Revision

- Have you stated clearly and specifically the purpose of the report?
- Have you put into the report everything required? Do you have sufficient supporting evidence? Have you stated the implications of your information clearly?
- Are all your facts and numbers accurate?
- Have you answered the questions your readers are likely to have?
- Does the report contain anything that you would do well to cut out?
- Does your organization suit the needs of your content and your audience?
- Are your paragraphs clear, well organized, and of reasonable length? Are there suitable transitions from one point to the next?
- Is your prose style clear and readable?
- Is your tone appropriate to your audience?
- Are there people you should share your draft with—for example, members of the target audience—before going on to a final draft?

Editing

- Have you checked thoroughly for misspellings and other mechanical errors?
- Have you included all the formal elements that your report needs?
- Are format items such as headings, margins, spacing, typefaces, and documentation consistent throughout the draft?
- Are your headings and titles clear, properly worded, and parallel? Do your headings in the text match those in the table of contents?
- Is your documentation system the one required? Have you documented wherever appropriate? Do the numbers in the text match those in the notes?
- Have you keyed the tables and figures into your text and have you sufficiently discussed them?
- Are all parts and pages of the manuscript in the correct order?
- Will the format of the typed or printed report be functional, clear, and attractive?
- Does your manuscript satisfy stylebook specifications governing it?
- Have you included required notices, distribution lists, and identifying code numbers?
- Do you have written permission to reproduce extended quotations or other matter under copyright? (Necessary only when your work is to be published or copyrighted.)
- While you were composing the manuscript did you have any doubts or misgivings that you should now check out?
- Have you edited your manuscript for matters both large and small?
- What remains to be done, such as proofreading final copy?

Reporting Technical Information

REPORTING TECHNICAL INFORMATION

7th Edition

Kenneth W. Houp
Late, The Pennsylvania State University

Thomas E. Pearsall
Emeritus, University of Minnesota

Contributing Author: Janice C. Redish
Vice President, American Institutes for Research, Washington, D.C.

MACMILLAN PUBLISHING COMPANY
New York

MAXWELL MACMILLAN CANADA
Toronto

MAXWELL MACMILLAN INTERNATIONAL
New York Oxford Singapore Sydney

Editor: Eben W. Ludlow
Production Supervisor: Marcia Craig
Production Manager: Nick Sklitsis
Text and Cover Designer: Pat Smythe
Cover Illustration: Salem Krieger
Photo Researcher: Chris Migdol
Illustrations: Precision Graphics

This book was set in Zapf International Medium by Ruttle, Shaw & Wetherill, Inc., and printed and bound by R. R. Donnelley & Sons Company. The cover was printed by Phoenix Color Corp.

Printed in the United States of America

Macmillan Publishing Company
866 Third Avenue, New York, New York 10022

Maxwell Macmillan Canada, Inc.
1200 Eglinton Avenue East
Suite 200
Don Mills, Ontario M3C 3N1

Library of Congress Cataloging-in-Publication Data

Houp, Kenneth W., 1913–
Reporting technical information / Kenneth W. Houp, Thomas E. Pearsall ; contributing author, Janice C. Redish.—7th ed.
p. cm.
Includes bibliographical references and index.
ISBN 0-02-393341-0 (paper)
1. Technical writing. I. Pearsall, Thomas E. II. Redish, Janice. III. Title
T11.H59 1992
808′.0666—dc20 91-7291
CIP

Printing: 1 2 3 4 5 6 7 Year: 2 3 4 5 6 7 8

PREFACE

In revising for this seventh edition of *Reporting Technical Information,* we were struck by a sense of both permanence and change. Permanence, in that our objectives have remained the same since the beginning:

> Our readers should be able to analyze a writing situation correctly; to find and organize material appropriate to audience, purpose, and situation; to design a functional report or letter that answers the needs of both writers and readers; and to write that report or letter correctly, clearly, and persuasively.

Though our objectives have remained fixed, the ways of reaching them have changed. We have been informed over the years by increasingly sophisticated research into technical communication and by changes in the nature of technical communication itself. This seventh edition is no exception to that on-going process.

Changes in Seventh Edition

Two significant changes in technical writing have brought about major revisions in this seventh edition of *Reporting Technical Information.* We, first of all, had to deal with the increasing use of word processing by technical writers and in the technical writing classroom. Word processing has put much more power in the hands of writers. They are now able to introduce typographical variation and graphics into their reports in a way that was previously possible only with expensive printing. With such new power has come the need for its responsible and efficient use.

We do not think that technical writing teachers or technical writing texts need to teach word processing technology, but, clearly, both need to impart insights in how to design documents using this technology. Recognizing that need, we turned to one of the foremost experts on document design in the United States: Dr. Janice C. Redish, Vice President, American Institutes for Research, Washington, DC. We asked her to contribute a chapter on document design, and she agreed to do so. Her contribution,

Chapter 10, "Document Design," explains the major principles of document design and shows how to couple these principles with word processing technology in creating powerful, accessible report formats.

Second, we recognize the increasing attention to collaborative writing in the technical writing classroom. Several writers or writers and editors working together to create reports and other documents has long been a common occurrence in the workplace. However, only recently have teachers turned to teaching collaborative techniques. To assist these teachers, we have introduced **Chapter 3, "Writing Collaboratively."** In Chapter 3 we explain how to organize and manage situational analysis, research, planning, and writing in a collaborative way. We include in the chapter a section on group conferencing techniques useful not only in collaborative writing but in other small group meetings as well.

Collaborative writing has also been partly responsible for the third major change. Collaborative groups need ways of staying on track through the composing process. Using checklists is a way of assuring that needed steps are not overlooked. To that end we have provided **planning and revision checklists** following every chapter except Chapter 1, "An Overview of Technical Writing." The checklist following Chapter 2, "Composing," is the most comprehensive; it, combined with some elements from Chapter 3, "Writing Collaboratively," is also printed on the endpapers inside our front cover. Such checklists are, of course, useful to individual writers as well as those working in groups.

Other important changes we have made are these:

- **A new more logical grouping of the chapters, working from the basics of technical writing to the applications of technical writing.**
- **An increased use of annotations.** We have increased significantly the annotation of our examples, making the example material more accessible and useful to the reader.
- **Individual chapter tables of contents.** Readability research clearly shows the effectiveness of such overviews in helping readers comprehend and retain material.[1]
- **New and rewritten chapter exercises that call for collaborative work and ask the student to deal directly with chapter material.**
- **An expanded section on documentation in Chapter 11, "Design Elements of Reports."** Technical writers need to know two systems of documentation: the numbered note system and the parenthetical author–date system. Word processing makes notes using superscript numbers and either footnotes or endnotes easy to do. As a result, many company and organizational reports have such notes. On the other hand, most scientific reports and articles use the parenthetical author–date system. We, therefore, provide instruction in both methods. We

have relied primarily on the documentation system of the Modern Language Association (MLA) in doing so.[2] In addition, we compare MLA style with American Psychological Association (APA) style. The MLA system, in our opinion, is now the simplest of all the major systems and, therefore, the easiest to learn. Once students have mastered basic principles of documentation using the MLA system, they should have little difficulty in using another system, such as APA, when the occasion calls for it.

- **A color insert on computer graphics.** The use of computer graphics, often in color, is becoming common in technical writing. However, not all computer graphics are equal. Many, while aesthetically pleasing, violate principles of good graphic design. To help students use computer graphics, we asked Dr. Redish to select and comment on some computer graphics that use design principles well. The color insert resulted from that request.
- **Additional material in Chapter 4, "Writing for Your Readers," on readability, the political implications of writing in an organization, and writing for a combined audience.**
- **An additional memorandum report and a letter report in Chapter 13, "Correspondence."** Our reviews showed that students needed additional help in this area. We also in this chapter revised other examples to bring them into keeping with contemporary business practice.
- **An expanded and reorganized "Handbook."** We have reorganized the "Handbook" using an alphabetical listing that makes information more accessible. We have added a section on acronyms and one on non-sexist usage.
- **A new and fresh design.** Macmillan's design team has designed a new format for this seventh edition. We think it improves both the accessibility and aesthetics of the book.

Plan of the Book

With all these changes, this is what you will find in the seventh edition.

Chapter 1, "An Overview of Technical Writing," defines technical writing and describes the world of workplace writing—the forms that technical reporting take and the problems that writers of technical information encounter.

Part I, "The Basics of Technical Writing," contains five chapters that take the student from the very beginning of the composing process in situational analysis, through the gathering of information, to the final polishing of a report's style.

Part II, "The Techniques of Technical Writing," contains three chapters that describe the basic techniques of organizing material to inform, to define and describe, and to present persuasive arguments.

Part III, "Document Design in Technical Writing," covers document design, the elements of reports from title page to documentation, and the use of graphics.

Part IV, "Applications of Technical Writing," covers advanced and extended applications of the basic principles and techniques. Here we show how to apply the principles explained in Parts I, II, and III to correspondence, instructions, proposals, progress reports, feasibility reports, and empirical research reports.

Part V, "Oral Reports," discusses the planning and presentation of oral reports and places heavy emphasis on the use of graphics in such reports.

Part VI, "Handbook," provides a ready reference when questions of grammatical usage, punctuation, and mechanics arise.

The **Appendixes** include an extended student report, a guide to technical reference sources, and a bibliography intended to lead readers to other sources for the many subjects covered in this book.

Teaching Ancillaries

With this 7th edition we offer more teaching aids than ever before.

- A greatly expanded Instructor's Manual with Transparency Masters and Tests. The new manual, co-authored with Professor Richard Raymond of Armstrong State College, includes advice on course planning, sample syllabi, suggested examination questions, and, for each chapter, teaching objectives, teaching hints, a discussion of exercises and assignments, a chapter quiz, transparency masters, and exercise solutions.
- A set of thirty-nine two-color transparency acetates. The transparencies display both existing text figures and new student examples.
- "RTI Writer" computer software available in both Macintosh and IBM versions using a program developed at the University of Minnesota by Professors Victoria Mikelonis and Ann Hill Duin and by Deborah Hansen. We have provided tutorials covering all the chapters in Part IV, Applications for Technical Writing, plus a master planning and revision checklist based on the checklists in the front endpapers of this text. This software should be especially useful in laboratory situations.

For information on the Instructor's Manual, the transparencies, and the computer software call your Macmillan representative, or write to the College English Editor, Macmillan Publishing Company, 866 Third Avenue, New York, New York 10022.

Acknowledgments

Detailed acknowledgments to the many sources we have drawn upon for this book are found in the **Chapter Notes** on pages 667–675. However, we acknowledge here a number of people who have been particularly helpful. We thank our many friends in academia, the business world, and in the Society for Technical Communication who have furnished us sample materials, in particular, Ann Duin, University of Minnesota; Mary Fran Buehler, Jet Propulsion Laboratory, California Institute of Technology; Diana Lutz and James M. Lufkin, Honeywell; and Charles R. Pearsall, Deere and Company.

We thank the colleagues who took the time to review our work and make so many useful suggestions: Carol Barnum, Southern College of Technology; Virginia Book, University of Nebraska-Lincoln; and Agricultural and Mechanical College; Sherry Little, San Diego State University; Carol Shehadeh, Florida Institute of Technology; Henrietta Nickels Shirk, Northeastern University; Elizabeth Tebeaux, Texas A & M University; Chester Tillman, University of Florida; Arthur Walzer, University of Minnesota; Thomas L. Warren, Oklahoma State University; Gloria Jaffe, University of Central Florida; Larry Beason, Eastern Washington University; Peter Dusenbury, Bradley University; Michael Keene, University of Tennessee at Knoxville; Joan McCoy, Southern College of Technology; Jean Vining, Houston Community College; Dean Hall, Kansas State University; Gerald Harris, Devry Institute of Technology; Judith Rosenberg, College of Lake County; Elbert F. Jones, University of Maryland; Kristin R. Woolever, Northeastern University.

Donald J. Barrett, Chief Reference Librarian, United States Air Force Academy, has once again revised Appendix B, "Technical Reference Books and Guides" for us, and Professor James Connolly, University of Minnesota, has again contributed the section on visuals found in Part V, "Oral Reports."

As previous users of this text know, Professor Kenneth W. Houp died shortly before the publication of the sixth edition. Despite the loss of Ken, the use of the pronoun "we" will continue throughout this book. Ken's ideas and work are still much in evidence throughout the book. As well, any book is a collaboration whether there is a single author or many. There are the reviewers who provide the insights that the author likely will not come to unaided. Most important of all are the editors, who, bless them, can point out a book's deficiencies without implying that somehow the author shares those deficiencies. *Reporting Technical Information* has had the good fortune to have three of Macmillan's best: Johnna Barto, Marcia Craig, and Eben W. Ludlow. As Ken did, they believe all writing is subject to infinite improvement.

Finally, I express my love and gratitude to my wife Anne for her loving and loyal support.

Thomas E. Pearsall

BRIEF CONTENTS

CONTENTS

Reporting Technical Information

CHAPTER 1

An Overview of Technical Writing

SOME MATTERS OF DEFINITION

THE SUBSTANCE OF TECHNICAL WRITING

THE NATURE OF TECHNICAL WRITING

THE ATTRIBUTES OF GOOD TECHNICAL WRITERS

THE QUALITIES OF GOOD TECHNICAL WRITING

A DAY IN THE LIFE OF TWO TECHNICAL WRITERS

Marie Enderson: Computer Specialist and Occasional Technical Writer
Ted Freedman: Technical Writer and Company Editor

This first chapter is intended to give you the broadest possible view of technical writing. Beginning with Chapter 2, we go into details, but for the details to be meaningful, they have to be seen against the background given here.

Some Matters of Definition

As you work your way through this book, you will see that technical writing is essentially a problem-solving process that involves the following elements at one or more stages of the process:

- A technical subject matter that is peculiar to or characteristic of a particular art, science, trade, technology, or profession
- A recognition and accurate definition of the communication problem involved
- The beginning of the solution through the establishment of the role of the communicator and the purpose and audience (or audiences) for the communication
- Discovery of the accurate, precise information needed for the solution of the problem through thinking, study, investigation, observation, analysis, experimentation, and measurement
- The arrangement and presentation of the information thus gained so that it achieves the writer's purpose and is clear, useful, and persuasive to the intended audience or audiences

The final product of this problem-solving process is a piece of technical writing that may range in size and complexity from a simple memorandum to a stack of books. To expand our overview of technical writing, we discuss it under these five headings:

The Substance of Technical Writing
The Nature of Technical Writing
The Attributes of Good Technical Writers
The Qualities of Good Technical Writing
A Day in the Life of Two Technical Writers

The Substance of Technical Writing

Organizations produce technical writing for both internal and external use. Internally, documents such as feasibility reports, technical notes, and memorandums go from superiors to subordinates, from subordinates to superiors, and between colleagues at the same level. If documents move in more than one direction, they may have to be drafted in more than one version. Company policy, tact, and the need-to-know are important considerations for intracompany paperwork.

Many examples come to mind. The director of information services must study and report on the feasibility of providing middle management with personal computers. The research department must report the results of tests on new products. The personnel department must instruct new employees about company policies and procedures. In fact, the outsider cannot conceive of the amount and variety of paperwork a company must generate simply to keep its internal affairs in order. Survey research indicates that college-educated employees spend about 20% of their time on the job writing.[1] In fact, most college-educated workers rank the ability to write well as "very important" or "critically important" to their job performance.[2]

Externally, letters and reports of many kinds go to other companies, the government, and to the users of the company's products. Let us cite a few of the many possibilities: A computer company must prepare instructional manuals to accompany its computers. A university department prepares a proposal to a state government offering to provide research services. An architectural firm prepares progress reports to report to clients the status of contracted building programs. An insurance company must write letters accepting or denying claims by its policyholders.

The manufacture of information has become a major industry in its own right. Much of that information is research related. Many government agencies, scientific laboratories, and commercial companies make research their principal business. They may undertake this research to satisfy their internal needs or the needs of related organizations. The people who conduct the research may include chemists, computer scientists, physicists, mathematicians, psychologists—the whole array of professional specialists. They record and transmit much of this research via reports. The clients for such research may be government agencies or other institutions that are inadequately equipped to do their own research. Reports may, in fact, be the only products of some companies and laboratories.

Much technical writing goes on at universities and colleges. Professors have a personal or professional curiosity that entices them into research. If they believe that their findings are important, they publicize the information in various ways—books, journal articles, papers for professional societies. Students assigned research problems to further their education present what they have done and learned in laboratory reports, monographs, and theses.

Many reports are prepared for public consumption. For example, a state department of natural resources is entrusted not only with conserving our woodlands, wetlands, and wildlife but also with making the public aware of these resources. State and federally supported agricultural extension services have as a major responsibility the preparation and dissemination of agricultural information for interested users. Profit-earning companies must create and improve their public image and also attract customers and applicants for employment. Airlines, railroads, distributors of goods and services, all have to keep in the public view. Pamphlets, posted notices, and radio and television announcements are commonly used to meet all these needs.

Myriad applications such as these—company memos and reports, government publications, research reports, public relations releases—create a great flood of paperwork. Some of it is only of passing interest; some of it makes history. Some of it is prepared by full-time professional writers, but most of it is prepared by professionals in a technical field who are writing about their own work.

The Nature of Technical Writing

Technical writing is a specialty within the field of writing as a whole. Beginning technical writers must serve an apprenticeship. They must gain a working knowledge of their new subject matter and its terminology. They must learn to develop a prose style that is clear, objective, and economical. They must learn report types, variations in format, standards for abbreviations, the rules that govern the writing of numbers, the uses of tables and graphs, and the kinds of people who read technical reports and their expectations. That is, novices must learn the whole special business of being technical writers.

And yet a broad and sound foundation in other writing is a tremendous asset to technical writers, for it gives them versatility both on and off the job. They can write a good letter, prepare a brochure, compose a report. In this comprehensive sense, they are simply *writers*. The same writing skills that are important in a college classroom are important on the job. Surveys show that workers rank writing skills in this order of importance:[3]

1. Clarity
2. Conciseness
3. Organization
4. Grammar

They understand, too, that not all writing is done in the same tone and style. As writers they have not one style but a battery of them:

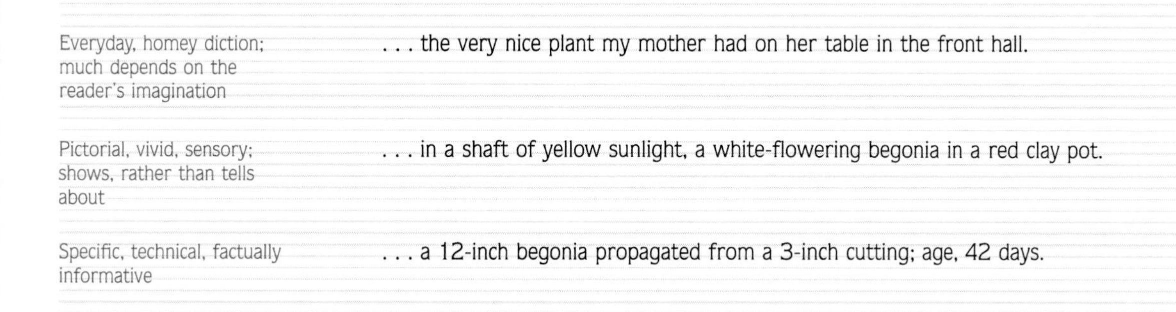

Everyday, homey diction; much depends on the reader's imagination
. . . the very nice plant my mother had on her table in the front hall.
Pictorial, vivid, sensory; shows, rather than tells about
. . . in a shaft of yellow sunlight, a white-flowering begonia in a red clay pot.
Specific, technical, factually informative
. . . a 12-inch begonia propagated from a 3-inch cutting; age, 42 days.

As a technical writer, whether part-time or full-time, you may have to use all of these "languages," for your job will be to convey your message to your intended readers. By playing the right tune with these languages in different combinations, and by adding other writing skills in generous measure, you can produce leaflets, proposals, brochures, sales literature, reports to stockholders, and a great variety of letters.

In writing intended for your professional colleagues, you will be nearer to the closing begonia example than the first two. Your diction will be objective and accurate. By relying on this style, you can produce operating manuals, feasibility reports, research reports, progress reports, and similar materials. When your audience is appropriate for this style, your writing is likely to show these characteristics:

- Your purpose is usually spelled out in the opening paragraph or two. All included information bears upon the accomplishment of the stated purpose. For example, a technical paper on smoke detectors may set forth only one major objective: to determine the relative effectiveness of photoelectric and ionization chamber types in detecting smoldering fires, flaming fires, and high temperatures. Other major topics would be reserved for other papers.
- The vocabulary tends to be specialized. Some of the terms may not appear in general dictionaries. If the audience shares the writer's professional specialization, the specialized terms may not be defined within the text, on the assumption that professional colleagues will be familiar with them. At other times, the terms may be listed and defined in accompanying glossaries.
- Sentences are highly specific and fact filled.
- When appropriate to the material, numbers and dimensions are numerous.

- Signs, symbols, and formulas may pepper the prose.
- Graphs and tables may substitute for prose or reinforce and expand upon the surrounding prose. Figures and illustrations of all sorts are widely used, sometimes to supplement prose, sometimes to replace it.
- Documentation and credits appear in notes and bibliographies.

As these listed characteristics perhaps make clear, audience analysis is tremendously important to successful technical writing. What is appropriate for your professional colleagues may be wholly inappropriate for the general public. In matters of definition, for example, terms are not normally defined if the audience can be expected to know them. But the indispensable corollary to that proposition is that terms *must be defined* when your audience, for whatever reason, cannot be expected to know them. Or, using another characteristic, sentences can be fact filled when the audience is highly professional and highly motivated. However, when your readers do not share your motivation, profession, and enthusiasm, you must slow your pace and make your prose less dense. In technical writing, you must know your audience as well as your objectives and adapt your style and material to both.

The Attributes of Good Technical Writers

To write clear and effective reports, you must build upon the natural talents you have in communicating ideas to others. But how can you build successfully? What skills, characteristics, and attitudes are of most value to the technical writer? From experience, we can summarize some of the major attributes that will stand you in good stead:

- You must be reasonably methodical and painstaking. Plan your work for the day and for the rest of the week. Look up from time to time to take stock of what you and others are doing, so that you do not squander your time and energy on minor tasks that should be put off or dispensed with altogether. File your correspondence. Keep at your desk the supplies you need to do your work. Keep a clear head about ways and means for accomplishing your purpose.
- Be objective. Try not to get emotionally attached to anything you have written; be ready to chuck any or all of it into the wastebasket. While reading your own prose or that of your colleagues, do not ask whether you or they are to be pleased but whether the intended audience will be pleased, informed, satisfied, and persuaded.

- As a professional, keep in mind that most of what you do will eventually have to be presented in writing. Do your work so that it will be honestly and effectively reportable. Keep a notebook or a deck of note cards. Record what you do and learn.
- Remind yourself frequently that *clarity* is your most important attribute. Until the sense of a piece of writing is made indisputably clear, until the intended reader can understand it, nothing else can profitably be done with it.
- As someone who must write, understand that writing is something that can be learned, even as chemistry, physics, and mathematics can be. The rules of writing are not as exact, perhaps, as those of science, but they can never be thrown overboard if you are to bring your substance home to your reader.

One writer, who knew well the nature and substance of technical writing, summed up the way to be successful with three imperatives that undergird much of this book:

> (1) Know your reader; (2) know your objective; (3) be simple, direct, and concise.[4]

The Qualities of Good Technical Writing

Because the qualities of good technical writing vary, depending upon audience and objective, we cannot offer you a list that applies equally to everything you write. However, some qualities are apparent more often than not in good technical writing:

Good technical writing . . .

- Arrives by the date it is due.
- Makes a good impression when it is picked up, handled, and flipped through.
- Has the necessary preliminary or front matter to characterize the report and disclose its purpose and scope.
- Has a body that provides essential information and that is written clearly without jargon or padding.
- Has—when needed—a summary or set of conclusions to reveal the results obtained.
- Has been so designed that it can be read selectively: for instance, by some users, only the summary; by other users, only the introduction and conclusions; by still other users, the entire report.

- Has a rational and readily discernible plan, such as may be revealed by the table of contents and a series of headings throughout the report.
- Reads coherently and cumulatively from beginning to end.
- Answers readers' questions as these questions arise in their minds.
- Conveys an overall impression of authority, thoroughness, soundness, and honest work.

Beyond all these basic characteristics, good technical writing is free from typographical errors, grammatical slips, and misspelled words. Little flaws distract attention from the writer's main points.

A Day in the Life of Two Technical Writers

By way of summary, let us describe two representative writers, whom we shall identify as Marie Enderson and Ted Freedman.

Marie Enderson: Computer Specialist and Occasional Technical Writer

Marie has a bachelor's degree in engineering technology. She works in the information services division of a small electronics company that employs some 400 people. Marie has been with the company for a little over a year. Since her childhood, she has been recognized as a whiz at mathematics. In college, she was drawn to the use and design of computing systems. Her major responsibility is to provide technical support for computer systems users in the company.

Marie's first project with the company was a design for an office automated system for the shipping department. She interviewed the supervisors and workers in the department to establish the department's needs. She then matched the needs to available "off-the-shelf" equipment and programs and designed a system to automate much of the department's work. After finishing her design she had to prepare a written report and oral briefing describing it for the shipping department and her boss. She had a ghastly time the next two weeks. She found, as do many novices at writing, that she knew *what* she wanted to say but not *where* or *how* to say it. But the 10-page report did somehow get written and, after a thorough overhaul by Ted Freedman (whom we'll meet next), was presented. Her oral report was a summary of the written report, and it was well received. Her system design was accepted and will be implemented in several months.

Marie's first experience with on-the-job technical writing taught her four important things: (1) an engineer is not simply a person whose only product is a new design or a gadget that works; (2) things that go on in your head and hands are lost unless they are recorded; (3) writing about what you have thought and done is a recurring necessity; and (4) technical writing, strange and difficult as it may seem at first, is something that can be learned by anyone of reasonable intelligence and perseverance.

Marie's present project is a set of instructions for the accounting department to help them use an automated system that was installed for them over a year ago. Marie's predecessor had installed the equipment and furnished the accountants a set of the manuals produced by the computer and program manufacturers. The manuals are well written, but because they are written by different manufacturers for a general audience they do not integrate the components of the system in a way meaningful to the accountants. Marie has studied the system and interviewed the users to determine their needs. She has drafted a 20-page booklet that supplements the manufacturers' manuals and shows the accountants how to use the new equipment and programs in their work. She has sent the draft to Ted Freedman for his comments.

Ted Freedman: Technical Writer and Company Editor

Ted Freedman was hired three years ago by the company as a technical writer–editor. He holds a bachelor's degree in technical communication. His office is a sparsely furnished cubicle down the hall from the publications and mailing departments. His office possessions include an old typewriter, a brand-new microcomputer and printer, a four-foot shelf of dictionaries and reference manuals, and an extra-large wastepaper basket.

At 8:45 this morning Ted is scheduled for a project review session in the company auditorium. He arrives at the auditorium with five minutes to spare. For the next hour he studies flip charts, slide projections on the huge screen, chalk-and-blackboard plans for company reorganization (minor), and staffing proposals for three new projects totaling $578,400. From the platform, Chief Scientist Muldoon requests that Ted develop research timetables and preview reporting needs.

At 10:20 he meets with a commercial printer to examine the artwork and layout for a plush report the company is preparing for a state commission. The work looks good but is a little lacking in typographical variety, he suggests.

At 12:55, back from lunch in the company cafeteria, Ted glances over the memos that collected on his desk during the morning—nothing urgent. Then he opens the manila envelope lying in his mail rack. In it is a computer disk that contains Marie's instructional booklet and a printout of the booklet.

At 1:30 he gets Marie on the telephone and arranges for a meeting at 3:00 so that they can run through the draft together. In the meantime he looks over the printout. He notices some computer jargon. He is pretty sure the accountants would not have a clear idea of the distinction between "on-line" and "batch environments." He circles both phrases. Reading on, he finds a spot where the text should be supported by a graphic. He makes a note of it in the margin. He realizes that the booklet would be more accessible to the reader with more headings in it. He puts Marie's disk into his microcomputer and, scrolling through her text on the screen, inserts headings that fit her arrangement and material. Thus, the afternoon wears on.

At 3:00 Marie arrives and the two confer, make changes, and plan later alterations in the draft. As before, they work amicably together. They intersperse their writing and editing with an occasional trip to the water cooler, a chat with a department head, and a trip to the library to consult a specialized reference work.

Ted is good at his work and is considered to have a great future with the company.

Marie and Ted are roughly representative of many thousands of technical writers–editors, most of them, like Marie, part-time as the need arises. To gain a more rounded understanding of their duties and behavior, we would have to pay them many additional visits; however, certain things are evident even from this brief visit. Like most writers on the job, they work in collaboration with others. Also, much of the time they are not writing at all, in the popular sense. Some of the time they are simply listening hard to what people are saying to one another—trying to clarify, simplify, and translate into other terms. A generous portion of their time is spent on tasks that have little direct connection with writing but eventually provide grist for the writing mill. The techniques, tools, and processes that writers such as Ted and Marie need to accomplish their work are the subject matter of this book.

Exercises

1. As your instructor directs, bring to class one or more magazine or newspaper clippings that you believe to be technical. In what respects is the writing technical? Subject matter? Purpose? Tone? Vocabulary? Other?
2. Rewrite a brief paragraph of technical prose (perhaps a clipping submitted in Exercise 1) to substantially lower its technical level. Explain what you have done and why.
3. With the help of *Ulrich's International Periodicals Directory* (see page 655), find several periodicals in your professional field. Examine one or more copies of them. In what ways and to what extent does your examination of such periodicals confirm or change your first impressions of technical writing?
4. On a two-column page, list your present *assets* and *limitations* as a technical writer.
5. Turn to the job advertisements section of a large metropolitan newspaper such as the Sunday *New York Times*. What advertisements for technical writers do you find? What qualifications are demanded of them?
6. Talk with a professional person of your acquaintance, if possible, one in a field you would like to enter. Ask how much writing he or she does and what kinds. Ask how much importance is attached to good writing on the job. Write a short report of what you learn for your instructor.

PART I

The Basics of Technical Writing

Composing

Writing Collaboratively

Writing for Your Readers

Gathering and Checking Information

Achieving a Readable Style

CHAPTER 2

Composing

SITUATIONAL ANALYSIS

DISCOVERY

ARRANGEMENT

DRAFTING AND REVISING

The Rough Draft
Revision

EDITING

Checking Mechanics
Checking Documentation and Format
Checking Graphics

REVISING AND EDITING WITH WORD PROCESSING

Spelling Checkers
Grammar and Style Checkers

The composing process is similar to all high-level reasoning processes in that we do not understand it completely. Most researchers in artificial intelligence—the use of computers to replicate human reasoning—have concluded that no general rule of reasoning works for all problems. As one of them put it, "The human brain . . . has an incredibly large processing capacity, much greater than several Cray computers, and it is beyond our understanding in its ability to connect, recall, make judgments, and act. Thus, all experiments to discover a generalized problem-solving system were, in a practical sense, failures."[1]

Just how complex is the human brain that is "much greater than several Cray computers" (one of the new generation of supercomputers)? As one authority points out, "There are more neurons in the human brain than stars in the Milky Way—educated estimates put the number of neurons at about 10^{12} or one trillion. Each of those cells can 'talk' to as many as 1,000 other cells, making 10^{15} connections."[2] Given that level of complexity and those kinds of numbers, no one can map out completely how any complex, high-level, problem-solving process works. And the composing process is precisely that: a complex, high-level, problem-solving process.

But, since classical times we have understood some things about the composing process. Aristotle, for example, recognized the wisdom of taking one's audience into account. In recent years, empirical research has revealed additional useful facts about the process. What we tell you in this chapter is based upon both those classical concepts that have stood the test of time and modern research. We don't pretend to have all the answers or even that all our answers are right for you. But we can say that the process we describe is one that draws upon the actual practices of many experienced writers, and it works for them.

For most skilled and experienced writers, the composing process breaks up roughly into five parts. The first part involves **situational analysis.** It's that time when you're trying to bring a thought from nowhere to somewhere. It's a time when you think about such things as your audience, your topic, and your purpose. In a second stage, you "discover" the material you need to satisfy your purpose and your audience. That **discovery** process may go on completely within the trillion cells of your brain or, as is often the case in technical writing, in libraries, laboratories, and workplaces as well.

When the discovery stage is almost complete, you pass into a stage where you **arrange** your material. That is, before writing a draft you may rough out a plan for it or even a fairly complete outline.

With your arrangement in hand, you are ready for the fourth part of the composing process, the **drafting** and **revising** of your document. For many competent writers, drafting and revising are separate steps; for others, the two go on almost concurrently.

In the final stage of the writing process, you **edit** your work to satisfy the requirements of standard English and to meet the requirements of good format.

Time spent on these five parts is usually not equal. Situational analysis, discovery, and arrangement for a complicated piece of work may take 80% or more of the time you have for the project. For an easy piece of routine writing, these first three stages may take a few minutes, and drafting and revising may take up the bulk of the time. Some situations call for careful, scrupulous editing; others do not.

The process is often not linear. If the drafting bogs down, you may have to return to the situational analysis stage to resolve the problem. Drafting and revising may alternate as you write for a while, then stop to read and revise. But, in rough outline, what we have described for you is the competent writer's composing process. Throughout this book, we frequently deal with the process. We remind you again and again of the needs of your audience and provide you ways to discover material to satisfy different purposes and topics. In the rest of this chapter we provide some strategies that you can use to develop a competent writing process of your own. Because any part of the process can be done in cooperation with others, we, in Chapter 3, provide information on how to write collaboratively.

Situational Analysis

In this section we discuss situational analysis, dealing first with topic and purpose and, second, with audience and persona.

Topic and Purpose

The topics and purposes of technical writing are found in the situations of technical writing. The topics are many. You may have a mechanism or process to explain; that is your topic. You may have to define a term or explain a procedure. You may have to report the results and conclusions of a scientific experiment or a comparison shopping study. New research has to be proposed. Delayed work has to be explained. All these and many more are the topics of technical writing.

Although the topics of technical writing are varied, the purposes are more limited. Generally, your purpose is either to inform or to argue. Most topics can be handled in one of these two ways, depending upon the situation you find yourself in. Often, you are simply informing. For example, the situation may call for you to describe a mechanism so that someone can understand it. As you will see when you read Chapter 8, "Defining and Describing," mechanism description will often call for you to divide the mechanism into its component parts and then describe these parts, perhaps as to size, shape, material, and purpose. As another example, you may have to define a term from your discipline. In your definition, you may tell what category of things the thing defned belongs to and what distinguishes it from other members of the same category. You may give an example of the thing described.

On the other hand, when dealing with your mechanism or definition, you may really be mounting an argument. You may not merely be describing a mechanism; you may be attempting to demonstrate its superiority to other mechanisms of the same type. To do so, you'll need to argue, perhaps by showing how your mechanism is more economical and easier to maintain than other mechanisms. In the same way, you may not simply be defining a term; you may be arguing that your definition is more comprehensive or more correct than previous definitions of the same term.

Be sure to have your topic and purpose in hand before you proceed in your writing project. It's good practice to write them down, something like this:

> I will define alcoholism in a way that reflects recent research. Further, I will demonstrate that my definition, which includes the genetic causes of alcoholism as well as the environmental ones, is more complete and accurate than definitions that deal with environmental causes alone.

Will topic and purpose change as you proceed with your project? That depends on the situation. Frequently, the situation may call for you to stick closely to a narrow topic and purpose: *We have to explain to our clients our progress (or lack of progress) in installing the air conditioning system in their new plant.* Or, in another typical situation, *We have to provide instructions for the bank tellers who will use the computer consoles we have installed at their stations.* While the way you handle such topics and purposes is subject to change as you explore them, the topics and purposes themselves really are not subject to change. On the other hand, the situation may call for you to explore a topic, perhaps the effect that the rising age of the American population may have on the restaurant business. While you have defined your topic well enough to begin your exploration, the precise topic and the purpose may have to wait until you discover more information about your subject.

Audience and Persona

Writers make important decisions about both content and style based upon consideration of audience and the persona the writer wants to project. **Persona** is a term that refers to the role the writer has or assumes when writing. It relates to, among other things, the position of the writer and his or her relationship to the audience and the situation. For example, a bank lending officer might assume one persona when writing to a loan applicant and a different persona when writing to a supervisor to justify a loan that has been made.

Professional people consider both audience and persona seriously when composing, as this quote from a hydrology consultant at an engineering firm indicates:

> We write about a wide range of subject matters. Some things are familiar to a lay audience. Most people can understand a study about floods. They can understand a study that defines a 100-year flood plane. They can imagine, say, water covering a street familiar to them. But other subjects are very difficult to communicate. We work with three-dimensional models of water currents, for example, that are based upon very recondite hydrolic movements. We also have a wide audience range. Some of our reports are read by citizen groups. Sometimes we write for a client who has a technical problem of some sort and is only interested in what to do about it. And sometimes we write for audiences with high technical expertise like the Army Corps of Engineers. Audiences like the Army Corps expect a report to be written in a scientific journal style, and they even want the data so they can reanalyze it. A lot of times the audience is mixed. A regulatory agency may know little about the subject of one of our reports, but they may have a technically trained person on the staff who does. In any case, we must understand what it is that the client wants, and we must be aware of what he knows about the subject. We must convince clients that we know what we are doing. We depend upon return business and word-of-mouth reputation, and we must make a good impression the first time. Much of the technical reputation of this company rides on how we present ourselves in our technical reports.[3]

Here are some questions you need to ask about your audience and persona when you are preparing to write.

What Is the Level of Knowledge and Experience of Your Readers? In technical writing, the knowledge and experience your readers possess are key factors. Do your readers, for instance, understand your professional and technical language? If they do, your task is easier than if they do not. When they do not, you'll have to be particularly alert to your word choice, choosing simpler terms when possible, defining terms when simpler choices are not possible. It goes beyond word choice. There are whole concepts that a lay audience may not have. Geologists, for example, thoroughly understand the concept of plate tectonics and can assume that geologists in their audience understand it equally well. When addressing a lay audience, however, the geologist writer would be wise to assume little understanding of the concept. If the geologist wishes to use the concept, he or she will have to take time to explain it in a way that the audience can grasp.

What Will the Reader Do with the Information Presented? This question relates to the reader's purpose and concerns. Suppose, for instance, that you are writing about a procedure. People may read about procedures for many reasons. In one case, the reader may wish to perform the procedure. In another, the reader may have to make a decision about whether to adopt the procedure. In yet another, the readers may simply want generalized information about the procedure, perhaps because they find it interesting.

Each case calls for a different selection of content and a different style. Readers wishing to perform the procedure need a complete set of step-by-step instructions. The decision maker needs to know by what criteria the procedure has been evaluated and why under these criteria it is a better choice than other procedures. Readers for interest want the general concept of the procedure explained in language they can understand.

What Is Your Relationship to the Reader? Are the readers your bosses, clients, subordinates, peers, or students? If you are a public employee, are you writing to a taxpayer who contributes to your salary? Writers in the workplace when interviewed about how they write reveal that they pay a good deal of attention to the effect of such relationships on tone, as these quotations demonstrate:[4]

- Writing to *my boss* I try to pinpoint things a little more.
- *When you have something as personal as a phone call and a conversation back and forth* . . . I feel free to use "I" rather than "we."
- We always want them *to realize* they can call on us if they have any questions.
- This [referring to a statement] is a bit more on *a personal level. . . . The other* [statement] is much too formal.
- Just to say "Send his address," would, I think, be *a little* too authoritarian.

The roles writers find themselves in also affect their choice of content. Imagine the difference in approach between that of a Chevrolet sales representative trying to sell a fleet of Chevrolets to a company and that of a young executive of the same company reporting to his or her superior that the results of a feasibility study demonstrate Chevrolets to be the best purchase. In the first instance, only the advantages of Chevrolets are likely to be emphasized. In the second, a more balanced appraisal will be expected.

What Is Your Reader's Attitude about What You Are Going to Say? Audiences can be suspicious and hostile. Or they may be apathetic. Of course, they may be friendly and interested. Their attitude should affect how you approach them. If you have an unfriendly audience, you must take particular care to explain your position carefully in language that is understandable but not patronizing. You may need more examples than you would with a friendly audience. A friendly audience may be persuaded with less information. With a friendly audience, you may present your conclusions first and then support them. With an unfriendly audience, it's a sound idea to present your support first and then your conclusions.

Readers may have attitudes about the language you use. For example, public health officials have had a difficult time expressing how to avoid exposure to AIDS. Such advice, to be effective, must refer very explicitly to various sexual practices. Newspapers have had to change their usual practice to allow such language to be printed, and some readers have found the language offensive. In most cases, protection against AIDS has won out over reader sensibilities, but the problem illustrates well the social context of audience analysis.

What Persona Do You Wish to Project to Your Audience? If you have read many scientific journals, you have probably already realized that they have a certain tone about them, a tone to which words like *objective, formal,* and *restrained* seem readily to apply. Scientists, to find acceptance in such journals, must adopt such a tone. A breezy, light journalistic style, though it might be just as clear, would not be acceptable. In the same way bankers, for example, must present themselves in a careful, formal way. We're not likely to give our money for safekeeping into the hands of someone who comes on like a television used car dealer. Young executives writing to their bosses are likely to be a bit deferential. The bosses, in turn, want to sound firm but reasonable and not authoritarian. What has come to be called "corporate culture" plays a role in the persona that a writer may adopt. In writing, you must project the values and attitudes of the organization you work for. To do so, you may look over past correspondence and reports to see what practices have been used, what sort of tone writers in the organization have adopted.

Taking on a persona when you write is something like taking on a persona when you dress. The student who exchanges his blue jeans and running shoes for a business suit and leather shoes when he reports for a job interview is slipping out of one persona into another. The teacher who exchanges a comfortable sweater and skirt for a businesslike dress when she leaves the classroom to consult in industry is exchanging one persona for another. It's a common enough experience in life, and you should not be surprised to find such experiences in writing situations. Both dressing and writing have their own rhetorics. However, don't misinterpret anything we have said here as a rationale for your being obscure or jargony. You can be clear no matter what persona you adopt.

Discovery

At some point in your writing process, you must "discover" the material you will use in your writing. Discovery is the teasing out of your mind the information you will use and modify to meet the needs of your topic, purpose, audience, and persona. Discovery is making connections. It's the putting together of two pieces of information to create a third piece that didn't exist before the connection. A mind that is well stocked with information in the subject matter to be discovered is probably a mind that will be successful at discovery. Those trillion neurons need something to work with; so the more you read, observe, and experience, the better writer you are likely to be.

Of course, all the material you need may not be in your mind when you begin. Discovery is using libraries and laboratories to fill in the gaps in your knowledge. You may also use interviews, on-site inspections, letters of inquiry, and the like to gather information. We discuss such information-gathering techniques in Chapter 5, "Gathering and Checking Information." The techniques we discuss here are those you can use to explore your own mind.

Brainstorming

In brainstorming you uncritically, without thought of organization, jot down every idea about a subject that pops into your head. The key to successful brainstorming is that you do not attempt to evaluate or arrange your material at the first stage. These processes come later. Evaluation or arrangement at the first critical stage may cause you to discard an idea that may prove valuable in the context of all the ideas that the brainstorming session produces. Also, avoiding evaluation at this point prevents the self-censorship that often blocks a writer.

We illustrate this process with excerpts from a brainstorming session on an essay you can see in finished form—Chapter 4, "Writing for Your Readers." We have put the ideas down here exactly as they appear on the brainstorming pad—abbreviations, fragmented ideas, and all:

Purpose vs. audience analysis

purpose of writing—purpose of reading

on the job vs. in school

need long known
  Aristole defined generation gap
  Roman water commissioner
  typographical experiments

Swain quote—use!

ping pong balls and mousetraps

technicians, lay people (executives), operators, engineers, scientists, jr. engineers, 1st echelon technicians, 2nd ech.—how to break down?

why do I read scientific articles/what's in it for me? What does development mean for me? practical interest

Instructions can be written for lay people or technician—or operator

Westinghouse stuff—Jim Souther

The brainstorming session for Chapter 4 filled four sheets of 22-inch by 17-inch newsprint with closely packed writing. We paid no attention to arrangement at this point. Neither did we evaluate. Some of the ideas made it through to the finished product; some did not. (Some show up in other chapters.) All we tried to do was to put our ideas and facts out into the open where we could use them and think about them.

Because brainstorming is a fairly painless process, it's frequently a good device to break down the normal resistance most of us have to hard thinking. It can result in your writing down a good deal more information than you ever thought you possessed. It can also reveal rather quickly to you those places where you have holes that will need to be filled at a later time with information you gather.

Using Arrangement Patterns for Discovery

Although the discovery stage is not the stage when you arrange your material, you can use familiar arrangement patterns as aids in discovering your material.[5] For example, suppose your purpose is to describe a procedure for a reader who wishes to perform the procedure. If you were familiar with writing instructions, as you will be after reading Chapter 14, "Instructions," you would know that a set of instructions often lists and sometimes describes the tools that must be used to perform the procedure. Furthermore, instructions describe the steps of the procedure, normally in chronological order. Knowing what is normally required for a set of instructions, you can, in effect, brainstorm your material in a more guided way.

You can begin by writing down the tools that will be needed for the procedure. Think about what you know about your audience. Are they experienced with the tools needed? If so, stop with listing the tools. If they are not experienced, jot down some information they'll need to use the tools properly.

When done with the tools, think about the steps of the procedure. As you do so, write down the key steps of the procedure. Keep your readers in mind. Are there some steps so unfamiliar to your readers that you need to provide additional information to help them perform the steps? If so, rough out what that information might be.

As in brainstorming, in very little time you can get information out where you can see it. Also, as in brainstorming, if there are gaps in your knowledge, you can discover them early enough to fill them.

Another task frequently encountered in technical writing is arguing to support an opinion. In discovering an argument, you can begin by stating that opinion clearly, perhaps something like "Women should get equal pay for equal work." Next, you can turn your attention to the subarguments that might support such an opinion. For example, first, you would have to establish that in many instances women are not getting equal pay for equal work, and, therefore, a problem really does exist. Then you might think of a philosophical argument: Ethically, women have a right to equal pay for equal work. You might think of an economic argument: Women's needs to support themselves and their families equal those of men. And so forth. As you think about subarguments, you will begin to think about the information you will need to support them. Some of it you may have; some you may find you will need to research. The very form and needs of your argument serve as powerful tools to help you discover your material.

Other Successful Discovery Techniques

Most experienced writers develop their own discovery techniques. In the workplace, writers often use past documents of a similar nature to jog their minds. Many professional writers keep journals that they can mine for ideas and data. Scientists keep laboratory notebooks that can be invaluable when it's time to write up the research. In the workplace, people talk to each other to discover and refine ideas.

Asking questions, particularly from the reader's point of view, is a powerful discovery technique. Suppose you were describing the use of computers for word processing. What questions might the reader have? *Do I have to be an expert typist to use a word processor? What are the advantages of word processing? The disadvantages? How do I judge the effectiveness of a word processor? What do word processors cost? Will word processing make writing and revising easier?* As you ask and answer such questions, you are discovering your material.

When you have established your topic and purpose, analyzed your audience, and discovered your material, it's time to think about arrangement.

Arrangement

When you begin your arrangement, you should have a good deal of material to work with. You should have notes on your audience, purpose, and persona. Your discovered material may be in various forms. It may, for instance, be a series of notes produced by brainstorming or other discovery techniques. You may have cards filled with notes taken during library research or notebooks filled with jottings made during laboratory research. You may have previous reports and correspondence on the same topic you are writing about. In fact, you may have so much material that you may not know where to begin.

You can save yourself much initial chaos and frustration if you remember that certain kinds of reports (and sections of reports) have fairly standard arrangements. The same arrangement patterns that helped you discover your material can now serve you as models of arrangement. For instance, you might divide your subject into a series of topics, as we have done with the chapters of this book. If you're describing a procedure and know that your readers wish to perform that procedure, you may use a standard instruction arrangement: introduction, tool list and description, and steps of the procedure in chronological order. If you are arranging an argument, you have your major opinion, often called the major thesis, and your subarguments, often called minor theses. You'll probably want to consider the strength of your minor theses when you arrange your argument. Generally, you want to start and finish your argument with strong minor theses. Weaker minor theses you'll place in the middle of your argument.

Documents such as progress reports, proposals, and empirical research reports have fairly definite arrangements that we describe for you in Part IV, "Applications of Technical Writing." Not all the arrangements described will fit your needs exactly. You must be creative and imaginative when using them. But they do exist. Use them when it is appropriate.

Let us illustrate the processs by using again the development of our

Chapter 4, "Writing for Your Readers." When we began arranging that chapter, we started by confirming for ourselves our purpose, which was to explain to college students how to write for the kinds of audiences they might someday meet in the workplace. We first looked for key ideas that directly supported our purpose and that were major enough that we could group lesser ones under them. Looking at our discovered information, we saw that types of audiences seemed to be the dominant idea. This suggested classification by audience type as the arrangement scheme. (See page 51.) First ideas may not always work, but in this case ours did. Once we had hit upon the classification scheme, our arranging went well but not without a good deal of trial-and-error adjustments.

First, we had to decide how to classify our audience by type. We took a large sheet of paper and wrote the various audiences on it:

technicians
lay people (executives)
engineers
scientists
jr. engineers
operators

As an afterthought, we added "history of audience analysis" to the list.

Next, we took a large sheet of paper and wrote "Scientists" across the top of it. Under this heading we listed all the things from our information that seemed to apply to the scientist. Part of the list read like this:

definition by education
definition by experience
limited background information needed
OK for theoretical calculations
show all facts
OK to use math, formulas, scientific terms
tables OK
be honest

We did the same for "engineers" and for "lay people (executives)." We looked at our lists for a while. They didn't work as they should. Some of the things listed under "Scientists" fit just as well under "Engineers." We didn't want a lot of needless repetition. Some of the things we had to say about lay people and executives did not seem to belong in the same section. Some of the things we had to say did not fit under any of the headings.

We had to look for new headings. After several more trial-and-error runs, we ended up with these:

History of Audience Analysis
A Lay Audience

Executives
Experts
Technicians
The Combined Audience

Under these headings we made a more complete outline, filling in facts and ideas under the major headings. No one but ourselves was to see the outline, so we skipped the formality of Roman numerals and other outline apparatus.

When we looked over our outline, we were pretty satisfied but had a nagging doubt about beginning with the history of audience analysis. For one thing, we were violating our basic classification scheme. We could probably justify that, but were we also violating our own rules of audience analysis? Would you be as fascinated by Aristotle's ideas on audience analysis as we were? Or by the fact that Sextus Julius Frontinius, water commissioner of Rome in 97 A.D., invented the information report? Reluctantly, we decided that you wouldn't be. We scrapped that section. We kept just a few examples to illustrate the introduction to the chapter.[6] When our rough but thorough outline was finished, we were ready to write our first draft of the chapter.

How thorough you are at this stage depends upon such things as the complexity of the material you are working with and your own working habits. Simple material does not require complicated outlines. And perhaps nowhere else in the writing process does personality play such a prominent role as it does at this arrangement stage. Some people prepare fairly complete outlines; others do not.

An article by Dr. Blaine McKee, "Do Professional Writers Use an Outline when They Write?" explores the outlining practices of professional technical writers.[7] Figure 2–1 is a table from his article showing the types of outlines used by the professionals. Only 5% used no outline at all, and only 5% reported using the elaborate sentence outline. Most reported using some form of the topic outline or a mixture of words, phrases, and sentences. Most kept their outlines flexible and informal, warning against getting tied to a rigid outline too early. But all but 5% did feel the need

Figure 2–1 Types of Outlines Used

Type		Percentage
Topic outline		60
Word	8.75%	
Word and phrase	51.25%	
Combination of word, phrase, and sentence		30
Sentence outline		5
No outline		5
Total		100

to think through their material and to get some sort of arrangement pattern down on paper before beginning the first draft of a report.

Experienced writers seldom go beyond the thorough but informal outlining procedure described earlier. However, if you do need a formal outline and need instruction in preparing one, look in Part VI, our alphabetically arranged "Handbook," under "Outlining."

Drafting and Revising

When you have finished arranging your material, you are ready for the drafting and revising of your report. We do not try to convince you that writing is an easy mechanical job. It is not. But we do give you suggestions that should make a tough job easier.

The Rough Draft

Writing a rough draft is a very personal thing. Few writers do it exactly alike. As you have seen, most write from a plan of some sort; a few do not. Some write at a fever pitch; others write slowly. Some writers leave revision entirely for a separate step. Some revise for style and even edit for mechanics as they go along, working slowly, trying, in effect, to get it right the first time. All we can do is to describe in general the practices of most professonal writers. Take our suggestions and apply them to your own practices. Use the ones that make the job easier for you and revise or discard the rest.

Probably our most important suggestion is to begin writing as soon after the arrangement stage as possible. Writing is hard work. Most people, even professionals, procrastinate. Almost anything can serve as an excuse to put the job off: one more book to read, a movie that has to be seen, anything. The following column by Art Buchwald describes the problem of getting started in a manner that most writers would agree is only mildly exaggerated.

> MARTHA'S VINEYARD—There are many great places where you can't write a book, but as far as I'm concerned none compares to Martha's Vineyard.
>
> This is how I managed not to write a book and I pass it on to fledgling authors as well as old-timers who have vowed to produce a great work of art this summer.
>
> The first thing you need is lots of paper, a solid typewriter, preferably electric, and a quiet spot in the house overlooking the water.
>
> You get up at 6 in the morning and go for a dip in the sea. Then you come back and make yourself a hearty breakfast.
>
> By 7 a.m. you are ready to begin Page 1, Chapter 1. You insert a piece of paper in the typewriter and start to type "It was the best of times . . ." Then you look out the window and you see a sea gull diving for a fish. This is not an ordinary sea gull. It seems to have a broken wing and you get up from the desk

to observe it on the off chance that somewhere in the book you may want to insert a scene of a sea gull with a broken wing trying to dive for a fish. (It would make a great shot when the book is sold to the movies and the lovers are in bed.)

It is now 8 a.m. and the sounds of people getting up distract you. There is no sense trying to work with everyone crashing around the house. So you write a letter to your editor telling him how well the book is going and that you're even more optimistic about this one than the last one which the publisher never advertised.

It is now 9 a.m. and you go into the kitchen and scream at your wife, "How am I going to get any work done around here if the kids are making all that racket? It doesn't mean anything in this family that I have to make a living."

Your wife kicks all the kids out of the house and you go back to your desk. . . . You look out the window again and you see a sailboat in trouble. You take your binoculars and study the situation carefully. If it gets worse you may have to call the Coast Guard. But after a half-hour of struggling they seem to have things under control.

Then you remember you were supposed to receive a check from the *Saturday Review* so you walk to the post office, pause at the drugstore for newspapers, and stop at the hardware store for rubber cement to repair your daughter's raft.

You're back at your desk at 1 p.m. when you remember you haven't had lunch. So you fix yourself a tuna fish sandwich and read the newspapers.

It is now 2:30 p.m. and you are about to hit the keys when Bill Styron calls. He announces they have just received a load of lobsters at Menemsha and he's driving over to get some before they're all gone. Well, you say to yourself, you can always write a book on the Vineyard, but how often can you get fresh lobster?

So you agree to go with Styron for just an hour.

Two hours later with the thought of fresh lobster as inspiration, you sit down at the typewriter. The doorbell rings and Norma Brustein is standing there in her tennis togs looking for a fourth for doubles.

You don't want to hurt Norma's feelings so you get your racket and for the next hour play a fierce game of tennis, which is the only opportunity you have had all day of taking your mind off your book.

It is now 6 p.m. and the kids are back in the house, so there is no sense trying to get work done any more for that day.

So you put the cover on the typewriter with a secure feeling that no matter how ambitious you are about working there will always be somebody on the Vineyard ready and eager to save you.

But you must begin and the sooner the better. Find a quiet place to work, one with few distractions. Choose a time of day when you feel like working, and go to work. What writing tools should you use? Our tools, for instance, for the first draft of almost any piece of writing are a yellow, legal-sized pad and a can filled with pencils of 2.5 hardness. Other writers compose on typewriters or microcomputers. One writer we know insists upon using white paper and a fountain pen filled with blue ink. The moral of all this is that the tools used are a matter of individual choice. There is something a bit ritualistic about writing, and most competent writers insist upon their own rituals.

Where should you begin? Usually, it's a good strategy to begin not with the beginning but with the section that you think will be the easiest to write. If you do so, the whole task will seem less overwhelming. As you write one section, ideas for handling others will pop into your mind. When you finish an easy section, go on to a tougher one. In effect, you are writing a series of short, easily handled reports rather than one long one. Think of a 1,500-word report as three short, connected, 500-word reports. You will be amazed at how much easier this attitude makes the job. We should point out that some writers do prefer to begin with their introductions and even to write their summaries, conclusions, and recommendations, if any, first. They feel this sets their purpose, plan, and final goals firmly in their minds. If you like to work that way, fine. Do remember, though, to check such elements after you have written the discussion to see if they still fit.

How fast should you write? Again, this is a personal thing, but most professional writers write rapidly. We advise you not to worry overmuch about phraseology or spelling in a rough draft. Proceed as swiftly as you can to get your ideas on paper. Later, you can smooth out your phrasing and check your spelling, either with your dictionary or your word processor's spell checker. However, if you do get stalled, reading over what you have written and tinkering with it a bit is a good way to get the flow going again. In fact, two researchers of the writing process found that their subjects spent up to a third of their time pausing. Generally, the pauses occurred at the ends of paragraphs or when the writer was searching for examples to illustrate an abstraction.[8]

Do not write for more than two hours at a stretch. This time span is one reason why you want to begin writing a long, important report at least a week before it is due. A report written in one long five- or six-hour stretch reflects the writer's exhaustion. Break at a point where you are sure of the next paragraph or two. When you come back to the writing, read over the previous few paragraphs to help you collect your thoughts and then begin at once.

Make your rough draft very full. You will find it easier to delete material later than to add it. Nonprofessional writers often write thin discussions because they think in terms of the writing time span rather than the reading time span. They have been writing on a subject for perhaps an hour and have grown a little bored with it. They feel that if they add details for another half-hour they will bore the reader. Remember this: At 250 words a minute, average readers can read an hour's writing output in several minutes. Spending less time with the material than the writer must, they will not get bored. Rather than wanting less detail, they may want more. Don't infer from this advice that you should pad your report. Brevity is a virtue in professional reports, but the report should include enough detail to demonstrate to the reader that you know what you're talking about. The path between conciseness on one hand and completeness on the other is often something of a tightrope.

As you write your rough draft, indicate where your references will go. Be alert for paragraphs full of numbers and statistics and consider presenting such information in tables and graphs. (See Chapter 12, "Graphical Elements of Reports.")

Whether your planning has been detailed or casual, keep in mind that writing is a creative process. Discovery does not stop when you begin to write. The reverse is usually true. For most people, writing stimulates discovery. Writing clarifies your thoughts, refines your ideas, and leads you to new connections. Therefore, be flexible. Be willing to revise your plan to accommodate new insights as they occur.

Revision

For some writers, revision goes on while they are writing. Therefore, for them, revising as a separate step is little more than minor editing, checking for misspellings and awkward phrases. For other writers, particularly those who write in a headlong flight, revising is truly rewriting and sometimes even rearranging the rough draft. Naturally, there are many gradations between these two extremes. We, for example, do some revising while we write, but we save most of it for a separate step. Most of our revising occurs when we type from our yellow legal pad manuscript into our word processor. It's at this stage that we pause and ponder the most and rework our language, content, and arrangement.

Whether your revising occurs while you write or as a separate step, you should be concerned about arrangement, content, logic, and style.

Arrangement and Content In checking your arrangement and content, try to put yourself in your reader's place. Does your discussion take too much for granted? Are questions left unanswered that the reader will want answered? Are links of thought missing? Have you provided smooth transitions from section to section, paragraph to paragraph? Do some paragraphs need to be split, others combined? Is some vital thought buried deep in the discussion when it should be put into an emphatic position at the beginning or end? Have you avoided irrelevant material or unwanted repetitions?

In checking content, be sure that you have been specific enough. Have you quantified when necessary? Have you stated that, "In 1990, 52% of the workers took at least 12 days of sick leave," rather than, "In a previous year, a majority of the workers took a large amount of sick leave"? Have you given enough examples, facts, and numbers to support your generalizations? And, conversely, have you generalized enough to unify your ideas and to put them into the sharpest possible focus? Have you adapted your material to your audience?

Is your information accurate? Don't rely on even a good memory for facts and figures that you are not totally sure of. Follow up any gut feeling you have that anything you have written seems inaccurate, even if it means a trip back to the library or laboratory. And check and double-check your math and equations. You can destroy an argument (or a piece of machinery) with a misplaced decimal point.

Logic Be rigorous in your logic. Can you really claim that A caused B? Have you sufficiently taken into account other contributing factors? Examine your discussion for every conceivable weakness of arrangement and content and be ready to pull it apart. All writers find it difficult to be harshly critical of their own work, but you will find it a necessary attitude.

Style After you have revised your draft for arrangement and content, read it over for style. (We treat this as a separate step, which it is. But, of course, if you find a clumsy sentence while revising for arrangement and content, rewrite it immediately.) Use Chapter 6, "Achieving a Readable Style," to help you. Rewrite unneeded passive voice sentences. Cut out words that add nothing to your thought. Cross out the pretentious words and substitute simpler ones. If you find a cliché, try to express the same idea in different words. Simplify; cut out the artificiality and the jargon. Be sure the diction and sentence structures are suitable to the occasion and the audience. Remember that you are trying to write understandably, not impressively. The final product should carry your ideas to the reader's brain by the shortest, simplest path.

In actual workplace situations, writers often share their drafts with colleagues and ask for their opinions. Often, someone who is not as close to the material as the writer can spot flaws far more quickly than can the writer. As you'll see when we discuss revising and editing with word processing, you may also share your work with your microcomputer.

When you are writing instructions, it's an excellent idea to share an early draft with people who are similar in aptitude and knowledge with the people for whom the instructions are intended. See if they can follow the instructions. Ask them to tell you where they had trouble carrying out your instructions or where poor vocabulary choice or insufficient content threw them off track.

Editing

Editing is a separate step that follows drafting and revising. It's the next-to-final step before releasing your report to its intended audience. When drafting and revising a manuscript, you may have to backtrack to the discovery or the arranging stage of the process. But when you are editing, it's either because you are satisfied with your draft or because you have run out of time. It's in the editing stage that you make sure your report is as mechanically perfect as possible, that it meets the requirements of standard English and whatever format requirements your situation calls for. If you are working for a large organization or the government, you may have to concern yourself with things such as stylebook specifications, distribution lists, and code numbers.

Checking Mechanics

Begin by checking your mechanics. Are you a poor speller? Check every word that looks the least bit doubtful. Some particularly poor spellers read their draft backwards to be sure that they catch all misspelled words. Develop a healthy sense of doubt, and use a good dictionary or the spelling checker of your word processing program. Do you have trouble with subject–verb agreement? Be particularly alert for such errors. In Part VI, we have provided you with a handbook that covers some of the more common mechanical problems. Word processing can help here by providing a check for some of the errors a computer program is able to detect.

Checking Documentation and Format

When you are satisfied with your mechanics, check your documentation. Be sure that all notes and numbers match. If you have not already done so, complete format requirements such as your title page and table of contents (TOC). (For help with such things, see Chapter 11, "Design Elements of Reports.") Check your TOC for accuracy. Be sure the headings in the report and the TOC match.

Checking Graphics

Check your graphics for accuracy, and be sure you have mentioned them at the appropriate place in the report. If they are numbered, be sure that their numbers and the numbers you use in referring to them match.

Again, word processing makes all these tedious little tasks easier. With word processing you don't have the agony of retyping an entire page to correct for one or two errors. When you are satisfied you have done all that needs to be done, print or type your final draft. Whether you, a typist, or a word processor types your final draft, as a final step, proofread it one more time before you turn it over to your audience. The author of a report is responsible for all errors no matter how the report is typed.

Revising and Editing with Word Processing

The advent of word processing has eased revision considerably since the time when revising reports called for writing between the lines and in the margins and even the cutting up of reports and splicing them together in new arrangements. Word processing offers, in effect, unlimited space for revision, allowing the writer to insert new sentences, paragraphs, and headings at any point in the report. With word processing, sentences and paragraphs can be easily moved from one place to another. Writers can get a cleanly typed printout at any stage of the writing and revising process. Because a clean text is easier to read than one full of arrows and written-over sentences, writers working with word processing can check their arrangement, content, logic, and style more easily than ever before. Because the machine does the retyping that even small changes sometimes entail, writers are far less reluctant to make needed changes than in the past. To further help the writer with editing and revising, spelling checkers and grammar and style checkers are available.

Spelling Checkers

You may have a spelling checker in your word processing program or one that you have bought separately. Spelling checkers work by looking up every word in your text in the dictionary that is a part of the program. Most spelling checkers have no sense of grammar or usage. They will stop on a word spelled correctly if that word does not happen to be in the program's dictionary, as is true of many technical terms. More importantly, they will not stop at a word that is in the program's dictionary when that word is used incorrectly in context.

A spelling checker won't catch errors like these:

The *student's* all came to class today.
They wanted to *here* your speech.
They wanted to hear *you* speech.

Use a spelling checker first, but then make sure to proofread as well.

Grammar and Style Checkers

You can get programs that check your work for grammar, punctuation, and style. Some will flag passive sentences, long sentences, wordy phrases, double words (like *and and*), unmatched pairs of quotation marks, and other problems. Some also flag problems with subject–verb agreement, incorrect possessives, and other grammatical faults. Some give your text a "readability rating" according to one or more readability formulas.

Grammar and style checkers can be helpful. They can make you more aware of your writing style. If you tend to write in the passive voice, they'll press you to change to active voice. If you tend to use wordy phrases or unnecessarily long words, you'll see shorter, crisper alternatives.

Use grammar and style checkers with great caution, however. Some current text-analysis programs are too rule-bound to be flexible, and some of the rules may be of doubtful validity. As one authority says, "Syntax writers that can truly evaluate a user's writing style await future breakthroughs in artificial intelligence."[9]

Think about the advice they give you in light of the purpose and audience for your document. Not every passive voice sentence should be rewritten as an active sentence. Not every sentence more than 22 words is too long. Grammar and style checkers work only at the sentence and word level, but the most serious problems with many documents are in their content and overall organization. If you change words and sentences here and there without considering larger issues of content and arrangement, you may actually be making your document less useful and understandable.

Planning and Revision Checklists

The following questions are a summary of the key points in Chapter 2, and they provide a checklist when you are composing.

SITUATIONAL ANALYSIS

- What is your topic?
- Why are you writing about this topic? What is your purpose (or purposes)?
- What are your readers' educational levels? What are their knowledge and experience in the subject matter area?
- What will your readers do with the information? What is their purpose?
- Do your readers have any expectations as to style and tone? Serious, light, formal, what?
- What is your relationship to your readers? How will this relationship affect your approach to them?
- What are your readers' attitudes about what you are going to say?

DISCOVERY

- What discovery approach can you use? Brainstorming? Using arrangement patterns? Other?
- Are there documents similar to the one you are planning that would help you?
- Do you have notes or journal entries available?
- What questions are your readers likely to want answered?
- Do you have all the information you need? If not, where can you find it? People? Library? Laboratory research?

ARRANGEMENT

- Are there standard arrangement patterns that would help you, for example, instructions, argument, proposals?
- Will you need to modify any such standard pattern to suit your needs?
- Do you need a formal outline?
- When completed, does your organizational plan fit your topic, material, purpose, and audience?
- Is everything in your plan relevant to your topic, purpose, and audience?
- If you have a formal outline, does it follow outlining conventions? Entries grammatically parallel? Each section divided at least into two parts? Correct capitalization? Entries substantive?

DRAFTING

- Do you have a comfortable place to work?
- Where in your organizational plan can you begin confidently?

REVISION

- Have you stated clearly and specifically the purpose of the report?
- Have you put into the report everything required? Do you have sufficient supporting evidence? Have you stated the implications of your information clearly?
- Are all your facts and numbers accurate?
- Have you answered the questions your readers are likely to have?
- Does the report contain anything that you would do well to cut out?
- Does your organization suit the needs of your content and your audience?
- Are your paragraphs clear, well organized, and of reasonable length? Are there suitable transitions from one point to the next?
- Is your prose style clear and readable?
- Is your tone appropriate to your audience?
- Are there people you should share your draft with—for example, members of your target audience—before going on to a final draft?

EDITING

- Have you checked thoroughly for misspellings and other mechanical errors?
- Have you included all the formal elements that your report needs?
- Are format items such as headings, margins, spacing, typefaces, and documentation consistent throughout draft?
- Are your headings and titles clear, properly worded, and parallel? Do your headings in the text match those in the table of contents?
- Is your documentation system the one required? Have you documented wherever appropriate? Do the numbers in the text match those in the notes?
- Have you keyed the tables and figures into your text and have you sufficiently discussed them?
- Are all parts and pages of the manuscript in the correct order?
- Will the format of the typed or printed report be functional, clear, and attractive?

- Does your manuscript satisfy stylebook specifications governing it?
- Have you included required notices, distribution lists, and identifying code numbers?
- Do you have written permission to reproduce extended quotations or other matter under copyright? (Necessary only when your work is to be published or copyrighted.)
- While you were composing the manuscript did you have any doubts or misgivings that you should now check out?
- Have you edited your manuscript for matters both large and small?
- What remains to be done, such as proofreading final copy?

Exercises

1. In a report, describe accurately and completely your current writing process. Be prepared to discuss your report in class.
2. Interview someone who has to write frequently (as, for example, many of your professors have to). Ask about the person's writing process. Base your questions upon the process described in this chapter, that is, ask about situational analysis and arranging, drafting, revising, and editing techniques. Take good notes during the interview, and write a report describing the interviewee's writing process.
3. Choose some technical or semitechnical topic you can write about with little research—perhaps a topic related to a hobby or some school subject that you enjoy. Decide upon a purpose and audience for writing about that topic. For example, you could instruct high school seniors in some laboratory technique. Or you could explain some technical concept or term to someone who doesn't understand it, to one of your parents, perhaps. Analyze your audience and persona following the suggestions on pages 17–20. With your purpose, audience, and persona in mind, brainstorm your topic. After you complete the brainstorming, examine and evaluate what you have. Reexamine your topic and purpose to see if information you have thought of during the brainstorming has changed them. Keeping your specific topic, purpose, audience, and persona in mind, arrange your brainstorming notes into a rough outline. Do not worry overmuch about outline format, such as Roman numerals, parallel headings, and so forth.
4. Turn the informal outline you constructed for Exercise 3 into a formal outline. (See "Outlining" in Part VI, the "Handbook.")
5. Write a rough draft of the report you planned in Exercises 3 and 4. Allow several classmates to read it and comment upon it. Revise and edit the rough draft into a final, well-written and well-typed draft. Submit all your outlines and drafts to your instructor.

CHAPTER 3

Writing Collaboratively

REVISING AND EDITING

Revising
Editing

GROUP CONFERENCES

Conference Behavior
Group Roles

As we pointed out to you on page 15, you can write collaboratively as well as individually. Organizations conduct a good deal of their business through group conferences. In a group conference, people gather together, usually in a comfortable setting, to share information, ideas, and opinions. Organizations use group conferences for planning, disseminating information, and, most of all, for problem solving. As a problem-solving activity, writing lends itself particularly well to conferencing techniques. And, in fact, collaborative writing is common in the workplace. In this section, we discuss some of the ways that people can collaborate on a piece of writing. We conclude with a brief discussion of group conferencing skills, skills that are useful not only for collaborative writing, but for any conference situation you are likely to find yourself in.

Collaborative writing can be two people working together or five or six. Writing groups with more than seven members are likely to be unwieldy. In any case, all the elements of composing—situational analysis, discovery, arrangement, drafting and revising, and editing—generally benefit by having more than one person working on them.

There is a downside in that groups sometimes digress and wander off the point of the discussion. Therefore, it helps to have some set procedures that guide discussion without stifling it. To that end, we have provided a planning and revision checklist at the end of this chapter and many others. The checklist provided on the front endpapers of this book combines the Chapter 2, "Composing," checklist with key elements from the checklist that follows this chapter. Following these checklists will help you stay on track. The checklists raise questions about topic, purpose, and audience that will guide either the individual or the group to the answers needed.

Planning

The advantage of working in a group is that you are likely to hit upon key elements that working alone you might overlook. The collaborative process greatly enhances situational analysis and discovery. Shared information about audience is often more accurate and complete than individual knowledge. By being forced to hammer out a purpose statement that satisfies all its members, a group heightens the probability that the purpose statement will be on target.

The flow of ideas in a group situational analysis and in a discovery brainstorming session will come so rapidly that you risk losing some of them. One or two people in the group should serve as recorders to capture the thoughts before they are lost. It helps if the recording is done so that all can see—on a blackboard, a pad on an easel, or a computer screen. During the brainstorming, remember to accept all ideas, no matter how outlandish they may appear. Evaluation and selection will follow.

The group can take one of the more organized approaches to discovery. For instance, if instructions are clearly called for, the group can use the arrangement pattern of instructions to guide discovery. If discovery includes gathering and checking information, the subject of Chapter 5, working in a group can really speed the process. The group can divide the work to be done, often assigning an area of the work to someone with expertise in that area.

When the brainstorming and other discovery techniques are finished, the group must evaluate the results. This is a time when trouble can occur. When everyone is brainstorming, it's fun to listen to the flow. There is a synergy working that helps to produce more ideas than any one individual is likely to develop working alone. But when it comes time to evaluate and select ideas, which means that some ideas will be rejected, tension in the group may result. Feelings may be ruffled. Keep your discussion at this point as open but as objective as you can. Divorce as much as possible the ideas from those who offered them. Evaluate the ideas on their merits—on how well they fit the purpose and the intended audience. Whatever you do, don't attack people for their ideas. Again, someone should keep track of the discussion in a way that the group can follow.

In collaborative writing, a good way of evaluating the ideas and information you are working with is to arrange them into an organizational plan. The act of arranging will likely show those items that work to meet the situation without shining too bright a spotlight on those that don't. A formal outline is not always necessary, but, in general, a group needs a tighter, more detailed organizational plan than does an individual. (See "Outlining" in Part VI, the "Handbook.") When you have reached such a plan, the planning is finished.

You may want to take one more step, however. Collaborative writing, like individual writing, can profit through networking with individuals or groups outside your immediate working group. You may, for example, want to seek comments about your organizational plan from people with particular knowledge of the subject area. If you're writing in a large organization, it might pay to seek advice from people senior to you who may see political implications your group has overlooked. In writing instructions, you would be wise to discuss your plan with several members of the group to be instructed. Be ready to go back to the drawing board if your networking reveals serious flaws in your plan.

Drafting

In the actual drafting of a document, a group can choose from one of several possible approaches.

Dividing the Work

For lengthy documents, perhaps the most common procedure is to divide the work among the group. Each member of the group takes responsibility for a segment of the organizational plan and writes a draft based upon the group plan. It's always possible, even likely, that in the writing a writer will alter the plan to some degree. If the alterations are slight enough that they do not cause major problems for group members following other segments of the plan, such alterations are appropriate. However, if such changes will cause problems to others, the people affected should be consulted about them.

Before beginning to write, a group should take two other steps that can save a lot of hassle and bother later on. First, they should agree on as many format features as possible, for example, spacing, typefaces, table and graph design, and the form of headings and footnotes. Our Part III, "Document Design in Technical Writing," provides help in this area. Second, the group should set deadlines for completed work and stick to them. The deadlines should allow ample time for the revising stage and for the delays that seem inevitable in writing projects. Even when a group has good agreements about design features, there will likely be many stylistic differences in the first drafts. A group that divides the work must be prepared to spend a good deal of time in revising and polishing to get a final product in which all the segments fit together smoothly.

Drafting in Collaboration

In a second method of collaboration, a group may want to actually draft the document in collaboration, rather than dividing up the work. Word processing, in particular, makes such close collaboration possible. Two or three people sitting before a keyboard and a screen will find that they can write together. Generally, one person will control the keyboard, but the collaborators not on the keyboard can provide immediate feedback and revision to what appears on the screen. Such close collaboration is not as cumbersome as it may sound, and experience shows that students and others can do it.[1]

One Person Doing the Drafting

The third method is to have one person draft the entire document. This produces a uniformity of style but has the obvious disadvantage in the classroom that not everyone will get needed writing experience. An alternative approach is to divide the work but then have one writer put the segments together, blending the parts into a stylistic whole. In large organizations you will find all of these methods or combinations of them in use.

Revising and Editing

Collaboration works particularly well in revising and editing. People working in a group frequently will see problems in a draft, and solutions to those problems, that a person working alone will not see.

Revising

In revising concern yourself primarily with content, organization, style, and tone. Be concerned with how well a draft fits purpose and organization. When the group can work together in the same location, everyone should have a copy of the draft, either on paper or on the computer screen. Comments about the draft should be both criterion based and reader based.[2]

Criterion-Based Comments Criterion-based comments measure the draft against some standard. For example, the sentences may violate good sentence standards by being too long or by containing pretentious language. (See Chapter 6, "Achieving a Readable Style.") Or perhaps in classifying information, the writer has not followed good classification procedures. (See pages 151–156.) Whatever the problem may be, approach it in a positive manner. Say something like "The content in this sentence is good. It says what needs to be said. But maybe it would work better if we divided it in two sentences. A sixty-word sentence may be more than our audience can handle."

Reader-Based Comments Reader-based comments are simply your reaction as a reader to what is before you. Compliment the draft whenever you can: "This is good. You really made me understand this point." Or you can express something that troubles you: "This paragraph has good factual content, but perhaps it could explain the implications of the facts more clearly. At this point, I'm asking what does it all mean. Can we provide an answer to the 'so-what' question here?"

Word Processing Word processing offers an attractive technique for revising, particularly when geography or conflicting schedules keep group members apart. Each member can do a draft on a disk and then send a copy of the disk to one or more coauthors. The coauthor can make suggested revisions on the disk and send it back to the original author. It helps if the revisions are distinguished in some way, perhaps through the use of asterisks or brackets. The original author can react to the changes in a way he or she thinks appropriate. Collaborators can use electronic mail in a similar way. If the collaborators can get together, they can slip the disk into the word processor and work on it side by side.

Comments from outside the Group As with the organizational plan, you should consider seeking comments on your drafts from people outside the group. People senior to you in your organization can be particularly useful to you in advising about tone and as to whether your work reflects the values and attitudes of the organization.

Problems in the Group Although effective, collaborative revision can cause problems in the group. We all get attached to what we write. Criticism of it can sting as much as adverse comments about our personality or our habits. Therefore, all members of the group should be particularly careful at this stage. Support other members of the group with compliments whenever possible. Try to begin any discussion by saying something good about a draft. As in discussing the plan, keep comments objective and not personal. Be positive rather than negative. Show how a suggested change will make the segment you are discussing stronger—for instance, by making it fit audience and purpose better.

If you are the writer whose work is being discussed, be open to criticism. Do not take criticism personally. Be ready to argue for your position, but also be ready to listen to opposing arguments. Really *listen.* Remember that the group is working toward a common goal—a successful document. You don't have to be a pushover for the opinions of others, but be open enough to recognize when the comments you hear are accurate and valid. If convinced that revision is necessary, make the changes gracefully and move on to the next point. If you react angrily and defensively to criticism, you poison the well. Other group members will feel unable to work with you and may find it necessary to isolate you and work around you. Harmony in a group is important to its success. Some debate is appropriate and necessary, but all discussions should be kept as friendly and positive as possible.

Editing

Make editing a separate process from revision. In editing, your major concerns are format and standard usage. Editing by a group is more easily accomplished than is revision. Whether a sentence is too long may be debatable. If a subject and verb are not in agreement, that's a fact. Use the handbook provided as Part VI of this text to help you to find and correct errors. Final editing should also include making the format throughout the document consistent. This is a particularly important step when the work of drafting has been divided among the group. Even with good agreements beforehand about format, inconsistencies will no doubt crop up. Be alert for them. All the equal headings should look alike. Margins and spacing should be consistent. Footnotes should all be in the same style, and so forth. See Part III, "Document Design in Technical Writing," for help in this important area.

The final product should be seamless. That is, no one should be able to tell where Mary's work leaves off and John's begins. To help you reach such a goal, we now provide you with some principles of conferencing.

Group Conferences

Collaborative writing is valuable as a means of writing and learning to write. In a school setting, collaborative writing is doubly valuable because it also gives you experience in group conferencing. You will find group conferencing skills necessary in the workplace. Most organizations use the group conference for training, problem solving, and other tasks. In this section we briefly describe good conference behavior and summarize the useful roles conferees can play. You'll find these principles useful in any conference and certainly in collaborative writing.

Conference Behavior

A good group conference is a pleasure to observe. A bad conference distresses conferees and observers alike. In a bad group conference, the climate is defensive. Conferees feel insecure, fearing a personal attack and preparing to defend themselves. The leader of a bad conference can't talk without pontificating; advice is given as though from on high. The group punishes those members who deviate from the majority will. As a result, ideas offered are tired and trite. Creative ideas are rejected. People compete for status and control, and they consider the rejection of their ideas as a personal insult. They attack those who reject their contributions. Everyone goes on the defensive, and energy that should be focused on the group's task flows needlessly in endless debate. As a rule, the leader ends up dictating the solutions, perhaps what was wanted all along.

In a good group conference, the climate is permissive and supportive. Members truly listen to one another. People assert their own ideas, but they do not censure the opinions of others. The general attitude is, "We have a task to do; let's get on with it." Members reward each other with compliments for good ideas and do not reject ideas because they are new and strange. When members do reject an idea, they do it gently with no hint of a personal attack on its originator. People feel free to operate in such a climate. They come forward with more and better ideas. They drop the defensive postures that waste so much energy and put the energy instead into the group's task.

How do members of a group arrive at such a supportive climate? To simplify things, we present a list of **dos** and **don'ts.** Our principles cannot, of course, guarantee a good conference, but if they are followed they can contribute to a successful outcome.

Dos

- Do be considerate of others. Stimulate people to act rather than pressuring them.
- Do be loyal to the conference leader without saying yes to everything. Do assert yourself when you have a contribution to make or when you disagree.

- Do support the other members of the group with compliments and friendliness.
- Do be aware that other people have feelings. Remember that conferees with hurt feelings will drag their feet or actively disrupt a conference.
- Do have empathy for the other conferees. See their point of view. Do not assume you know what they are saying or are going to say. Really listen and hear what they are saying.
- Do conclude contributions you make to a group by inviting criticism of them. Detach yourself from your ideas and see them objectively as you hope others will. Be ready to criticize your own ideas.
- Do understand that communication often breaks down. Do not be shocked when you are misunderstood or when you misunderstand others.
- Do feel free to disagree with the ideas of other group members, but never attack people personally for their ideas.
- Do remember that most ideas that are not obvious seem strange at first. Yet they may be the best ideas.

Don'ts

- Don't try to monopolize or dominate a conference. The confident person feels secure and is willing to listen to the ideas of others. Confident people do not fear to adopt the ideas of others in preference to their own, giving full credit when they do so.
- Don't continually play the expert. You will annoy other conferees with constant advice and criticism based upon your expertise.
- Don't pressure people to accept your views.
- Don't make people pay for past mistakes with continuing punishment. Instead, change the situation to prevent future mistakes.
- Don't let personal arguments foul a meeting. Stop arguments before they reach the personal stage by rephrasing them in an objective way.

Perhaps the rule "Do unto others as you would have them do unto you" best summarizes all these *dos* and *don'ts*. When you speak you want to be listened to. Listen to others.

Group Roles

You can play many roles in a group conference. Sometimes you bring new ideas before the group and urge their acceptance. Perhaps at other times you serve as information giver and at still others as harmonizer, resolving differences and smoothing ruffled egos. We describe these useful roles that you as a conference leader or member can play. We purposely do not distinguish between leader and member roles. In a well-run conference, an observer would have difficulty knowing who the leader is. We divide the roles into two groups: **task roles**—roles designed to move the group toward the accomplishment of its task—and **group maintenance roles**—roles designed to maintain the group in a harmonious working condition.

Task Roles When you play a task role, you help the group accomplish its set task. Some people play one or two of these roles almost exclusively, but most people slide easily in and out of most of them.

- **Initiators** are the idea givers, the starters. They move the group toward its task, perhaps by proposing or defining the task, or by suggesting a solution to a problem or a way of arriving at the solution.
- **Information seekers** see where needed facts are thin or missing. They solicit the group for facts relevant to the task at hand.
- **Information givers** provide data and evidence relevant to the task. They may do so on their own or in response to the information seekers.
- **Opinion seekers** canvass group members for their beliefs and opinions concerning a problem. They might encourage the group to state the value judgments that form the basis for the criteria of a problem solution.
- **Opinion givers** volunteer their beliefs, judgments, and opinions to the group or respond readily to the opinion seekers. They help set the criteria for a problem solution.
- **Clarifiers** act when they see the group is confused about a conferee's contribution. They attempt to clear away the confusion by restating the contribution or by supplying additional relevant information, opinion, or interpretation.
- **Elaborators** further develop the contributions of others. They give examples, analogies, and additional information. They might carry a proposed solution to a problem into the future and speculate about how it would work.
- **Summarizers** draw together the ideas, opinions, and facts of the group into a coherent whole. They may state the criteria that a group has set or the solution to the problem agreed upon. Often, after a summary, they may call for the group to move on to the next phase of work.

Group Maintenance Roles When you play a group maintenance role, you help to build and maintain the supportive group climate. Some people are so task oriented that they ignore the feelings of others as they push forward to complete the task. Without the proper climate in a group, the members will often fall short of completing their task.

- **Encouragers** respond warmly to the contributions of others. They express appreciation for ideas and reward conferees by complimenting them. They go out of their way to encourage and reward the reticent members of the group when they do contribute.
- **Feeling expressers** sound out the group for its feelings. They sense when some members of the group are unhappy and get their feelings out in the open. They may do so by expressing the unhappiness as their own and thus encourage the others to come into the discussion.
- **Harmonizers** step between warring members of the group. They smooth ruffled egos and attempt to lift conflicts from the personality level and objectify them. With a neutral digression, they may lead the group away from conflict long enough for tempers to cool, allowing people to see the conflict objectively without further help from them.
- **Compromisers** voluntarily withdraw their ideas or solutions in order to maintain group harmony. They freely admit error. With such actions, they build a climate within which conferees do not think their status is riding on their every contribution.
- **Gatekeepers** are alert for blocked-out members of the group. They subtly swing the discussion away from the forceful members to the quiet ones and give them a chance to contribute.

Planning and Revision Checklist

The following questions are a summary of the key points in this chapter, and they provide a checklist for composing collaboratively. To be most effective, the questions in this checklist should be combined with the checklist questions following Chapter 2, "Composing." To help you use the two checklists together, we have combined Chapter 2 questions with the key questions from this list and printed them on the front endpapers of this book.

Planning

- Is the group using appropriate checklists to guide discussion?
- Has the group appointed a recorder to capture the group's ideas during the planning process?

Revision

- Are format items such as headings, margins, spacing, typefaces, and documentation consistent throughout the group's documents?
- Does the group have criteria with which to measure the effectiveness of the draft?

- When planning is completed, does the group have an organizational plan sufficiently complete to serve as a basis for evaluation?
- How will the group approach the drafting stage?
  - By dividing the work among different writers?
  - By writing together as a group?
  - By assigning the work to one person?
- Has the group agreed on format items such as spacing, typography, table and graph design, headings, and documentation?
- Has the group set deadlines for the work to be completed?
- Are there people you should share your draft with? Supervisors? Peers? Members of the target audience?
- Do people phrase their criticisms in an objective, positive way, avoiding personal and negative comments?
- Are the writers open to criticism of their work?
- Is the climate in the group supportive and permissive? Do members of the group play group maintenance roles as well as task roles, encouraging one another to express their opinions?

Exercises

By following the techniques outlined in this chapter, groups could do most of the writing exercises in this book collaboratively. For a warm-up exercise in working collaboratively, work the following problem:

- Divide into groups of three to five people. Consider each group to be a small consulting firm. An executive in a client company has requested a definition of a technical term used in a document the firm has prepared for that company.
- The group plans, drafts, revises, and edits an extended definition (see pages 164–168) in a memo format (see page 382) for the client.
- Following the completion of the memo, the group critiques its own performance. Before beginning the critique, the group must appoint a recorder to summarize the critique.
  - How well did the members operate as a group?
  - What methods did the group use to work together to analyze purpose and audience and to discover its material?
  - What technique did the group use to draft its memo?
  - Was the group successful in maintaining harmony while carrying out its task?
  - What trouble spots emerged?
  - What conclusions has the group reached that will help future collaborative efforts?
- The recorders report to the class the summaries of the groups. Using the summaries as a starting point, the class discusses collaborative writing.

CHAPTER 4

Writing for Your Readers

TECHNICIANS

Point of View
Background

THE COMBINED AUDIENCE

Illustrative Example
Other Situations

In Chapter 2, "Composing," we tell you how and why to analyze your readers. In this chapter we provide you with some ways to use that analysis to plan strategies for communicating successfully with your readers.

Perhaps in no other kind of writing is this business of matching a particular piece of writing to a particular audience as important as it is in technical writing. Engineers of the nineteenth century were aware of the problems facing audiences of different levels. In 1887, the American engineer Arthur M. Wellington had different parts of his book, *The Economic Theory in the Location of Railways*, set in three different type sizes: large type for the lay reader, medium-sized type for the reader who could understand some technical data, and small type for the reader who needed the most detailed scientific data. Readers, knowing their own limitations and interests, had clear signals as to what to read and what not to read.[1]

In this chapter we break down audiences a bit differently. We tell you how to write for lay people, executives, experts, and technicians. In Figure 4–1, we show you the major concerns and characteristics of these audiences. We also tell you how to put together a combined report for an audience in which these groups are mixed. However, no matter the audience, four basic imperatives are always the same:

- Consider your readers' knowledge and experience.
- Determine your readers' point of view, that is, their concerns and interests.
- Include only the content that the readers need.
- Organize, write, and format that content so that your readers can access it, understand it, and use it.

Before we discuss the four audiences, let us caution you that no audience is uniform, falling readily into a neat category. We speak of a lay audience, or an executive audience, but such audiences are by no means totally homogeneous units. An audience might be compared to an aggregate of rocks of all shapes and sizes as opposed to a mass of smooth marble. This chapter, therefore, is not audience adaptation made simple, foolproof, and mechanical. Rather, it contains a series of generalizations that should help you understand the process, but only you can adapt for your own audience.

AUDIENCE	CONCERNS AND CHARACTERISTICS
Lay Persons	■ Read for learning and interest ■ Have more interest in practice than theory ■ Need help with science and mathematics ■ Enjoy and learn from human interest ■ Require background and definitions ■ Need simplicity ■ Learn from simple graphics
Executives	■ Read to make decisions ■ Have more interest in practice than theory ■ Need plain language ■ Learn from simple graphics ■ Need information on people, profits, and environment ■ Expect implications, conclusions, and recommendations expressed clearly ■ Read selectively—skimming and scanning ■ Have self-interests as well as corporate interests
Technicians	■ Read for How-To Information ■ Expect emphasis on practical matters ■ May have limitations in mathematics and theory ■ May expect theory if higher level
Experts	■ Read for how and why things work ■ Need and want theory ■ Will read selectively ■ Can handle mathematics and terminology of field ■ Expect graphics to display results ■ Need new terms defined ■ Expect inferences and conclusions to be clearly but cautiously expressed and well supported
Combined	■ One person may combine the attributes of several audiences ■ Readers may consist of representatives of several audiences

Figure 4–1 Audience Concerns

A Lay Audience

We discuss the lay audience at some length for two reasons. First, many technical experts really do not understand the special needs of this important audience. In a free society, citizens who do not understand science and technology are likely to use science and technology badly. Second, some of the techniques used for lay people have their uses for other audiences as well. For example, in the same way that lay people do, even experts sometimes need definitions and background to comprehend what they are reading.

Who are **lay people?** They are fourth graders learning how the moon causes solar eclipses. They are the bank clerk reading a Sunday newspaper story about genetic engineering and the biologist with a Ph.D. reading an article in *Scientific American* entitled "The Nature of Metals." In short, we are all lay people once we are outside our own particular fields of specialization. Most lay people have at least a high school diploma. In 1987, 75.7% of the U.S. population over 25 years of age had at least four years of high school; 19.9% had at least four years of college.[2] Despite these encouraging statistics, some studies indicate that 60 million of 180 million adult Americans read at only marginal levels. They have difficulty reading things such as leases and newspapers; some even have difficulty with road signs and menus in fast-food restaurants.[3] Many of the high school graduates and even college graduates have only a smattering of mathematics and science and are a little vague about both subjects.

Point of View

Lay people read for interest. They read to understand the world in which they live. They are not very expert in the field, or they would not be reading an article written for a lay audience. In these days of environmental concern and consumerism, they may be reading as a prelude to action. When such is the case you cannot ignore the politics of the situation. The experience of the Battelle Memorial Institute provides a case in point. The U.S. Department of Energy chose Battelle to identify sites for a national repository for high-level nuclear wastes. As Neal E. Carter, a vice president of Battelle, explains, Battelle made its plans and schedules "based upon scientific methods for achieving such resolutions."[4] Battelle soon found that public reaction and opposition threw its plans and schedules awry. Mr. Carter tells what happened:

> Let me describe in detail how these developments have significantly altered the ivory tower atmosphere of our scientists. Our technical staff had to become directly involved in providing the background studies needed by the staff of the United States Congressmen involved in the development of the Nuclear Waste Policy Act. Testimony before Congress was required and most importantly our scientists had to become directly involved in almost 200 briefings, public meetings, and public hearings in over 23 states in which we were doing our research studies. Here is where the new challenge set in. During the public meetings, the scientists had to convert their technical jargon into terminology that would be understandable to the lay public, and they had to withstand the consequences of meetings with a disturbed public. For example, I can recall one public meeting where, after we had finished our testimony, we returned to find our rented cars had been covered with manure. Public meetings were filled with irate citizens carrying signs with vocabulary not well known to Ph.D. scientists. There were threats, personal threats aimed at scientists giving testimony on their siting studies. There was general disbelief and disrespect by the audience. The educated Ph.D. scientist had not been trained to deal with these threats, and disbeliefs, and disrespect by individuals that are not trained scientifically as he is. We did

> not like having the content of our research program and its schedule highly influenced by the general public.

The hostile audience reaction to the Battelle scientists clearly demonstrates the mistake of approaching an audience in a way they cannot understand or relate to. The Battelle scientists had to regroup, to analyze their audience, and to develop strategies that would achieve their purposes. The Battelle solution was a series of reports, written and oral, simplifying extremely complex matters for a nontechnical lay audience. While remaining simple and understandable, the reports must be scientifically accurate and persuasive. The scientists involved have had to recognize that their roles go beyond being scientific investigators. They had to be salespeople, in Mr. Carter's words, "constantly communicating the benefits of research to the public and political sector."

A lay audience's reason for reading technical information may not always be as dramatic as in the Battelle situation, but it is almost always practical. Lay people, for instance, are generally much more concerned with what things do than how they work. Their interest is personal. What impact will this new development have on them? They are more interested in the fact that widespread computer networks may invade their privacy than the fact that computers work on a binary number system. They are more interested in the safety, efficiency, and cost of nuclear waste disposal than in the theory of such disposal.

Lay people, more than any other audience, present a bewildering complexity of interests, skills, educational levels, prejudices, and concerns. Can we define exactly how to write for them? The truth is that we cannot—completely. But we can suggest some broad avenues of approach that—to paraphrase Lincoln—apply to all of the lay people some of the time, some of the lay people all of the time, but *not* all of the lay people all of the time.

Human Interest

Human interest serves two purposes when you are writing for lay people: It motivates people to read and seems to help them retain more of what they read.

Reader Motivation To begin with, you often have to motivate lay people to read what you have written. Frequently, this means introducing human interest into your writing. Most of us, at any educational level, have an interest in other human beings and in human personalities. Most writers for lay audiences recognize this interest and use it to gain acceptability for their subject matter. For example, an article in *Time* about farm problems will give us statistical information about the number of farmers losing their farms. But the writer of the article knows that many of us do not relate very well to bare, abstract statistics. Therefore, the

Time writer will also usually introduce people into the article. Perhaps Bill and Mary Gould and their two children, a "typical" farm family living in Iowa, will be described. We'll learn what effect losing their farm has had on their lives. We are interested in learning about what happens to real people, and through such knowledge we can better understand the farm problem.

In the following example, notice how the introduction to an article on hurricane prediction uses human drama to capture the reader's interest:

> Today, some 45 million Americans live in coastal areas vulnerable to hurricanes. Many of them have seen merely the fringes of a hurricane and believe it capable of nothing more dangerous than tearing off shingles or flattening large trees. Only a fraction have endured the unbelievable fury at the storm's center.
>
> How bad, in fact, can a hurricane be? Bob Sheets, a forecaster with the National Hurricane Center, presents visitors with two photographs of an apartment house in Pass Christian, Mississippi. The first shows a solid, three-story brick structure, separated from the Gulf of Mexico by an eight-foot seawall, a four-lane highway, a row of substantial oak trees, a generous front lawn, and a swimming pool. When Hurricane Camille was predicted in August 1969, 25 people felt confident enough to stage a "hurricane party" there. The second photo shows the same site, after Camille passed through. Only the swimming pool remained. Of the partygoers, one managed to cling to the upper branches of a tree. Another was swept out to sea and cast back 12 hours later and four miles down the beach, semi-conscious, but somehow still alive. The other 23 died.[5]

The technique involved in this excerpt is explained by the editor of a science magazine designed for a lay audience:

> We take out some of the content whenever it gets in the way of the story telling. We stress readability and the quality and freshness of the writing over content because we think the first imperative is that people actually read the article—not a trivial task when you are trying to interest two million very different people in the complexities of cosmology or molecular biology. We also think that what most people carry away from a popular magazine article is a rough sense of the subject, not the details. So we emphasize the cultural context, the human impact, the anecdotal example, precisely because they contribute more to that lingering impression and are more important to the lives of our readers than the detailed physics or the viral mechanisms, however scientifically elegant.[6]

Increased Retention In addition to motivating people to read, human interest also seems to improve readers' retention of what they read. In one experiment, *Time-Life* editors were asked to rewrite a passage from a high school history text. They introduced what they called "nuggets" into the passage, that is, anecdotes and stories about historical people. The result: student recall rose from 20% on the original passage to 60% on the rewritten one. Other research indicates that examples, questions, elaborations, and summaries all improve reader recall.[7]

While providing human interest and perhaps even human drama, be careful not to exaggerate scientific achievements. Newswriters, when writing stories about scientific achievements, sometimes forget this need for caution—to the dismay of the scientists involved. A newspaper story concerning research on the skin ailment psoriasis carried this headline: "Psoriasis Cure Breakthrough Seen." The lead of the story announced, "Scientists Wednesday announced a breakthrough in treating psoriasis, the skin disease which causes misery for about 6 million Americans." The scientist involved criticized the story, saying that nowhere in the report presented by the scientists was the word *breakthrough* used. He concluded by saying, "The last sentence of our writeup said cure of psoriasis is probably 50 years away. Yet the title of this article you sent says 'Psoriasis Cure Breakthrough Seen.' All I can say is [censored]!"[8]

Background

Without sufficient background, readers will not be able to comprehend and absorb your material. For example, most American and Canadian readers of this book know enough about baseball to comprehend a sentence such as, "Casey hit Cohen's high hard one down the right field line, moving Morrisey from first to third." American and Canadian readers have what reading experts call a *schema* upon which to hang the sentence. That is, they can visualize the baseball field. They know that Casey has to be the batter at home plate and Cohen has to be the pitcher. They know where the right field line is and that it is far enough away from third base to allow Morrisey to gain a few steps on the throw from right field. All that is a lot to know, but readers with the appropriate baseball schema can easily organize and integrate the new information in the sentence into their knowledge. Readers without the appropriate schema can make little sense of the sentence and will not absorb the information in it.

As a writer on a technical or scientific subject, you are likely to be an expert. Your biggest difficulty, perhaps, will be to remember how little you knew about your subject matter before you studied and experienced it, before you had a schema for it. Do try to remember and to provide your lay audience with the background information they'll need to comprehend and absorb your material. Provide such information in terms your reader can understand. In the following example, the writer introduces his subject, dietary control of salt, by discussing the role of salt (sodium chloride) in the human body. Such information would be known to the nutritionist but not to the lay reader.

Beginning with the familiar

Proceeding to the unfamiliar

Sodium chloride, most frequently encountered in the food supply as common table salt, is an essential part of the human diet. As it dissolves in water, it dissociates into two ions—one of sodium and the other of chloride. In all mammals, including humans, the sodium ion is required to maintain the pressure and volume of blood. It is also essential in controlling the passage of water into and out of the body's cells, and the relative volumes of fluids inside and outside those cells. In addition, sodium is needed for the transmission of nerve impulses and for the metabolism of carbohydrates and proteins.

The chloride ion, too, is essential, and is involved in maintaining the acid–base balance in the blood, and in tissue osmolarity (the passage of water across cell walls to maintain proper concentrations of various chemical entities). It is necessary for activating certain essential enzymes, and for the formation of hydrochloric acid in the stomach, needed in the digestive process.

Thus, both sodium and chloride are normal and necessary constituents of body tissues and fluids, and must be provided for in the diet.[9]

In providing background, the writer begins with the familiar, "common table salt," and works his way to the unfamiliar. Analogy is a special way of comparing the familiar to the unfamiliar and as such is often very useful in lay writing. Notice its use in this excerpt from an article on laser technology:

The main laser bay is a giant clean room two stories high and nearly the size of a football field. To take the analogy a step further, a single low-power laser beam begins at one end zone, is amplified, and is divided by mirrors into six separate beams. Each beam is then amplified further and directed through one of six long tunnel-like chains of optical components. Just beyond the 50-yard line, each beam is divided again, this time into four beams, and all 24 are directed toward a four-foot stainless steel sphere right about where the opposite goal line would be.[10]

Analogy is a powerful device to help the lay reader. Of course, you must be sure your readers have the necessary schema to understand your analogy. For instance, the example above will do little for readers who do not possess a schema for football. In the everyday world around us there are countless things—such as light bulbs, radios, garden hoses, faucets, windows, mirrors, trees, tennis rackets, baseballs, clay, loam, granite, ocean waves—known and somewhat understood by everyone that the writer can use to explain about every law of science. It's a question of using one's imagination and knowing how to talk in lay terms without being condescending. (See also pages 149–151.)

In addition to providing a schema, background information is often used to stress the importance of the subject at hand. A *Harvard Medical School Health Letter* on alcoholism begins this way:

> Consider that half of all deaths in automobile accidents, half of all homicides, and a fourth of all suicides are related to alcohol abuse; that persons with a "drinking problem" are seven times more likely to be separated or divorced than those in the general population; that the total cost of alcohol abuse in this country may exceed 44 *billion* dollars; that an alcoholic's life span is shortened (on average) by 10–12 years; and that at least ten million persons in this country abuse alcohol. No wonder some have labeled alcoholism the most devastating sociomedical problem faced by human society short of war and malnutrition.[11]

Give your readers, then, a grounding in your subject. When possible, use familiar things as points of comparison. If your background also wakens your readers' concerns and interest, so much the better.

Definitions

Although we learn our expert language long after we have acquired our common, everyday language, expert language becomes such a part of our life that we often forget that others don't share it with us. We forget that we are, in effect, bilingual, possessing both a common language that we share with others and an expert language that we share with a much smaller group. In reaching out to lay people, we must remember that they need specialized terms and words defined. You are the host when you write. You have invited your readers to come to you. You owe them every courtesy, and defining difficult terms is a courtesy. If you force your readers to the dictionary every fourth line, their interest will soon flag.

Depending upon the needs of both writer and audience, terms can be defined either briefly, usually by the substitution of a more familiar term, or at length. The use of brief definitions is seen in this segment of *The Harvard Medical School Health Letter* on alcoholism:

Brief definitions

Brief definitions

> Alcoholics have a much higher incidence of *peptic ulcers* and *pancreatitis* (inflammation of the pancreas) than nonalcoholics. In addition, many suffer from repeated episodes of nausea, vomiting, and abdominal distress—most often due to superficial *gastritis* (inflammation of the lining of the stomach).[12]

Here, the lay definition substitutes simpler, more familiar language for the physician's technical terms. In the same article is this more extended definition:

Extended definition

If a group of experts attempts a definition of "alcoholism," as many definitions as experts usually emerge, which, of course, points out the complexity of this particular "-ism." Some argue the relative merits of biological (it is a "disease") versus social–psychological (it is a "behavior disorder") theories, but most settle for descriptive definitions—such as the one from the Rutgers University Center of Alcohol Studies: "An alcoholic is one who is unable consistently to choose whether he shall drink or not, and who, if he drinks, is unable consistently to choose whether he shall stop or not." Ultimately, all will agree to descriptive definitions which point out that for the person with a serious drinking problem, tremendous disruption occurs in terms of health, interpersonal relationships, and the basic activities of life—eating, sleeping, and working.[13]

Notice how this definition is used to stress one of the major points of the article—that alcoholism, however caused, disrupts the lives of alcoholics and the lives of those around them. If the writer had left the readers to define *alcoholism* for themselves, the emphasis wanted by the writer would have been lost. Likewise, a dictionary definition may not emphasize the points important to the writer. Compare this definition of *alcoholism* from *The American Heritage Dictionary of the English Language* with the one from *The Harvard Medical School Health Letter:*

Dictionary definition

1. A chronic pathological condition, chiefly of the nervous and gastroenteric systems, caused by habitual excessive alcoholic consumption. 2. Temporary mental disturbance, muscular incoordination, and paresis caused by excessive alcoholic consumption.

The dictionary definition deals primarily with the physiological aspects of alcoholism and overlooks the social aspects that concern the writer.

Be careful not to distort the true meaning of terms if you substitute more common terms for technical language. One researcher, for example, felt his work was distorted by this lead in a newspaper story:

> A research group reported Friday that marijuana causes chimpanzees to overestimate the passage of time, and a single dose can keep them befuddled for up to three days.

The researcher commented:

> The term "befuddle" was not employed in our scientific report, and the statement in the news article "and a single dose can keep them befuddled for up to three days" is erroneous and misleading. Three days were required to recover normal baseline performance following administration of high doses.[14]

Scientists choose words very precisely and for good reason. Although their findings must be interpreted for lay readers, to distort or to sensationalize their work is a disservice both to them and to the reader. Define, then, for clarity and understanding and to aid your own exposition, but do it with care. Refer also to the section on definition on pages 164–169.

Simplicity

There are several ways to keep an article simple for lay people. Two of them we have already discussed: give needed background and define those specialized terms that you must use. Most often, avoid specialized words for which you can find simple substitutes. Experts can read certain meanings into the word *homeostasis,* and you should use it for them, but for a lay audience *stable state* or *equilibrium* will serve as well. But another caution here: Most people like to enrich their vocabularies. So don't avoid technical terms altogether. Just don't put them together in incomprehensible strings with a reader-be-damned attitude.

Be Directive Another way to simplify your writing is to be directive. That is, provide your readers with a roadmap through your writing by including sufficient introductions, transitions, headings, summaries, and the like. We have a good deal more to say about designing documents for lay people and other audiences in Chapter 10, "Document Design." In Chapter 11, "Design Elements of Reports," we provide the information you need to prepare introductions, summaries, conclusions, headings, and the like.

Use Plain Language Some scientific specialities are loaded with mathematics. Others, such as biochemistry, are full of formulas, complicated charts, and diagrams incomprehensible to lay people. Mathematics, formulas, and diagrams are useful shorthand expressions for experts. Through them, experts find a precision impossible to obtain in any other way. But what experts sometimes forget are the years spent learning how to handle such precise tools. The average person, lacking those years of training, rarely can handle them. When you write for lay people, you must force yourself to express your ideas in plain language. As we have seen in the background examples, it can be done.

Avoid Overcomplex Sentences Pay some attention to your sentence length. Research confirms a fact that is perhaps intuitively obvious: Sentences that are too long or too complex cause difficulties for readers.[15] A reader familiar with a subject can comprehend sentences of greater difficulty than can a reader unfamiliar with the subject. In effect, the reader unfamiliar with the subject is fighting two battles at one time: one with the complex sentence structure, another with the subject matter. Therefore, when dealing with lay readers be ever conscious of the need to keep your sentence structures, as well as your vocabulary, at a simpler level than might be appropriate for a more expert audience.

In Chapter 6, "Achieving a Readable Style," we discuss how to achieve a style that aids rather than impedes understanding. The two paragraphs that follow illustrate that style. They are from a Public Health Service pamphlet that explains noninsulin-dependent diabetes to a lay audience. To give background information about the disease, the writer uses a simple vocabulary and well-constructed sentences of a reasonable length.

> Ten million Americans—about 1 in 20—have diabetes. Over 90 percent of these people have a form of the disease called *noninsulin-dependent diabetes.* This kind of diabetes begins in adulthood. Because the symptoms of this disease can be mild, about half of the people who have noninsulin-dependent diabetes don't know it. Over time, however, diabetes can slowly damage the heart, blood vessels, kidneys, and nerves. The damage can occur even in someone who is not aware that he or she has the disease.
>
> Treament can control noninsulin-dependent diabetes, and, most experts feel, reduce the chances that these harmful changes or complications will threaten health. In many cases, weight control alone can control diabetes. However, the treatment that exists now is not a cure. Until a cure is found, treatment must continue throughout a person's life.[16]

Graphics

In writing for a lay audience, you use graphics to interest your readers as well as to inform and explain. You can use bar charts in place of equations to explain mathematical concepts or in place of formulas for chemical concepts. Or you can combine tables and pictographs that establish facts, formulas, or definitions quickly, clearly, and in a way that interests readers.

Be sure to use graphics suited to a lay audience. For example, tables should be simple, as in Figure 4–2, not complex as in Figure 4–3. Use simple bar graphs, as in Figure 4–4, not complex line graphs, as in Figure 4–5.

If your readers are among the 60 million Americans only marginally literate, you may use graphics to carry the major points of your message. Most people are familiar with the skull and crossbones poison symbol used on products like Drano to warn those who can't read the written

Table E. **Percent of Women Divorced After First Marriage, by Age at First Marriage: 1970 and 1985**

Age at first marriage	1985	1970
Total	23.2	14.2
Under 20 years	32.4	19.6
20 to 24 years	18.2	10.9
25 to 29 years	13.6	9.2
30 years and over . . .	11.8	9.1

Figure 4–2 A Simple Table
Source: Bureau of the Census, *Studies in Marriage and the Family* (Washington, DC: GPO, 1989) 4.

Table A. **Family Groups with Children Under 18, by Type and Race and Hispanic Origin of Householder or Reference Person: 1988**
(Numbers in thousands)

	All family groups		Family households		Related subfamilies		Unrelated subfamilies	
Race and group	Number	Percent	Number	Percent	Number	Percent	Number	Percent
ALL RACES								
Family groups with children . . .	34,345	100.0	31,920	100.0	1,998	100.0	427	100.0
Two-parent	24,977	72.7	24,600	77.1	366	18.3	11	2.6
One-parent	9,368	27.3	7,320	22.9	1,632	81.7	416	97.4
Mother only	8,146	23.7	6,273	19.7	1,480	74.1	393	92.0
Father only	1,222	3.6	1,047	3.3	152	7.6	23	5.4
WHITE								
Family groups with children . . .	28,104	100.0	26,618	100.0	1,167	100.0	319	100.0
Two-parent	22,013	78.3	21,699	81.5	304	26.0	10	3.1
One-parent	6,090	21.7	4,918	18.5	863	74.0	309	96.9
Mother only	5,100	18.1	4,066	15.3	743	63.7	291	91.2
Father only	990	3.5	852	3.2	120	10.3	18	5.6
BLACK								
Family groups with children . . .	5,057	100.0	4,195	100.0	766	100.0	96	100.0
Two-parent	2,055	40.6	2,016	48.1	39	5.1	–	–
One-parent	3,003	59.4	2,180	52.0	727	94.9	96	100.0
Mother only	2,812	55.6	2,020	48.2	701	91.5	91	94.8
Father only	191	3.8	160	3.8	26	3.4	5	5.2
HISPANIC[1]								
Family groups with children . . .	3,321	100.0	2,991	100.0	291	100.0	39	(B)
Two-parent	2,205	66.4	2,123	71.0	77	26.5	5	(B)
One-parent	1,116	33.6	868	29.0	214	73.5	34	(B)
Mother only	977	29.4	754	25.2	193	66.3	30	(B)
Father only	139	4.2	114	3.8	21	7.2	4	(B)

– Represents zero.
B Base less than 75,000.
[1] May be of any race.
Note: Family groups comprise family households, related subfamilies, and unrelated subfamilies.

Figure 4–3 A Complex Table
Source: Bureau of the Census, *Studies in Marriage and the Family* (Washington, DC: GPO, 1989) 13.

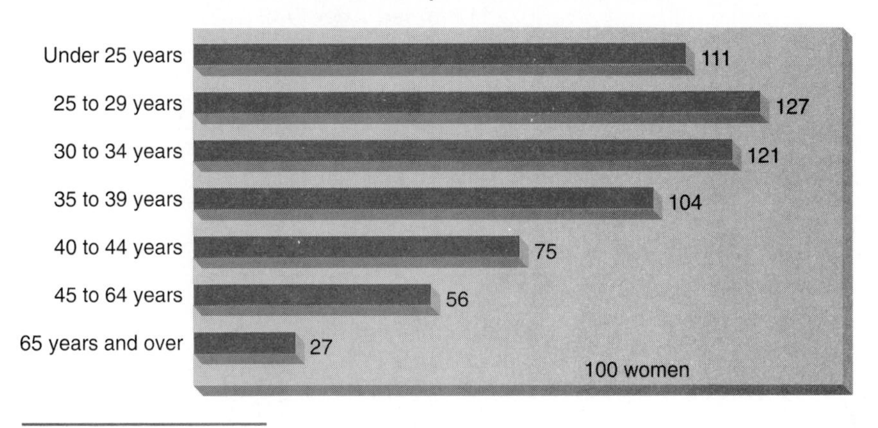

Figure 4–4 A Simple Bar Chart
Source: Bureau of the Census, *Studies in Marriage and the Family* (Washington, DC: GPO, 1989) 5.

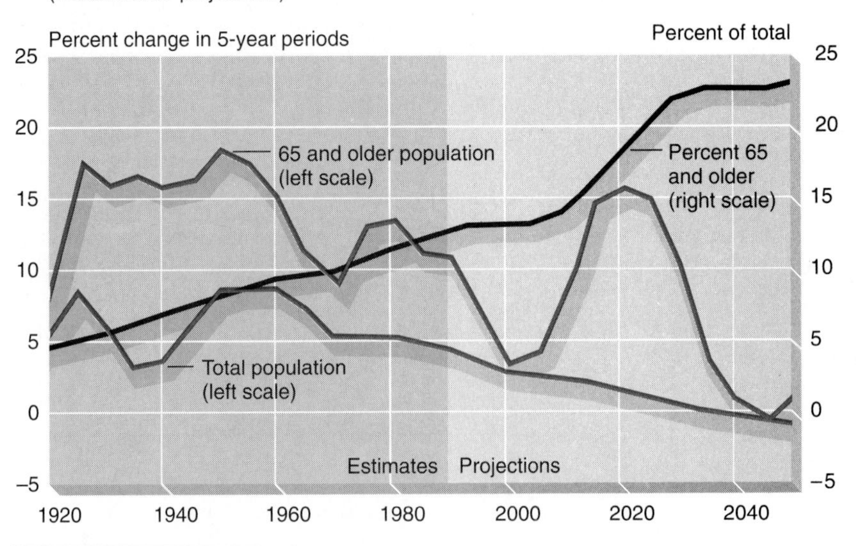

Figure 4–5 A Complex Line Graph
Source: Bureau of the Census, *Studies in Marriage and the Family* (Washington, DC: GPO, 1989) 6.

warning on the label. The Puget Sound Power and Light Company was faced with the problem of newly arrived Asian immigrants unfamiliar with both electricity and the English language. To solve the problem of educating these new customers, the company put their warnings in pictures and in several Asian languages, as shown in Figure 4–6.

We have a good deal more to say about graphics in Chapter 12, "Graphical Elements of Reports."

Lay people present a challenging audience to write for because their needs and capabilities are so difficult to pin down and define. We can say, in general, that in writing for them we should provide ample background (without mathematics or complicated formulas), definitions, illustrations, and simple charts. Remember, lay people are reading mainly for interest, and their interest is mainly practical: How will this scientific development affect our lives? So your cardinal rule in writing for lay people should be to keep things uncomplicated, interesting, human, practical, and personal and, if possible, to provide a touch of drama.

Examples of good writing for lay people are easy to find. Magazines such as *Time, Newsweek,* and *National Geographic* regularly print articles on scientific subjects for their readers. Reading and analyzing articles in such magazines is an excellent way to learn how to write for a lay audience.

If your children put things in wall outlets,

Nếu trẻ em nhét đồ vật vô các lỗ cắm điện,

ຖ້າລູກຂອງທ່ານເອົາສິ່ງຂອງໄປຍັດໃສ່ບ່ອນສຽບໄຟຟ້າ,

如果你的孩子把任何東西放入牆上電線插穴内

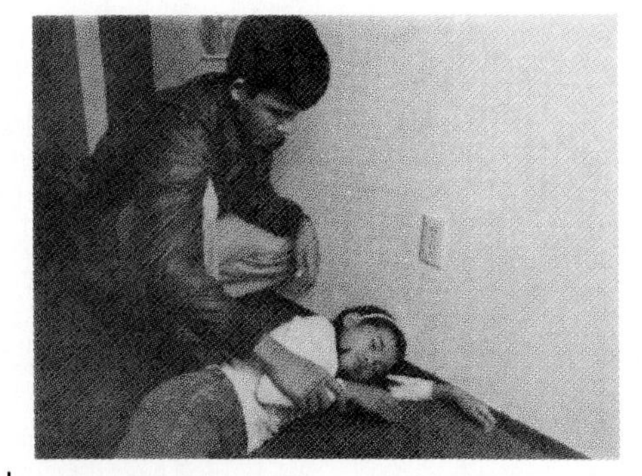

they could get shocked.

thì chúng có thể bị giựt điện

ພວກເຂົາອາດຖືກໄຟຊັອດ.

他們會受到電震的傷害

Figure 4–6 A Graphic Warning
Source: Reprinted by permission of the Puget Sound Power and Light Company.

Executives

Executives are the managers, the supervisors, the administrators, the decision makers of an organization. Legislators, granting agency reviewers, high school principals, and farmers, as well as Lee Iacocca, function as executives. Much of what we have said about lay people also applies to executives. You cannot assume that executives possess very much knowledge in the field you are writing about. While most executives have college degrees and many have technical experience, they represent many disciplines—not necessarily including the one you are writing about. Some executives may have training in management, accounting, a social science, or the humanities, but little or no technical background.

Point of View

Even more than lay people, executives' chief concerns are with practical matters—what things do as opposed to how they work. Executives want to know how technological development will affect the development of their companies, and they probably can use more technical background than lay people. Both of these attributes are evident in this passage from the *Bell Laboratories Record,* a journal published, as it states, "for the management and technical staff of the Bell System":

Implications of SDP design

The SDP [a new computer] components were chosen to be not only fast but flexible, so the processor is programmable. The same computer, therefore, can be used for other tasks that require fast processing and great memory capacity. It is already being used in several Navy applications where custom-designed equipment would otherwise have had to be developed. The resultant cost savings, and the logistics advantage of having the same spare parts inventory for a variety of applications, are two of the attractive features of the SDP.

What SDP does, not how it does it

Quite different applications are possible. For example, the SDP could also be used to process signals from radio telescopes. Because these signals have a low signal-to-noise ratio, analysis requires the processing of a great deal of data collected over a long period of time. Millions of data from several hours may have to be stored and repetitively processed.

The SDP might also be used in telecommunication systems where large quantities of data must be processed quickly. It could be used to recognize speech and synthesize spoken answers, cancel echoes, or multiplex signals by time assignment speech interpolation. Its ability to process many signals simultaneously might be useful in a variety of applications in central offices.[17]

Notice that throughout the passage the emphasis is on *what* the computer does, its functions and potential capabilities, not on *how* it does what it does.

When writing for executives, define important technical terms that you must use, but avoid shop jargon altogether. Executives are busy people; don't force them to use a dictionary any more than you would lay people.

When writing for executives, write in plain language. Do not use mathematics beyond their grasp. Use simple graphics of the types suitable for lay people: bar graphs, pie charts, and pictographs.

Although executives resemble lay readers in many ways, there is also a significant difference between them. While what lay people read obviously influences their lives and their decisions, they only occasionally have to act directly upon it. Executives, however, must often make decisions based upon what they read. People and profits figure largely in executive decisions.

Executives must also consider the social, economic, and environmental effects of their decisions upon the community at large. Aesthetics, public health and safety, and conservation are key decision-affecting factors today, and few executives would consider a report complete if it did not deal with them. All this means that executives are usually more interested in the implications of the data than in the data themselves. Never fail to give executives the *so-whats* of the information you provide.

What questions do executives want you to answer in a report written for them? They want to know how a new process or piece of equipment can be used. What new markets will it open up? What will it cost, and why is the cost justified? What are the alternatives?

Why did you choose the new equipment over the other alternatives? Give some information about the also-rans. Convince the executive that you have explored the problem thoroughly. For all the alternatives include comments on cost, size of the project, time to completion, future costs in upkeep and replacement, and the effects on productivity, efficiency, and profits. Consider such aspects as new staffing, competition, experimental results, and problems likely to arise. What are the risks involved? What environmental impact will this new development have? Figure 4–7, "What Managers Want to Know," illustrates what information, according to one research study, executives feel they need in a report.

Experts, in particular, often find writing for executives a difficult task. Experts are often most interested in methodology and theory. As Professor Mary Coney points out, executives are more interested in function. Excerpts from a salmon study done for the Alaskan Fish and Game Department illustrate the frame of mind the expert researcher should have while writing for the executive. In the introduction, the researcher poses the questions that will be answered in the report. They are the questions an executive would ask:

> Why have they gone? Can the runs be restored to any significant degree? Is it reasonable to base a large industry on the harvest cycle of a wild resource? What should be done? What should be done now?[18]

Problems
What is it?
Why undertaken?
Magnitude and importance?
What is being done? By whom?
Approaches used?
Thorough and complete?
Suggested solution? Best? Consider others?
What now?
Who does it?
Time factors?

New Projects and Products
Potential?
Risks?
Scope of application?
Commercial implications?
Competition?
Importance to Company?
More work to be done? Any problems?
Required manpower, facilities, and equipment?
Relative importance to other projects or products?
Life of project or product line?
Effect on Westinghouse technical position?
Priorities required?
Proposed schedule?
Target date?

Tests and Experiments
What tested or investigated?
Why? How?
What did it show?
Better ways?
Conclusions? Recommendations?
Implications to Company?

Materials and Processes
Properties, characteristics, capabilities? Limitations?
Use requirements and environment?
Areas and scope of application?
Cost factors?
Availability and sources?
What else will do it?
Problems in using?
Significance of application to Company?

Field Troubles and Special Design Problems
Specific equipment involved?
What trouble developed? Any trouble history?
How much involved?
Responsibility? Others? Westinghouse?
What is needed?
Special requirements and environment?
Who does it? Time factors?
Most practical solution? Recommended action?
Suggested product design changes?

Figure 4–7 What Managers Want to Know
Source: James W. Souther, "What to Report," *IEEE Transactions on Professional Communication* PC-28 (1985): 6.

The stated purpose of the report further reassures the executive that the researcher is on the right track:

> Our approach has been first to gather and understand as much relevant information as could reasonably be found; and then to organize, interpret, and project toward the goal of defining a conceptual framework for successful actions by the State of Alaska through its Department of Fish and Game.[19]

Here it is obvious that scientific findings will be wedded to executive needs. Function—"successful action"—lies at the heart of the report.

Be honest. Remember that if your ideas are bought, they are *your* ideas. Your reputation will stand or fall on their success. Therefore, don't overstate your case. Qualify your statements where necessary.

Give your conclusions and recommendations clearly. In writing a report for an executive, remember that you must interpret your material and present its implications, not merely give the facts. Professor James Souther, who made an extensive study of the reading habits of Westinghouse executives, points out that in executive reports the professional judgment of the writer should be the focal point. "True," Professor Souther writes, "it is judgment based on objective study and evaluation of the evidence; but it is judgment nevertheless."[20] The researcher who amasses huge amounts of detail but neglects to state the implications, conclusions, and recommendations that follow from the facts has failed to do the complete job. Many technical people who have aspired to executive rank have not made it because they failed to grasp the importance of this point.

Reading Habits

Organize your report around executive reading habits. Professor Souther reports that executive reading habits are "surprisingly similar."[21] "All managers said they read the *abstract;* most said they read the *introduction* and *background* sections as well as the sections containing *conclusions* and *recommendations* to gain a better perspective of the material being reported and to find an answer to that all important question—'What do we do next?' "[22] Managers seldom read the body and appendixes of reports. When they do, it is because, as Professor Souther reports, they are "especially interested . . . deeply involved . . . forced to read by the urgency of the problem," or "skeptical" of the writer's conclusions.[23] Figure 4–8 illustrates how Westinghouse managers read reports.

Readability authorities have classified the way people read into five methods:[24]

- **Skimming.** Going through a document very quickly, mainly to get a general idea of its nature and contents
- **Scanning.** Reading rapidly to find specific needed facts or conclusions, for example, looking only for financial information in a report

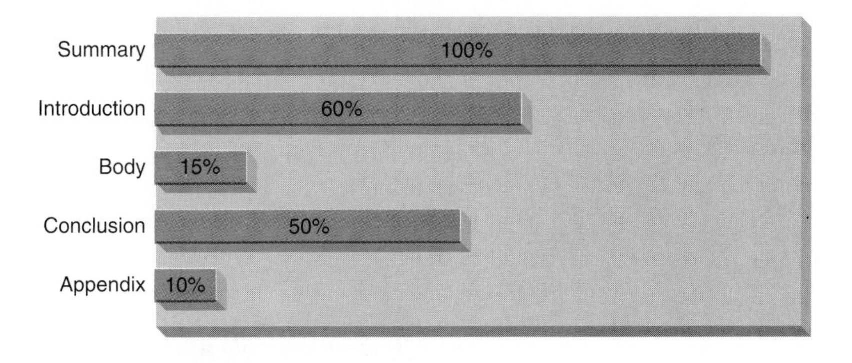

Figure 4–8 How Managers Read Reports
Source: James W. Souther, "What to Report," *IEEE Transactions on Professional Communication* PC-28 (1985): 6.

- **Search reading.** Scanning but slowed down so as to pick up more of the content
- **Receptive reading.** Reading at whatever speed is necessary for high comprehension
- **Critical reading.** Reading to evaluate the document and its contents

Professor Souther's research seems to indicate that executives most often skim, scan, and search read. They are likely to read receptively or critically only in such sections as introductions, summaries, and conclusions. Because of this tendency, you should be sure to organize your executive reports so that your readers can easily find the sections they want. As with the lay reader, you should be directive, providing good introductions, transitions, headings, and summaries. Executives expect to see conclusions and recommendations early in a report with the justification following. If you feel you must report a large amount of technical data, put the data in an appendix where executives can read them if they want to (or assign experts to read them).

Writing styles suitable for executives are well illustrated in magazines intended for middle- and upper-management audiences, such as *Business Week* and *Forbes*. Our chapters on proposals, progress reports, and feasibility reports deal with the arrangement and formats of specialized executive reports. Appendix A provides an example of an executive report. See also Part III, "Document Design in Technical Writing," for ways of designing reports for various audiences.

Organizational Politics

In writing for executives, you must be acutely aware of politics and personalities in matters both large and small. Be loyal to your supervisors and to people who work for you. If they make mistakes, tell them privately, but don't broadcast their shortcomings in company memos. Be aware that people reading your reports and correspondence have their own self-interest at heart. Self-interest can take many forms. It may be an interest in money or power or in protecting the reader and his or her department. Reports that seem to threaten such interests may receive a hostile reception, even when they are otherwise well prepared.

The organization for which you work may have a corporate style. You will belong to what communication experts call a **discourse community.** Members of a discourse community hold certain things in common among themselves: what information is important to them, what style is most acceptable, what lines of argument are most persuasive.[25] Most people belong to more than one discourse community. For example, an accountant who works for an insurance company may also in her spare time help to keep the books for her church. On the insurance job, she may find it necessary to cultivate a courteous but brisk style. She will safely assume that most of her accounting terminology will be generally known by her fellow accountants and that insurance concepts will be known to all who are likely to read her reports. She will know that the emphasis in all her reports will be on a straightforward statement of her data and the implications of that data. At church she may find it necessary to use a more indirect, less business-like approach, perhaps more value oriented than money oriented. She may have to define and explain her professional words and concepts.

As a college student, you may lack experience in business and, therefore, not yet have a schema on which to hang further information about organizational politics. However, we do want you to know that writing in a large organization, particularly for executives, is a tricky business. When you enter an organization after college, you will be wise to find yourself a mentor, a senior person, to whom you can go with your writing. Don't expect that person to do your work for you, but you can hope for advice about when your writing has violated your organization's (or supervisor's) values and beliefs.

Experts

Who are experts? For our purposes, we will define **experts** as people with either a master's degree or a doctorate in their fields or a bachelor's degree and years of experience, for example, college professors, industrial researchers, and engineers who design and build. They know their fields intimately. When they read in their own fields, they seldom look for background information. You may offer background information that you feel is particularly pertinent to the narrow subject at hand—such as a review of the experiments leading up to the one you have conducted. But this background will not be presented in simple terms, as it would be in an article for nonexperts. In some cases, rather than give background, you can refer experts to other sources—books or articles—where they can pursue background if they need to.

Point of View

Experts are very concerned with how and why things work. They want to see the theoretical calculations and the results of basic research. They want your observations, your facts—what you have seen, what you have measured. In general, they expect you to work your way through your data and your interpretation of those data to your conclusions. In reporting such things, be as complete as time, space, and human patience allow. Modern technology and science depend upon many people's cooperatively working upon an accretion of facts, many of which seem trivial standing alone.

Writing for Experts

Despite the tolerance experts have for data, the enormous amounts of information available in most fields have made scanners and search readers of experts as well as of executives. Research indicates that experts review reports quickly to see if the reports contain needed information. In their review, the experts depend upon, in order of importance, the *summary, conclusions, abstract, title page,* and *introduction.*[26] Therefore, these components must be carefully written and constructed in expert reports. As in executive reports, make information accessible through the use of headings.

Shorthand Methods When writing for experts, you may use any shorthand methods such as abbreviations, mathematical equations, chemical formulas, and scientific terms that you are sure your audience can comprehend. Complicated formulas and equations needed to support the conclusions, but not essential for understanding them, are often placed in an appendix rather than the body of the report.

Graphics In writing for experts, you will use graphics to display your results and to support your claims. Graphics will be an integral part of your text. Tables provide an excellent way to lift classifications and groups of closely related facts out of the text and display them clearly. Because they best portray the relationship between variables, line graphs are frequently used. Scientific concepts are also frequently expressed in drawings, maps, and photographs. A line drawing combined with a prose explanation is often used to aid the expert reader to understand new equipment or new concepts. Figure 4–9 illustrates the technique.

Abbreviations and Symbols You do not normally have to define terms unless you have used them in some new or unusual way. However, a word of caution here. Where possible, abbreviations and symbols should be the standard ones for a given field, as defined by the authorities in that field. When they are not standard, symbols should be defined as in this example:

From Einstein's mass–energy equation, one can write the relations:

$$E(\text{Btu}) = m(\text{lb}) \times 3.9 \times 10$$

m being the loss of mass.[27]

The writer does not define *E, Btu,* and *lb* because these are standard symbols and abbreviations for *energy, British thermal units,* and *pounds.* He does define *m* because it is not standard. Writers who do not define nonstandard symbols cannot expect readers to know what they are. In fact, in a few years, returning to their own reports, *they* may not know

Hinged-Blade, Vertical-Shaft Windmill

The blade profile is reduced when moving against the wind.

Marshall Space Flight Center, Alabama

A vertical-shaft windmill concept calls for hinged, flapping blades to increase energy-conversion efficiency by reducing the wind-energy loss. When a blade is moving on the downwind half cycle, the wind pressure forces the blade halves to unfold from each other and present a larger cross section to the wind, thereby capturing more wind energy (see figure). When a blade is moving on the upwind half cycle, the wind folds the blade halves together, thereby reducing the cross section to wind and therefore the wind resistance.

Of course, the windmill design must be made resistant to the vibrations caused by the flapping of the blades. The sound of the flapping may also be a consideration.

This work was done by Bennie Shultz, Jr., of **Marshall Space Flight Center.** *No further documentation is available.* *MFS-25980*

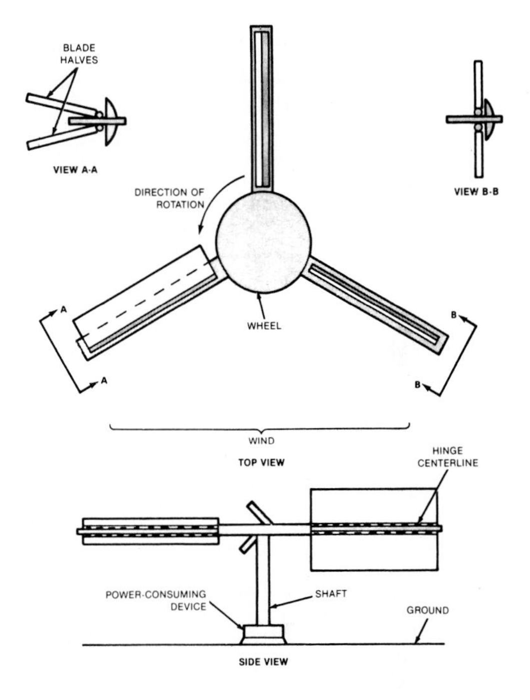

Hinged Blade Halves unfold to catch the wind when moving with it, then fold away from the wind when moving against it.

Figure 4–9 Prose and Graphic Explanation
Source: NASA Tech Briefs 9 (1985): 134.

what the symbols mean. Symbols and abbreviations may be defined as they are used, or if there are many of them, they can be defined in a glossary at the front of the report. (See pages 164–169.)

Background and Definitions The fact that most professional and scientific fields have grown increasingly specialized leads to another caution. Experts in the same field but who have specialized in different aspects of the field may find they have difficulty in understanding one another. Not too long ago, we sat in on a meeting of agronomists being addressed by a fellow agronomist—one who specializes in plant genetics. The question-and-answer period that followed the talk made it painfully obvious that few of the agronomists had understood the speaker. He had not taken into account how far removed they were from his specialization. In this and similar cases, some of the same techniques recommended for the lay audience, such as giving background and defining terms, would be appropriate.

Qualifying Statements Expert reports are, of course, not merely calculations and facts. Experts as well as executives want your inferences and conclusions. When you do draw inferences from your facts and observations, be sure to make no unwarranted leaps. Stay within the bounds of the scientific method. In presenting your conclusions, be careful that your language shows where you are certain and where you are in doubt. Because of this need for scientific honesty and caution, most expert discussions and conclusions do contain qualifications as well as positive statements. The following excerpt is an example of the balance between fairly positive statements and qualified statements that is typical of scientific discussion. (We have italicized the qualifiers.)

Positive statement

Qualified statement

Positive statement

Qualified statement

Positive statement but promising, not a certainty

Use of a LANDSAT intensity image provided additional information on varnish distribution on the alluvial fans, thus permitting discrimination of two fan gravel units that were not distinguishable in the radar data. The use of all four bands of LANDSAT *should* add the capability to distinguish compositions. We thus have seven dimensions of imagery available for computer processing techniques.

Surely, particulate alluvial and colluvial regoliths, evaporites, and eolian deposits represent widespread geologic units on terrestrial planetary surfaces. On cloud covered planets (e.g., Venus), multiconfiguration radar data *appear* capable of discriminating many of the surficial geologic units likely to exist. On other planets, additional use of optical sensors will permit a more thorough compositional mapping. LANDSAT/radar combinations offer considerable *promise* for making high-quality geological maps of lithologic-textural units.[28]

Experts know that certainty in science is a hard-won achievement. They are content with probabilities until thorough experimentation and observation remove all reasonable doubt. Their style reflects their basic caution and honesty. Our Chapter 18, "Empirical Research Reports," describes the most predominant type of expert report. You can find good samples of expert reports in the journals that cover your particular discipline. For examples of expert reports that aim at a somewhat more generalized but still expert scientific audience, we recommend to you *Science*, the journal of The American Association for the Advancement of Science.

Technicians

Technicians are at the heart of any operation. They are the people who build, maintain, and operate equipment. Technicians' educational levels vary widely. Most typically, technicians will range anywhere from a high school to a college graduate. They may have been trained in one of the many vocational schools. The high school graduate may have a great deal of on-the-job-training and experience. The college graduate may be better educated in theory but have less practical experience.

Point of View

Because technicians build, maintain, and operate equipment, their major concern is with "how-to" information. How do I use this word processing program? How do I operate this lathe? How do I replace the fuel pump in this model engine? And so forth.

While, generally speaking, technical audiences care more about "how-to" information and more about the practical application of a theory than about the theory itself, they will appreciate some theory. How much theory you give, and how complex you make it, depends often on both their education and their point of view. College-educated technicians, or those with long experience, border on being experts, and you can treat them much as you would an expert. Less expert technicians or people operating as technicians in areas outside their major interest require much less theory. For example, a college professor installing a new washer in her kitchen sink faucet would probably not desire much background theory about washers.

Some technicians may not be able to follow complicated mathematics, and they'll grow restive with too much theory. Be careful to avoid excessive sentence complexity. (see pages 119–123.)

Background

You can assume a good deal of knowledge on the part of the technician audience but not as much as with the expert audience. For technicians you will need to supply some background information and some definitions. An article from *Bell Laboratories Record* begins this way:

> Waveguide systems are often loosely called "plumbing." The name implies a network of empty pipes where electrical energy flows unimpeded. Actually a waveguide is a precisely designed, electrically tuned structure for propagating electromagnetic waves. It transmits certain determinable frequencies well, does not transmit some frequencies at all, and transmits others only with large losses.[29]

The author of this article has assumed general knowledge in the field on the part of his audience. For example, he does not define "electrically tuned structure" or "electromagnetic waves." His audience will recognize these terms. But he begins his article with background information on the subject: waveguide systems. He uses analogy—"plumbing"—and later describes in great detail what waveguides look like.

For the most part, keep theory explanations simple and fairly nonmathematical. While college-educated technicians normally can handle calculus, others may not be able to. For both groups, the wisest choice may be to put most of the necessary math in graph form. If you feel equations must be given the technical audience, definitely put them in an appendix. The equations should supplement the graphical presentations. Equations should provide further depth for the reader who can use it. They should not be necessary for understanding the basic text.

Depend upon analogy when writing for the technical audience. Analogy bridges the gap between a reader's general information and the particular object or theory you are trying to explain. An excerpt from the *Bell Laboratories Record* article on waveguides illustrates the principle:

> Every electron orbiting about an atomic nucleus gives rise to magnetic fields. In some materials the field comes largely from the motion of the electron around the nucleus, but in ferromagnetic materials it depends more on the spin associated with the electron itself. The spin creates a small magnetic movement that is precisely aligned with the axis of spin (this can be visualized as similar to the alignment of the earth's magnetic field between the north and south poles).[30]

Keep in mind that technicians read primarily for "how-to" information. Therefore, technical manuals and operating instructions are the most common applications of writing for technicians. Both are described in Chapter 14, "Instructions." The magazines *Popular Science* and *Popular Mechanics* and many manuals offer fine examples of writing for the technician.

The Combined Audience

We have discussed in isolation each of the major audiences you are likely to have. And sometimes you will write for them in isolation, for instance, write exclusively for a lay or an executive audience. But frequently you will have a combined audience. Sometimes, one person can combine the aspects of several audiences. For instance, someone may be reading about stereo speaker systems that can be installed in his house. He may be a lay person in that he doesn't understand all the electronic jargon. He may be an executive in the sense that he may make a decision about buying such equipment. He may be a technician in that he may install the equipment himslf.

At other times your audience may be a team of people—perhaps composed of executives, experts, and technicians. Many, perhaps most, organizational documents have more than one reader. In this age of photocopiers and fax machines, many reports, letters, and memos are copied for people other than the primary reader. Also, reports and correspondence may have a longer life than you anticipate. They go into filing cabinets, where they become a part of organizational history. They may be read years later, perhaps as background to a new development or even as evidence in a trial. All this means that your life is frequently complicated by the need to consider several audiences at once. As with any communication situation, the strategy you choose must suit your purpose, audience, and material.

Illustrative Example

To illustrate how all this may work, imagine for a moment you are a computer expert working for a large investment company. You have just completed a study to decide which of two computer systems to install at the company. Let's call them Brand X and Brand Y. The system is to be purchased for the use of the Research Division of the company. The Information Systems Division (ISD), for which you work, will buy, install, and maintain the system.

You have decided, based upon your research, that Brand X, even though it costs slightly more, is the system to buy. You have to write a report stating your conclusions and recommendations and justifying them. You have a combined audience of three people, all of them senior to you:

Jack Anderson. A vice president of the company. The Research Division is one of his responsibilities. You have heard that his major concern is always cost, and that he doesn't like to read long reports. He is the decision maker in this matter, but he is a team player. He will certainly confer with the other people concerned before making his decision.

Sally Kroger. Head of the ISD. She has a Ph.D. in computer science. She has been with ISD for 10 years and head for the last five. She is a stickler for detail and will want full justification for your conclusions and recommendations. She doesn't mind long reports, even seems to prefer them.

Bruce Hogan. Head of the Research Division. He has been complaining that his staff do not have adequate computer support for their research activities. It is because of his complaints that you conducted your study. He has a master's degree in business administration. You talked to him a good deal when you were assessing the computer needs in the Research Division. You know that he has used computers and has a practical interest in them but that he has no interest in computer theory.

How to satisfy this diverse audience? You have stacks of information in front of you: masses of statistics concerning input, output, power, remote consoles, ease of use, available programs, and cost. You have brainstormed a series of pros and cons for each system. The pros and cons are helpful, but they don't seem to lend themselves to a coherent report. Also, they get you into more detail than either Jack Anderson or Bruce Hogan is likely to want.

You analyze the situation. You think about your audience again and what their true concerns are. Jack Anderson will want to be sure that Bruce Hogan's complaints are taken care of. Also, he will need to be persuaded that the higher costing Brand X is the better choice. Bruce Hogan's major concern is that all his needs will be satisfied. He will not care that the cost is higher for Brand X. Sally Kroger will be interested in cost, but it's not a vital factor for her. She will want to see that Hogan's needs are satisfied. She also will be quite interested in a comparison of the maintenance requirements of the two systems.

Now you're getting somewhere. The major concerns for your audience seem to be cost, needs satisfaction, and maintenance. It's because Brand X meets the needs of the Research Division so well and because of its superior maintenance record that, despite its higher cost, you have chosen it over Brand Y.

Your organization based upon your situational analysis begins to fall into place. You decide that your report will have three sections: *cost, needs satisfaction,* and *maintenance.* The cost section can be straightforward and short, probably based largely on some comparison tables. The needs section will show how you assessed the needs of the Research Division and how Brand X satisfies those needs far better than does Brand Y. You have some good quotes from Bruce Hogan you can include in this section to show that he has been consulted. You remind yourself to keep this section free of computer jargon so that Bruce can read it, and Jack Anderson as well if he should decide to read it.

The maintenance section is of major concern to Sally Kroger and will probably not even be read by Jack Anderson or Bruce Hogan. It's the section in which you can let yourself go both with technical computer language and in satisfying Sally's penchant for detail.

When the three sections are complete, you'll top them off with an *executive summary* that states your conclusions and recommendations and succinctly supports them. (See pages 291–292.) You suspect that the summary may be the only section that Jack Anderson reads, so you remind yourself to make sure that the support for buying the more expensive Brand X is adequate to meet his skepticism. You will also direct your readers through all parts of the report with a good introduction, transitions, and headings.

Step back for a moment and see what you have done. As shown in Figure 4–10, you have considered the major concern of each reader and made sure that you have addressed it. You have satisfied the needs of the *decision maker,* Jack Anderson; *the user,* Bruce Hogan; and *the expert,*

Sally Kroger. Because you know that executives scan reports and read them very selectively, you have organized and will format your report to make that easily done.

Other Situations

Other situations may be more or less complex than the one we have described for you. In another situation you might need, for instance, to write a short memo to a primary reader, but you know there will be several secondary readers. To satisfy the primary reader, you would have to adjust the level of technicality to the appropriate level. If the level of technicality is low, you run the risk of boring more technically competent secondary readers, but, at least, you won't confuse them. If the level of technicality is high, you might have to offer the secondary readers some help in the way of definitions, analogies, and so forth. Often, such help can be kept out of the way of a more technically competent primary reader by putting it in a graphic or an attachment. Because the graphic and attachment stand a little apart from the body of the memo, the primary reader can ignore them.

Sometimes needed definitions can be placed in footnotes or glossaries where, again, a more expert reader can bypass them. (See pages 168–169.) We have much more to say about such matters in Chapters 8, 10, 11, and 12, where we discuss definition, document design, and graphics. But remember that all such devices and techniques are used to satisfy the needs of your readers. Therefore, always begin by considering your audience and consider it at every stage of the writing process from preliminary research to final formatted product. Perhaps the advice that new staffers on *Life* magazine used to hear best sums up the attitude you should maintain toward all the readers you may want to reach: "*Never underestimate the intelligence of your readers and never overestimate their knowledge.*"

Figure 4–10 Writing for a Combined Audience

WRITING FOR A COMBINED AUDIENCE

Reader	Main Concern	Style	Target Audience
Jack Anderson (Decision maker)	Cost	Cost section: straightforward and short, based on comparison tables	*Executive*
Bruce Hogan (User)	Needs satisfaction	Needs satisfaction section: free of jargon, includes quotes	*Technician, Executive*
Sally Kroger (Information systems expert)	Maintenance	Maintenance section: detailed with technical language	*Expert, Executive*
Anderson, Hogan, Kroger	Satisfying Jack Anderson	Executive summary: conclusions, recommendations, with supporting evidence	*Executive, Expert, Technician*

Planning and Revision Checklist

The following questions are a summary of the key points in this chapter, and they provide a checklist when you are planning and revising any document for your readers.

Planning

- How many readers will you have?
- Is any one person (or any one group) your primary reader? Are there secondary readers? Consider both primary and secondary readers when answering the questions that follow.
- What is the relationship of the primary readers to you? The secondary readers?
- What do you hope to accomplish with your document? What is your primary purpose? Secondary? Other?
- Why will the primary readers read your document? For enjoyment and interest? To learn a skill? To do some task? To make a decision? To gain knowledge in the subject area? Other? Secondary readers?
- What are the primary readers' interest, knowledge, and experience in the subject of your document? Do they have the necessary schema and background to assimilate your material easily? If not, what can you provide to help them? Explanations? Definitions? Graphics? Analogies? Examples? Anecdotes? Secondary readers?
- What are the primary readers' attitudes toward your subject matter? Secondary readers?
- What are your primary readers' attitudes toward you? Secondary readers?
- If your document includes conclusions and recommendations, how will your primary readers feel about them? Enthusiastic? Friendly? Hostile? Skeptical? Indifferent? Secondary readers?
- Is self-interest involved in the primary readers' attitudes? For instance, will the primary readers benefit or lose as a result of your document? Will the primary readers feel threatened or be made angry by the document? Secondary readers?
- If the primary readers have negative attitudes (hostile, skeptical, indifferent), what can you do to overcome those attitudes? Provide more support? Point out long-term benefits? Soften your language? Make your presentation more interesting? Secondary readers?
- Are you writing within a recognizable discourse community such as a corporation or a professional group? If so, are you familiar and comfortable with the customs and voice of that community? Is there a prevailing style and tone? What arguments are most persuasive? Do you have access to models of similar documents you can use to help you?
- How are the primary readers likely to read the report? Skim? Scan? Search read? Read for comprehension? Read critically? How will the secondary readers read? Given how your readers will read your document, what can you do to help your readers get the most from it? Will some of the following help? Table of contents? Graphics? Questions? Introductions and transitions? Summaries? Checklists?

Revision

- Have you written your text at a level appropriate for your readers, both primary and secondary? Do you need more or fewer definitions, examples, descriptions, and explanations? Have you provided analogies and metaphors when they would be useful?
- Have you met your objectives for each audience addressed?
- Will your readers feel that their objectives have been met?
- Is your purpose clearly stated in your introduction?
- Are your conclusions and recommendations clearly stated? Can your readers easily find them?
- Does your format allow your reader to read selectively? Do your introductions and transitions forecast adequately what is to come? Have you furnished a table of contents if needed? Do you have enough headings and subheadings to guide your readers? Would review questions, summaries, or checklists be useful?
- Have you furnished good graphic support? Have you helped your reader when possible by using pictures, flowcharts, and diagrams?
- Have you considered adequately the politics and personalities involved in the situation you are addressing? Have you achieved the proper tone? Have you inadvertently said anything rude or misleading?
- Does your language distort or sensationalize your material in any way?

Exercises

1. Think about some concept in your discipline that you understand thoroughly. Write a letter explaining that concept to someone you know who does not understand it: a good friend, a parent, a sister, or a brother. In your planning, use the checklist provided on page 78. Include graphics if you think they will help. Your goal is to interest your reader and teach him or her something about a world you are coming to understand and enjoy.
2. Take an article from a journal in your field that discusses some development or concept in your field that may have practical value. For example, it may suggest a new method of producing some product, a new market for an old product, or some service that could be performed for a fee. Imagine an executive who might find a discussion of this concept or development useful. Assume this executive is intelligent and college educated but that he or she knows little about the subject matter involved. Write that executive a memorandum in which you explain the concept or development, and fully discuss its implications to the executive's organization. Use graphics if they will help. Use the memorandum format shown on page 382.
3. Compare and contrast an article or report written for one kind of audience with another written for a different audience—perhaps an expert article with an executive one or an executive article with a technician one. Try

to get articles on similar subjects and in your field. Ask and answer some of these questions about the articles:

a. What stylistic similarities do you see between the two articles? What dissimilarities? Which article has the greater complexity? What indicates this complexity? What is the average sentence length in each article? Paragraph length? Which article defines the most terms?
b. What is the author's major concern in each article? How do these concerns contrast?
c. What similarities and dissimilarities of format and arrangement do you see?
d. Are there differences in the kinds of graphics used in the two articles?
e. Which article presents the most detail? Why? What kind of detail is presented in each article?
f. How much background information are you given in each article? Does either article refer you to other books or articles?

4. Perhaps you are already planning a long paper for your writing course. Using the Planning Checklist on pages 78–79, write an analysis of the anticipated audience. Prepare your analysis as a memorandum addressed to your instructor. Use the memorandum format shown on page 382.

CHAPTER 5

Gathering and Checking Information

PREPARING AND ADMINISTERING A QUESTIONNAIRE

Drafting the Questions

Administering the Questionnaire

CHECKING READERS' ATTITUDES AND REQUIREMENTS

CONDUCTING INTERVIEWS

WRITING LETTERS OF INQUIRY

PERFORMING CALCULATIONS AND ANALYSES

REVIEWING INFORMATION ALREADY GATHERED

Information gathering, whether in the library or elsewhere, is usually an important part of any major writing project. In addition to information-gathering techniques, you will need some techniques for checking and reviewing your information as you gather it, both to stay on course and to correct your course as necessary. To illustrate these techniques, in this chapter we take you through an extended example.

Let's suppose that the subject of the aging of America interests you. Further imagine that your major is restaurant management. Project yourself into the role of a young executive managing one restaurant in a small restaurant chain. The new owner of the chain, Jane Lewis, is new to the restaurant field. She bought the chain after retiring from a successful career as an insurance sales representative. She is 45 years old with a college degree in history. You know she is intelligent, well educated, and understanding about business, but not too well informed about the restaurant business. You cannot assume that she will possess a thorough grounding in restaurant management techniques. Therefore, you would need to justify fully any recommendations about restaurant management that you might make to her.

The restaurant you manage caters to a preteen and teenage clientele and serves mainly hot dogs, hamburgers, gigantic ice cream concoctions, and the like. The food is served with a lot of hoopla, and the decor is bright and colorful. You know from a newspaper article you read that while in 1970 the 15- to 17-year-old group greatly outnumbered the 22- to 34-year-old group, by 1990, the two groups were almost equal. Furthermore, the older group is projected to soon outnumber the younger group. You wonder if your new boss might not be interested in this change. You decide to research the matter. You suspect that the information gathered may lead to a report recommending significant changes in the decor, menu, and manner of service of the restaurant.

Calling upon Your Memory

In information gathering it's usually wise to begin with yourself. What do you already know about your subject? You may already possess all the information you need; then again, you may not. Using a brainstorming technique, write down what you already know. Here are some assorted facts you may have in your head:

- America's birthrate is declining and people are living longer. Therefore, the average age of the American population is increasing.

- The birthrate during the 15 years or so after the end of World War II in 1946 was higher than normal. The people born then are now something of a bulge moving through the population. An article in *Newsweek* had useful information about this.
- In a restaurant management course, the professor referred to a highly successful restaurant chain in Chicago. The chain has a tricky name, Lettuce Entertain You, and its customers are primarily young urban professionals.
- The Cornell University School of Hotel and Restaurant Management publishes a journal that contains excellent articles on hotel and restaurant management. The journal should be worth browsing through.
- Somewhere recently an article appeared on how to evaluate the effect a restaurant's environment had on its customers. Could it have been in the Cornell journal? The article said that lights too dim had a negative effect on customers, and bright lights increased the speed with which people ate and left the restaurant.

If you are normally observant, you could extend this list several times over. Someone interested in restaurant management could no doubt extend the list to several dozens of items. Here, near the beginning of the prewriting stage, you can get more down in writing than you realize. Even some notions about arrangement may appear.

Searching the Literature

After you have searched your own brain for information on your subject, the next logical step is to search the brains of others. The library is the best starting point to learn what others have found out and recorded about your subject. In a very real way, the library is our collective memory. Because technical writing is an advanced writing course, we assume you have the basic knowledge needed to use a library. However, you may not be aware of the large number of publications and computerized bibliographies that exist to help you find needed technical information. To assist you with such research, we recommend to you our Appendix B, "Technical Reference Books and Guides." There you will find guides to reference books, periodicals, report literature, computerized information retrieval, and U.S. Government publications. An hour spent browsing through Appendix B will likely save you from many hours of wasted effort in the library.

If you are a college student, you may have a choice of three or more libraries: your town library, the main library on campus, the library maintained by each school of your college and, sometimes, by a department within a school.

Because of its nearness to your living quarters, or some equally practical consideration, you decide to try the main library on central campus. After an hour of searching through the subject catalog and a few indexes and consulting with the reference librarian, you are pleased to have located six items as a start on your "working bibliography." (See Figure 5–1.)

Eventually, you will be able to extend this initial working bibliography to several times its present length. But at the start, you have evidence that a sizeable quantity of reading matter will be available to you.

Now you have an important decsion to make—which information source to read first. In general, it is best to select a published source that is large, recent, authoritative. On the basis of these criteria, the fourth source probably should be your first choice. You find the book and start to open it. But hold on a moment! How are you going to extract and use any pertinent information it contains? Having a method of doing this will prove critically important to you later as you turn to organizing and writing your report.

Many researchers and report writers find that converting recorded data into prose reports is unbearably torturous and time-consuming. It is true, of course, that there are no magical shortcuts or miracle methods that

Backus, Harry. Designing Restaurant Interiors. New York: Lebhar-Friedman Books, 1977.

Lawson, Fred. Restaurant Planning and Design. New York: Van Nostrand Reinhold, 1973.

Davern, Jeanne M., ed. Places for People. New York: McGraw-Hill, 1976.

U.S. Bureau of the Census. Statistical Abstract of the United States: 1989. 109th ed. Washington, DC: U.S. Government Printing Office, 1989.

Lambert, Carol U. "Environmental Design: The Food-Service Manager's Role." The Cornell Hotel and Restaurant Administration Quarterly 22 (1981): 62-68.

Levine, Joshua. "Lessons from Tysons Corner." Forbes 30 Apr. 1990: 186–188.

Figure 5–1 Working Bibliography

will produce "instant reports." However, there is no sense whatsoever in plunging blindly into the information-recording-and-compiling process or in continuing with methods that have proved cumbersome in the past. Has it been, perhaps, your practice to copy research information onto the sheets of a notebook? If so, you know the battle that ensues when your individual notes have to be sorted and somehow placed in usable order for compilation into a report. You have to leaf back and forth through the notebook to identify and locate notes bearing upon the topic of present interest. Some notes may escape your attention; others may accidentally be used twice. Still worse, the resulting text is likely to be disjointed and badly organized.

Using Note Cards

Abandon the notebook practice for taking notes. Instead, take your notes on cards of four- by six-inch size, or even larger. On one set of cards, you will keep the bibliographic data you need, one card per reference. By having each reference on a separate card, you'll be able, when it comes time to do your final bibliography or references, to sort your cards into any sequence, alphabetical or otherwise, that you need.

On a second set of cards, keep the notes that you extract from your sources. Limit the information on each card to a single narrow topic—one that you may label with a few identifying words such as *Installation Cost, Service Procedures,* or *Maintenance*. Place the topic identifier prominently on each card, perhaps at the upper-left corner. You thus make it possible to sort through the entire collection of cards and organize into subpacks all cards bearing the same topic identifier. Now you can, at any stage of the process, copy onto a sheet of paper all of the topic identifiers and thus be able to inspect the extent of your information in topical array.

When you reach the final organization stage, you will be able to draft one or more tentative plans to hold and arrange the information for the body of your report—and at the same time be able to detect omissions, duplications, overlaps, and irrelevancies.

Once you have created the subpacks, you can make a still finer sorting of cards in each of the subpacks until you have them arranged in order for actual composition. It is a relatively straightforward process to pick up card 1 of subpack 1 and consume its information in order to write the first paragraph of your first body chapter. Prior sorting, permitted by the card note-taking process and the use of topic identifiers, will have freed your mind of the necessity to hunt and sort while composition is in process. Hence, you can give your full attention to the digestion of information and to its clear and coherent expression in prose.

If you do not have a note-taking and note-using method that works for you, try the method we have outlined. The method works for thousands of researchers and report writers, both student and professional.

Reading and Taking Notes

To return to our example problem: at this point, you have selected your first reference, *Statistical Abstract of the United States*. Make a bibliographic identification card for it, as in Figure 5–2. Notice the brief identifier placed in the upper-right corner. Such bibliographical identifiers will save you much time and writing later, as you will shortly see. Be sure to record all the bibliographical information you'll need to document your report (see pages 304–321).

Now let your reading and note taking begin. In the *Statistical Abstract* you are looking for demographic data to confirm that the average age of the American population is increasing. Looking through the table of contents, you note a table labeled "Resident Population, by Age, Sex, and Race: 1970–1987." Turning to it, you find that you don't need the entire table but that you can extract from it exactly the information you are seeking. You copy the information on a note card, as illustrated in Figure 5–3. The label, *Stat ab,* in the upper-right corner of the card in Figure 5–3 identifies the source as that fully described by the bibliographic identification card you made out earlier. The label in the upper-left corner, *Age groups,* is the topic identifier. The information at the bottom of the card is the number and name of the table and the page it is found on. You will need this information to document your source when you write your report or to find the source again should you need to.

You search on through the *Statistical Abstract* extracting further useful information. You place each piece of information on a separate card with appropriate labels and identifiers. The more careful you are at this stage, the easier your job will be when it comes time to write the report.

When you have extracted the information you need from the *Statistical Abstract,* you turn your attention to the article "Lessons from Tysons Corners." You read through it and decide it contains information you

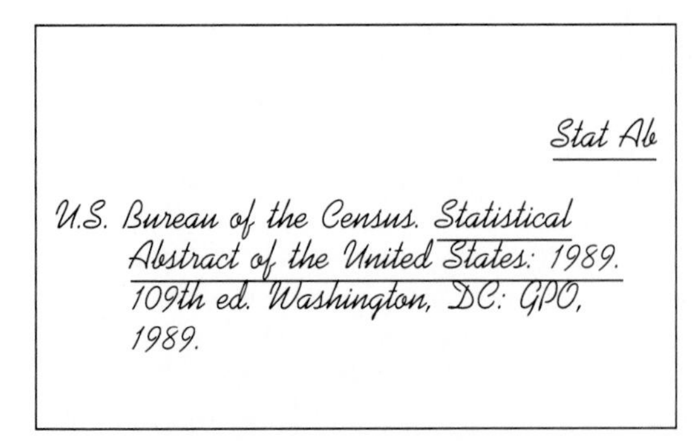

Figure 5–2 Bibliographic Identification Card

Age groups Stat ab

Year	Age Groups					
	10-14	15-19	20-24	25-29	30-34	35-39
1970	20,804	19,084	16,383	13,486	11,437	11,113
1980	18,242	21,168	21,319	19,521	17,561	13,965
1987	16,485	18,459	19,793	21,980	21,335	18,738

(Numbers in thousands)

Table 21, "Resident Population by Age, Sex, and Race: 1970–1987," p. 17.

Figure 5–3 Note Card

need. As a first step before extracting information, you prepare a bibliographic card for it, as illustrated in Figure 5–4.

In this article you find a passage that you think may lend authority to your paper. Because the language of the passage is economical and informative, you decide to copy it verbatim onto a card as shown in Figure 5–5. Note again the use of labels and identifiers on the note card. The label "p. 186" in the lower-right corner indicates the page from which the extract was taken. Because this is a verbatim quote, you have put quotation marks around it. Also, you have noted with an ellipsis (see page 602) where you have omitted some words.

Further on in the paragraph, you find valuable information. In this case, you don't need the entire passage, so you extract only the information you do need, as shown in Figure 5–6. When you condense or paraphrase material taken from a source, avoid using the sentence structures of the original. In that way, you eliminate any hint of plagiarism, always a troublesome thing in research reports.

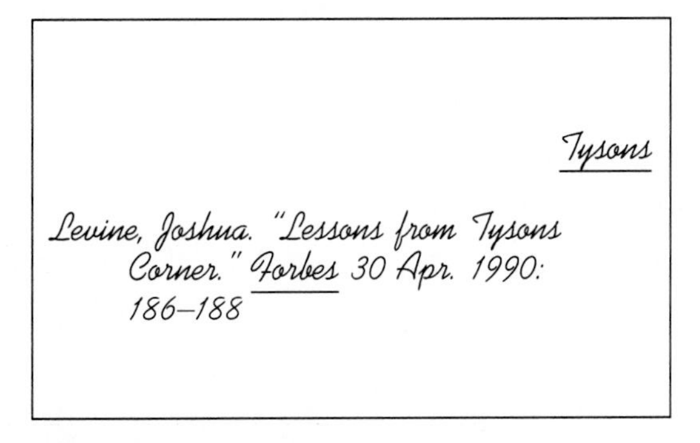
Tysons

Levine, Joshua. "Lessons from Tysons Corner." Forbes 30 Apr. 1990: 186–188

Figure 5–4 Bibliographic Identification Card

How did the shopping get so rich at Tysons Corner? Sure, location is a big part of the answer. But a competing mall across the street, the Galleria at Tysons II, with stores like Neiman Marcus and Saks, hasn't done as well. Storefronts stand empty at the Galleria, while the waiting list for Tysons now numbers 15 retailers.

The answer in part: superior marketing. Tysons knows its customers.

Customer analysis *Tysons*

"How did the shopping get so rich at Tysons Corner? . . . The answer in part: superior marketing. Tysons knows its customers."

p. 186

Figure 5–5 A Verbatim Abstract

Your reading and note taking continue. After finishing "Lessons from Tysons Corners," you read other books and articles. You return to the library to add to your original list of readings. You uncover additional sources referred to in your readings. Things go well. You read, take notes, and make plans.

Generalizing from Particulars and Particularizing from Generalizations

Now that you have done some reading in the literature, you may want to lean back in your chair and take a long hard look at what you have learned so far. You have learned, for example, that research data on how different restaurant environments affect people are distressingly scarce. What makes one environment work and another fail is somewhat unclear. You have a few facts that seem solid. People do not like restaurants to be

The answer in part: superior marketing. Tysons knows its customers. The average mall customer earns $36,000 a year. Tysons' shoppers earn $62,000 a year on average, and are younger and far less likely to have children than the average mall shopper. Only 10% of Tysons' customers are in their mid-to-late teens.

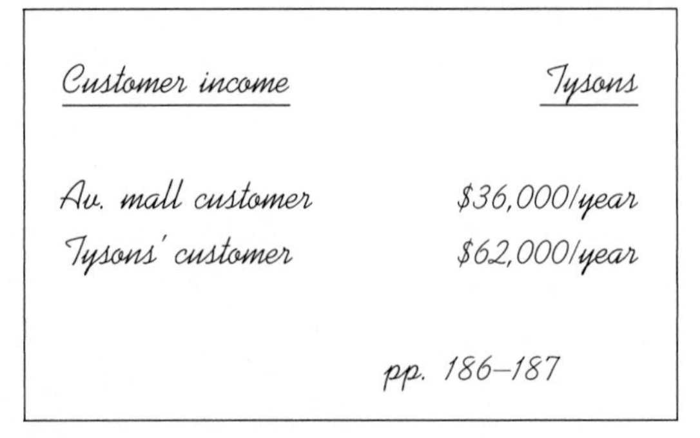

Customer income
Tysons
Av. mall customer
$36,000/year
Tysons' customer
$62,000/year
pp. 186–187

Figure 5–6 A Condensed Abstract

too dimly lit. They want to be able to read the menu without the help of a match. If the restaurateur's objective is a rapid turnover of customers, bright lights help. People seem to eat more quickly and leave under such conditions. You know that spacing is important to people and that Americans' reactions to space have been fairly well mapped out into four zones:

- Intimate distance: 0–18 inches
- Personal distance: 18 inches–4 feet
- Social distance: 4–12 feet
- Public distance: 12 feet and beyond

From that set of data, you generalize that most tables in a restaurant should be spaced so that diners are at least 18 inches away from diners at other tables. But how much over 18 inches? Would it have to be as much as 4 feet? Space is expensive. Furthermore, you have had some cheerful experiences in crowded restaurants. You remember one in particular. On your eighteenth birthday, your mother and father had given you your present while at dinner in a crowded restaurant where the tables were extremely close together. The people at the next table had asked to see the present and had passed it around, commenting on how nice it was. Rather than resenting the intrusion, you and your parents had enjoyed it. Somehow it made the occasion more festive.

You combine this personal experience with a researcher's report that in crowded situations people are more likely to talk to one another. Would conversations carried on between tables be desirable or undesirable for the young urban professionals you are beginning to see as your target clientele? You remember that in the *Statistical Abstract* you learned that the numbers of single people and people living alone have greatly increased since 1970. In 1970, 10.9 million people lived alone. By 1987, that number had increased to 21.1 million. Could some of the people living alone be lonely and therefore enjoy the chance for a more intimate restaurant atmosphere that encourages conversation?

You think and you ponder about these particular facts and draw generalizations from them. Some of your generalizations seem pretty firm; others you are doubtful about. They will need more checking.

The kinds of reasoning you have been following here have their formal names in logic. Reasoning from one or more particular facts to reach a covering generalization is called **induction.** Notice the flow of particulars in this excerpt from an article on how people use space in cities:

> When people stop to have a conversation, we wondered how far away do they move from the main pedestrian flow? People didn't move out of it. They stayed in or moved *into* it, and the great bulk of the conversations were smack in the center of the flow. The same gravitation characterized "traveling conversations"—the kind in which two people move about, alternating the roles of

"straight man" and principal talker. Although there is a lot of apparent motion, if you plot the orbits, they turn out to be quite restricted.

People also sit in the mainstream. At the Seagram plaza, the main pedestrian paths are on diagonals from the building entrance to the corners of the steps. These are natural junction and transfer points and the site of lots of activity. They are also a favored place for sitting and picnicking. Sometimes there will be so many people that pedestrians have to step carefully to negotiate the steps. The pedestrians rarely complain. While some will detour around the blockage, most will thread their way through it.[1]

The generalization to be drawn from these similar particulars is stated in the article: "What attracts people most, it would appear, is other people."[2]

The other kind of reasoning, called **deduction,** leads from the general to the particular. In deduction you apply a generalization to deduce a conclusion. For example, as the young restaurant executive, you have been musing about the spacing required between tables. The particulars of the research just cited as well as other research and your own personal experience have given you the generalization that people really don't mind being crowded a bit. In fact, urban people may even enjoy it. Using this generalization, you deduce that spacing between tables in an urban restaurant could be minimal, perhaps right at 18 inches. Your deduction has given you a conclusion. Rational thinking is based upon this constant flow of induction and deduction—generalizing from particulars and particularizing from generalizations. (See also pages 199–204.)

Our experience would also suggest another generalization: to continue your research you must delve into an additional source of information, namely, the local scene.

Inspecting Local Sites and Facilities

Library research is vital to most investigations. A few hours in the library can turn up data that represent months and even years of work. Library research can keep you from reinventing the wheel. Nonetheless, in most investigations there comes a time when you want to go and look and see, listen and hear, touch and feel for yourself.

On your tour of inspection, take along a notebook to jot down whatever you observe. At this time, do not attempt to be particularly logical; your on-the-spot notes can always be organized later. In this case, you decide to observe people in a local restaurant, Scarpelli's, that you know is popular with people in their 20s and 30s. You call the owner, Sandy Dolan, get his permission, and spend several hours people watching during the evening meal hour. Your jottings read as follows:

- Dress varies, perhaps in accordance with whether people came directly from work or not. Some people are in obvious business attire, others in jeans, open-necked shirts, and casual jackets. Footwear varies from well-shined black boots to scuffed jogging shoes. Despite the apparent casualness, people are not sloppy. They seem to care about how they look.

- The room is clean and well lighted, though not enough to produce a sense of glare. The walls are white. Tablecloths are white and heavy feeling. There are large mirrors in dark frames around all the walls. There are coat hooks between the mirrors, and along the walls tables are closely lined up between the coat hooks. Looks a bit like a restaurant in a French movie.
- People seem to know each other. There is some conversation between tables, and some people get up and speak to people at other tables. Atmosphere is relaxed and friendly.
- Waiters, male and female, are dressed in black slacks, white shirts, and short white aprons—kind of androgynous.
- Tables throughout the room are quite close together. Barely room for waiters to get through. People are cheerful about it. They hitch up their chairs almost without thought to make room when necessary. The noise level is fairly high. People are not shouting, but there is a steady sound of voices. People are talking as much as eating. They are enjoying themselves.
- Tables seem to fill and clear out at about 50-minute intervals. People have to wait for a table but not too long.

Depending upon your powers of observation and your knowledge of the facts of restaurant life, you might fill several notebook pages with such comments in an hour or two of observing. In any case, you will have familiarized yourself with a restaurant of the type you are beginning to envision yourself recommending. You still have a lot of questions, maybe more than answers, but that is entirely normal and healthy in the early stages of research.

To continue your research, several courses are now open to you. In this state of open-mindedness, it occurs to you that you would like to get some reactions from people about restaurants, perhaps your restaurant and the one you have just observed, Scarpelli's. A questionnaire administered to patrons might yield some interesting results. You decide to try it.

Preparing and Administering a Questionnaire

It is not easy to construct and administer questionnaires and tabulate their replies without years of study and experience. Even the "experts" do not pretend to have solved all the problems. We do not intend to make an instant expert of you, but we can offer some basic advice that should enable you to handle questionnaires of limited complexity and scope.[3]

It is always an imposition to ask someone to fill out a questionnaire. You should keep this fact in mind when the time comes to design and administer a questionnaire yourself. Fortunately, you can usually approach your subjects in such a way that the questionnaire does not seem an imposition.

First, be sure that a questionnaire is really necessary. Questionnaires should be used only when the desired information cannot be obtained from another source—interviews, personal observation, searches through daily newspapers, courthouse and library files, and so on. Time, money, and effort can be saved by avoiding the questionnaire process whenever possible.

Whether or not you decide a questionnaire is indicated, the research you do before making that decision is not wasted. If you find no questionnaire is needed, that means your research yielded a lot of the answers you thought might require a questionnaire, and you're farther along than you had thought. If you *do* need to use the questionnaire approach, the information you've obtained about your topic will be invaluable in designing that questionnaire. Among other things, your research should help you to limit the number of questions. As a general rule of thumb, the fewer questions you ask, the more likely you are to get cooperation from your respondents. Ask only for pertinent information and only for information that you can't get through other research avenues.

The extent of your preliminary research will vary, of course, depending on the amount of time involved. If the questionnaire is part of a project for a one-semester or one-term course, you will have to tailor your topic to suit the available time and plan your research accordingly. With a relatively narrow, straightforward topic (like our restaurant example), a week or so of study should suffice to establish your credentials as an authority and enable you to write an authoritative questionnaire. But if the questionnaire is part of a graduate-level thesis, your topic will undoubtedly be much broader—and you may need to spend six months or more simply acquiring the information that will serve as a basis for the questionnaire.

Drafting the Questions

When you've thoroughly studied your topic and determined that a questionnaire is definitely the best way to obtain at least some of the data you need, it's time to make up a list of questions. For several reasons, this part of the project is somewhat more difficult than it sounds.

Reliability First, your questions have to take into account the factor of **reliability.** That is, your questions should be worded so that they measure what you want to measure. In our restaurant problem, if you wanted to measure patrons' reactions to the environment of a restaurant, you would word your questions so that their reactions, good or bad, to the food do not color their answers. For instance, a general question such as "Did you enjoy your experience here tonight?" might measure overall customer satisfaction very well. But you could not use it by itself to interpret what part environment, food, or service played in the level of satisfaction. You have to word your questions specifically to measure what you hope to measure. As we'll see, our restaurant researcher managed to word the questionnaire to obtain fairly reliable results.

Validity Second, your questions must be worded to ensure a **valid** interpretation of each question by the respondents—that is, an interpretation that matches your own. This requirement overlaps the first to some extent but also goes beyond it. In trying to write questions that express your meaning so that respondents cannot possibly misinterpret it, you must try to read each question through their eyes. Remember that although you have studied the topic in depth, the questionnaire may be their first introduction to it. Some people will know more than others, of course, but the idea is to formulate questions that will be clear to *everyone*. Returning to the restaurant problem, you would not learn much from the question, "Did you enjoy the ambiance of the restaurant?" Some of your respondents might not know what is meant by *ambiance*. Even if they do know the meaning of the word (the atmosphere or environment surrounding one), they may not know what to include about ambiance in their answers. For instance, should they include the lighting, the waiters' attitudes, the presentation of the food, and so forth? Too many decisions are thus left to the respondents. With such a question both validity and reliability fly out the window. As in ensuring *reliability*, to ensure *validity* you have to word your questions very specifically, and you must use words that you know your respondents will interpret in the same way you do.

Question Types Third, you have several different *types* of questions to choose among in drafting your questionnaire. Once you have selected the type that best suits your purpose, in general, stick to it throughout the questionnaire. Changing from one type of question to another in midstream may confuse your respondents. The six types of questions most commonly used on questionnaires are as follows:

1. **Dichotomous.** A *dichotomy* is a division into two separate parts, and a dichotomous question offers the respondent a choice between two answers—*yes* or *no*, *before* or *after*, *manual* or *electric*, and so on. Thus, this type of question provides a relatively limited span of responses and is primarily useful when the questionnaire topic can be analyzed in black-and-white terms.

2. **Multiple-choice.** A multiple-choice question allows the respondent to choose among several possible answers:

In preparing copy for a printer, if you must attach printed pages to a backing sheet, which of these methods do you prefer?

stapling ___ taping ___ gluing ___ using paper clips ___

If you think that there may be other possible answers, you can add a blanket category, followed by a line for the respondent's answer: other ________. This approach is probably the most common for questionnaires; it permits a fairly wide range of responses, but the choices are sufficiently controlled by the administrator to permit easy interpretation of results.

3. **Ranking.** In ranking, the respondent is asked to rank several possibilities in order of personal preference:

If your company required you to transfer to one of the following cities, what would be your order of preference? Place the appropriate number on each line, 1 for first choice, 2 for second choice, and so on:

Boston ___ Chicago ___ Denver ___ Miami ___

4. **Rating.** In rating, the respondent is asked to rate the relative importance of an item:

By placing a check at the appropriate place, indicate the importance of the following items in measuring your satisfaction with a hotel:

	Very Important	Somewhat Important	Not Important
Front desk service			
Room service			
Lounge service			
Restaurant service			

5. **Fill-in-the-blank.** This short-answer approach may be used to elicit either factual answers or opinions:

How many children do you have? (If none, write "none.") ____

How long have you held your present job? years __________
months __________

What is your gross annual income (to the nearest thousand)?

____

6. **Essay.** An essay questionnaire allows the respondent maximum freedom in composing an answer. Questions are somewhat general and followed by an inch or more of space for the answers:

Do you approve of the president's fiscal policies. Why or why not?

____

____

Despite the obvious advantage of complete and detailed answers, the essay method is the least efficient of the six described here, and it is therefore the least suitable for short-term projects. (By the same token, it could be the best for a long-term project in which thorough individual responses are important.) Essay questionnaires are time-consuming for both administrator and respondent, handwriting may be illegible and difficult to decipher, and replies may be very difficult to tabulate.

Administering the Questionnaire

When the questions are drafted to measure what you want to measure with no chance of reader misinterpretation and are formatted consistently, other problems may remain. How many dozens or thousands of questionnaires should you administer? Should the administration procedure involve mailing, telephoning, or personal interviews? Will the local telephone directory supply enough respondents, or should you use another source? Is there any guarantee that the respondents to your questionnaire will be representative of the population you hope to study?

There is no set formula for solving these problems. The answers will vary with each project, depending on the issue being investigated, the time period involved, any special circumstances that must be dealt with, and so on. For the sake of this discussion, however, let us assume that you as our restaurant researcher, with the help of an article in the *Cornell Hotel and Restaurant Administration Quarterly*, have drafted a satisfactory group of questions. You have decided to administer the questionnaire to the patrons of three restaurants: your own, Scarpelli's, and an expensive hotel dining room. You call the managers of the two restaurants you are not connected with and get their cooperation by agreeing to share the results of your questionnaire with them. The questionnaires and sufficient pencils will be given to everyone sitting at a table by the waiter when he or she presents the check. The questionnaire is shown in Figure 5–7.

Even the most intelligently devised questionnaires sometimes miss their mark, and the responses prove unrewarding or hard to tabulate. Therefore, we suggest that you administer the first version of any questionnaire

CUSTOMER ATTITUDE SURVEY

Completing this survey will take only a few moments of your time. Your answers will help us measure your reaction to this restaurant's physical environment (color, lighting, decoration, spacing, furnishings, and so forth). Please do not let your answers be influenced by the quality of food or service you have received tonight. Please complete the survey and return it to your waiter or cashier when he or she asks for it. Thank you very much for your help and cooperation.

1. Please tell us your age and sex. Age ____ Sex ____

2. In each pair of statements below, please circle the statement that most accurately describes your response to the physical environment of this restaurant.

appealing / unappealing
attractive / unattractive
bright / dull
cheerful / gloomy
comfortable / uncomfortable
distinctive / ordinary
expensive / inexpensive
formal / informal
good lighting / poor lighting
inviting / repelling
ornate / plain
pleasant / unpleasant
tasteful / tasteless

Figure 5–7 Questionnaire
Source: Adapted from Carol U. Lambert, "Environmental Design: The Food Service Manager's Role," *The Cornell Hotel and Restaurant Administration Quarterly* 22 (1981): 62–68.

you devise on a trial basis. Then make whatever revisions seem to be indicated.

Do not expect questionnaires to produce miracles. If you mail them (stamped return-address envelope included, of course), do not be shocked if only 35% respond. House-to-house canvassing may or may not produce a higher percentage of responses, depending on the time of day and the mood of the community among other things. Administering a questionnaire by telephone is a tedious procedure, but it permits you to cover a large geographical territory without pounding the streets. A combination of these means of administration may prove to be the best overall answer. Whatever method you choose, try to select your subjects on a random basis. By doing so, you can be surer of obtaining a representative sample.

In our restaurant situation, having licked the questionnaire problem, you stop to take stock again. It begins to seem likely that you will recommend some drastic changes in restaurant environment to your boss, Jane Lewis. You wonder if it would not be a good idea to check out her attitude toward such changes. You decide you had better talk to her.

Checking Readers' Attitudes and Requirements

In writing on the job, you usually have some notion of who your readers are, and you may even have close and daily contact with them. They may have a client relationship with you in that they may have hired you to do the research for them. Or, as in our restaurant example, there may be a boss–subordinate relationship. Whatever, if you do have such contact, you should check with your readers from time to time. It's another way of keeping your project on track.

In a large formal undertaking, you may stay in contact through progress reports. (See Chapter 16, "Progress Reports.") In a smaller, more informal situation, a few words of conversation or a short memo may do the job. Suppose, for example, your project is to write a set of instructions for calibrating a television set. You have reached the point in gathering information where you are investigating needed test equipment. The thought occurs to you that the company technicians who will be your readers may not know how to operate the test equipment. Do not continue in the dark. Go see or call the technicians, and ask them what they know. If they are unfamiliar with the test equipment, you will have to include operating instructions for it. If they are familiar with it, you can refer to it and move on. In such ways, you can shape your report to your readers.

In our restaurant example, you may now be realizing that switching a restaurant from a teenaged clientele might be an expensive proposition. It might call for extensive remodeling. For various reasons, your boss may not want to undergo such expense. Call her or send her a memo. (See page 382.) Summarize what you know to date, and ask if you should continue. If so, are there restrictions upon what you should be planning or thinking about? You do, and she gives you a green light. "Go ahead," she says. "Tell me what you find out and what you think I ought to do about it. We need to plan for the future."

Good writing is an act of constant refining of your material to suit your purpose and audience. Some of this refining goes on during the writing and revising stages. But much of it, and by no means the least important, goes on during the stage when you are gathering and checking your material.

Conducting Interviews

You have begun to sift through the answers to your questionnaire concerning the physical environment in your selected restaurants. Some preliminary answers are becoming obvious. The young adults who answered the questionnaires definitely favor the environment of Scarpelli's. They do not care very much for the environment in either your restaurant or the expensive hotel dining room. There are interesting mixes of adjectives for each restaurant, but the pointers are becoming clear. For instance, young adults find your restaurant "bright" but "unappealing" and "ordinary." They find the hotel dining room "distinctive" but "gloomy" and "unappealing." They find Scarpelli's "bright," "cheerful," and "appealing." You decide that it's time to flesh out what you have learned through the literature, on-site inspections, and questionnaires with some interviews.

You list some people who may be able to help you:

Carl Strickland, Associate Professor of Food Management
Virginia Book, Professor of Interior Design
Sandy Dolan, owner-manager of Scarpelli's

Your instincts suggest that Sandy Dolan, who has run a series of successful restaurants for 20 years, may well have the broadest view of what it takes to please customers. You call him, and he agrees to talk to you on Thursday, three days hence.

If you have watched interviews on TV you probably have noticed that most experienced interviewers keep a tablet or clipboard in front of them and from time to time glance at their prepared questions. This tactic is helpful for the expert—it will be essential for you.

Prepare a list of questions in advance. In general, stay clear of "dead end" questions that can be answered with a simple yes or no. (The experts call such questions *bipolar.*) Instead, phrase your questions in a way that invites the person being interviewed to expand on his or her first answers. Pass the ball to the interviewee and let that person do practically all the carrying. Phrase your questions clearly and economically; avoid 10-line "essay" questions that will eat up valuable time when your interviewee should be doing the talking. On the other hand, do not insist upon asking every question on your prepared list. If the interviewee discourses productively, let the discourse continue until he or she runs out of things to say. Also, do not insist upon posing the questions in their prepared order; instead, use the cues provided by what has just been said to guide you in selecting a reasonably relevant next question. If your interviewee permits it, tape-record the interview for later reference.

When you arrive for the interview, introduce or reintroduce yourself, giving your name, business or professional connections, subject matter for discussion, reasons for the interview, and the use to be made of the information gained. You should both sense the mood of the occasion and try to color the mood. For a half-hour interview, the opening five minutes may be given to generalities and personalities apparently far afield from the real reason for your interview. Once rapport has been established, take some key remark to lead into one of your prepared questions. What follows may read, in part, like this:

Interview

Q. Mr. Dolan, I notice that your color scheme in Scarpelli's is simple, basically mirrors with dark frames and white walls. Why have you chosen this scheme?

A. Well, yes, I guess that's right. We try to keep it simple. I dislike busy decoration if you know what I mean—all candles and swords and capes and phony ancestral portraits. Keep it simple, bright, and cheerful is what we try to do.

Q. Why the mirrors?

A. Well, that's part of it, really. In a restaurant, in a very real sense, your customers are your decoration. They come in here dressed casually but pleasingly. They're informal but not grubby. They add the needed color to the room. The mirrors reflect all that back and intensify it. Think of all the different scenes played out in these mirrors—much better than static paintings.

Q. I sense that the room is French in feeling. Is that accidental?

A. No, not really accidental, probably more subliminal.

Q. I'm not sure I understand what you mean by that.

A. OK. I mean I probably didn't consciously design the room to look like a French café. But I've traveled in France, and I like French restaurants. I associate them with good food. I was just following my own instincts in here, and the design, probably in a subliminal way, reflected my admiration for things French.

Q. Do you think your customers react favorably to the French look?

A. I think so. They keep coming back anyway. Actually, if you were to analyze it, an ethnic look that's not overdone, that's kept simple, is probably a successful look today. People expect ethnic food—French, Italian, Mexican, Middle Eastern, what have you—to be good. Yes, maybe an ethnic feel in the decoration helps. But keep it simple. For instance, you could catch an Arabian look with stucco in a sand color and maybe a discreet use of brass and leather. But don't for heaven's sake, fill up the place with burnooses on the wall and camel saddles for seats. See what I mean?

Q. Yes, I do. Thank you.

And so it goes. During an interview like this, you are looking for cues and clues. Most well-informed people, like Sandy Dolan, will talk at length if you let them. But do not expect blueprints and detailed estimates. Those will come later. You plan other interviews, perhaps with Professor Strickland and Professor Book. You also realize that it would be a good idea to follow up your questionnaires with some in-depth interviews of patrons of Scarpelli's. In the meantime, you drop a note of thanks to Sandy Dolan for his candid and helpful replies.

Writing Letters of Inquiry

Like questionnaires and interviews, letters of inquiry can provide you with useful information that may not be available in a library. When writing a letter of inquiry, keep in mind that you are imposing upon the other person, asking for time and attention. Therefore, brevity without curtness is in order. Sometimes a list of specific questions is appropriate—plus a general solicitation for advice and information. But be careful. If you ask for too much, you may get nothing at all. If there is any favor you can offer in return, introduce this thought toward the end of the letter. For additional information on how to write letters of inquiry, see pages 387–391.

As the young restaurateur of our example, you have used an article by Professor Carolyn U. Lambert in setting up your questionnaire. You have found the article quite useful. In one place, the article describes in a general way some research that might measure the effect of various table spacings and sizes in a restaurant. You wonder if a letter to Professor Lambert might not obtain more specific advice on how to go about such research. You write her as shown in Figure 5–8.

You mail your letter of inquiry and turn to a problem that has been much on your mind. Given the cost of remodeling for a different clientele, will the payback be sufficient to make it worthwhile? You sit down with calculator, pencil, and paper to see.

Performing Calculations and Analyses

A month ago when you began your research into a new clientele for a restaurant, you did not think very much about what the cost could be to carry out any recommendation you might make. With the passing days, however, your ideas have begun to take shape. You're excited about the possibility of converting one of the restaurants in the chain to serve a young adult clientele. It now occurs to you to consider the financial aspects of your project. What will be the cost? How much will it cost to borrow needed money? Will the return on investment be worthwhile? You decide to make some rough calculations.

Dear Professor Lambert:

As a manager in the Jane E. Lewis, Inc., restaurant chain, I am studying the possibility of remodeling one of the restaurants in the chain. Specifically, we are exploring the possibility of shifting the restaurant's environment and menu to suit a young-adult clientele rather than the present teenaged clientele.

Your article, "Environmental Design: The Food Manager's Role," has been most useful. A questionnaire based upon your sample adjective checklist elicited precisely the information we were looking for.

One of the problems still facing us is deciding what size tables to use and how to space them. In your article, you state that "Initial research in food-service settings could identify and compare the space required to eat and the space actually occupied by patrons, as well as determine the impact of making changes in a given space on conversation levels, turnover, and other patterns of behavior."

The questions you identify in this passage are the very questions we seek to answer. Have you researched this area further? Could you, perhaps, specify more exactly how such research could be accomplished?

I'll be happy to share the results with you of any research I do in this area. I enclose for you now a summary of the preliminary results obtained using your adjective checklist.

Thank you for your attention and for your very useful article.

Sincerely,

Figure 5–8 Letter of Inquiry

Several articles about restaurant renovation in one of the leading professional journals in the field, *Restaurant Hospitality*, have convinced you that $200,000 is about the minimum you can expect to spend for the kind of changes you have in mind. You decide to use a figure of $250,000 to allow for inflation and the extra costs that always seem to come up. A call to your boss, Jane Lewis, reveals that the figure doesn't frighten her off. She could put up $50,000 for a worthwhile project. The firm would have to borrow the remaining $200,000.

A call to a local bank reveals that a loan of $200,000 for business purposes would have to be for no more than five years at an interest rate of 11.5%. Because your boss is a good customer of the bank, it will allow a "balloon" (that is, a *lump sum*) repayment of the principal at the end of five years. You calculate the cost of the $200,000 loan. There would be 10 semiannual interest payments of $11,500 for a total of $115,000 which added to the $250,000 gives the tidy sum of $365,000.

The remodeled restaurant should seat 150 people. With good turnover, you should be able to count on two lunch sittings each business day of the week, that is, 300 people a day. On Saturdays, you plan on only 200 for lunch. For dinner, evidence from similar restaurants suggests you can serve approximately three sittings a night and thus serve 450 for dinner, Monday through Friday. The experience of the Lettuce Entertain You group that you have read about in *Restaurant Hospitality* indicates that you could plan on serving 900 people on Saturday night.

Let's see:

5 × 300 lunches =	1,500	
Saturday lunch =	200	
	1,700	lunches per week
5 × 450 dinners =	2,250	
Saturday dinner =	900	
	3,150	dinners per week

Sunday brunch business is booming in cities, and experience in restaurants such as R. J. Grunts in the Lettuce Entertain You group indicates that 900 for Sunday brunch would be a reasonable estimate. You plan to close the restaurant on Sunday nights.

Now let's figure possible gross revenue and net profits for all these meals. Drawing upon information gleaned from *Restaurant Hospitality* and your interview with Sandy Dolan of Scarpelli's, you have some numbers to work with:

Lunch checks average $7.50 per person:	1,700 × $ 7.50 =	$12,750
Dinner checks average $12.50 per person:	3,150 × $12.50 =	$39,375
Brunch checks average $9.75 per person:	900 × $ 9.75 =	$ 8,775
	Total	$60,900

Your calculations show that a weekly gross revenue of $60,900 would be reasonable. You know that with good management the net should be about 7% of the gross or about $4,300 a week.

Next, you'll compare this projected net profit against the current net of the teenaged operation. After figuring in other expenses that you can project over the five-year period, you'll be able to get a grasp of whether the planned new restaurant will repay the $365,000 invested in it and, if so, how long it will take.

Because our example is an executive report, your entire research, calculations, and analysis are heading toward a single point—a decision. Will remodeling for a new clientele be worthwhile—yes or no? If yes, what are the risks involved, and what are the first steps to be taken to get the project under way? In essence, your calculations and analysis are creating new knowledge based on what you have been able to discover in your research.

Reviewing Information Already Gathered

At many stages in your research, perhaps weekly, you should take stock of your situation. What have you learned? Where are you now? What step should you take next? Without an occasional review you will most assuredly waste much of your time and go off on tangents. Keep track of your purpose and your audience's purpose. They may quite properly shift a bit as you gain more insight into the problem, but don't let the shift be accidental. Know what you are doing and where you are heading.

In our restaurant problem, your review might reveal that you have established the following points:

- The arguments for shifting from a teenaged clientele to young adults seem real and convincing. Your preliminary **calculations** seem to indicate that the investment will pay off. You know, though, that you must also compare the investment in the restaurant with an alternative investment. That is, how well will the $365,000 invested in the restaurant do against the same sum invested in, say, a mutual fund or a money market fund. You make a note to do the comparison.
- Somewhat to your surprise, you find you already possess substantial information on American age groups and on restaurant management. Much of it needs to be checked and verified, but you are gaining in knowledge every day.

- Most if not all the information you need is available from individuals and **libraries.**
- You now have an extensive set of note cards filled with information on population growth, age groups, the post-World War II babyboom bulge, and restaurant environments. You have a much firmer grip on the financial side of things than you had a few weeks ago. You wonder if you can find some information on income by age groups. You decide to have a look at the *Statistical Abstract of the United States* again.
- You have moved back and forth between **particulars** and **generalizations.** You have concluded that a restaurant of the type you envision can seat people more densely than you first expected.
- Your **on-site inspection** of Scarpelli's confirms your generalization about spacing and gives you many ideas about environment.
- Your **questionnaire** has given you many useful insights into how young adults react to different restaurant environments. Simplicity seems to be a key factor. Decor that is overelaborate or too formal seems to be out. Some whimsy without excessive cuteness seems to be OK. That needs some checking.
- You learned a lot in your **interview** with Sandy Dolan. He confirms many of your notions about taste, particularly about "keeping it simple."
- You still await a reply to your **letter of inquiry.** You hope it will aid you in further research into table size and spacing.
- You realize that your purpose has changed as you have gained additional information. Originally, you had planned, somewhat vaguely, simply to report on demographic changes and to indicate what these changes might mean to the restaurant chain. Now you see quite clearly that you have enough evidence to recommend a change. Furthermore, you are beginning to shape what the changes should be and are gathering information to demonstrate how the changes will pay off. You have kept in touch with your boss, and she has agreed to this shifting and tightening of your purpose.

The larger the project and the longer it takes, the more such stocktaking as we have demonstrated here is essential. Notice, too, that research has a snowballing effect. Information begets information; the more you learn, the more adroit you will be in devising new means of adding to your knowledge. It's a comforting thought.

Understand that the material under each heading of this chapter is simply illustrative. We have shown you one questionnaire; you may need to prepare and administer several of them, or none at all. You may need to arrange for a dozen interviews, or none at all. In other words, you will have to cycle back and forth through the information gathering and checking methods we have shown you. In some projects you will use many of the methods extensively; in others you will need only a few.

How do you know when enough is enough, when you have gathered all the information you need? There is no easy answer to that question. Sometimes you'll find you are taking notes and suddenly you will realize that you already have the information in notes taken earlier. If that happens often enough, it's a good sign that your research may be complete. Perhaps a better sign is when a weekly review reveals that you have a firm purpose for what you are doing and that you possess the information you need to fulfill that purpose. When you reach that happy point, it will be time to move into the next stage, arranging the report.

You will, of course, really be arranging and even writing your report all through the information-gathering and -checking process. As you gather information from whatever source, you will take notes under headings that will suggest organizational units to you. Gradually, as your purpose, audience, and information take shape, the overall arrangement may become clear. In this way, you may skip from gathering and checking your information directly into arranging it without even realizing it. In the same way, you will be writing bits and pieces of your report as you go along. Many of your notes, whether based upon your research or your own thinking, with a little scrubbing and polishing will go right into your report. The results of your mathematical calculations may need only to be formalized into tables or graphs to be presentable. Writing is truly difficult if you sit down empty headed except for the thought you must fill a certain number of pages with writing. If you build toward that writing moment with careful planning, intelligent information gathering, and even some early writing and graphing, you will find that many of the terrors you may have concerning the task will disappear.

Checklist for Gathering and Checking Information

The following questions are a summary of the key points in this chapter, and they provide a checklist for gathering and checking information.

- What do you already know about your subject?
- To satisfy your purpose, what information do you need from the library?
- What documentation system will you use in your report? (See pages 304–321.) What bibliographic information do you need for complete entries in that system?

- What are the implications of your information? What generalizations can you draw from it? How can you apply your generalizations to new material?
- What sites and facilities should you be investigating?
- Would a questionnaire be useful in your research? What types of questions would best suit your investigation: Dichotomous, multiple choice, ranking, rating, fill-in-the-blank, or essay?
- Are your questions worded to obtain reliable and valid responses? Will your questions measure what you want to measure? Will your respondents interpret your language in the same way you do?
- With what group can you give your questionnaire a trial run?
- What should you check with your readers?
- What could you learn by interviewing people? Who are the people who would be useful interviewees? What do you want to learn from them? What questions will get that information for you?
- Are there people you can write to for information? What questions do you want to ask them? How soon do you need an answer?
- How will you analyze your information? How do your calculations and analyses affect the final outcome of your report? Which of your calculations would make good tables or graphs?
- Are you still on track for your purpose and your reader's purpose? Has either purpose shifted in any way? If so, do you understand where you are heading? Are you satisfied with the direction?
- What questions remain to be answered? Where can you find the answers? Are your notes in good order?

Exercises

Perhaps by this time you have a term writing project. If so, using that project, complete the following exercises.

1. Calling upon your memory, construct a list of things that you already know about your subject.
2. Go to the library and, using the reference tools available to you, construct a working bibliography. We suggest that you write your bibliography on cards.
3. Consult one of the sources listed in the bibliography you constructed for Exercise 2. On a note card, write a direct quotation from the source. On another card, condense or paraphrase the quote on the first card.
4. Using the subject area of your projected report, complete one or more of the following assignments.
   - **a.** Inspect a local site or facility. Write up what you learn in a report of about 500 words.
   - **b.** Construct a questionnaire that you could use with people who can provide you information in the subject area.
   - **c.** Interview some expert on the subject. Write up the interview in a report of about 500 words.

d. Write a letter of inquiry to some expert in the subject area. Ask only for information that you cannot obtain in any other way.

5. Using the various review techniques—such as induction, deduction, calculations, and analysis—discussed in this chapter, conduct regular reviews of your projected writing project. When requested, write up these reviews in a manner given by your instructor.

CHAPTER 6

Achieving a Readable Style

SPECIFIC WORDS

POMPOSITY

Empty Words
Elegant Variation
Pompous Vocabulary

GOOD STYLE IN ACTION

A readable text is one that an intended reader can comprehend without difficulty. Many things can make a text difficult to read. For example, the content may include unexplained concepts that the reader does not understand. The material in the text that is new to the reader may not be explained in terms of material already familiar to the reader. The material may not be arranged or formatted in a way to make it accessible to the reader. We cover such aspects of readability elsewhere, notably in Chapters 2, "Composing"; 4, "Writing for Your Readers"; and 10, "Document Design." In this chapter, we deal with those elements of your style at the paragraph, sentence, and language level that can contribute to your text being readable and clear.

Examples of unclear writing style are all too easy to find, even in places where we would hope to find clear, forceful prose. Read the following sentence:

> While determination of specific space needs and access cannot be accomplished until after a programmatic configuration is developed, it is apparent that physical space is excessive and that all appropriate means should be pursued to assure that the entire physical plant is utilized as fully as feasible.

The murky sentence quoted comes from a report issued by a state higher education coordinating board. Actually, it's better than many examples we could show you. While difficult, the sentence is probably readable. Others are simply indecipherable. When you have finished this chapter, you should be able to analyze a passage like the one just cited and show why it is so unclear. You should also know how to keep your own writing clear, concise, and vigorous. We discuss paragraphs, lists, clear sentence structure, specific words, and avoiding pomposity. We have broken our subject into five parts for simplicity's sake. But all the parts are closely related. All have but one aim—achieving readability.

If you have a style checker available to you in your computer software, it will incorporate many of the principles we discuss in this chapter. Nevertheless, use it with great care. Style checkers used without understanding the principles involved in good style can be highly misleading. (We discuss style checkers and the problems with them more fully in Chapter 2, "Composing." See pages 32–33.)

The Paragraph

In Chapters 7, 8, and 9, we discuss the ways that you can inform a reader, define and describe, and argue. These strategies may be used not only to develop reports but also to develop paragraphs within reports. Thus, paragraphs will vary greatly in arrangement and length, depending upon their purpose.

The Central Statement

In technical writing, however, the **central statement** of a paragraph more often than not appears at the beginning of the paragraph. This placement provides the clarity of statement that good technical writing must have. In a paragraph aimed at persuasion, however, the central statement may appear at the close, where it provides a suitable climax for the argument. No matter where a central statement is placed, unity is achieved by relating all the other details of the paragraph to the statement, as in this paragraph of inference and speculation:

We have italicized the central statement concerning "the electronic library." The rest of the paragraph presents facts and speculation in support of the central statement.

The electronic library is coming. Publishers are preparing for this eventuality. Already electronic storage of text is more economical and more compact than storage of paper. Whereas the conventional library expends much effort in managing the whereabouts of physical objects—books, maps, serials, and so forth—the electronic library will have no such objects to keep track of. Circulation control will be a thing of the past because the electronic book will always be available to users. Books will still be available for convenient reading and study; they will not become obsolete. They will, however, be printed mainly on demand and on the user's electronic printer. I expect to see speciality firms offering leather-bound volumes for scholars to order, and in this way all the familiar book formats will be preserved. Like the paperback publisher, the electronic publisher will distribute much more information than was previously available, and the information will be far less expensive to distribute and maintain.[1]

Paragraph Length

Examination of well-edited magazines such as *Scientific American* reveals that their paragraphs seldom average more than 100 words in length. The editors of magazines know that paragraphs are for the reader. Paragraphing breaks the materials into related subdivisions for the reader's better understanding. When paragraphs run on too long, the central statements that provide the generalizations needed for reader understanding are either missing or hidden in the mass of supporting details.

In addition to considering the need of the reader for clarifying generalizations, editors also consider the psychological effect of their pages on the reader. They know that large blocks of unbroken print have a forbidding appearance that intimidates the reader. If you follow the practices of experienced editors, you will break your paragraphs whenever your presentation definitely takes a new turn. As a general rule, paragraphs in reports and articles should average 100 words or fewer. In letters and memorandums, because of their page layout, you should probably hold average paragraph length to fewer than 60 words.

Transitions

Most often, a paragraph presents a further development in a continuing sequence of thought. When this is true, the paragraph's opening central statement will be so closely related to the preceding paragraph that it provides most of the **transition** you need. When a major transition between ideas is called for, consider using a short paragraph to guide the reader from one idea to the next.

The following five paragraphs provide an excellent example of paragraph development and transition:

Hyperactivity is the key word. It or *hyperactive* appears six times. Its repetition provides a major transitional device. The second sentence leads into the central theme of the five paragraphs—that hyperactivity may sometimes be "in the eye of the beholder."

Hyperactivity is essentially a symptom, which may be the result of a child's basic personality, a temporary state of anxiety, or subclinical seizure disorders; or it may reflect a true hyperkinetic state. It may also, according to the Council of Child Health of the American Academy of Pediatrics, be strictly "in the eyes of the beholder" (AAP Council on Child Health, 1975).

The central statement in the paragraph shows its relationship to the first paragraph through the use of the words *hyperactivity* and *descriptions*. The second half of the statement leads into the subject of the paragraph—the behavioral activities that provide a measure of hyperactivity. Many central statements provide transition by looking both backward and forward.

Descriptions of hyperactivity are generally given in behavioral terms, such as motor activity, attention span, frustration tolerance, excitability, impulse control, irritability, restlessness, and aggressiveness. Although these behaviors are measurable, they may not adequately reflect the kind of problems that different children have. Objective measures of attention span have been developed and have been used in some studies to help make the diagnosis, usually during tasks requiring continuous performance by the child. Standard questionnaires have also been designed, to be used by parents and teachers as they make observations at different times; such questionnaires provide useful evaluation information.

Researchers provides the new element that will be considered in this paragraph. The words *measuring* and *interpreting* alert the reader that this paragraph continues the central theme of all five paragraphs.

Facts to support the difficulty of measurement are given.

Researchers, however, acknowledge the difficulties of measuring these attributes and of interpreting the measurements. For example, in a study at the University of North Carolina designed to measure the motor activity and interest span of children diagnosed as hyperkinetic, Routh (1975) reported that of 78 referrals from physicians, teachers and parents, only 47% of the children were judged to be overactive—despite the fact that all of the children were considered to be "problem children" by those referring them to the testing service. In a study of teacher ratings of hyperactivity conducted at the University of Iowa, older teachers rated more children hyperactive than did young teachers (Johnson, 1974).

Clinical experience shifts the reader from *researchers* as a source of information to a new source. The word *also* shows that the central idea of the previous paragraph is still being pursued. The clause beginning *many factors* introduces the subject matter of the paragraph.

Clinical experience also indicates that many factors alter the activity patterns of such children or the perception of the patterns by parents. Such factors range from the presence or absence of breakfast, weather conditions and resultant seasonal cycles that alter activities during cold seasons, to the interpersonal family relationships or existence of disruptive family problems.

Age of the child introduces the central subject of this paragraph. The words *another, determinant, detection*, and *hyperactivity* all announce the relationship of the age of the child to the central theme of the five paragraphs.

The final sentence rounds off the five paragraphs by returning, in more formal language, to the central theme that hyperactivity may sometimes be "in the eye of the beholder."

Age of the child appears to be another determinant of detection of the syndrome of hyperactivity. The majority of cases are noted at school, and there is usually a gradual diminution in the hyperactivity at puberty or slightly thereafter. While some observers cite isolated cases of identifiable hyperkinesis in older individuals, it is unusual. Also, in some areas of the world, the syndrome is not recognized as a problem by school authorities—again, an example of differences in perception or of occurrence.[2]

The five paragraphs illustrate that you will develop paragraphs coherently when you keep your mind on the central theme. If you do so, the words needed to provide proper transition will come rather naturally. More often than not, your transitions will be repetitions of key words and phrases supported by such simple expressions as *also, another, of these four, because of this development, so, but,* and *however*. When you wander away from your central theme, no amount of artificial transition will wrench your writing back into coherence.

Lists and Tables

One of the simplest things you can do to ease the reader's chore is to break down complex statements into lists. Visualize the printed page. When it appears as an unbroken mass of print, it intimidates readers and makes it harder for them to pick out key ideas. Get important ideas out into the open where they stand out. Lists help to clarify introductions and summaries. You may list by (1) starting each separate point on a new line, leaving plenty of white space around it, or (2) using numbers within a line, as we have done here. Examine the following summary from a student paper, first as it might have been written and then as it actually was:

> The exploding wire is a simple-to-perform yet very complex scientific phenomenon. The course of any explosion depends not only on the material and shape of the wire but also on the electrical parameters of the circuit. In an explosion the current builds up and the wire explodes, current flows during the dwell period, and "postdwell conduction" begins with the reignition caused by impact ionization. These phases may be run together by varying the circuit parameters.

Now, the same summary as a list:

> The exploding wire is a simple-to-perform yet very complex scientific phenomenon. The course of any explosion depends not only on the materials and shape but also on the electrical parameters of the circuit.
>
> An explosion consists primarily of three phases:
>
> 1. The current builds up and the wire explodes.
> 2. Current flows during the dwell period.
> 3. "Postdwell conduction" begins with the reignition caused by impact ionization.
>
> These phases may be run together by varying the circuit *parameters.*

The first version is clear, but the second version is clearer, and readers can now file the information in their minds as "three phases." They will remember it longer.

Some writers avoid using lists even when they should use them, so we hesitate to suggest any restrictions on the technique. But, obviously, there are some subjective limits. Lists break up ideas into easy-to-read, easy-to-understand bits, but too many can give your page the appearance of a laundry list. Also, journal editors sometimes object to lists in which each item starts on a separate line. Such lists take space, and space costs money. So use lists when they clarify your presentation, but use them discreetly.

Tables perform a function similar to lists. You can use them to present a good deal of information—particularly statistical information—in a way easy for the reader to follow and understand. We discuss tables and their functions in Chapter 12, "Graphical Elements of Reports."

Clear Sentence Structure

The basic English sentence structure appears in two patterns, *subject–verb–object* (SVO) and *subject–verb–complement* (SVC):

> John(S) hit(V) the ball(O).
> The baby(S) cried(V) lustily(C).

Around such simple sentences as "John hit the ball," the writer can hang a complex structure of words, phrases, and clauses that serve to modify and extend the basic idea. In this section on clear sentence structure, we discuss how to go about that task. In order, we discuss sentence length, sentence order, sentence complexity and density, active verbs, active and passive voice, and first-person point of view.

Sentence Length

Over the years, many authorities have thought that sentence length is an indicator of how difficult or easy a sentence is. More recent research has led to the conclusion that while sentence length and word length may, indeed, be indicators they are not the primary causes of the difficulty. Rather, the true causes may relate more to the use of difficult sentence structures and words unfamiliar to the reader. This position is summed up well in this statement:

> A sentence with 60, 100, or 150 words needs to be shortened; but a sentence with 20 words is not necessarily more understandable than a sentence with 25 words. The incredibly long sentences that are sometimes found in technical, bureaucratic, and legal writing are also sentences that have abstract nouns as subjects, buried actions, unclear focus, and intrusive phrases. These are the problems that must be fixed, whether the sentence has 200 words or 10.
>
> Similarly, short words are not always easier words. The important point is not that the words be short, but that your readers know the words you are using.[3]

In general we agree with such advice. Sentence density and complexity likely cause readers more grief than does sentence length alone. Nevertheless, it's probably worth keeping in mind that most professional writers average only slightly more than 20 words per sentence. Despite that average, their sentences, of course, may range from short to fairly long, but, for the most part, they avoid sentences like this one from a bank in Houston, Texas:

> You must strike out the language above certifying that you are not subject to backup withholding due to notified payee underreporting if you have been notified that you are subject to backup withholding due to notified payee underreporting, and you have not received a notice from the Internal Revenue Service advising you that backup holding has terminated.

Sentence Order

What is the best way to order a sentence? Is a great deal of variety in sentence structure the mark of a good writer? One writing teacher, Francis Christensen of the University of Southern California, looked for the answers to those two questions. He examined large samples from 20 successful writers, among them John O'Hara, John Steinbeck, William Faulkner, Ernest Hemingway, Rachel Carson, and Gilbert Highet. In his samples, he included 10 fiction writers and 10 nonfiction writers.[4]

What Christensen discovered seems to disprove any theory that good writing requires extensive sentence variety. The professionals whose work was examined depended mostly on basic sentence patterns. They wrote 75.5% of their sentences in plain **subject–verb–object (SVO)** or **subject–verb–complement (SVC)** order, as in these two samples:

> Doppler radar increases capability greatly over conventional radar. (SVO)
> Doppler radar can be tuned more rapidly than conventional radar. (SVC)

Another 23% of the time, the professionals began a sentence with a short **adverbial opener:**

> As with any radar system, Doppler does have problems associated with it.

These adverbial openers are most often simple prepositional phrases or single words such as *however, therefore, nevertheless,* and the other conjunctive adverbs. Generally, they provide the reader with a transition between thoughts. Following the opening, the writer usually continued with a basic SVO or SVC sentence.

These basic sentence types—*SVO(C)* or *adverbial* + *SVO(C)*—are used 98.5% of the time by the professional writers in Christensen's sample. What did the writers do with the remaining 1.5% of their sentences? For 1.2%, they opened the sentence with **verbal clauses** based upon participles and infinitives such as "*Breaking* ground for the new church," or "*To see* the new pattern more clearly." The verbal opener was again followed most often with an SVO or SVC sentence, as in this example:

> Looking at it this way, we see the radar set as basically a sophisticated stopwatch that sends out a high-energy electromagnetic pulse and measures the time it takes for part of that energy to be reflected back to the antenna.

Like the adverbial opener, the verbal opener serves most of the time as a transition.

The remaining 0.3% of the sentences (about 1 sentence in 300) were **inverted constructions,** in which the subject is delayed until after the verb, as in this sentence:

> No less important to the radar operator are the problems caused by certain inherent characteristics of radar sets.

What can we conclude from Christensen's valuable study? Simply this: The professionals were interested in getting their content across, not in tricky word order. They conveyed their thoughts in a clear container not clouded by extra words. You should do the same.

Sentence Complexity and Density

Research indicates that sentences that are *too* complex in structure or *too* dense with content are difficult for many readers to understand.[5] Basing our observations on this research, we wish to discuss four particular problem areas: openers in front of the subject, too many words between the subject and the verb, noun strings, and multiple negatives.

Openers in Front of the Subject As the Christensen research indicates, professional writers place an adverbial or verbal opener before their subjects about 25% of the time. When these openers are held to a reasonable length, they create no problems for readers. The problems occur when the writer stretches such openers beyond a reasonable length. What is *reasonable* is somewhat open to question and depends to an extent on the reading ability of the reader. However, most would agree that the 27 words and 5 commas before the subject in the following sentence make the sentence difficult for many readers:

Opening phrase too dense

Because of their ready adaptability, ease of machining, and aesthetic qualities that make them suitable for use in landscape structures such as decks, fences, and retaining walls, preservative-treated timbers are becoming increasingly popular for use in landscape construction.

The ideas contained in this too dense sentence become more accessible when spread over two sentences:

Puts central idea before supporting evidence

Preservative-treated timbers are becoming increasingly popular for use in landscape construction. Their ready adaptability, ease of machining, and aesthetic qualities make them highly suited for use in landscape structures such as decks, fences, steps, and retaining walls.

The second version has the additional advantage of putting the central idea in the sequence before the supporting information.

The conditional sentence is a particularly difficult version of the sentence with the subject too long delayed. You can recognize the conditional by its *if* beginning:

Subject too long delayed

If heat (20°–35° C or 68°–95° F optimum), moisture (20%+ moisture content in wood), oxygen, and food (cellulose and wood sugars) are present, spores will germinate and grow.

To clarify such a sentence, move the subject to the front and the conditions to the rear. Consider the use of a list when you have more than two conditions:

List helps to clarify

Spores will germinate and grow when the following elements are present:

- Heat (20°–35° C or 68°–95° F optimum)
- Moisture content (20%+ moisture content in wood)
- Oxygen
- Food (cellulose and wood sugars)

Words between Subject and Verb In the following sentence, too many words between the subject and the verb cause difficulty:

Subject and verb too widely separated

Creosote, *a brownish-black oil composed of hundreds of organic compounds, usually made by distilling coal tar, but sometimes made from wood or petroleum,* has been used extensively in treating poles, piles, cross-ties, and timbers.

The sentence becomes much easier to read when it is broken into three sentences and first things are put first:

Revised

Creosote has been used extensively in treating poles, piles, cross-ties, and timbers. It is a brownish-black oil composed of hundreds of organic compounds. Creosote is usually made by distilling coal oil, but it can also be made from wood or petroleum.

You might wish to break down the original sentence into only two sentences if you had an audience that you thought could handle denser sentences:

Revised

Creosote, a brownish-black oil composed of hundreds of organic compounds, has been used extensively in treating poles, piles, cross-ties, and timbers. It is usually made by distilling coal tar, but it can also be made from wood or petroleum.

Noun Strings Noun strings are another way that writers sometimes complicate and compress their sentences beyond tolerable limits. A noun string is a sequence of nouns that serves to modify another noun: for example, *multichannel microwave radiometer,* where the nouns *multichannel* and *microwave* serve to modify *radiometer*. Sometimes the string may also include an adjective, as in *special multichannel microwave radiometer*.

Nothing is grammatically wrong with the use of nouns for modifiers. Such use is an old and perfectly respectable custom in English. Expressions such as *fire fighter* and *creamery butter* in which the modifiers are nouns go unremarked and virtually unnoticed. The problem occurs when writers either string many nouns together in one sequence or use many noun strings in a passage. Both tendencies are evident in this paragraph:

Seven noun strings in one paragraph is excessive

We must understand who the initiators of *water-oriented greenway efforts* are before we can understand the basis for *community environment decision making processes. State government planning agencies and commissions* and *designated water quality planning and management agencies* have initiated such efforts. They have implemented *water resource planning and management studies* and have aided *volunteer group greenway initiators* by providing *technical and coordinative assistance.*[6]

In many such strings, the reader has great difficulty in sorting out the relationships among the words. In *volunteer group greenway initiators*, does *volunteer* modify *group* or *initiators?* The reader has no way of knowing.

The solution to untangling difficult noun strongs is to include the relationship clues such as prepositions, relative pronouns, commas, apostrophes, and hyphens. For instance, placing a hyphen in *volunteer-group* would clarify that *volunteer* modified *group*. The strung-out passage just quoted was much improved by the inclusion of such clues:

Relationship clues help to clarify noun strings

We must understand who the initiators of efforts *to* promote water-oriented greenways are before we can understand the process *by which* a community makes decisions *about* environmental issues. Planning agencies and commissions *of the* state government and agencies *which* have been designated *to* plan and manage water quality have initiated such efforts. They have implemented studies *on* planning and managing water resources and have aided volunteer groups *that* initiate efforts *to* promote greenways by providing them with technical advice and assistance *in* coordinating their activities.[7]

The use of noun strings in technical English will no doubt continue. They do have their uses, and technical people are very fond of them. But perhaps it would not be too much to hope that writers would hold their strings to three words or fewer and not use more than one per paragraph.

Multiple Negatives Writers can introduce excessive complexity into their sentences through the use of multiple negatives. By *multiple negative*, we do not mean the grammatical error of the *double negative:* for instance, "He does *not* have *none* of them." We are talking about perfectly correct constructions that include two or more negative expressions, like these, for example:

Negative statements

- We will *not* go *unless* the sun is shining.
- We will *not* pay *except* when the damages exceed $50.
- The lever will *not* function *until* the power is turned on.

The positive statements of all of these are better and clearer than the negative versions:

Positive statements

- We will go only if the sun is shining.
- We will pay only when the damages exceed $50.
- The lever functions only when the power is turned on.

Most research shows that readers have difficulty sorting out passages that contain multiple negatives. If you doubt the research, try your hand at interpreting this government regulation (italics for negatives are ours):

Excessive use of negatives

§928.310 Papaya Regulation 10.

Order. (a) *No* handler shall ship any container of papayas (*except* immature papayas handled pursuant to §928.152 of this part): (1) During the period January 1 through April 15, 1980, to any destination within the production area *unless* said papayas grade at least Hawaii No. 1, *except* that allowable tolerances for defects may total 10 percent *Provided*, that *not* more than 5 percent shall be for serious damage, *not* more than 1 percent for immature fruit, *not* more than one percent for decay: *Provided further*, that such papayas shall individually weigh *not* less than 11 ounces each.[8]

Active Verbs

The verb determines the structure of an English sentence. Many sentences in technical writing falter because the finite verb does not function properly; that is, (1) comment upon the subject, (2) state a relationship about the subject, or (3) relate an action that the subject performs. Look at the following sentence:

Action in a noun

Sighting of the ground was accomplished by the pilot at 7 A.M.

English verbs can easily be changed into nouns, but sometimes—as we have just seen—the change can lead to a faulty sentence. The writer has put the true action into the subject and subordinated the pilot and the ground as objects of prepositions. The sentence should read:

Action in a verb

At 7 A.M. the pilot *saw* the ground.

The poor writer can ingeniously bury the action of a sentence almost anywhere. With the common verbs *make, give, get, have,* and *use*, the writer can bury the action late in the sentence in an object:

Action in an object

The punch card operator *has* the job of translating language symbols into machine symbols.

or,

The speaker did not *give* a satisfactory explanation of his technique.

Properly revised, the sentences put the action where it belongs, in the verb:

Action in a verb

The punch card operator *translates* language symbols into machine symbols.

The speaker did not adequately *explain* his technique.

The poor writer can even bury the action in an adjective:

Action in an adjective

A new discovery produces an *excited* reaction in a scientist.

Revised:

Action in a verb

A new discovery *excites* a scientist.

There is constructions may trap the writer into inefficient sentences. For example:

There are some technical editors who prefer the passive voice.

Revised:

Some technical editors prefer the passive voice.

When writing, and particularly when rewriting, you should always ask yourself: "Where's the action?" If the action does not lie in the verb, rewrite the sentence to put it there, as with this sample:

Action in nouns

Music therapy is the scientific *application* of music to accomplish the *restoration, maintenance,* and *improvement* of mental health.

This sentence provides an excellent example of how verbs are frequently turned into nouns by the use of the suffixes *ion, ance* (or *ence*), and *ment*. If you have sentences full of such suffixes, you may not be writing as actively as you could be. Rewritten to put active ideas into verb forms, the sentence reads this way:

Action in verbs

> Music therapy *applies* music scientifically *to restore, maintain,* and *improve* mental health.

The rewritten sentence defines "music therapy" in one-third less language than the first sentence, without any loss of meaning or content.

Active and Passive Voice

We discuss active and passive voice sentences on pages 183–184, but let us also quickly explain the concept here. In an active voice sentence, the subject performs the action and the object receives the action, as in "Mary hit the ball." In a passive voice sentence, the subject *receives* the action, as in "The ball has been hit." If you want to include the doer of the action, you must add this information in a prepositional phrase as in "The ball has been hit *by Mary*."

We urge you to use the active voice more than the passive, but we do not wish to imply that you should ignore the passive altogether. The passive voice is often useful. You can use the passive voice to emphasize the object receiving the action. The passive voice in "Influenza may be caused by any of several viruses" emphasizes *influenza*. The active voice in "Any of several viruses may cause influenza" emphasizes the *viruses*.

Often the agent of action is of no particular importance. When such is the case, the passive voice is appropriate because it allows you to drop the agent altogether:

Appropriate passive

> Edward Jenner's work on vaccination was published in 1796.

Be aware, however, that inappropriate use of the passive voice can cause you to omit the agent when knowledge of the agent may be vital. Such is often the case in giving instructions:

Poor passive

> All doors to this building will be locked by 6 P.M.

This sentence may not produce locked doors until it is rewritten in the active voice:

Active

> The night manager will lock all doors to this building by 6 P.M.

While the passive voice has its uses, most editors feel that too much of it produces lifeless and wordy writing. They, therefore, advise against using it except when it is clearly appropriate. The *Council of Biology Editors Style Manual* succinctly expresses the reasons for this advice:

> Use the active voice except where you have good reason to use the passive. The active is the natural voice, the one in which people usually speak and write, and its use is less likely to lead to wordiness and ambiguity. Avoid the "passive of modesty," a device of writers who shun the first-person singular. "I discovered" is shorter and less likely to be ambiguous than "it was discovered." When you write "Experiments were conducted," the reader cannot tell whether you or some other scientist conducted them. If you write "I" or "we" ("we" for two or more authors, never as substitute for "I"), you avoid dangling participles, common in sentences written in the third-person passive voice.[9]

As the biology editors point out, the passive voice does, indeed, lead to many dangling participles, as in this sentence:

Passive

While conducting these experiments, the chickens were seen to panic every time a hawk flew over.

Chickens conducting experiments? Not really. The active voice straightens out the matter:

Active

While conducting these experiments, we saw that the chickens panicked every time a hawk flew over.

(See also "Dangling Modifier," in Part VI, "Handbook.")

Write in the active voice whenever appropriate. When revising, find and rewrite inappropriate passive voice sentences into the active.

First-Person Point of View

Once, many reports and scientific articles were written in the third person—"This investigator has discovered"—rather than first person—"I discovered." Because this practice introduces needless complexity, it is now much less common. Along with the *Council of Biology Editors Style Manual,* many other style manuals for scientific journals now advise against the use of the third person on the grounds that it is wordy and confusing. We agree with this advice.

The judicious use of *I* or *we* in a technical report is entirely appropriate. Incidentally, such usage will seldom lead to a report full of *I*'s and *we*'s. After all, there are many agents in a technical report other than the writer. In describing an agricultural experiment, for example, researchers will report how *the sun shone, photosynthesis occurred, rain fell, plants drew nutrients from the soil,* and *combines harvested.* Only occasionally will researchers need to report their own actions. But when they must, they should be able to avoid such roundabout expressions as "It was observed by this experimenter," (See also page 550.)

A Caution about Following Rules

We must caution you before we leave this section on clear sentence structure. We are not urging upon you an oversimplified primer style, one often satirized by such sentences as "Jane hit the ball" and "See Dick catch the ball." Mature styles have a degree of complexity to them. Good writers, as Christensen's research shows, do put information before the subject. Nothing is wrong with putting information between the subject and verb of a sentence. You will find many such sentences in this book. What we have done, however, is alert you to the fact that research does show that sentences that are too long, too complex, too dense, for whatever reason, cause many readers difficulty. Keep in mind that, despite increasingly good research into its nature, writing is a craft and not a science. Be guided by the research available, but do not be simplistic in applying it.

Specific Words

The semanticists' abstraction ladder is a ladder composed of rungs that ascend from very specific words such as *table* to abstractions such as *furniture, wealth,* and *factor.* The human ability to move up and down this ladder enabled us to develop language, on which all human progress depends. Because we can think in abstract terms, we can call a moving company and tell it to move our furniture. Without abstraction we would have to bring the movers into our house and point to each object we wanted moved. But like many helpful writing techniques, abstraction is a device you should use carefully.

Stay at an appropriate level on the abstraction ladder. Do not say "inclement weather" when you mean "rain." Do not say "overwhelming support" when you mean "62% of the workers supported the plan." Do not settle for "suitable transportation" when you mean "a bus that seats 32 people."

Writing that uses too many abstractions is lazy writing. It relieves writers of the need to observe, to research, and to think. They can speak casually of "factors," and neither they nor their readers really know what they are talking about. Here is an example of such lazy writing. The writer was setting standards for choosing a desalination plant to be used at Air Force bases.

Too abstract

- The cost must not be prohibitive.
- The quality of water must be sufficient to supply a military establishment.
- The quality of the water must be high.

The writer here thinks he has said something. He has said little. He has listed slovenly abstractions when with a little thought and research he could have listed specific details. He should have said:

Use of specific detail

- The cost should not exceed $3 per thousand gallons.
- To supply an average base with a population of 5,000, the plant should purify 750,000 gallons of water a day (AFM 88-10 sets the standard of 150 gallons a day per person.)
- The desalinated water produced should not exceed the national health standard for potable water of 500 parts per million of dissolved solids.

Abstractions are needed for generalizing, but they cannot replace specific words and necessary details. Words mean different things to different people. The higher you go on the abstraction ladder, the truer this is. The abstract words *prohibitive, sufficient,* and *high* could have been interpreted in as many different ways as the writer had readers. No one can misinterpret the specific details given in the rewritten sentences.

Abstractions can also burden sentences in another way. Some writers are so used to thinking abstractly that they begin a sentence with an abstraction and *then* follow it with the specific word, usually in a prepositional phrase. They write,

Poor — The problem of producing fresh water became troublesome at overseas bases.

Instead of,

Revised — Producing fresh water became a problem at overseas bases.

Or,

Poor — The circumstance of the manager's disapproval caused the project to be dropped.

Instead of,

Revised — The manager's disapproval caused the project to be dropped.

We do not mean to say you should never use high abstractions. A good writer moves freely up and down the abstraction ladder. But when you use words from high on the ladder, use them properly—for generalizing and as a shorthand way of referring to specific details you have already given.

Pomposity

When writing, state your meaning as simply and clearly as you can. Do not let the mistaken notion that writing should be more elegant than speech make you sound pompous. Writing *is* different from speech. Writing is more concise, more compressed, and often better organized than speech. But elegance is not a prerequisite for good writing.

A sign at a service station where one of us gets his gasoline reads, "No gas will be dispensed while smoking." Would the attendants in that service station speak that way? Of course not. They would say, "Please put out that cigarette" or "No smoking, please." But the sign had to be elegant, and the writer sounds pompous, and illiterate as well.

If you apply what we have already told you about clear sentence structure, you will go a long way toward tearing down the fence of artificiality between you and the reader. We want to touch on just three more points: empty words, elegant variation, and pompous vocabulary.

Empty Words

The easiest way to turn simple, clear prose into elegant nonsense is to throw in empty words, like these phrases that begin with the impersonal *it:* "It is evident," "It is clear that," or most miserable of all, "It is interesting to note that." When something is evident, clear, or interesting, readers will discover this for themselves. If something is not evident, clear, or interesting, rewrite it to make it so. When you must use such qualifying phrases, at least shorten them to "evidently," "clearly," and "note that." Avoid constructions like "It was noted by Jones." Simply say, "Jones noted."

Many empty words are jargon phrases writers throw in out of sheer habit. You see them often in business correspondence. A partial list follows:

to the extent that
with reference to
in connection with
relative to
with regard to
with respect to
is already stated

in view of
inasmuch as
with your permission
hence
as a matter of fact

We could go on, but so could you. When such weeds crop up in your writing, pull them out.

Another way to produce empty words is to run an abstract word in tandem with a specific word. This produces such combinations as

20 in number *for* 20
wires of thin size *for* thin wires
red in color *for* red

When you have expressed something specifically, do not throw in the abstract term for the same word.

Elegant Variation

Elegant variation will also make your writing sound pompous. (H. W. Fowler and F. G. Fowler invented the term "elegant variation" in their book, *The King's English.* H. W. Fowler used it again in *A Dictionary of Modern English Usage.* Writers should own both books and use them often.) **Elegant variation** occurs when a writer substitutes one word for another because of an imagined need to avoid repetition. This substitution can lead to two problems: (1) The substituted word may be a pompous one. (2) The variation may mislead the reader into thinking that some shift in meaning is intended. Both problems are evident in the following example:

Elegant variation

Insect damage to evergreens varies with the condition of the plant, the pest species, and the hexapod population level.

Confusion reigns. The writer has avoided repetition, but the reader may think that the words *insect, pest,* and *hexapod* refer to three different things. Also, *hexapod,* though a perfectly good word, sounds a bit pompous in this context. The writer would better have written,

Revised

Insect damage to evergreens varies with the condition of the plant, the insect species, and the insect population level.

Remember also that intelligent repetition provides good transition. Repeating key words reminds the reader that you are still dealing with your central theme (see pages 114–115.)

Pompous Vocabulary

Generally speaking, the vocabulary you think in will serve in your writing. Jaw-breaking thesaurus words and words high on the abstraction ladder will not convince readers that you are intellectually superior. Such words will merely convince readers that your writing is hard to read. We are not telling you here that you must forego your hard-won educated vocabulary. If you are writing for readers who would understand words like *extant* or *prototype,* then use them. But use them only if they are appropriate to your discussion. Don't use them to impress people.

Nor are we talking about the specialized words of your professional field. At times these are necessary. Just remember to define them if you feel your reader will not know them. What we are talking about is the desire some writers seem to have to use pompous vocabulary to impress their readers.

The following list is a sampling of heavy words and phrases along with their simpler substitutes.

accordingly: so
acquire: get
activate: begin
along the lines of: like
assist: help
compensation: pay
consequently: so
due to the fact that: because
facilitate: ease, simplify
for the purpose of: for
in accordance with: by, under
in connection with: about
initiate: begin
in order to: to
in the event that: if
in the interests of: for
in this case: here
make application to: apply
nevertheless: but, however
prior to: before
subsequent to: later, after
utilize: use

You would be wise to avoid the word-wasting phrases on this list and other phrases like them. You really don't need to avoid the single words shown, such as *acquire* and *assist.* All are perfectly good words. But to avoid sounding pompous, don't string large clumps of such words together. Be generous in your writing with the simpler substitutes we have shown you. If you don't, you are more likely to depress your readers than to impress them. Don't be like the pompous writers who seek to bury you under the many-syllabled words they use to express one-syllable ideas, as in this example from the U.S. Department of Transportation:

Pompous

The purpose of this PPM [Policy and Procedure Memorandum] is to ensure, to the maximum extent practicable, that highway locations and designs reflect

> and are consistent with Federal, State and local goals and objectives. The rules, policies, and procedures established by this PPM are intended to afford full opportunity for effective public participation in the consideration of highway location and design proposals by highway departments before submission to the Federal Highway Administration for approval. They provide a medium for free and open discussion and are designed to encourage early and amicable resolution of controversial issues that may arise.

We urge you to read as much good writing—both fiction and nonfiction—as time permits. Stop occasionally as you do and study the authors' choice of words. You will find most authors to be lovers of the short word. Numerous passages in Shakespeare are composed almost entirely of one-syllable words. The same holds true for the King James Bible. Good writers do not want to impress you with their vocabularies. They want to get their ideas from their heads to yours by the shortest, simplest route.

Good Style in Action

A final example will summarize much that we have said. Insurance policies had for so long been verbal bogs that most buyers of insurance had given up on finding one clearly written. However, the St. Paul Fire and Marine Insurance Company decided that perhaps it was both possible and desirable to simplify the wording of its policies. The company revised one of its policies, eliminating empty words and using only words familiar to the average reader. In the revision, the company's writers avoided excessive sentence complexity and used predominantly the active voice and active verbs. They broke long paragraphs into shorter ones. The insurance company became *we* and the insured *you*. Definitions were included where needed rather than segregated in a glossary.

The resulting policy is wonderfully clear. Compare a paragraph of the old with the new.[10]

Old:

Cancellation

Passive voice

> This policy may be cancelled by the Named Insured by surrender thereof to the Company or any of its authorized agents, or by mailing to the Company written notice stating when thereafter such cancellation shall be effective. This

Average sentence length: 29 words

Empty words

Unfamiliar words

All one paragraph

A 66-word sentence

policy may be cancelled by the Company by mailing to the Named Insured at the address shown in this Policy written notice stating when, not less than thirty (30) days thereafter, such cancellation shall be effective. The mailing of notice as aforesaid shall be sufficient notice and the effective date of cancellation stated in the notice shall become the end of the policy period. Delivery of such written notice either by the Named Insured or by the Company shall be equivalent to mailing. If the Named Insured cancels, earned premium shall be computed in accordance with the customary short rate table and procedure. If the Company cancels, earned premium shall be computed pro rata. Premium adjustment may be made at the time cancellation is effected or as soon as practicable thereafter. The check of the Company or its representative, mailed or delivered, shall be sufficient tender of any refund due the Named Insured. If this contract insures more than one Named Insured, cancellation may be effected by the first of such Named Insureds for the account of all the Named Insureds; notice of cancellation by the Company to such first Named Insured shall be deemed notice to all Insureds and payment of any unearned premium to such first Named Insured shall be for the account of all interests therein.

New:

Can This Policy Be Canceled?

Active voice

Average sentence length: 15 words

Clear, specific language

Short paragraphs

Yes it can. Both by you and by us. If you want to cancel the policy, hand or send your cancellation notice to us or our authorized agent. Or mail us a written notice with the date when you want the policy cancelled. We'll send you a check for the unearned premium, figured by the short rate table—that is, pro rata minus a service charge.

If we decide to cancel the policy, we'll mail or deliver to you a cancellation notice effective after at least 30 days. As soon as we can, we'll send you a check for the unearned premium, figured pro rata.

Examples that substitute specific, familiar words for the high abstractions of the original policy are used freely. For instance:

> You miss a stop sign and crash into a motorcyle. Its 28-year-old married driver is paralyzed from the waist down and will spend the rest of his life in a wheelchair. A jury says you have to pay him $1,300,000. Your standard insurance liability limit is $300,000 for each person. We'll pay the balance of $1 million.

Or:

> We'll defend any suit for damages against you or anyone else insured even if it's groundless or fraudulent. And we'll investigate, negotiate and settle on your behalf any claim or suit if that seems to us proper and wise.
>
> You own a two-family house and rent the second floor apartment to the Miller family. The Millers don't pay the rent and you finally have to evict them. Out of sheer spite, they sue you for wrongful eviction. You're clearly in the right, but the defense of the suit costs $750. Under this policy we defend you and win the case in court. The whole business doesn't cost you a penny.

Incidentally, there is no fine print in the policy. It is set entirely in 10-point type, a type larger than that used in most newspapers and magazines. Headings and even different-colored print are used freely to draw attention to transitions and important information. Most states now require insurance policies sold within their borders to meet "plain language" requirements. We can hope, therefore, that the impossible-to-read insurance policy is a thing of the past.

You can clean up your own writing by following the principles discussed in this chapter and demonstrated in the revised insurance policy. Also, if you exercise care, your own manner of speaking can be a good guide in writing. You should not necessarily write as you talk. In speech, you may be too casual, even slangy. But the sound of your own voice can still be a good guide—in this way. When you write something, read it over; even read it aloud. If you have written something you know you would not speak because of its artificiality, rewrite it in a comfortable style. Rewrite so that you can hear the sound of your own voice in it.

Planning and Revision Checklist

You will find the planning and revision checklists that follow Chapter 2, "Composing," (pages 33–35 and inside the front cover) and Chapter 4, "Writing for Your Readers," (pages 78–79) valuable in planning and revising any presentation of technical information. The following questions specifically apply to style. They summarize the key points in this chapter and provide a checklist for revising.

Planning

You can revise for good style, but you can't plan for it. Good style comes when you become aware of the need to avoid the things that cause bad style: ponderous paragraphs; dense, clotted sentences; excessive use of passive voice; pomposity; and the like. Good style comes when you write to express your thoughts clearly, not to impress your readers. Good style comes when you have revised enough writing that the principles involved become ingrained in your thought process.

Revision

- Have you a style checker in your word processing software? If so, use it, but exercise the cautions we advocate on pages 32–33.
- Are the central thoughts in your paragraphs clearly stated? Do the details in your paragraphs relate to the stated central thought?
- Have you broken up your paragraphs sufficiently to avoid long, intimidating blocks of print?
- Have you guided your reader through your paragraphs with the repetition of key words and with transition statements?
- Have you used listing or tables when such use would help the reader?
- Are your sentences of reasonable length? Have you avoided sentences of 60 or even 100 words? Does your average sentence length match that of professional writers—about 20 words?
- Professional writers begin about 75% of their sentences with the subject of the sentence. How does your percentage of subject openers compare to that figure? If your average differs markedly, do you have a good reason for the difference?
- When you use sentence openers other than the subject, do they provide good transitions for your readers?
- Have you held your sentence openers before the subject to a reasonable length?
- Have you avoided large blocks of words between your subject and your verb?
- Have you used noun strings to modify other nouns? If so, are you sure your readers will be able to sort out the relationships involved?
- Have you avoided the use of multiple negatives?
- Are your action ideas expressed in active verbs? Have you avoided burying them in nouns and adjectives?
- Have you used active voice and passive voice appropriately? Are there passive voice sentences you should revise to active voice?
- Have you used abstract words when more specific words would be clearer for your readers? Do your abstractions leave interpre-

tations open to the reader that you do not intend? When needed, have you backed up your abstractions with specific detail?
- Have you avoided empty jargon phrases?
- Have you chosen your words to express your thoughts clearly for your intended reader? Have you avoided pompous words and phrases?

Exercises

1. You should now be able to rewrite the example sentence on page 112 into clear, forceful prose. Here it is again; try it:

   > While determination of specific space needs and access cannot be accomplished until after a programmatic configuration is developed, it is apparent that physical space is excessive and that all appropriate means should be pursued to assure that the entire physical plant is utilized as fully as feasible.

2. Following are some expressions and sentences that the Council of Biology Editors believes should be rewritten.[11] Using the principles you have learned in this chapter, rewrite them:
   - an innumerable number of tiny veins
   - at this point in time
   - bright green in color
   - we conducted inoculation experiments on
   - due to the fact that
   - during the time that
   - fewer in number
   - for the reason that
   - goes under the name of
   - if conditions are such that
   - in the event that
   - in view of the fact that
   - it is often the case that
   - it is possible that the cause is
   - it would appear that
   - lenticular in character
   - oval in shape
   - plants exhibited good growth
   - prior to
   - serves the function of being
   - subsequent to
   - the fish in question
   - the treatment having been performed
   - throughout the whole of this experiment
   - the tube which has a length of 3 m
   - a process for the avoidance of waste
   - judging by present standards, these trees are
   - If we interpret the deposition of chemical signals as initiation of

courtship, then initiation of courtship by females is probably the usual case in mammals.

- A direct correlation between serum vitamin B_{12} concentration and mean nerve conduction velocity was seen.
- It is possible that the pattern of herb distribution now found in the Chilean site is a reflection of past disturbances.
- Following termination of exposure to pigeons and resolution of the pulmonary infiltrates, there was a substantial increase in lung volume, some improvement in diffusing capacity, and partial resolution of the hypoxemia.
- Some in the population suffered mortal consequences from the lead compound in the flour.

3. Turn the following sentence into a paragraph of several sentences. See if listing might be a help. Make the central idea of the passage its first sentence.

> If, on the date of opening of bid or evaluation of proposals, the average market price of domestic wool of usable grades is not more than 10 percent above the average of the prices of representative types and grades of domestic wools in the wool category which includes the wool required by the specifications (see (f) below), which prices reflect the current incentive price as established by the Secretary of Agriculture, and if reasonable bids or proposals have been received for the advertised quantity offering 100 percent domestic wools, the contract will be awarded for domestically produced articles using 100 percent domestic wools and the procedure set forth in (e) and (f) below will be disregarded.

4. Lest you think all bad writing is American, here are two British samples quoted in a British magazine devoted to ridding Great Britain of gobbledygook.[12] Try your hand with them.

- The garden should be rendered commensurate with the visual amenities of the neighborhood.
- Should there be any intensification of the activities executed to accomplish your present hobby the matter would have to be reappraised.

5. Write a short report, or revise one you did earlier, using the stylistic principles of this chapter.

PART II

The Techniques of Technical Writing

Informing

Defining and Describing

Arguing

CHAPTER 7

Informing

CHRONOLOGICAL ARRANGEMENT

TOPICAL ARRANGEMENT

EXEMPLIFICATION

ANALOGY

CLASSIFICATION AND DIVISION

Our purpose in this and the next two chapters is to show you how writing techniques are tools to be chosen to fulfill the various purposes you have when you write. In this chapter, we discuss those techniques that are useful when your primary purpose is to *inform:* chronological arrangement, topical arrangement, exemplification, analogy, and classification and division. You can use any of these techniques as an overall arranging principle for an entire paper. For example, an entire paper could be arranged in topical order. But you will also use these techniques as subordinate methods of development within a larger framework. For example, within a paper arranged topically you might have small sections based upon chronology, exemplification, classification, and so forth. The two uses of informing techniques are mutually supportive and not in conflict with one another. Also, as you will see in the next two chapters, you have to incorporate many of these informing techniques into your overall strategy when your purpose is to define, describe, or argue.

Chronological Arrangement

When you have reason to relate a series of events for your readers, arranging the events **chronologically,** that is, by time, is a natural way to proceed. In a chronological narrative, keep your readers informed as to where they are in the sequence of events. In the example that follows—a narration about the Three Mile Island nuclear accident—we have printed in boldface the phrases that the author uses to orient his readers.

Begins by announcing time at the beginning of the accident

In the early morning of March 28, 1979, when workers were repairing a malfunction of the secondary loop's water purification system, the main feedwater supply pumps were suddenly, for some unknown reason, shut down **(the first incident, 4:00:37 A.M.).**

This incident triggered a cascade of unexpected events and finally caused a core-melting accident.

With the main feedwater cut off, the temperature in the steam generator began to rise. The emergency feedwater pumps would normally have supplied enough water to cool down the steam generator, but the two valves in the secondary loop's emergency feedwater pipes had been left in the wrong position—instead of open, they were closed **(the second incident, which must**

have happened sometime during the preceding two day's maintenance work).

Thus the temperature and pressure in the primary loop started rising.

Keeps readers on track with transitions

At 3 seconds after the accident started, the relief valve on the pressurizer opened to relieve the accumulated pressure in the primary loop, **and at 8 seconds** fission in the reactor was suspended by insertion of control rods. Water temperature and pressure soon began to fall, as expected, **and at about 12 seconds** returned to normal.

Up to this point all safety redundant systems had worked well. The trouble which started with the first incident would have been subdued and recorded only as one of the day's minor malfunctions if the relief valve had closed automatically as it should.

But it failed to do so **(the third incident),** for reasons still unknown, and the plant operators were unaware of this fact. With the relief valve open, the pressure in the primary system continued to fall, and this allowed the unpressurized water to boil, triggering the automatic injection of emergency cooling water into the core.

Because this safety feature worked properly, there was again a chance of averting subsequent damage to the reactor core. Instead the operators, who believed the relief valve was closed, observing the pressurizer water level going up (due to the increase of steam bubbles in the primary cooling system) were completely misled.

Uses parentheticals effectively to indicate sequence

Here a major misstep occurred; the operators, believing the cooling system was being overfilled with water, blocked the emergency core-cooling water flow **(the fourth incident).** More and more of the water in the core boiled to steam and the reactor coolant pumps began to vibrate excessively **(the fifth incident)** since they had to pump a mixture of water and steam. The operators shut off two reactor coolant pumps **at 1 hour 13 minutes into the accident** and the other two **at 1 hour 40 minutes (the sixth incident).**

This was the most critical point during the accident; the core began to overheat and the fuel cladding started melting. A small fraction of the radioactive gases escaped through leaks into the environment.

At 2 hours 18 minutes after the accident began, the operators finally discovered the stuck-open relief valve and successfully blocked it off to stop the water flow through the valve from the primary cooling system. This marked the end of the critical phase of the accident.[1]

Such a chronological narrative is frequently used in technical writing to provide a historical overview that will aid the reader in understanding some explanation or argument that is to come. The passage just quoted serves such a use. For the most part, such narratives are related in a businesslike way. Sometimes, however, the writer may choose to use a narration more dramatically—perhaps in brochures or advertisements for a lay audience—to simplify complex ideas and provide human interest.

The following prologue to a National Aeronautics and Space Administration publication for the general public fulfills both purposes. The publication reports the exploration of Mars by the Viking spacecraft. We have again presented time and sequence markers in boldface to draw your attention to them:

Dramatic descriptions

Sunlight glints on metal as a robot lander separates from its carrier spacecraft high over Mars and begins **the final day** of a 440-million-mile journey from Earth. The two craft drift apart, joined only by the invisible thread of a radio link. The lander crouches within a great white shell, a heat shield that will protect it and help reduce its speed during its hot descent through the atmosphere.

Exact time shows sequence

Seven minutes and one second later, a compact 52-pound computer in the lander sends a series of commands to small liquid-fuel rockets on the heat shield. **The rockets fire for 22 minutes and 16 seconds,** reducing the craft's velocity, and the robot slowly falls out of its orbit toward a Martian basin called *Chryse Planitia.* Two radar altimeters lock onto the planet and feed altitude and velocity information to the computer. It controls the methodical landing sequence, navigating second by second, curving the descent toward the Martian surface. **Touchdown is just over three hours away.**

Two hundred twenty million miles away, in a control room on Earth, engineers and scientists watch a maze of numbers flashing across video screens, detailing the descent. These men and women have worked a sizable fraction of their professional careers preparing for this moment. Now they wait and watch. Their lander is on its own, controlled by transistor and microchip, dependent on wire and solder. It will descend slowly to success or crash in failure on commands stored in the computer's memory and on the intricate workings of computers, radars, and rocket engines.

Humans have ceded control to the machine, because as Hamlet said in a different context, "The time is out of joint." For space travel, the quotation is apt.

At the lander time is now, but on Earth lander time is past. A speed-of-light radio report from the lander takes 19 minutes to reach controllers on Earth. What they read on their screens has already happened, and if anything goes wrong during the landing they cannot alter the past.

The lander descends, and it feels the first traces of Mars' atmosphere impinging on the heat shield. A sensor sniffs and tastes the alien air. The computer responds with a command, and a tape recorder whirs into action, storing atmospheric data that will be relayed later to Earth.

The alien robot from Earth drops deeper into the atmosphere of Mars, slowing as molecules of air impact the heat shield. The computers signal ignition of the terminal-descent rocket engines.

Three minutes and 59 seconds after the lander first senses atmosphere, when it has reached an altitude of 19,273 feet, a small mortar fires, freeing a great red-and-white parachute to blossom in the thin air. The craft slows rapidly now—from 519 to 118 miles an hour—and the curving descent path approaches vertical. **Seven seconds after the mortar fires,** the computer breaks small explosive bolts and the heat shield falls rapidly away to crash, unnoticed, somewhere on the surface below. Three landing legs extend and lock, and the craft turns in midflight so it will be aligned to provide lighting for photography after the landing.

The parachute flutters away in the thin, carbon dioxide air and falls in a crumbled heap far from the landing site. **For 60 seconds the machine descends,** its rockets breaking the fall. Velocity is but five miles an hour when the robot touches the ground. Shock absorbers on the three legs take the strain. The rockets cut off. **The computer, after three hours, 12 minutes, and 50 seconds** in total control, passes stewardship of the lander back to its human partners. **It is July 20, 1976,** and Viking is ready to explore Mars.

Describes emotional reactions to add human interest

On Earth, the sun is edging above the horizon at the Jet Propulsion Laboratory in California when the lander's signals tell the scientists that their machine has landed safely. The men and women of the Viking team grin and cheer as the tension breaks. They slap each others' backs. They clasp hands. It has been a long night, a night climaxing years of preparation for the first landing on another planet. The scientists have spent the last month poring over photographs from the orbiter and agonizing over a site safe enough for a landing. **No one knows it now, but their work will continue for six more years.** They will overcome mechanical problems. They will be excited by some scientific results and baffled by others. **The years will pass** and they will develop a deep, new understanding of the strange red planet, Mars.[2]

The events the writer describes are dramatic in and of themselves. The writer heightens the drama with language such as "Sunlight glints on metal," "The lander crouches within a great white shell," and "Their lander is on its own, controlled by transistor and microchip, dependent on wire and solder." The writer uses human interest when, for example, he describes the emotional reaction of the scientists to Viking's successful landing. As we pointed out to you in Chapter 4, "Writing for Your Readers," the use of human interest is often appropriate when writing for a lay audience.

Chronological narratives don't have to be restricted to past events. They can be used to forecast future events, as in this segment where the writer projects the future flight plans of the Voyager spacecraft:

After Neptune, the Voyager spacecraft planetary encounters will be over. Only one planet in the solar system remains unvisited—Pluto—and neither of the Voyagers can change its course to visit that planet.

But the Voyager missions will continue as the two spacecraft hurtle onward through space, one above the ecliptic and one below, searching for the edge of the heliosphere—the heliopause, which is the outer boundary of the sun's energy influence. Crossing the heliopause, perhaps early in the next century, they will enter true interstellar space. These spacecraft may give us the first direct measurements of the environment outside our solar system, including interstellar magnetic fields and charged particles.[3]

Because the arrrangement of your material follows the chronology of the events you relate, arrangement is not a particular problem for you when you use chronological order. Choosing the level of detail you need may be a problem. Obviously, the narrator of the events of the Three Mile Island accident could have used more or less detail than he did. As in most kinds of writing, purpose and audience are your best guides in these matters. If your purpose, for example, is to give a broad overview for a lay or executive audience, you will limit the amount of detail. If you have, on the other hand, an expert audience who will wish to analyze carefully the sequence you describe, you will need to provide considerable detail.

A major application of chronology in technical writing is describing process, that is, describing a sequence of events that progresses from a beginning to an end and results in a change or a product. Process descriptions are written in one of two ways:

- *For the interested observer*—to provide an understanding of the process.
- *For the doer*—to provide instruction for performing the process.

We cover the first type of process description in Chapter 8, "Defining and Describing," the second in Chapter 14, "Instructions."

Topical Arrangement

Technical writing projects often begin with a topic, say something like "Christmas tree farming." One way to deal with such topics is to look for subtopics under the major topic. These should serve as umbrella statements beneath which you can gather yet smaller subsubtopics and related facts. In the case of the Christmas tree topic, the subtopics might very well be "production" and "marketing." "Production" could be broken down further into "planting," "maintaining," and "harvesting." "Marketing" could be broken down into "retail," "wholesale," and "cut-your-own." With some thought, you can break most topics down into umbrella-sized subtopics.

While you are settling upon the topic and subtopics for your paper, you should also be aware of the need for topic limitation. Students, in particular, often hesitate to limit their topic sufficiently. They fear, perhaps, that if they limit their material too severely they will experience difficulty in writing essays of a sufficient length. The truth of the matter is really the reverse. You will find it easier to write a coherent, full essay of any length if you *limit the scope of your topic.* With your scope limited and your purpose and audience clearly defined, you can fill your paper with specific facts and examples. When your scope remains broad and your purpose and audience vague, you must deal in abstract generalizations.

Suppose you wish to deal with the subject of robotics in about a thousand words. If you keep the purpose simply as "explaining robotics," what can you say in a thousand words? Probably just a few simple-minded generalizations about the theory, history, and applications of robotics.

But suppose you define your purpose and audience in relation to your subject matter. Perhaps your audience is a group of executives who may have an interest in the use of robotics in their industries. You know that their major interest in robotics will be in application, not in history or theory. You limit yourself, therefore, to application. You refine your purpose. You decide you want to give them some idea of the scope of robotics. To do so, you reach into your knowledge of robotics and choose several applications that can serve as illustrative examples.

You decide upon the use of robotics in performing the following three tasks:

- installing windows in automobiles as the automobiles pass on an assembly line
- arc welding in an airplane plant
- mounting chips in a computer factory

By using these three applications as your subtopics, you can illustrate a wide range of current robotics practices. Within a few minutes, you can limit your topic to manageable size and make a good start on arrangement.

Figure 7–1 reproduces two pages from a pamphlet on the radioactive gas radon. The pamphlet uses a topical arrangement, each topic marked for the reader by a question, "What is radon?" "Where does radon come from?" and so forth. The use of questions shows that the pamphlet's writer understands that headings phrased as questions are an effective device to increase reader understanding. (See also pages 248–258.)

What is radon?

Radon is a radioactive gas which occurs in nature. You cannot see it, smell it, or taste it.

Where does radon come from?

Radon comes from the natural breakdown (radioactive decay) of uranium. Radon can be found in high concentrations in soils and rocks containing uranium, granite, shale, phosphate, and pitchblende. Radon may also be found in soils contaminated with certain types of industrial wastes, such as the byproducts from uranium or phosphate mining.

In outdoor air, radon is diluted to such low concentrations that it is usually nothing to worry about. However, once inside an enclosed space (such as a home) radon can accumulate. Indoor levels depend both on a building's construction and the concentration of radon in the underlying soil.

How does radon affect me?

The only known health effect associated with exposure to elevated levels of radon is an increased risk of developing lung cancer. Not everyone exposed to elevated levels of radon will develop lung cancer, and the time between exposure and the onset of the disease may be many years.

Scientists estimate that from about 5,000 to about 20,000 lung cancer deaths a year in the United States may be attributed to radon. (The American Cancer Society expects that about 130,000 people will die of lung cancer in 1986. The Surgeon General attributes around 85 percent of all lung cancer deaths to smoking.)

Your risk of developing lung cancer from exposure to radon depends upon the concentration of radon and the length of time you are exposed. Exposure to a slightly elevated radon level for a long time may present a greater risk of developing lung cancer than exposure to a significantly elevated level for a short time. In general, your risk increases as the level of radon and the length of exposure increase.

How certain are scientists of the risks?

With exposure to radon, as with other pollutants, there is some uncertainty about the amount of health risk. Radon risk estimates are based on scientific studies of miners exposed to varying levels of radon in their work underground. Consequently, scientists are considerably more certain of the risk estimates for radon than they are of those risk estimates which rely solely on studies of animals.

To account for the uncertainty in the risk estimates for radon, scientists generally express the risks associated with exposure to a particular level as a *range* of numbers. (The risk estimates given in this booklet are based on the advice of EPA's Science Advisory Board, an independent group of scientists established to advise EPA on various scientific matters.)

Despite some uncertainty in the risk estimates for radon, it is widely believed that **the greater your exposure to radon, the greater your risk of developing lung cancer.**

How does radon cause lung cancer?

Radon, itself, naturally breaks down and forms radioactive decay products. As you breathe, the radon decay products can become trapped in your lungs. As these decay products break down further, they release small bursts of energy which can damage lung tissue and lead to lung cancer.

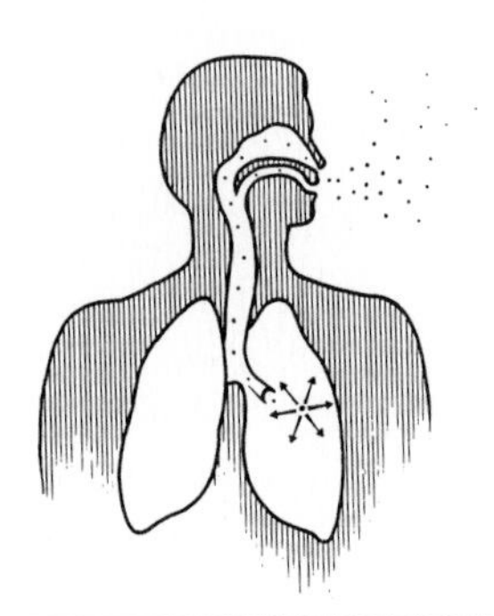

Figure 7–1 A Topical Arrangement
Source: U.S. Environmental Protection Agency, *A Citizen's Guide to Radon* (Washington, DC: GPO, 1986) 1–2.

Exemplification

Technical writing sometimes consists largely of a series of generalizations supported by examples. The writer makes statements such as this one by science essayist Lewis Thomas:

Central statement

> The tools possessed by NASA for this kind of year-round scrutiny [of the earth] have become simply flabbergasting, and a lot better ones are still to come if the research program can be adequately funded.[4]

At this point Dr. Thomas could have left the generalization stand unsupported, or he could have provided examples to give it credibility and interest. As most experienced writers would, he chose to provide examples:

Supporting examples

> Already instruments in space can make quantitative records of the concentrations of chlorophyll in the sea (and, by inference, the density of life), the acre-by-acre distribution of forests and fields and farms and deserts and human living quarters, the seasonal movement of ice packs at the poles, the distribution and depth of snowfalls, the chemical components in the outer and inner atmosphere, and the upwelling and downwelling regions of the waters of the earth. It is possible, now, to begin monitoring this planet, spotting early the evidences of trouble ahead for all ecosystems and species including ourselves.[5]

There are two common ways to use examples. One way is to give one or more extended, well-developed examples. We have used this method in showing you a large sample of Dr. Thomas's writing. The other way is to give a series of short examples that you do not develop in detail. Dr. Thomas himself used that approach. Like most everything else in writing, the use of examples calls for judgment on your part. Too few examples and your writing will lack interest and credibility. Too many examples and your key generalizations will be lost in excessive detail.

Analogy

Analogies are comparisons: they compare the unfamiliar to the familiar to make the unfamiliar more understandable for the reader. You should frequently use short, simple analogies, particularly when you are writing for lay people. For example, many people have difficulty in understanding the immense power released by nuclear reactions. A completely technical explanation of $E = mc^2$ probably would not help them very much. But suppose you tell them that if one pound of matter—a package of butter, for instance—could be converted directly to energy in a nuclear reaction, it would produce enough electrical power to supply the entire United States for 45 hours (that is, over 11 billion kilowatt hours). Such a statement reduces $E = mc^2$ to an understandable idea.

Scientists recognize the need for analogy when called upon to explain difficult concepts. A scientist working with microelectronic integrated circuits, that is, microchips, when called upon to explain how small the circuits are, said, "You grope for analogies. If you wanted to draw a map of the entire United States that showed every city block and town square, it would obviously be a *very* big map. But with the feature sizes we're working with to create microcircuits right now, I could draw that entire map on a sheet of paper not much larger than a postage stamp."[6]

A writer looking for a way to explain the immense age of the universe relative to humankind put it this way:

> Some 12 to 20 billion years ago, astronomers think a "primeval atom" exploded with a big bang sending the entire universe flying out at incredible speeds. Eventually matter cooled and condensed into galaxies and stars. Eons after life began to develop on Earth, humans appeared. If all events in the history of the universe until now were squeezed into 24 hours, Earth wouldn't form until late afternoon. Humans would have existed for only two seconds.[7]

Analogies serve particularly well in definitions or descriptions. If, after describing a diode, you tell a lay audience that it is similar to a water faucet in that you can use it to control the flow of electrons or shut them off completely, you make the concept more understandable.

Besides being practical, analogies can liven up your writing. Here is a writer having fun with some far-fetched analogies that, nevertheless, help you to grasp the enormousness of the quantities he is discussing:

> If all the Coca-Cola ever produced were dumped over Niagara Falls in place of water, the falls would flow at a normal rate for 16 hours and 49 minutes. . . . Two ships the size of the Queen Elizabeth could be floated in the ocean of Hawaiian Punch Americans consume annually. . . . There are 30,000 peanut butter sandwiches in an acre of peanuts, and 540 peanuts in a 12-ounce jar of peanut butter.[8]

Analogies are sometimes fairly extended. Here is a famous extended analogy of the astronomer Sir James Jeans that explains why the sky is blue.

Extended analogy

> Imagine that we stand on an ordinary seaside pier, and watch the waves rolling in and striking against the iron columns of the pier. Large waves pay very little attention to the columns—they divide right and left and reunite after passing each column, much as a regiment of soldiers would if a tree stood in

their road: it is almost as though the columns had not been there. But the short waves and ripples find the columns of the pier a much more formidable obstacle. When the short waves impinge on the columns, they are reflected back and spread as new ripples in all directions. To use the technical term they are "scattered." The obstacle provided by the iron columns hardly affects the long waves at all, but scatters the short ripples.

We have been watching a sort of working model of the way in which sunlight struggles through the earth's atmosphere. Between us on earth and outer space the atmosphere interposes innumerable obstacles in the form of molecules of air, tiny droplets of water, and small particles of dust. These are represented by the columns of the pier.

The waves of the sea represent the sunlight. We know that sunlight is a blend of many colors—as we can prove for ourselves by passing it through a prism, or even through a jug of water, or as Nature demonstrates to us when she passes it through raindrops of a summer shower and produces a rainbow. We also know that light consists of waves, and that the different colors of light are produced by waves of different lengths, red light by long waves, and blue light by short waves. The mixture of waves which constitutes sunlight has to struggle through the obstacles it meets in the atmosphere just as the mixture of waves at the seaside has to struggle past the columns of the pier. And these obstacles treat the light waves much as the columns of the pier treat the seawaves. The long waves which constitute red light are hardly affected, but the short waves which constitute blue light are scattered in all directions.

Thus the different constituents of sunlight are treated in different ways as they struggle through the earth's atmosphere. A wave of blue light may be scattered by a dust particle, and turned out of its course. After a time a second dust particle again turns it out of its course, and so on, until finally it enters our eyes by a path as zigzag as that of a flash of lightning. Consequently the blue waves of sunlight enter our eyes from all directions. And that is why the sky looks blue.[9]

Sir James has used analogy in a very imaginative way. By comparing light waves (an unfamiliar concept) to ocean waves (a concept familiar to his English readers), he has made it easy for his readers to grasp his meaning. Such extended analogies are most useful when you are writing for lay people.

Throughout your writing, use analogy freely. It is one of your best bridges to the uninformed reader.

Classification and Division

Classification and division, like chronological and topical arrangement, are useful devices for bringing order to any complex body of material. You may understand classification and division more readily if we explain them in terms of the **abstraction ladder.** We borrow this device from the semanticists—people who make a scientific study of words. We will construct our ladder by beginning with a very abstract word on top and working down the ladder to end with a specific term:

- **Factor:** Almost anything can be a factor. You could be a factor in someone's decision. So could wealth, climate, and geography, to name but a few significant concepts. Without more specific references, we cannot determine what is specifically meant at this rung of the ladder.
- **Wealth:** Now we have moved down a rung on the ladder. We have added specific information. Wealth could be money. It also could be stocks and bonds, land, furniture, or any of the other numerous material objects that people value highly.
- **Furniture:** We now have become much more specific. We can mean beds, tables, chairs, desks, lamps, and so on.
- **Table:** Now we are zeroing in on the object. However, *table* still refers to a huge class of objects: coffee tables, kitchen tables, dining room tables, library tables, end tables, and so on.
- **Kitchen table:** Now we know a good deal more. We know the function of the object. People eat at kitchen tables. Cooks mix cakes on them. We can even generalize somewhat about their appearance. Kitchen tables are usually plain objects, compared to end tables, for example. Many of them are made of wood and have four legs.
- **John Smith's kitchen table:** Now we know precisely what we are talking about. We can describe the size, weight, shape, and color of this table. We know the material of which it is made. We know that John Smith's family eats breakfast at this table but not dinner. In the evenings John Smith's daughter does her homework there.

We must keep one important distinction in mind: even "John Smith's kitchen table" is not the table itself. As soon as we have used a word for an object, the abstraction process has begun. Beneath the word is the table *we see* and beneath that is the table *itself,* consisting of paint, wood, and hardware that consist of molecules that consist of atoms and space.

In **classification** you move *up* the abstraction ladder, seeking higher abstractions under which to group many separate items. In **division** you move *down* the abstraction ladder, breaking down higher abstractions into the separate items contained within them. We will illustrate classification first.

Suppose for the moment that you are a dietitian. You are given a long list of foods found in a typical American home and asked to comment on the value of each. You are to give such information as calorie count, carbohydrate count, mineral content, vitamin content, and so forth. The list is as follows: onions, apples, steak, string beans, oranges, cheese, lamb chops, milk, corn flakes, lemons, bread, butter, hamburger, cupcakes, and carrots.

If you try to comment on each item in turn as it appears on the list, you will write a chaotic essay. You will repeat yourself far too often. Many of the things you will say about milk will be the same things you say about cheese. To avoid this repetition and chaos you need to classify the list, to move up the abstraction ladder seeking groups that look like the following:

Food

I. Vegetables
   A. Onions
   B. String beans
   C. Carrots
II. Fruit
   A. Apples
   B. Oranges
   C. Lemons
III. Meat
   A. Steak
   B. Lamb Chops
   C. Hamburger
IV. Cereal
   A. Corn Flakes
   B. Bread
   C. Cupcakes
V. Dairy
   A. Milk
   B. Cheese
   C. Butter

By following this procedure, you can use the similarities and dissimilarities of the different foods to aid your organization rather than have them disrupt it.

In division you move down the abstraction ladder. Suppose your problem now to be the reverse of the foregoing. You are a dietitian and someone asks you to list examples of foods that a healthy diet should contain. In this case you start with the abstraction, food. You decide to divide this abstraction into smaller divisions such as vegetables, fruit, meat, cereal, and dairy. You then subdivide these into typical examples such as cheese, milk, and butter for dairy. Obviously, the outline you could construct here might look precisely like the one already shown. But in classification we arrived at the outline from the bottom up; in division, from the top down. Figure 7–2 shows such an outline from a brochure on diet.

Very definite rules apply in using classification and division.

- **Keep all headings equal.** In the preceding example, you would not have headings of "Meat," "Dairy," "Fruit," "Cereal," and "Green Vegetables." "Green Vegetables" would not take in a whole class of food as the other headings do. Under the heading of "Vegetables," however, you could have subheadings of "Green Vegetables" and "Yellow Vegetables."

The key is following a Choose More Often approach. It doesn't mean giving up your favorite foods. It means taking steps to choose more often foods that are low in fat and high in fiber. For example, if you enjoy eating steak, choose a low-fat cut such as round steak, trim off the excess fat, broil it, and drain off the drippings. Pizza? To try a low-fat version that is rich in fiber, use a whole-grain English muffin or pita bread topped with part-skim mozzarella, fresh vegetables, and tomato sauce. And cookies or other desserts? In many recipes you can reduce the fat, and substitute vegetable oils or margarine for butter. To increase fiber, use whole wheat flour in place of white flour.

Here's how the Choose More Often approach works:

Choose More Often:

Low-fat meat, poultry, fish
Lean cuts of meat trimmed of fat (round tip roast, pork tenderloin, loin lamb chop), poultry without skin, and fish, cooked without breading or fat added.

Low-fat dairy products
1 percent or skim milk, buttermilk; low-fat or nonfat yogurt; lower fat cheeses (part-skim ricotta, pot, and farmer); ice milk, sherbet.

Dry beans and peas
All beans, peas and lentils—the dry forms are higher in protein.

Whole grain products
Breads, bagels, and English muffins made from whole wheat, rye, bran, and corn flour or meal; whole grain or bran cereals; whole wheat pasta; brown rice; bulgur.

Fruits and vegetables
All fruits and vegetables (except avocados, which are high in fat, but that fat is primarily unsaturated). For example, apples, pears, cantaloupe, oranges, grapefruit, pineapple, peaches, bananas, carrots, broccoli, Brussels sprouts, cabbage, kale, potatoes, tomatoes, sweet potatoes, spinach, cauliflower, and turnips, and others.

Fats and oils high in unsaturates
Unsaturated vegetable oils, such as canola oil, corn oil, cottonseed oil, olive oil, and soybean oil, and margarine; reduced-calorie mayonnaise and salad dressings.

Figure 7–2 Classification
Source: U.S. Department of Health and Human Services, *Eating for Life* (Washington, DC: GPO, 1988) n. pag.

- **Apply one rule of classification or division at a time.** In the preceding example, the classification is done by food types. You would not in the same classification include headings *equal* to the food types for such subjects as "Mineral Content" or "Vitamin Content." You could, however, include such subheadings under the food types.
- **Make each division or classification large enough to include a significant number of items.** In the preceding example, you could have many equal major headings, such as "Green Vegetables," "Yellow Vegetables," "Beef Products," "Lamb Products," "Cheese Products," and so forth. In doing so, however, you would have overclassified or overdivided your subject. Some of the classifications would have included only one item.

- **Avoid overlapping classifications and divisions as much as possible.** In the preceding example, if you had chosen a classification scheme that included "Fruits" and "Desserts," you would have created a problem for yourself. The listed fruits would have to go in both categories. You cannot always avoid overlap, but you can keep it to a minimum.

As long as writers observe the rules, they are free to classify and divide their material in any ways that best suit their purposes. An accountant, for example, who wished to analyze the money flow for construction of a state's highways might choose to classify them by source of funding: federal, state, county, city. An engineer, on the other hand, concerned with construction techniques, might choose to classify the same group of roads by surface: concrete, macadam, gravel.

In a fact sheet about controlling the insects that prey on trees, the author chose to classify the insects according to damage produced:

Types of Damage

Leaf Chewers

Classification by types of damage

A number of insects, mostly caterpillars and beetles, damage foliage by consuming all or parts of leaves or needles. This may take the form of skeletonizing, leaf mining, or free feeding. The skeletonizers consume the soft leaf tissues leaving a lacy pattern of the veins and sometimes the upper or lower leaf surface. The miners work between the leaf surfaces causing brownish or papery blotches or winding narrow trails. The free feeders eat the complete leaf and sometimes fold or roll the leaves or web leaves together.

Broad-leafed trees in otherwise good condition can withstand a complete defoliation without serious injury. They will produce a second crop of leaves the same season if defoliation occurs before August. Extensive defoliation for several successive seasons can kill the branches (called die-back) and possibly the tree.

Most evergreens, however, will die if completely defoliated. Any branch that is stripped of all its needles should be pruned out.

Sap-Sucking Insects and Mites

Aphids, plant bugs, scales, thrips, and mites extract the sap from buds, leaves, twigs, or stems. Their attacks frequently cause curling, spots, galls, yellowing, mottling, or deformed leaves and flowers. A healthy, well-established tree will withstand most infestations without permanent damage. However, some die-back may follow heavy infestations and ornamentals are frequently disfigured. This detracts from their value in the landscape.

Some aphids and scale insects also secrete a sweet sticky material called honeydew. During periods of heavy infestations, honeydew coats the leaves and branches of the trees as well as sidewalks, lawn furniture, and other objects under the trees. A black sooty-mold fungus grows on this honeydew, giving the trees a sooty appearance.

Disease Carriers

Some insects are vectors of disease organisms. The most common of these is Dutch elm disease, which is spread by elm bark beetles. Chemicals are sometimes needed to control these insects, which carry diseases but which may not otherwise be harmful to trees.

Borers

Trees and shrubs may be attacked by several kinds of insects which bore or tunnel inside trunks, branches, or twigs. Borers generally infest weakened or dying trees rather than healthy, vigorous ones. Proper pruning, fertilizing, and watering will help prevent borer problems. There are few practical chemical controls for borers. Some species may be controlled in individual specimen trees by injecting the burrows or tunnels with carbon tetrachloride and then plugging the treated burrows with clay or putty.[10]

Using this classification scheme, the author teaches his readers how to recognize whatever pests are plaguing their gardens at any given moment. The rest of the article explains how to deal with each type of pest. Obviously, insects can be classified in many ways; the types of damage they produce suggested the best classification scheme for this author, in view of the article's purpose and the needs of the audience.

Planning and Revision Checklists

You will find the planning and revision checklists that follow Chapter 2, "Composing" (pages 33–35 and inside the front cover), and Chapter 4, "Writing for Your Readers" (pages 78–79), valuable in planning and revising any presentation of technical information. The following questions specifically apply to the techniques discussed in this chapter. They summarize the key points of the chapter and provide a checklist for planning and revising.

CHRONOLOGICAL ARRANGEMENT

Planning

- Do you have a reason to narrate a series of events? Historical overview? Background information? Drama and human interest for a lay audience? Forecast of future events?
- What are the key events in the series?

Revision

- Is your sequence of events in proper order?
- Are all your time references accurate?
- Have you provided sufficient guidance within your narrative so that your readers always know where they are in the sequence?

Planning

- In what order do the key events occur?
- Do you know or can you find out an accurate timing of the events?
- How much detail does your audience need or want?

Revision

- Is your level of detail appropriate to your purpose and the purpose and needs of your audience?

TOPICAL ARRANGEMENT

Planning

- What is your major topic?
- What is your purpose?
- What is your audience's interest in your topic? How does their interest and purpose relate to your purpose?
- Given your purpose and your audience's purpose, how can you limit your topic?
    - What subtopics would be appropriate to your purpose and your audience's purpose?
    - Can you divide your subtopics further?

Revision

- Do your topics and subtopics meet your purpose and your audience's purpose and interests?
- Did you limit your subject sufficiently so that you could provide specific facts and examples?
- Do you have headings?
    - Do your headings accurately reflect how your readers will approach your subject matter?
    - Are your headings phrased as questions? If not, would it help your readers if they were?

EXEMPLIFICATION

Planning

- Do your generalizations need the support of examples?
- Do you have or can you get examples that will lend interest and credibility to your document?

Revision

- Have you left any generalizations unsupported? If so, have you missed a chance to interest and convince your readers?
- Have you provided sufficient examples to provide interest and credibility to your material?

ANALOGY

Planning

- What is your audience's level of understanding of your subject matter?
- Would the use of analogy provide your readers with a better understanding of your subject matter?
- Are there things familiar to your readers that you can compare to the unfamiliar concept, for example, water pressure to voltage?

Revision

- Have you provided analogies wherever they will help reader understanding?
- Do your analogies really work? Are the things compared truly comparable?

CLASSIFICATION AND DIVISION

Planning

- Where is your subject matter on the abstraction ladder?
  - Are you moving up the ladder, seeking higher abstractions under which you can group your subject matter: classification?
  - Are you moving down the ladder, breaking your abstractions down into more specific items: division?
- What is your purpose in discussing your subject matter?
- What is your audience's purpose and relationship to your subject matter?
- What classification or division will best meet your purpose and your audience's purpose?

Revision

- Are all the parts of your classification equal?
- Have you applied one rule of classification and division at a time?
- Is each classification or division large enough to include a significant number of items?
- Have you avoided overlapping classifications and divisions?
- Does your classification or division meet the needs of your purpose and your audience's purpose?

Exercises

1. Write a memo to an executive. The purpose of the memo is to inform the executive about the subject matter of the memo. Base the arrangement of the memo on one of the techniques described in this chapter. Accompany your memo with a short explanation of why you chose the arrangement technique you did. Your explanation must show how your purpose and your reader's purpose and interests led to your choice. For instruction on the format of memos, see page 382.
2. Write a chronological narrative of several paragraphs that is intended to serve as either a historical overview or a dramatic prologue for a larger report. Choose as a subject for your narrative some significant event in your professional field. Accompany your narrative with a description of your audience and an explanation of how their purpose and yours led you to the level of detail you use in your narrative.
3. Write an extended analogy of several paragraphs that will make some complicated concept in your discipline comprehensible to a fourth-grade student.
4. Reproduced in this exercise is an excerpt from a longer article about the Earth's crust. The article is an informational piece written primarily for the "employees and stockholders" of a large corporation. Therefore, the audience is a mixed one that will include, certainly, lay people, technicians, and executives, and perhaps even experts. The writer's task is to explain things so that lay people will find the article both interesting and understandable, but at the same time include facts and inferences that the other readers may find useful. The excerpt uses some of the techniques

described in this chapter. Working in a group, discuss where, how, and why the author has used these techniques. Reach a judgment as to how well he has used them. In your discussion, use concepts from Chapter 4, "Writing for Your Readers," as well as the concepts in this chapter. (*A helpful hint:* see, in particular, pages 149–151.) When the group has finished its discussion, each member of the group should write a report that presents his or her own views on the questions raised in this exercise.

> Typically, the Earth's crust is 12 miles thick, though it varies from three to 60 miles in depth depending on location. The distance from the surface to the very center of the globe is about 3,950 miles.
>
> If the Earth were a beachball exactly 12 inches in diameter, the entire, quasi-solid layer of rock and dirt that we call the crust would literally be less than 2/100 of an inch thick in most areas and less than one tenth of an inch at the very most.
>
> The crust is the only part of the Earth we know much about, and the only part with which we've had any practical, albeit limited, experience. How limited? Take an apple and prick its skin as delicately as you possibly can with a sharp needle. If the Earth were that apple—forget the beachball—and you had a very gentle touch, you'd have just made a hole roughly as deep as the deepest well ever drilled into the Earth's crust.
>
> Seismograph Service Corporation (SCC), a subsidiary of Raytheon based in Tulsa, Okla., knows all about such wells. SSC tickles and probes, thumps, and sounds the Earth's crust with a kind of rock radar to help oil and gas drillers decide where to punch those impudent holes.
>
> "It's much like the problems that doctors have imaging the human body," explains Dale Stone, head of SSC's geophysical research department. "Especially since seismologists use the same kind of ultrasonic energy."
>
> The anomaly a doctor might be looking for is a cancer. Our anomaly is oil. And for us, the equivalent of the differing densities of bone, flesh, and muscle are things like sandstone, shale, and limestone.
>
> Not all that long ago—the debate was still going on in the 1960s—geologists were doing well to have figured out that the continents all came from one mother mass and somehow moved monumentally hither and grandly yon across the Earth's surface—the theory of continental drift. Then earth scientists discovered that it wasn't neat, tear-along-the-dotted-line continents that were drifting, but entire "plates" of crustal matter, huge slabs of Earth's cracked eggshell upon which nations and oceans ride like so many puddles and piles of dirt.
>
> Plate tectonics—one of science's great breakthroughs—was born and became the mechanism by which the earth "moves." The theory of plate tectonics holds that the Earth's entire crust is made up of somewhere between a dozen and 20 slabs that fit together like a badly matched jigsaw puzzle with some pieces a bit too large for each other. And not only that, but some of those pieces are growing, new matter being squeezed from deep within the Earth like sheets of hot pasta.
>
> So one plate pushes against and even overlaps another here and there, and the plates are constantly making room for each other. The result? The edges chafe and grind and create earthquakes. One plate rams another and

eventually mountains wrinkle upward when the rammed plate folds like a throw rug kicked into a corner. Plates crack from the strain and fault lines form—more earthquake potential—and magma from the bowels of the planet leaks up through rifts, ridges, and hot spots to create volcanoes and geothermal springs. We live upon a seething machine that turns "solid as a rock" into a bad joke.

Nothing quite so grand concerns Seismograph Service, which looks for oil and gas, not geological enlightenment. Still, as Carol Rorschach, SSC's communications manager, puts it, "Were it not for naturally occurring phenomena such as the movements of plates, presumably the subsurface would be more nearly homogeneous, and there would be no oil left—no places where the hydrocarbons would have been trapped."

Fortunately for the energy industry, oil and gas deposits usually gather conveniently near the surface in a variety of spots after being formed by the rare series of coincidences that turn "rotten dinosaurs," as Dale Stone puts it, into hydrocarbons. "What we look for is porous rock," Stone explains. "You can't put oil into hard rock. You need something like sand or limestone that has become porous, dead underground river channels, reefs from ancient oceans. We also look for formations such as salt domes that trap the hydrocarbons, or fault planes that can be lateral traps."

SSC "looks" by twitching the Earth's crust with short pulses of energy—with explosives, truckmounted hydraulic rams that massage the earth, air guns that blast underwater bubbles, and a variety of other tools—and recording the vectors of energy reflected from whatever is underground.

"It's like throwing a rock in a pond," says Stone. "Ripples of energy radiate outward, and wherever that energy strikes a lack of continuity, whether it's a lily pad in the pond or a layer of limestone underground, some of it scatters back toward the focal point."

It would be nice if seismograms were neat snapshots of the crust under our feet, for the equipment used lacks little in range or penetration.

"We can look down 12 miles or more," Stone says, "and image as much of the Earth's crust as you'd be interested in seeing.

But seismograms aren't photographs, for the energy waves ricochet and refract, rock formations "migrate" and reappear in the wrong place, the crust under Oklahoma reacts differently than does the ground beneath Brazil.

SSC was one of several companies that pioneered the use of computer processing to clean up and amplify seismic data, beginning in the mid-1960s, but it was the only company to process that data on site.

"By 1970, everybody was processing seismic by computer," Stone explains, "but the only computers programmed to do it were in Dallas, Houston, and Tulsa. If you recorded the data in Borneo, it had to be shipped all the way back here to then stand in line waiting for computer time."

So in 1970, the subsidiary designed, assembled, and encoded an entire computer package powerful, durable, and transportable enough—by trailer, off-road truck or helicopter sling—to do the job in jungle or desert, on offshore platform or North Slope vastness.

"Some of the data processing techniques we've learned have been applied to earthquake research," says Stone.

"We're particularly interested in folds and fault patterns and pressures, because they're natural oil traps, and that's also where earthquakes occur."

Safely warehoused in packing crates at the subsidiary's Tulsa headquarters are some of the world's more detailed records of the Earth's crust—files full of not only digitized data from the computer-processed era of seismology, but also long manila envelopes that shelter laboriously made magnetic analog recordings and even primitive paper traces stretching all the way back to the 1930s.

It's not too far-fetched to wonder, in fact, if some of the motivating secrets of the machine we call Earth aren't already in our hands, awaiting interpretation, in a file cabinet in Tulsa.[11]

CHAPTER 8

Defining and Describing

DEFINITION

DESCRIPTION

Visual Language
Analogy
Place Description
Mechanism Description
Process Description

As a writer about technical subjects you will have a constant need to define terms and to describe places, mechanisms, and processes. Most often, your definitions and descriptions will be part of a larger effort; for instance, you often need to define the terms used in an argument. But there will be occasions when your major objective will be to define a term or to describe some mechanism or process. The techniques of defining and describing are closely related, and we take up both in this chapter.

Definition

Everyone with a trade or a profession has a specialized vocabulary to suit that occupation. Plumbers know the difference between a *globe valve* and a *gate valve.* Electrical engineers talk easily about *gamma rays* and *microelectronics.* Statisticians understand the mysteries of *chi-square tests* and *one-way analyses of variance.* In fact, learning a new vocabulary is a major part of learning any trade or profession. Unfortunately, as you grow accustomed to using your specialized vocabulary, you may forget that others don't share your knowledge—your language may be incomprehensible to them. So, the first principle in understanding definition is to realize that you will have to do it frequently. You should define any term you think is not in your reader's normal vocabulary. The less expert your audience is, the more you will have to define. But, sometimes, when you use a new specialized term or an old term used in a new way, you will even need to define for your fellow specialists.

Definitions range in length from a single word to long essays or even books. Sometimes, but not usually, a **synonym** inserted into your sentence will do, as in this example:

> The oil sump, *that is, the oil reservoir,* is located in the lower portion of the engine crankcase.

Synonym definition serves only when a common interchangeable word exists for some bit of technical vocabulary you wish to use.

Sentence Definitions

Most often you will want to use at least a one-sentence definition containing the elements of a **logical definition:**

> *term = genus or class + differentia*

Although you may not have heard of the elements of a logical definition, you have been giving and hearing definitions cast in the logical pattern most of your life. In the logical definition, you state that something is a member of some genus or class and then specify the differences that distinguish this thing from other members of the class.

Term	=	Genus or Class	+	Differentia
An ohmmeter	is	an indicating instrument		that directly measures the resistance of an electrical circuit.
A legume	is	a fruit		formed from a single carpel, splitting along the dorsal and the ventral sutures, and usually containing a row of seeds borne on the inner side of the ventral suture.

The second of these two definitions, particularly, points out a pitfall you must avoid. This definition of a legume would satisfy only someone who was already fairly expert in botany. Real lay people would be no further ahead than before, because terms such as *carpel, ventral suture,* and so forth are not familiar to them. When writing for nonexperts, you may wish to settle for a definition less precise but more understandable:

> A legume is a fruit formed of an easily split pod that contains a row of seeds, such as a pea pod.

Here you have stayed with plain language and given an easily recognized example. Both of these definitions of a legume are good. The one you would choose depends on your audience.

Sometimes you may wish to begin a definition by telling what something is *not,* as in the following definition from *Chamber's Technical Dictionary:*

> *metaplasm.* Any substance within the body of a cell which is not protoplasm; especially food material, as yolk or fat, within an ovum.

You should, of course, avoid circular definition. "A botanist is a student of botany" will not take the reader very far. However, sometimes you may appropriately repeat on both sides of a definition common words you are sure your reader understands. In the following *Chamber's* definition it would be pointless to drag in some synonym for a word as readily understandable as *test.*

> *Gmelin's test.* A test for the presence of bile pigments; based upon the formation of various colored oxidation products on treatment with concentrated nitric acid.

Extended Definitions

To make sure you are understood, you will often want to extend a definition beyond a single sentence. The most common techniques for extending a definition are description, example, and analogy. However, any of the arrangement techniques, such as chronology, topical order, classification, and division, may be used. The following definition, again from *Chamber's*, goes beyond the logical definition to give a description:

Description

> *anemometer.* An instrument for measuring the velocity of the wind. A common type consists of four hemispherical cups carried at the ends of four radial arms pivoted so as to be capable of rotation in a horizontal plane, the speed of rotation being indicated on a dial calibrated to read wind velocity directly.

In our lay definition of legume, an example was given: "such as a pea pod." Often analogy is valuable:

Analogy

> A *voltmeter* is an indicating instrument for measuring electrical potential. It may be compared to a pressure gauge used in a pipe to measure water pressure.

Look for opportunities to enhance extended definitions with graphics. Figure 8–1 demonstrates how a graphic can clarify relationships for a reader.

The following definition of a hurricane is a good example of an extended definition intended for an intelligent lay audience. In it, the writer makes extensive use of both process and mechanism descriptions. Notice, also, that the writer begins by defining other terms needed in understanding hurricanes:

Defines related terms

A **hurricane** is defined as a rotating wind system that whirls counterclockwise in the northern hemisphere, forms over tropical water, and has sustained wind speeds of at least 74 miles/hour (119 km/hour). This whirling mass of energy is formed when circumstances involving heat and pressure nourish and nudge the winds over a large area of ocean to wrap themselves around an atmospheric low. **Tropical cyclone** is the term for all wind circulations rotating around an atmospheric low over tropical waters. A **tropical storm** is defined as a cyclone with winds from 39 to 73 mph, and a **tropical depression** is a cyclone with winds less than 39 mph.

Describes process

It is presently thought that many tropical cyclones originate over Africa in the region just south of the Sahara. They start as an instability in a narrow east-to-west jet stream that forms in that area between June and September as a result of the great temperature contrast between the hot desert and the cooler, more humid region to the south. Studies show that the disturbances generated over Africa have long lifetimes, and many of them cross the Atlantic. In the 20th century an average of 10 tropical cyclones each year whirl out across the Atlantic; six of these become hurricanes (1). The hurricane season is

1 Understanding GL:M

This chapter introduces the central concepts of McCormack & Dodge's general ledger system, General Ledger:Millennium (GL:M). It defines the purpose of GL:M, describes its functions and outstanding features, and shows how they can help meet business needs.

A sentence definition

GL:M: A Definition

GL:M is a general ledger software package that helps you perform accounting and reporting functions with accuracy and flexibility. It automates accounting processes, to make maintaining the ledger as easy as possible.

Extension by example

The general ledger stands at the center of an integrated financial information system. Data flows into it from all related subsystems, such as accounts receivable, accounts payable, purchase order, fixed assets, and human resources.

Figure 1-1 shows how GL:M draws data from all subsystems into a centralized base of financial information.

Graphical illustration of information flow

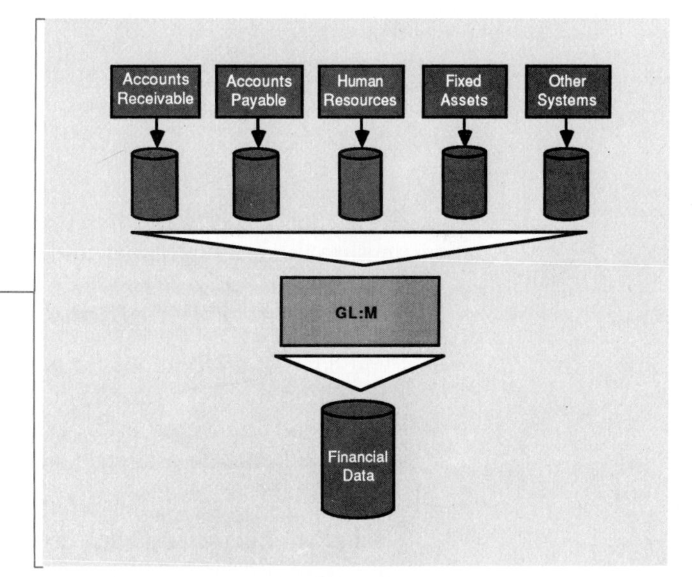

Figure 1-1. A Financial Information System

Figure 8–1 Graphic and Extended Definition
Source: GL: M General Ledger: Millenium—Overview (Atlanta, GA: Dun and Bradstreet Software Services, Inc., 1987) 1–1. Copyright © 1987 by Dun and Bradstreet Software Services, Inc. Reprinted by permission.

set as being June 1 through November 30. An "early" hurricane occurs in the 3 months before the season, and a "late" hurricane takes place in the 3 months after the season (2).

Describes mechanism

Hurricanes are well-organized. The 10-mile-thick inner spinning ring of towering clouds and rapid upper motion is defined as the hurricane's eyewall; it is here that condensation and rainfall are intense and winds are most violent. Harbored within the eyewall is the calm eye of the hurricane—usually 10–20 miles across—protected from the inflowing winds and often free of clouds. Here, surface pressure drops to a minimum, and winds subside to less than 15 mph. Out beyond the eyewall, the hurricane forms into characteristic spiral rain bands, which are alternate bands of rain-filled clouds. In the typical hurricane, the entire spiral storm system is at least 1,000 miles across, with hurricane-force winds of 100 miles in diameter and gale-force winds of 400 miles in diameter. A typical hurricane liberates about 100 billion kilowatts of heat from the condensation of moisture, but only about 3% of the thermal energy is transferred into mechanical energy in the form of wind. Sustained wind speeds up to 200 mph have been measured, but winds of about 130 mph are more typical. It is estimated that an average hurricane produces 200 billion tons of water a day as rain.[1]

As this writer has done, extend your definition as far as is needed to ensure the level of reader understanding desired.

Placement of Definitions

You have several options for placement of definitions within your papers: (1) within the text itself, (2) in footnotes, (3) in a glossary at the beginning or end of the paper, and (4) in an appendix. Which method you use depends upon the audience and the length of the definition.

Within the Text If the definition is short—a sentence or two—or if you feel most of your audience needs the definition, place it in the text with the word defined. Most often, the definition is placed after the word defined, as in this example:

Besides direct electric and magnetic induction, another source of power-frequency exposure is contact currents. Contact currents are the currents that flow into the body when physical contact is made between the body and a conducting object carrying an induced voltage. Examples of contact current situations include contacts with vehicles parked under transmission lines and contacts with the metal parts of appliances, such as the handle of a refrigerator.[2]

Sometimes, the definition is slipped in smoothly before the word is used—a technique that helps break down the reader's resistance to the unfamiliar term. The following definition of *supernova* is a good example of the technique:

> The vast majority of stars in a typical galaxy such as our own are extremely stable, emitting a remarkably steady output of radiation for millions of years. Occasionally, however, a star in an advanced stage of evolution will spontaneously explode, and for a few months it will be several hundred million times intrinsically more luminous than the sun. Such a star is a supernova, and at the time of its greatest brilliance it may emit as much energy as all the other stars in its galaxy combined.[3]

When you are using key terms that must be understood before the reader can grasp your subject, define them in your introduction.

In Footnotes If your definition is longer than a sentence or two, and your audience is a mixed one—part expert and part lay—you may want to put your definition in a footnote. A lengthy definition placed in the text could disturb the expert who does not need it. In a footnote it is easily accessible to the lay person and out of the expert's way, as demonstrated in Figure 8–2.

In a Glossary If you have many short definitions to give and if you have reason to believe that most members of your audience will not read your report straight through, place your definitions in a glossary. (See Figure 8–3 and pages 164–169.) Glossaries do have a disadvantage. Your readers may be disturbed by the need to flip around in your paper to find the definition they need. When you use a glossary, be sure to draw your readers' attention to it, both in the table of contents and early in the discussion.

In an Appendix If you need one or more lengthy extended definitions (say, more than 200 words each) for some but not all members of your audience, place them in an appendix. (See pages 164–169.) At the point in your text where readers may need them, be sure to tell readers where they are.

Description

In technical writing you will chiefly have to describe three things: places, mechanisms, and processes. After explaining the use of visual language and analogy in description, we deal with the three.

There are two types of electric and magnetic fields, those that travel or propagate long distances from their source (also called electromagnetic waves) and those that are confined to the immediate vicinity of their source. At distances that are close to a source compared to a wavelength,[1] fields are primarily of the confined type. Confined fields decrease in intensity much more rapidly with distance from their source than do propagating fields. Propagating fields dominate, therefore, at distances that are far from the source compared to a wavelength. The fields to which most people are exposed from radio broadcast antennas are examples of propagating fields since these sources are generally much more than one wavelength (1–100 meters) removed from inhabited areas. The power-frequency fields that people encounter are of the non-propagating type because power lines and appliances are much closer to people than one 60 Hz wavelength (several thousand kilometers). Only a very miniscule portion of the energy in power lines goes into propagating fields. Because the power-frequency fields of public health concern are not of the propagating type, it is technically inappropriate to refer to them as "radiation."

[1] A wavelength is the distance that a propagating field travels during one oscillatory cycle. For fields in air this distance is c/f where c is the velocity of light and f is the frequency of the oscillating field.

Figure 8–2 Definition in a Footnote
Source: U.S. Congress, Office of Technology Assessment, *Biological Effects of Power Frequency Electric and Magnetic Fields—Background Paper* (Washington, DC: GPO, 1989) 6.

Visual Language

The following brief descriptive passage shows how a combination of dimensions and a few words indicating shape and position can help you accurately visualize a canal in cross section:

What was a barge canal like? Engineering experience had settled the overall geometry of canals well before the end of the 18th century. They were trapezoidal in cross section; at the bottom their width was 20 to 25 feet and at water level it was 30 to 40 feet. The water in them was only three to four feet deep, so that the angle of the sides was quite flat. This gradual slope helped to minimize erosion due to wave action and slumpage due to seepage.[4]

We visualize things in essentially five ways—by shape, size, color, texture, and position—and have access to a wide range of comparisons and terms to describe all five.

Shape You can describe the shape of things with terms such as *cubical, cylindrical, circular, convex, concave, square, trapezoidal,* or *rectangular*. You can use simple analogies such as *C-shaped, L-shaped, Y-shaped,*

Terms printed in boldface

Gramatically parallel sentence fragments used for definitions (see pages 611–612)

When needed, definitons are extended in complete sentences

GLOSSARY

Alternator—a device which supplies alternating current.

Anemometer—a device for measuring wind speed.

Crossflow turbine—a drumshaped water turbine with blades fixed radially along the outer edge. The device is installed perpendicularly to the direction of stream flow.

DC to AC inverter—a device which converts electrical current from direct to alternating.

FERC—The Federal Energy Regulatory Commission, established by Congress to regulate non-federal hydroelectric projects.

Flow—the quantity of water, usually measured in gallons or cubic feet, flowing past a point in a given time.

Generator—a device that converts mechanical energy into electrical energy—a large number of conductors mounted on an armature that rotates in a magnetic field.

Head—the vertical height in feet from the headwater (with a dam) or where the water enters the intake (no dam) to where the water leaves the turbine.

Induction generator—an alternating current generator whose construction is identical to that of an AC motor.

Intake structure—a structure that diverts the water into the penstock; a small dam.

Isolation transformer—a device used to isolate the utility grid system from an earth-grounded electric generating system.

Net energy billing—an electric metering system in which the meter turns backwards when electricity flows from the generating system to the utility lines and forward when the utility is supplying electricity to the residence.

Pelton wheel—a water turbine in which the pressure of the water supply is converted to velocity by a few stationary nozzles, and the water jets then impinge on buckets mounted on the rim of the wheel.

Penstock—the pipe that carries pressurized water from the intake structure to the turbine.

Photovoltaic array—several photovoltaic modules connected together, usually mounted in a frame.

Photovoltaic module—several solar cells connected together on a flat surface.

PURPA—Public Utility Regulatory Act, a federal regulation requiring utilities to buy back power generated by small producers.

Solar easements—a written agreement with a person's neighbors that protects his/her access to the sun through the prohibition of any structures that might block their access.

Synchronous inverter—a device that links the output from a wind generator to the power line and the domestic circuit. The varying voltage and frequency generated by the windmill is instantly converted to exactly the same type of electricity distributed by a utility's power grid.

Figure 8–3 Definitions in a Glossary
Source: U.S. Department of Energy, *Homemade Electricity: An Introduction to Small-Scale Wind, Hydro, and Photovoltaic Systems* (Washington, DC: GPO, n.d.) 57.

cigar-shaped, or *spar-shaped.* You can describe things as *threadlike* or *pencillike* or as *sawtoothed* or *football-shaped.*

Size You can give physical dimensions for size, but you can also compare objects to coins, paper clips, books, typewriters, and football fields.

Color You can use familiar colors such as red and yellow and also, with some care, such descriptive terms as *pastel, luminous, dark, drab,* and *brilliant.*

Texture You have many words and comparisons at your disposal for texture, such as *pebbly, embossed, pitted, coarse, fleshy, honeycombed, glazed, sandpaperlike, mirrorlike,* and *waxen.*

Position You have *opposite, parallel, corresponding, identical, front, behind, above, below, right, left, north, south,* and so forth to indicate position.

Analogy

The use of analogy will aid your audience in visualizing the thing described. In the following example, analogies (which we have set in boldface) play a key role in helping the reader visualize the immense size of the Martian volcano, Olympus Mons:

> Olympus Mons is a great shield volcano, the largest volcano in the solar system. **It is similar to Hawaii's Mauna Loa but many times larger. Olympus Mons alone contains more lava than the entire Hawaiian Island chain.** Its summit rises 16.7 miles above the mean surface—**the Martian equivalent of sea level.** The great mountain covers an area about 435 miles across, and **would cover the state of Montana.** It is bounded by a huge escarpment, a near-vertical cliff, 3.7 miles high. Lava repeatedly flowed over the cliff, dropped nearly four miles to the plain below, and splashed far beyond. The upper slopes of Olympus Mons are terraced by the lava flows. Ridges that appear to be lava channels and roofed-over lava tubes course down the lower slopes. Impact craters are extremely rare on Olympus Mons, indicating that the volcano was active in recent geologic time, to cover the cratering that surely occurred long ago.[5]

Place Description

Writers of technical material must often describe places. Place descriptions appear in research reports when the location of the research has some bearing on the research. Highway engineers must describe the locations of projected highways. Surveyors have to describe the boundaries of parcels of property. Police officers have to describe the scene of an accident. In place description, you must pay particular attention to your point of view and your selection of detail.

Point of View When describing a place, you must be aware of your point of view—that is, where you are positioned to view the place being described. When describing a place, if you, in your mind's eye, position yourself physically before you begin to write, you'll avoid confusing and annoying your readers with an inconsistent jumping about in your description. You can shift your point of view, but be aware when you are making the shift. Have a good reason for doing it, and don't do it too often. Most often the reason will be to allow yourself to give an overview and then proceed to a close-up of selected details, as in the following passage, in which an archaeologist reconstructs through imagination and scientific knowledge a Roman outpost in second-century Britain:

Aerial point of view

Description by size and shape

A new fort, built of stone, rose on the eastern part of a level plateau. The enclosure was oblong; at the north and south ends the fort's massive walls, broken by central gates, ran 85 meters; on the east and west sides the length of the walls is just under 150 meters.

Shift to eye-level point of view

Use of analogy

Description by position

The overgrown ruins of the last wood fort at Vindolanda, abandoned 40 years earlier, stood just beyond the west wall of the new fort. Here the garrison engineers must have dumped many carloads of clay in the process of covering up the earlier structures and preparing a level site for the stone foundations of what we call Vicus I. On both sides of the main road leading from the west gate were erected a series of long barrackslike buildings; the masonry was dressed stone bound with lime mortar. The largest barracks was nearly 40 meters long, and all of them seem to have been about five meters wide. We deduce that the structures were one story high and probably served as quarters for married soldiers. They were divided into single rooms, one to a family, by partitions spaced about seven meters apart. There was probably storage space under the roof, and a veranda outside each room would have provided cooking space. The four barracks we have located so far could have housed 64 families. The British army in India offered almost identical housing to the families of its Indian soldiers. Smaller buildings, possibly housing for the families of noncommissioned officers, stood nearby.[6]

In the passage just cited, the author begins with an aerial point of view. He looks down upon the fort and describes it in general outline—its walls, roads, and buildings. That job done, he returns to ground level, and his point of view is now that of a well-informed guide walking you through the barracks. Notice the masonry, he says. Here are the rooms. One family lived in each. See how the rooms were partitioned off and where the storage space and cooking areas were. The writer uses the descriptive language of size, shape, and position. Physical description and function are gracefully combined.

The following passage concerns the details photographed by two spacecraft on Venus. The scene is visualized from the point of view of the spacecrafts' vantage points on the surface:

Details on the surface of Venus were virtually unknown until two Soviet Venera spacecraft landed on it in 1975 and returned pictures. Earlier speculations had ranged from steaming swamps to dusty deserts, from carbonated seas to oceans of bubbling petroleum. All were wrong, though the desert concept seems closest to what is now known. Photographs from the Venera spacecraft revealed a dry rocky surface fractured and changed by unknown processes.

Description by texture

There are rocks of many different kinds and a dark soil. The two spacecraft landed about 2000 km (1200 mi) from each other. One landed on an ancient plateau or plains area. At this site there are rocky elevations interspersed with a relatively dark, fine-grained soil. This soil seems to have resulted from a weathering of the rocks, possibly by a chemical action. The rocky outcrops are generally smooth on a large scale, with their edges blunted and rounded. The dark soil fills some of the cavities in the rocks. By contrast the other Venera landed at a site where there are rocks which look younger and less weathered, with not much evidence of soil between them.[7]

Description by texture

In this passage, the use of textured language is striking: *dry rocky surface fractured and changed by unknown processes; fine-grained soil; weathered; smooth; blunted and rounded.* Color is not mentioned, but the shade *dark* is a dominant note.

Selection of Detail Remember that in any description you are selecting details. You would find it quite impossible to describe everything. Very often, the details chosen support a thesis, perhaps an inference or chain of inferences. Notice how the selected details in the following passage about Jupiter's ring support the inferences (which we have set in boldface):

Voyager 2 took pictures of Jupiter's ring on the inbound leg, but more interesting were the pictures it took while behind Jupiter, looking back at the ring. Where Voyager 1's pictures were faint, the ring now stood out sharp and bright in the newest photos, telling scientists instantly that **the ring's particles scattered sunlight forward more efficiently than they scattered it back-**

ward, and therefore were tiny, dust-like motes. (Large particles backscatter more efficiently.)

While the dust particles of the ring appeared to extend inward toward Jupiter, probably all the way to the cloud tops, the ring had a hard outer edge, as if cut from cardboard.

Close examination of Voyager photos after the encounter revealed two tiny satellites, orbiting near the outer edge of the ring and herding the particles in a tight boundary. **The source of the ring's dust probably lies within the bright portion of the ring itself. The dust may be due to micrometeorites striking large bodies in the ring.**[8]

Mechanism Description

The physical description of some mechanism is perhaps the most common kind of technical description. It is a commonplace procedure with little mystery attached to it. For example, we see a friend having little success struggling to unplug a clogged sink drain by using a plunger.

"Look," we say to her, "you ought to buy yourself a plumber's snake. It would unclog that drain in a couple of minutes."

"Really. What's a plumber's snake?"

"Well, it's a tool for unplugging drains. Mostly, it's a flexible, springlike, steel cable about five feet long and with the diameter of a pencil. The cable has a football-shaped boring head on its working end. The head is about two inches long and at its widest point is twice the diameter of the cable. The whole business looks a bit like a snake, hence its name."

"That so? Anything to it besides the cable and head?"

"Uh-huh, there's a crank. It's a hollow steel tube in the shape of an opened up Z. It's about ten inches long, so you can get both hands on it. You slip the crank over the cable. With it, you can rotate the cable after you've inserted it in the drain, so that the head operates something like a drill to bore through the clog."

"Sounds like a handy gadget. I'll have to get one."

In this brief passage, we have used most of the techniques of good technical description. Our purpose has been to make you *see* the object and understand its function. We have done the following:

- Described the overall appearance of the plumber's snake and named the material with which it is made, steel.

- Divided the mechanism into its component parts—cable, boring head, and crank.
- Described the appearance of the parts, given their functions, and showed how they worked together.
- Pointed out an important implication, a *so-what,* of one of the descriptive facts: "It's about ten inches long, *so you can get both hands on it.*"
- Given you only information important in this description. For example, because it is of no consequence in this description, we have not told you the color of the mechanism.
- Used figurative language, *springlike* and *football-shaped,* and comparisons to familiar objects such as pencils, snakes, and drills to clarify and shorten the description.

In our discussion of mechanism description, we tell you how to plan one and provide several examples of typical mechanism descriptions.

Planning Despite certain elements that most mechanism descriptions have in common, there is no formula for writing them. Your judgment must be involved, weighing such matters as purpose and audience. As you plan your mechanism description, you'll need to answer questions like the ones that we provide next. The answers to these questions will largely determine how you arrange your description and the details and so-whats you elect to provide your readers.

What is the purpose of the description? To impart general knowledge? To sell the mechanism? To transfer technology?

Why will the intended reader read the description? To understand the mechanism? To consider buying the mechanism? To use the mechanism? What is the reader's level of experience and knowledge in regard to the mechanism? Would the reader, for example, understand the technical terminology involved, or would he or she need terms defined? Would the reader understand the implications of the facts presented about the mechanism or would you have to furnish the so-whats?

Consider these questions about the mechanism.

- What is the purpose and function of the mechanism?
- How can the mechanism be divided?
- What are the purpose and function of the parts?
- How do the parts work together?
- How can the parts be divided? Is it necessary to do so?
- What are the purpose and function of the subparts?
- How do the parts and subparts work together?

- Which of the following are important for understanding the mechanism and its parts and subparts? Construction, materials, appearance, size, shape, color, texture, position?
- Are there any so-whats you need to express explicitly for the reader?
- Would the use of graphics—photographs and drawings, for example—aid the reader?
- Are there any analogies that would clarify the description for the reader?

When you can answer these questions, you will be better able to select the details concerning the mechanism that you need and to arrange them in a way that will fulfill your purpose. Because purpose, audience, and mechanism vary from situation to situation, the arrangement and content of mechanism descriptions also vary. But they are all likely to include statements about function, some sort of division of the mechanism into its component parts, physical descriptions in words or graphics or some combination of the two, analogies, and so-whats.

Examples of Mechanism Descriptions In Figure 8–4, our annotations draw your attentions to the various features of a mechanism description as exemplified in the descriptive overview of the Hubble Space Telescope. Our Figure 8–5 is one of several graphics that accompany the telescope's description. As demonstrated in Figure 8–5, such graphics are frequently annotated and are "cut away" to show the interior of the mechanism described.

Professionals often describe mechanisms for reasons of technology transfer. *Technology transfer* is a rather fancy term for a common but important part of the technical and scientific life. Technical people exchange concepts through their technical journals. Sometimes the transfer is from one discipline to another. Very often the concept described concerns a mechanism. Figure 8–6 reproduces a mechanism description as it appeared in a journal devoted to technology transfer. Like most such descriptions, it uses a graphic to reduce the need for extensive visual description in words. So-whats are prominently featured. In fact, the head designed to attract the reader's attention is primarily a so-what.

As is done in most mechanism descriptions, the authors divide the mechanism into its parts—a base and a cover. They give details of construction, again not forgetting the so-whats. The purpose of the description is primarily to make the reader aware of the mechanism. Most people would need additional information to construct the mechanism for themselves. Therefore, the description provides a way for the reader to get "further information."

Description by size and analogy

The Hubble Space Telescope is just over 13 meters (43 feet) long and 4 meters (14 feet) in diameter, about the size of a bus or tanker truck. Upright, it is a five-story tower; carried inside the Space Shuttle for the trip to orbit, it fills the payload bay.

Division of telescope into its component parts

The Hubble Space Telescope is made up of three major elements: the Optical Telescope Assembly, the focal plane scientific instruments, and the Support Systems Module, which is divided into four sections, stacked together like canisters:

Subdivision of Support System Module

Functions of key parts of Support System Module

Aperture Door and Light Shield: protecting the scientific instruments from light of the sun, Earth, and moon and also from contamination

Forward Shell: enclosing the Optical Telescope Assembly mirrors

Equipment Section: girdling the telescope to supply power, communications, pointing and control, and other necessary resources

Aft Shroud: covering the five focal plane instruments and the three fine guidance sensors.

Miscellaneous details

Solar energy arrays and communications antennas are attached to the exterior shell. Doors allow astronauts to remove instruments and components from the equipment bays. Handrails and sockets for portable foot restraints attached to the external surface aid the astronauts in performing maintenance and repair tasks.

Space Telescope Vital Statistics

Use of list for ease of reference

Length:	*13.1 m (43.5 ft)*
Diameter:	*4.27 m (14.0 ft)*
Weight:	*11,000 kg (25,500 lb)*
Focal Ratio:	*f/24*
Primary Mirror	
Diameter:	*2.4 m (94.5 in)*
Weight:	*826 kg (1,825 lb)*
Reflecting Surface:	*Ultra-low expansion glass covered by aluminum with magnesium-fluoride coating*
Secondary Mirror	
Diameter:	*0.3 m (12 in)*
Weight:	*12.3 kg (27.4 lb)*
Reflecting Surface:	*Ultra-low expansion glass covered by aluminum with magnesium-fluoride coating*
Systems	
	Optical Telescope Assembly
	Support Systems Module
	Focal Plane Science Instruments
	Wide Field/Planetary Camera
	Faint Object Camera
	Faint Object Spectrograph
	Goddard High Resolution Spectrograph
	High Speed Photometer
	Fine Guidance Sensors (for astrometry)
Data Rate:	*Up to 1 mbps*

Figure 8–4 Mechanism Description of Hubble Space Telescope
Source: National Aeronautics and Space Administration, *Exploring the Universe with the Hubble Space Telescope* (Washington, DC: GPO, n.d.) 57.

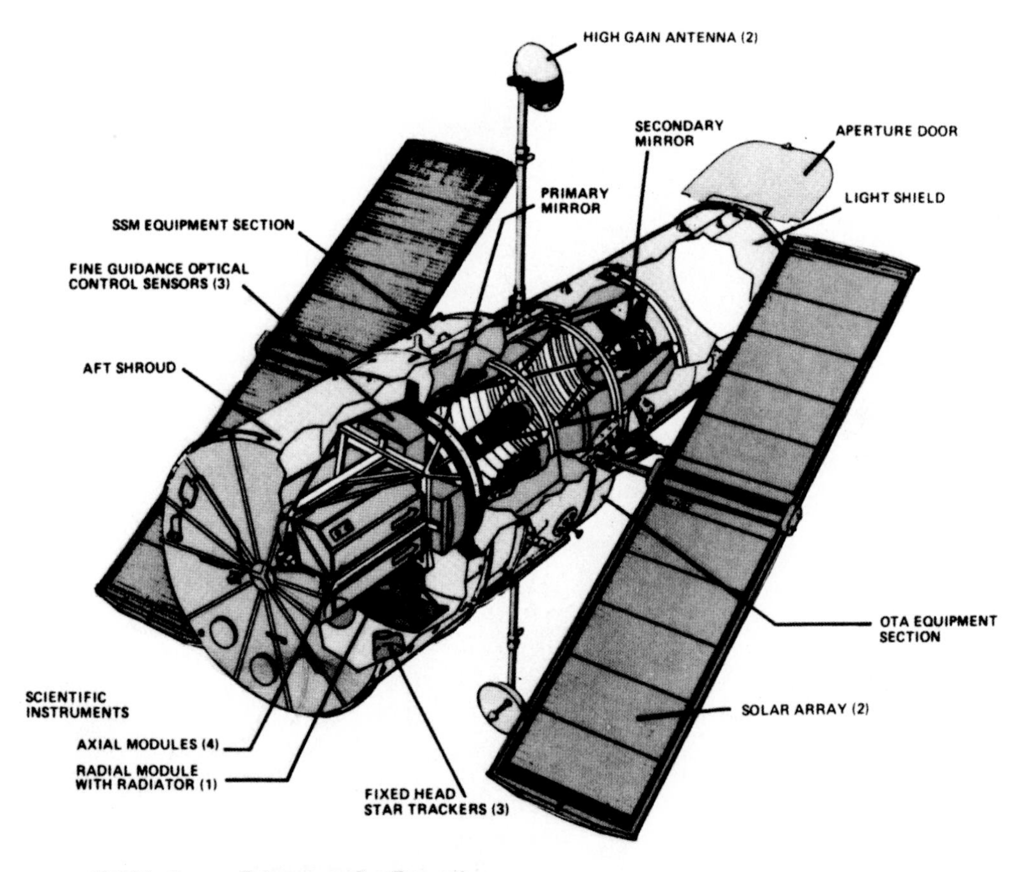

Figure 8–5 Graphic for Hubble Space Telescope
Source: National Aeronautics and Space Administration, *Exploring the Universe with the Hubble Space Telescope* (Washington, DC: GPO, n.d.) 58.

Some mechanism descriptions may be sales motivated, for example, catalog descriptions. In sales the implications of your facts—the so-whats—take on special importance, as in the next example:

Our English Steel Scraper Mat Is Guaranteed for Life!

This mat is assembled from strips of heavy-gauge steel which are woven and twisted together. It just can't come apart. The entire Scraper Mat has been dipped in a galvanizing solution which totally protects it from rusting out. Clay, mud, dirt, or snow scraped off your shoes or boots fall through the steel strips. Ordinary door mats simply move the grime from one part of your shoes to another; Scraper Mat actually removes it altogether.[9]

Protective Package for a Gamma-Ray Detector

Enclosure resists contamination,
voltage breakdown, and vibration.

NASA's Jet Propulsion Laboratory, Pasadena, California

Function of mechanism

A package for a germanium gamma-ray detector protects the semiconductor crystal from contamination, allows it to operate at high voltages, and isolates it from shock and vibration. The package seals the detector from its surroundings, whether in the atmosphere or in the vacuum of space.

Division into parts

The main parts of the package, a base and a cover, are made of aluminum. As shown in the figure, the cover is sealed to the base by a soft aluminum ring. Since no solder is used for the seal, there is no danger of contamination of the germanium by flux outgassing inside the package.

Construction details with so-whats

A high-voltage power connection and a low-voltage signal connection are soldered to the cover. Nitrogen is introduced into the sealed package through the evacuation port. The nitrogen gas inhibits high-voltage arcs in the package cavity, prevents oxidation of the crystal, and provides back pressure to minimize the release of gases from the package metal. A small amount of helium in the nitrogen charge allows leakage from the package to be measured with a sensitive helium leak detector.

Construction details with so-whats

A clamping ring surrounds the jaw fingers and ceramic jaws and clamps them to the germanium crystal. The device is thus supported without damage to sensitive crystal surfaces, even during high acceleration.

This work was done by Marshall Fong, Charles Lucas, Albert Metzger, Donald M. Moore, Robert Oliver, and Walter Petrick of Caltech for **NASA's Jet Propulsion Laboratory.** *For further information, Circle 28 on the TSP Request Card.*
NPO-16019

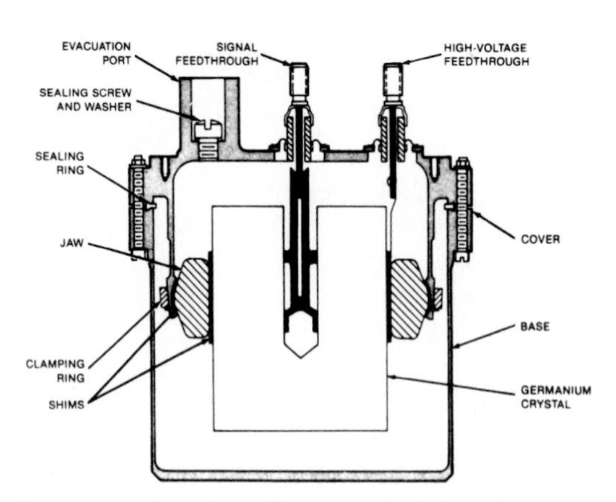

This **Housing for a Germanium Detector**, part of a gamma-ray spectrometer, holds the germanium crystal securely while protecting it from contamination. Although designed to hold nitrogen, the package can also be evacuated if necessary.

Figure 8–6 Description of a Mechanism
Source: NASA Tech Briefs Spring 1985: 72.

Rewritten without the so-whats, the description loses most of its force:

> This mat is assembled from strips of heavy-gauge steel which are woven and twisted together. The entire Scraper Mat has been dipped in a galvanizing solution.

Catalog descriptions illustrate how important purpose and audience analysis are in mechanism description. Lay audiences, such as those that read most catalogs, are primarily interested in the function of a mechanism and the so-whats concerning it. On the other hand, the more expert an audience is, the more they will be interested in the mechanism itself, as well as its functions and the so-whats.

Many of the principles found in mechanism description have nonmechanical applications. For example, we don't usually think of skin as a mechanism, but the same rational principles we have been discussing are found in Figure 8–7 in the passage about skin. The subject is divided, objective physical details are described, function is discussed, so-whats are given, and a graphic is provided.

Process Description

Process description is probably the chief use of chronological order in technical writing. By **process** we mean a sequence of events that progresses from a beginning to an end and results in a change or a product. The process may be humanly controlled, such as the manufacture of an automobile, or it may be natural—the metamorphosis of a caterpillar to a butterfly, for example.

Process descriptions are written in one of two ways:

- *For the doer*—to provide instructions for performing the process
- *For the interested observer*—to provide an understanding of the process

A cake recipe provides a good example of instructions for performing a process. You are told when to add the milk to the flour, when to reserve the whites of the eggs for later use. You are instructed to grease the pan *before* you pour the batter in, and so forth. Writing good instructions is a major application of technical writing, and we have devoted all of Chapter 14 to it. In this chapter we explain only the second type—providing an understanding of the process. We discuss planning and sentence structure and provide example process descriptions.

Planning Insofar as you have grasped the proper order of events, the organization of a process description will not present any particular

Skin

Function

The skin is the protective covering of the body; it also contains the sweat glands which help regulate body temperature. The skin consists of three layers, an external layer (epidermis), a deep layer or true skin (dermis), and the fat tissue layer (subcutaneous) (Figure 1-3). The skin is one of the most important organs of the body. The loss of a large part of the skin will result in death unless it can be replaced.

Division into parts

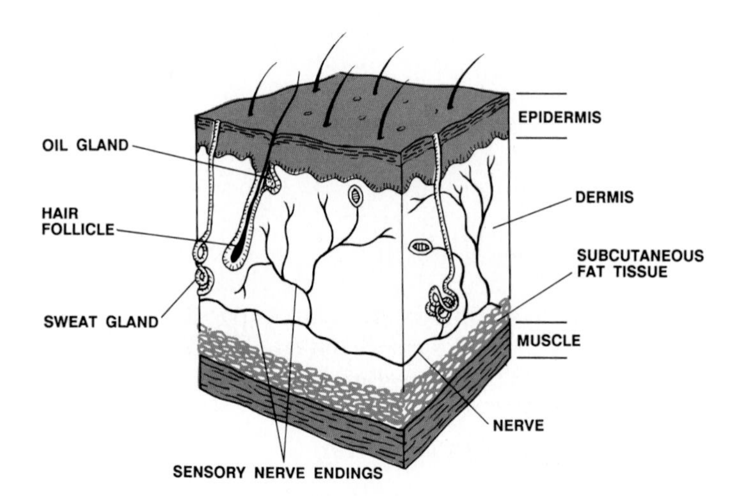

Function

The protective functions of the skin are many. Skin is watertight and keeps internal fluids in while keeping germs out. A system of nerves in the skin carries information to the brain. These nerves transmit information about pain, external pressure, heat, cold, and the relative position of various parts of the body.

Skin provides information to the first aider concerning the victim's condition. For example, pale, sweaty skin may indicate shock.

Figure 8–7 Skin Described as a Mechanism
Source: U.S. Department of Labor, *First Aid Book* (Washington, DC: GPO, 1988) 9.

problem to you. You simply follow the order of events as they occur. However, the amount of detail you include is likely to be a problem. A situational analysis that raises questions about purpose, reader, and the process itself will usually give you a good idea of the level of detail needed. The questions you should ask for a process description resemble those used for a mechanism description.

What is the purpose of the description? To describe a process you have followed? To describe in a general way how some process works? To describe in a detailed, highly technical way how some process works?

Why will the reader read the process description? To learn about the process in a general way? To make a decision, perhaps about using the process? To understand how the process might affect his or her work? To remain aware of activities in the discipline? To understand how an experiment was conducted?

What is the reader's level of experience and knowledge in regard to the process? Will the reader need some technical terms defined? Will the reader need some of the so-whats explained? How high is the reader's interest in the details of the process?

What is the purpose of the process?

Who or what does the process?

What are the major steps of the process? Do the steps divide up into subprocesses, some of which may be going on at the same time?

Are there graphics and analogies that would help the reader?

Sentence Structure The choice of sentence structure is important in process description. You have to make the correct choice between present and past tense and decide which voice and mood to use. Most process descriptions are written in the present tense (we have set the verbs in boldface):

Present tense

> Blood from the body **enters** the upper chamber, atrium, on the right side of the heart and **flows** from there into the lower chamber, the ventricle. The ventricle **pumps** the blood under low pressure into the lungs where it **releases** carbon dioxide and **picks up** oxygen.

The rationale for the use of present tense is that the process is an ongoing and continuing process, such as the heart's pumping of blood. Therefore, it is described as going on as you are observing it.

However, processes that have occurred in the past and are completed are described in the past tense. The major use of past tense in process description is found in empirical research reports where the procedures the researcher followed are described in past tense (verbs in boldface):

Past tense

> During the excavation delay we **accomplished** two tasks. First, we **installed** a temporary intake structure and **tested** the system's efficiency. Second, we **designed, built,** and **installed** a new turbine and generator.

In writing instructions, you will commonly use the active voice and imperative mood:

Imperative mood

> **Clean** the threads on the new section of pipe. **Add** pipe thread compound to the outside threads.

In following your instructions, it is the reader, after all, who is the doer. With its implied *you,* the imperative voice directly addresses the reader. But in the process description for understanding, where the reader is not the doer, the use of imperative mood would be inappropriate and even misleading.

In writing a process for understanding, therefore, you will ordinarily use the indicative mood in both active and passive voice:

Active voice — The size of the cover opening **controls** the rate of evaporation.

Passive voice — The rate of evaporation **is controlled** by the size of the cover opening.

In active voice the subject does the action. In passive voice the subject receives the action. Use of the passive emphasizes the receiver of the action while de-emphasizing or removing completely the doer of the action. (Incidentally, as the preceding examples illustrate, neither doer nor receiver has to be a human being or even an animate object.) When the doer is unimportant or not known, you should choose passive voice. Conversely, when the doer is known and important, you should choose the active voice. Because the active is usually the simpler, more direct statement of an idea, choose passive voice only when it is clearly indicated. We have more to say on this subject on pages 125–126.

Examples of Process Description Process description, like mechanism description, figures prominently in technology transfer. In their journals, scientists and technologists tell each other about useful processes just as they tell each other about useful mechanisms. Figure 8–8 describes an automatic coal-mining system. The description is particularly well done. It begins with an introduction that provides an overview of the process and the mechanisms involved and makes clear the type of coal mining for which this process is intended. In paragraphing, the author is guided by the steps of the process, one paragraph for each major step. He provides a graphic that makes the entire process clear at a glance.

Notice the use of present tense verbs throughout the description:

> The cut coal **falls** on loading ramps, where gathering arms **move** it to central screws.

In writing process descriptions to provide understanding, you'll find that extensive detail is not always necessary or even desirable. As in mechanism description, the amount of detail given relates to the technical level and technical interests of your readers. When *Time* magazine, for example, publishes an article about open-heart surgery, its readers do not expect complete details on how such an operation is performed. Rather, they expect their curiosity to be satisfied in a general way.

Our next example, from a book called *Wood Handbook,* illustrates how process descriptions are tailored to the needs and interests of the

Automatic Coal-Mining System

Coal cutting and removal would be done with minimal hazard to people.

NASA's Jet Propulsion Laboratory, Pasadena, California

Overview and function

A proposed automatic coal-mining system would cut coal, grind it, mix it with water to make a slurry, and transport the slurry to the surface. The system would include closed-circuit television monitoring, laser guidance, optical obstacle avoidance, proximity sensing, and other features for automatic control.

Use of system

The system is intended for longwall mining, in which the coal seam is divided into several blocks at least 600 feet (183 m) wide by corridors or "entries," which allow the movement of equipment and materials. Moving along an entry, an extracting machine cuts coal from the face of a block (see figure) to a depth of a few feet (about a meter). An extractor has two cutting heads, one at each end. The extractor moves on crawler tracks that are reversed at the end of each pass so that coal can be cut in the opposite direction. Thus, the extractor does not have to retreat in "deadhead" fashion to begin the next cut, and its productivity is increased.

One paragraph per major step

The cut coal falls on loading ramps, where gathering arms move it to central screws. The screws crush the coal and feed it to the transport tube.

Water is sprayed on the extractor drums as they cut to keep down explosive dust and to promote visibility. More water is added in the screws to convert the coal powder into a low-viscosity slurry.

A laser beam guides the extractor along the face, ensuring a straight cut. Connected to the slurry-transport subsystem, the laser and its reflector are moved forward automatically with each pass of the extractor. Any deviation from the path set by the beam represents a change in the direction of travel of the extractor; and an onboard controller changes the speed of one crawler track relative to the other, thereby adjusting the direction of travel.

A sonic or optical proximity sensor detects the approach of an entry and stops the extractor to prevent a collision with the far wall of the entry. The proximity sensor also guides the reversing maneuver and the start of a new cut.

The transport tube is made in repeating sections, with each section mounted on an individual skid to allow lateral motion. As the extractor moves along, it pulls each section toward it and connects with that section in turn.

The roof supports move forward in alternation, "walking" by being lifted off the floor by hydraulic pistons, then propelled by horizontal pistons. Each alternate support unit supports itself on adjacent neighbors and propels itself by pushing on its neighbors. The units that are not moving are holding up the roof. The movement of the roof supports may be controlled remotely by human operators watching via television, or they may be made to move automatically on the basis of the movement of the slurry-transport tube.

This work was done by Earl R. Collins, Jr., of Caltech for **NASA's Jet Propulsion Laboratory.** *For further information, Circle 114 on the TSP Request Card.*
NPO-15861

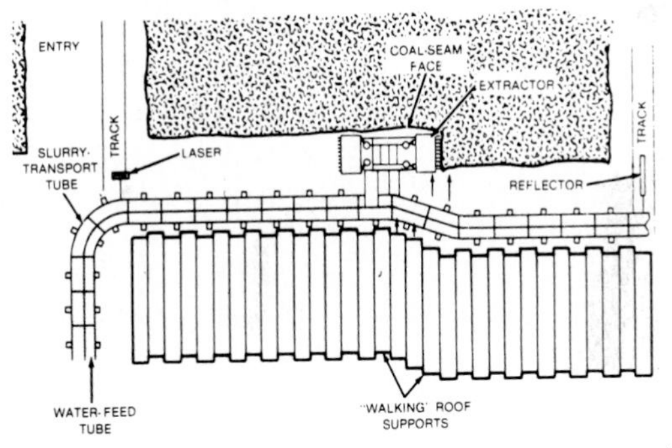

In the **Automatic Coal Mine,** the cutting, transport, and roof-support movement are all done by automatic machinery. The exposure of people to hazardous conditions would be reduced to inspection tours, maintenance, repair, and possibly entry mining.

Figure 8–8 Description of a Process
Source: NASA Tech Briefs Spring 1984: 398–99.

audience. The book's subtitle is *Wood as an Engineering Material,* and its preface makes clear that its primary audience comprises architects and engineers. One small section of it deals with shipworms, marine organisms that attack wood immersed in salt water:

> Shipworms are the most destructive of the marine borers. They are mollusks of various species that superficially are wormlike in form. The group includes several species of *Teredo* and several species of *Bankia,* which are especially damaging. These are readily distinguishable on close observation but are all very similar in several respects. In the early stages of their life they are minute, free-

Overview of process

Key so-what

swimming organisms. Upon finding suitable lodgement on wood they quickly develop into a new form and bury themselves in the wood. A pair of boring shells on the head grows rapidly in size as the boring progresses, while the tail part or siphon remains at the original entrance. Thus, the animal grows in length and diameter within the wood but remains a prisoner in its burrow, which it lines with a shell-like deposit. It lives on the wood borings and the organic matter extracted from the sea water that is continuously being pumped through its system. The entrance holes never grow large, and the interior of a pile may be completely honeycombed and ruined while the surface shows only slight perforations. When present in great numbers, the borers grow only a few inches before the wood is so completely occupied that growth is stopped, but when not crowded they can grow to lengths of 1 to 4 feet according to species.[10]

Notice that no attempt is made to give the full information about the shipworm that an entomologist might desire. We don't learn, for example, how the shipworm reproduces, nor do we even learn very clearly what it looks like. For the intended readers of the *Wood Handbook*, such information is not needed. They do need to know what is presented—the process by which the shipworm lodges on the wood and how it bores into it. They do need to know the key so-what expressed.

Knowing these facts, the engineers and architects can be alert to the potential for damage represented by the shipworm. Also, they will understand the need for the kinds of preventive measures, such as chemical treatment and sheathing, explained later in the text.

In another section of the *Wood Handbook*, the process of heat transfer through a wall is described:

Overview

Division of the process

Definitions and examples

Heat seeks to attain a balance with surrounding conditions, just as water will flow from a higher to a lower level. When occupied, buildings are heated to maintain inside temperature between inside and outside. Heat will therefore be transferred through walls, floors, ceilings, windows and doors at a rate that bears some relation to the temperature difference and to the resistance to heat flow of intervening materials. The transfer of heat takes place by one or more of three methods—conduction, convection, and radiation (see figure [our Figure 8–9]).

Conduction is defined as the transmission of heat through solid materials; for example, the conduction of heat along a metal rod when one end is heated in a fire. Convection involves transfer of heat by air currents; for example, air moving across a hot radiator carries heat to other parts of the room or space. Heat also may be transmitted from a warm body to a cold body by wave motion

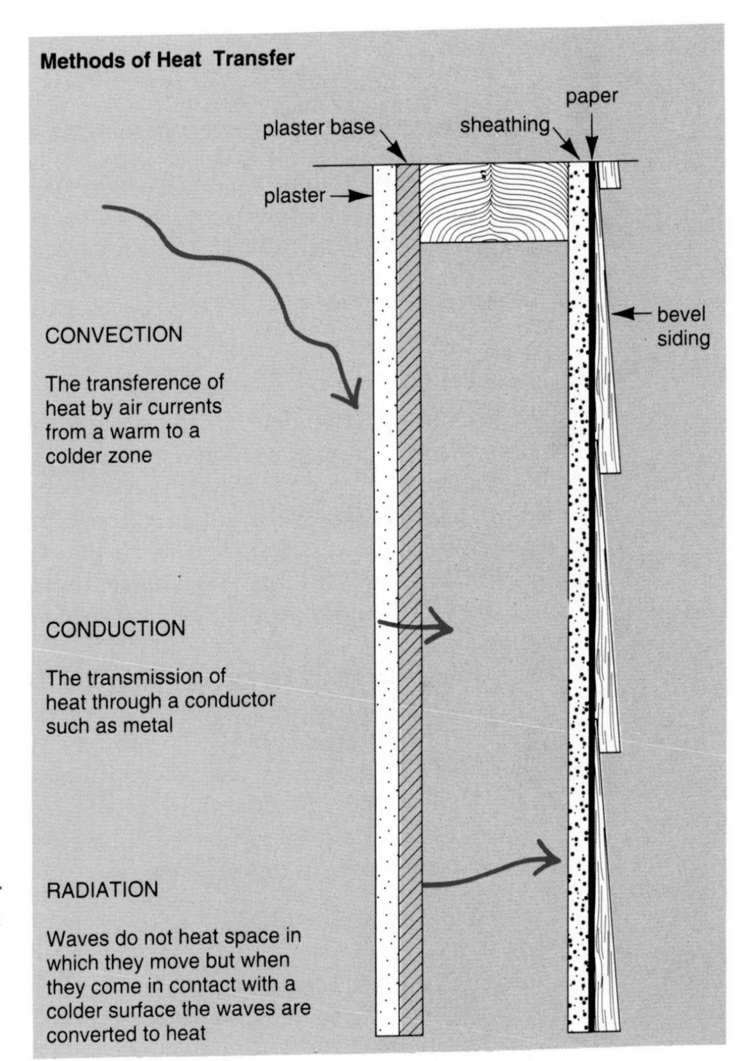

Figure 8–9 Illustration of Heat Transfer
Source: U.S. Department of Agriculture, *Wood Handbook* (Washington, DC: GPO, 1974) 20.

through space, and this process is called radiation because it represents radiant energy. Heat obtained from the sun is radiant heat.

Extended example of process

Heat transfer through a structural unit composed of a variety of materials may include one or more of the three methods described. Consider a frame house with an exterior wall composed of gypsum lath and plaster, 2- by 4-inch studs, sheathing, sheathing paper, and bevel siding. In such a house, heat is transferred from the room atmosphere to the plaster by radiation, conduction, and convection, and through the lath and plaster by conduction. Heat transfer across the stud space is by radiation and convection. By radiation, it moves from the back of the gypsum lath to the colder sheathing; by convection, the air warmed by the lath moves upward on the warm side of the stud space, and that cooled by the sheathing moves downward on the cold side. Heat transfer through sheathing, sheathing paper, and siding is by conduction. Some small air spaces will be found back of the siding, and the heat transfer across these spaces is principally by radiation. Through the studs from gypsum lath to sheathing, heat is transferred by conduction and from the outer surface of the wall to the atmosphere, it is transferred by convection and radiation.[11]

The architect and engineer readers are furnished far more detail in this description than they were in the description of the shipworm. Even an illustration is provided to make the process more understandable. The process of heat transfer, while still background information, is more useful to the architects and engineers than is the life cycle of a shipworm. Many factors of design depend upon a complete understanding of the process. Again, audience and purpose have determined the amount of detail presented.

In reports of empirical research, the section that describes the *methods and procedures* used by the researcher is a process description. Because the research has been completed by the time it is reported, procedures and methods are written in past tense. Writers of such reports include enough detail about the procedure so that readers equally expert to themselves could duplicate the research. Notice both the use of past tense and the narrative flow in this excerpt:

First-person active voice

Past tense

Because of the "real-dollar and real-time" exigencies that face researchers in corporations (Morehead, 1987), I could not focus upon both the writing of the first draft and upon the group brainstorming and editing processes. Since the purpose of the study was to look at group interaction, I chose the latter alternative. In order to describe these processes, I collected data nearly every working day during a 20-week internship at the site (10/86–3/87) and conducted four final interviews one month after my internship ended. Internship duties required two hours per day; the remaining four to six hours were spent gathering and analyzing data. Except for gathering data, I did not participate in the production of the executive letter. I observed two group brainstorming sessions and taped ten executive-letter editing sessions. In addition, I interviewed each participant in the editing process except for the CEO, to whom I was not given access. My account of the CEO's actions comes from statements from his secretary and three other participants. All 53 taped interviews and editing sessions were transcribed.[12]

Notice that the writer has in this description used primarily first person (*I*) and active voice. First-person, active voice is acceptable in most research journals, although some still insist upon the more anonymous passive voice, as used in this excerpt:

Passive voice

> Two teachers, one an English teacher at a high school and one a graduate teaching assistant, were trained to rate the compositions produced at each grade level. The compositions were coded for identification and mixed together before scoring.[13]

For more information about writing procedures in empirical research reports, see pages 548–549.

As with place and mechanism descriptions, there are no easy formulas to follow in writing process descriptions. You must exercise a good deal of judgment in the matter. As in all writing, you must decide what your audience needs to know to satisfy its purpose and yours. However, the checklists that follow should provide guidance to aid you in exercising your judgment.

Planning and Revision Checklists

You will find the planning and revision checklists that follow Chapter 2, "Composing" (pages 33–35 and inside the front cover), and Chapter 4, "Writing for Your Readers" (pages 78–79), valuable in planning and revising any presentation of technical information. The following questions specifically apply to defining and describing. They summarize the key points in this chapter and provide a checklist for planning and revising.

DEFINING

Planning

- Do your readers share the vocabulary you are using in your report? If not, make a list of the words you need to define.
- Do any of the words on your list have readily available synonyms known to your readers?
- Which words will require sentence definitions? Which words are so important to your purpose that they need extended definitions?

Revision

- In your sentence definitions, have you put your term into its class accurately? Have you specified enough differences so that your readers can distinguish your term from other terms in the same class?
- Will your readers understand all the terms you have used in your definitions?
- Have you avoided circular definition?

Planning

- How will you extend your definition? Description? Example? Analogy? Chronology? Topical order? Classification? Division? Graphics? Are there words within your definition that you need to define?
- Does everyone in your audience need your definitions? How long are your definitions? How many definitions do you have?
- Where can you best put your definitions: Within the text? In footnotes? In a glossary? In an appendix?

Revision

- Have you used analogy and graphics to help your readers? If not, should you?
- Does the placement of your definitions suit the needs of your audience and the nature of the definitions?

PLACE DESCRIPTION

Planning

- What is your purpose in describing this place?
- Who are your readers? What is their purpose for reading your description?
- What will be your point of view? From what vantage point can you best view the place to be described? Will you need more than one point of view?
- Are there analogies to places familiar to the readers that will help them visualize and understand this place? Would a graphic help the readers?

Revision

- Does your description satisfy your purpose and your readers?
- Is your point of view consistent? Do you have good reasons for any shifts in your point of view?
- Have you used well the visual language of shape, size, color, texture, and position?
- Are your inferences and implications clearly stated? Do the details you have selected support your inferences and implications?

MECHANISM DESCRIPTION

Planning

- What is the purpose of the description?
- Why will the intended reader read the description?
- What is the purpose and function of the mechanism?
- How can the mechanism be divided?
- What are the purpose and function of the parts?
- How do the parts work together?
- How can the parts be divided? Is it necessary to do so?
- What are the purpose and function of the subparts?

Revision

- Does your description fulfill your purpose?
- Does the level of detail in your description suit the needs and interests of your readers?
- Have you made the function of the mechanism clear?
- Have you divided the mechanism sufficiently?
- Do your descriptive language and analogies clarify the description?
- Have you clearly stated your so-whats?
- Have you provided enough graphic support? Are your graphics sufficiently annotated?

Planning

- Which of the following are important for understanding the mechanism and its parts and subparts? Construction? Materials? Appearance? Size? Shape? Color? Texture? Position?
- Are there any so-whats you need to express explicitly for the reader?
- Would the use of graphics aid the reader?
- Are there analogies that would clarify the description for the reader?

PROCESS DESCRIPTION

Planning

- What is the purpose of the description?
- Why will the reader read the description?
- What is the reader's level of experience and knowledge regarding the process?
- What is the purpose of the process?
- Who or what does the process?
- What are the major steps of the process?
- Are there graphics and analogies that would help the reader?

Revision

- Does your description fulfill your purpose?
- Does your description suit the needs and interests of your readers?
- Have you chosen the correct tense, past or present?
- Have you chosen either active or passive voice appropriately?
- Are the major steps of the process clear?
- Have you provided enough graphic support? Are your graphics sufficiently annotated?

Exercises

1. Reproduced in this exercise is an extended definition of the term *microelectronics*. The definition is an excerpt from an article on microelectronic integrated circuits. The article is an informational piece written primarily for the "employees and stockholders" of a large corporation. Therefore, the audience is a mixed one that will include, certainly, lay people, technicians, and executives, and perhaps even experts. The writer's task is to explain things so that lay people will find the article both interesting and understandable but at the same time to include facts and inferences that the other readers may find useful. Working in a group, discuss where, how, and why the author has used the techniques of definition. When the group has finished its discussion, each member of the group should write an analysis of the excerpt that shows what techniques the author has used and where, how, and why he has used them. In your analysis, use concepts from Chapter 4, "Writing for Your Readers," Chapter 7, "Informing," and this chapter.

> What then *is* microelectronics? It's the design and fabrication of increasingly compact, awesomely complex electrical circuits on tiny squares of crystalline material such as silicon or gallium arsenide, and the dimensions of those circuits are literally microscopic. The silicon chips that Raytheon

currently makes are typically 0.3 inches on a side, and the most advanced contain more than 100,000 "features," or individual components, *each* of them equivalent to something like one of those glowing vacuum tubes that helped snatch Jack Benny's voice out of the ether. Each feature can be etched upon the silicon in sizes as small as 1.25 microns in width, soon to be reduced to .8 micron and ultimately half a micron. (A micron is one *millionth* of a meter.)

"You grope for analogies," admits Dr. Bradford Becken of Raytheon's Submarine Signal Division, a dedicated user of microelectronics in its sonar systems. "If you wanted to draw a map of the entire United States that showed every city block and town square, it would obviously be a *very* big map. But with the feature sizes we're working with to create microcircuits right now, I could draw that entire map on a sheet of paper not much larger than a postage stamp."

Yet this tremendous sophistication, this incomprehensible tininess—"wires" etched in silicon so thinly that they are approaching a size too small for an electron to squeeze through—is all brought to market by techniques that approach the ultimate in mass production.

"You are able to apply manufacturing processes that are very repeatable, very cost-effective," explains Tom Shaw, manager of Raytheon's Missile System Division Laboratory in Bedford, Massachusetts—another heavy user of microcircuits in its missile guidance and control systems. "These repeatable processes make practical what most people would rightly assume is a highly complex operation."

It can cost $750,000 to design a single complex chip of the sort created and manufactured in Raytheon's Microelectronics Center, but the manufacturing process, while complex, is not that expensive. Some very complex devices can be sold for less than $5 apiece. Which is why grocery clerks wear watches that rival old Harrison's chronometer in accuracy, children play games on computers that exceed ENIAC in power, coffeemakers contain more computing power than a university lab of the 1950's, new generation airliners can automatically fly every phase of flight from takeoff to rollout, and pocket calculators with more power than those costing several hundred dollars just a decade ago have become giveaways.

The vacuum tube was turned into an antique by a device that at the time, in the 1950's, seemed the end of the line in electronic simplicity; the transistor. It was tough, didn't heat up, and took little space and less power. But transistors and their tiny teammates—diodes and capacitors—still had to be laboriously soldered onto circuit boards. This was fine for pocket radios and hobby kits, but a typical early-1960's computer had 100,000 diodes and 25,000 transistors. Solder that many connections and you end up with a price tag with even more digits.

When it was discovered that transistors and other semiconductors could simultaneously be made by the dozens on a wafer of pure silicon, which was then cut apart and wired to connectors and ultimately to circuit boards, it wasn't long before somebody realized how silly it was to cut them apart in the first place: why not make the entire circuitry—the whole "device"—right on the wafer? And thus was created the "integrated" circuit, soon to be miniaturized itself and made by the dozens on each silicon wafer.[14]

2. Write an extended definition of some term in your academic discipline. Use a graphic if it will aid the reader. In a paragraph separate from your definition, explain for your instructor your purpose and audience.
3. Describe some place that could be of importance to you in your professional field—a river valley, a power plant, a nurse's station, a business office, a laboratory. Have a definite purpose and audience for your description, such as supporting some inference about the place for an audience of your fellow professionals. Or you might emphasize some particular aspect of the place for a lay audience. Use at least two points of view, one that allows you to give an overview of the place and one that allows you to examine it in some detail. In a paragraph separate from your description explain for your instructor your purpose and audience.
4. Choose three common household tools such as a can opener, vegetable scraper, pressure cooker, screwdriver, carpenter's level, or saw. For each, write a one-paragraph description that could serve as a catalog description for a particular brand of the product, such as a Black and Decker can opener or Stanley saw.
5. Figure 8–6 is a mechanism description; Figure 8–8 is a process description. Both are intended to transfer technology from one technical person to another. Choose some new mechanism or process in your field that will lend itself to such a description. Using the appropriate figure as a model, write your description. Include at least one graphic as a part of your description. In a paragraph separate from your description, explain for your instructor your purpose and audience.
6. Write two versions of a process description intended to provide an understanding of a process. The first version is for a lay audience whose interest will be chiefly curiosity. The second version is for an expert or technical audience that has a professional need for the knowledge. The process might be humanly controlled such as buying and selling stocks, writing computer programs, fighting forest fires, giving cobalt treatments, or creating legislation. It could be the manufacture of some product—paint, plywood, aspirin, digital watches, maple syrup, fertilizer, extruded plastic. Or you might choose to write about a natural process—thunderstorm development, capillary action, digestion, tree growth, electron flow, hiccuping, the rising of bread dough. In a separate paragraph accompanying each version, explain for your instructor how your situational analysis guided your strategy.
7. Write a description of a mechanism that has moving parts, such as an internal combustion engine. You may choose either a manufactured mechanism—such as a farm implement, electric motor, or seismograph—or a natural mechanism, such as a human organ, an insect, or a geyser. Consider your readers to have a professional interest in the mechanism. They use it or work with it in some way. In a paragraph separate from your description, explain for your instructor your purpose and audience.

CHAPTER 9

Arguing

COMPARISON

Alternatives
Criteria

TOULMIN LOGIC

Discovering Flaws in Argument
Arranging Argument for Readers

Whenever you are exercising your professional judgment and expressing an opinion, you will need the tools and techniques of argument. In a business setting, you would argue for your recommendations and your decisions. As a scientist, you would argue in the discussion section of a research report to support the conclusions you have reached. Argument, in fact, is indispensable for the technical person.

You must present your argument in a persuasive way. The use of *induction, deduction,* and *comparison* is necessary in argument. *Toulmin logic,* named for its creator, Steven Toulmin, is often a good technique for discovering an argument and presenting it. We cover all these points in this chapter.

Persuasive Argument

In argument, you deal with opinions that lie somewhere on a continuum between verifiable fact on one hand and pure subjectivity on the other. Verifiable fact does not require argument. If someone says a room is 35 feet long and you disagree, you don't need an argument, you need a tape measure. Pure subjectivity cannot be argued. If someone hates the taste of spinach, you will not convince him or her otherwise with argument. The opinions dealt with in argument may be called propositions, premises, claims, conclusions, theses, or hypotheses, but under any name they remain opinions. Your purpose in **argument** is to convince your audience of the probability that the opinions you are advancing are correct.

Typically, an argument supports one major opinion, often called the **major proposition.** In turn, the major proposition is supported by a series of minor propositions. **Minor propositions,** like major propositions, are opinions, but generally they are nearer on the continuum to verifiable fact. Finally, the minor propositions are supported by verifiable facts and frequently also by statements from recognized authorities.

To understand how you might construct such an argument, imagine for the moment that you are the planning engineer for a new housing subdivision called Hawk Estates. Hawk Estates, like many such new subdivisions, is being built close to a city, Colorful Springs, but not in a city. The problem at issue is whether Hawk Estates should build its own sewage disposal plant or tap into the sewage system of Colorful Springs. (You have already ruled out individual septic tanks because Hawk Estates is built on nonabsorbent clay soil.) Colorful Springs will allow the tap-in. You have investigated the situation, thought about it a good deal, and have decided that the tap-in is the most desirable alternative. The heads of the company planning to build Hawk Estates are not convinced. It is their money, so you must write a report to convince them.

Major Proposition

In developing your argument, it helps to use a chart such as the one illustrated in Figure 9–1. The chart is a way of clearly separating and organizing your major proposition, minor propositions, and evidence. First you must state your major proposition: "Hawk Estates should tap in to the sewage system of the city of Colorful Springs."

Figure 9–1 Argument Arrangement Chart

MAJOR PROPOSITION	MINOR PROPOSITION	EVIDENCE
	Colorful Springs can handle Hawk Estates sewage.	• Estimate of waste from Hawk Estates. • City engineer's statement that Colorful Springs can handle estimated waste.
Hawk Estates should tap into sewage system of Colorful Springs.	Overall cost to Hawk Estates taxpayers only slightly higher if tapped into Colorful Springs.	• Initial cost of plant vs. cost of tap-in. • Yearly fee charged by Colorful Springs vs. operating cost of sewage lagoon. • Cost per individual tax payer.
	Proposed plant, a sewage lagoon, will be a nuisance to home owners.	• Well maintained lagoons Okay. • Lagoons hard to maintain, often smell bad, experts say. • Lagoon has to be located upwind of development.

Minor Propositions and Evidence

Now you must support your major proposition. Your first and clearly most relevant minor proposition is that Colorful Springs' sewage system can handle Hawk Estates' waste. Questions of cost, convenience, and so forth would be irrelevant if Colorful Springs could not furnish adequate support. As you did with your major proposition, you lead off this section of the report with your minor proposition. To support this proposition, you give the estimated amount of waste that will be produced by Hawk Estates, followed by a statement from the Colorful Springs city engineer that the city system can handle this amount of waste.

The minor proposition that states "the overall cost to Hawk Estates taxpayers will be only slightly more if they are tapped into the city rather than having their own plant" is a difficult one. It's actually a rebuttal of your argument, but you must deal with it for several reasons. First and foremost, you must be ethical and honest with your employers. Second, it would be poor strategy not to be. Should they find that you have withheld information from them, it would cast doubt upon your credibility. You decide to put this premise second in your argument. In that way you can begin and end with your strongest propositions, a wise strategy. To support this proposition you would list the initial cost of the plant versus the cost of the tap-in. You would further state the yearly fee charged by the city versus the yearly cost of running the plant. You might anticipate the opposing argument that the plant will save the taxpayers money. You could break down the cost per individual taxpayer, perhaps showing that the tap-in would cost an average taxpayer only an additional 10 dollars a year, a fairly nominal amount.

Your final minor proposition is that the proposed plant, a sewage lagoon, will represent a nuisance to the homeowners of Hawk Estates. Because the cost for the tap-in is admittedly higher, your argument will probably swing on this minor proposition. State freely that well-maintained sewage lagoons do not particularly smell. But then point out that authorities state that sewage lagoons are difficult to maintain. Furthermore, if not maintained to the highest standards, sewage lagoons emit an unpleasant odor. To clinch your argument, you show that the only piece of land in Hawk Estates large enough to handle a sewage lagoon is upwind of the majority of houses, during prevailing winds. With the tap-in, of course, all wastes are carried away from Hawk Estates and represent no problem of odor or unsightliness whatsoever.

Organization

When you draft your argument, you can follow the organization shown on the chart, adding details as needed to make a persuasive case. Although your major proposition is actually the recommendation that your argument leads to, you present it first, so that your audience will know where you are heading. In executive reports, which this one is, major conclusions and recommendations are normally presented first, as we point out in Chapter 4, "Writing for Your Readers."

When you sum up your argument, draw attention once again to your key points. Point out that in cost and the ability to handle the produced wastes, the proposed plant and the tap-in are essentially equal. But, you point out, the plant will probably become an undesirable nuisance to Hawk Estates. Therefore, you recommend that the builder choose the tap-in over the plant.

Throughout any argument you appeal to reason. In most technical writing situations, an appeal to emotion will make your case immediately suspect. Never use sarcasm in an argument. You never know whose toes you are stepping on or how you will be understood. Support your case with simply stated verifiable facts and statements from recognized authorities. In our example, a statement from potential buyers that "sewage lagoons smell bad" would not be adequate. But a statement to the same effect from a recognized engineering authority from a nearby university would be acceptable and valid.

Induction and Deduction

Much of your thought, whether you are casually chatting with friends or are on your most logical and formal behavior, consists of induction and deduction. In this section we cover both induction and deduction and discuss some of the fallacies you'll want to avoid in using them.

Induction

Induction is a movement from particular facts to general conclusions. It's a method of discovering and testing the inferences that you can draw from your information. Induction is the chief way we have of establishing causality, that A caused B. The inductive process consists of (1) looking at a set of facts, (2) making an educated guess to explain the facts, and (3) then investigating to see if the guess fits the facts. The educated guess is called a *hypothesis*. No matter how well constructed your hypothesis is, remember it is only a guess. Be ready to discard it in an instant if it doesn't fit your facts.

For example, shortly after they attend a church picnic, 40 people out of the 100 who attended fall ill. You look at the facts and form a loose hypothesis: Something they ate at the picnic caused the illness. Investigating further, you discover that there were two lines at the picnic serving table. All the people who became ill went through the left-hand line. You refine your hypothesis to the effect that the illness had something to do with the left-hand line. You conjecture that perhaps the food handlers on the left-hand line did not follow the proper sanitary precautions.

However, upon checking with the food handlers from the left-hand line, you find that they all swear that they were models of cleanliness. Furthermore, you discover that 10 people who went through the left-hand line did not become ill. Stymied for the moment, you drop the hypothesis about the unsanitary food handlers, but you still have good reason to suspect the left-hand line.

You form another left-hand line hypothesis. Some of the people on the left-hand line must have eaten something that those on the right-hand line did not. You discover that the left-hand line served Mrs. Smith's potato salad. The right-hand line served Mrs. Olson's potato salad. Furthermore, everyone who ate Mrs. Smith's potato salad became ill. Everyone who ate Mrs. Olson's potato salad or no potato salad at all did not become ill.

Your hypothesis has probably been upgraded in your mind at this point to a *conclusion:* Mrs. Smith's potato salad caused the illnesses. Unless you could obtain some samples of Mrs. Smith's salad for testing, your conclusion would have to remain, unfortunately, hypothetical. But, because of the large number of people involved, you feel quite secure in your conclusion. It is possible, of course, that all 40 people came down with an infection not related to the salad, but it seems unlikely.

The whole process of gathering evidence, making hypotheses, and testing hypotheses against the evidence is, of course, the scientific method at work. Looking for similarities and differences are major tools in testing hypotheses. Examining similarities and differences in the population has led medical authorities, including the Surgeon General of The United States, to declare that cigarette smoking is hazardous to your health. They looked at the population and saw a difference: There are those who smoke and those who don't. Within these two groups, they looked for similarities. Smokers had in common a high incidence of respiratory problems, including emphysema and lung cancer. Nonsmokers had in common a low incidence of such problems. The higher incidence of such problems in the smoking group when compared to the nonsmoking group was a significant difference.

Whenever you argue inductively to support a hypothesis, you must accept the possibility of new evidence proving you wrong. Nevertheless, far more judgments and decisions are made on inductive arguments than on direct evidence. Well-constructed inductive arguments are powerful. Yes, the Surgeon General could possibly be proven wrong, but more and more people are not betting their lives on it.

The actual practice of induction can be rather messy, as you chase down leads, retreat from blind alleys, and try out your hypotheses in a trial and error way. If for some reason, you wish to show your thought processes to your readers, you might reveal some of the messiness to them when you present your argument. But, more often, after you have reached your conclusions, you will probably wish to argue for them in a straightforward way, as we demonstrated with the Hawk Estates example. In the following example, the author argues inductively for the proposition that life does not exist on Mars. The argument is supported by a series of conclusions based upon evidence gathered by the Viking spacecraft.

Major proposition

The primary objective of Viking was to determine whether life exists on Mars. The evidence provided by Viking indicates clearly that it does not.

Three of Viking's scientific instruments were capable of detecting life on Mars:

Capability of cameras

Conclusion

- The lander cameras could have photographed living creatures large enough to be seen with the human eye and could have detected growth changes in organisms such as lichens. The cameras found nothing that could be interpreted as living.

Paragraph explaining function and capability of GCMS and presenting conclusion about its findings

- The gas chromatograph/mass spectometer (GCMS) could have found organic molecules in the soil. Organic compounds combine carbon, hydrogen, nitrogen, and oxygen and are present in all living matter on Earth. The GCMS searched for heavy organic molecules, those that contain complex combinations of carbon and oxygen and are either precursors of life or its remains. To the surprise of almost every Viking scientist, the GCMS, which easily finds organic matter in the most barren Earth soils, found no trace of any in the Martian samples.

Function of biology instrument

- The Viking biology instrument was the primary life-detection instrument. A one-cubic-foot box, crammed with the most sophisticated scientific hardware ever built, it contained three tiny instruments that searched the Martian soil for evidence of metabolic processes like those used by bacteria, green plants, and animals on Earth.

Capability of biology instruments

Appeal to authority

The three biology instruments worked flawlessly. All showed unusual activity in the Martian soil, activity that mimicked life. But biologists needed time to understand the strange chemistry of the soil. Today, according to most scientists who worked on the data, it is clear that the chemical reactions were not caused by living things.

Evidence for presence of oxidants

Conclusion

Furthermore, the immediate release of oxygen when the soil contacted water vapor in the instrument, and the lack of organic compounds in the soil, indicate that oxidants are present in the soil and the atmosphere. Oxidants—such as peroxides and superoxides—are oxygen-bearing compounds that break down organic matter and living tissue. Therefore, even if organic compounds were present on Mars, they would be quickly destroyed.

Analysis of the atmosphere and soil of Mars indicated that all the elements essential to life on Earth—carbon, nitrogen, hydrogen, oxygen, and phosphorus—are present on Mars. Liquid water is also considered an absolute requirement for life. Viking found ample evidence of water in two of its three phases—

vapor and ice—and evidence for large amounts of permafrost. But it is impossible for water to exist in its liquid state on Mars.

Conclusion

Restatement of major proposition

Possibility for future research

The conditions now known to exist on and just beneath the surface of Mars do not allow carbon-based organisms to exist and function. The biologists add that the case for life sometime in Mars' distant past is still open.[1]

This passage shows well the characteristics of an inductive argument. The argument begins with the major proposition. The major proposition is then supported by a series of conclusions that are in turn supported by evidence. Remember, whatever the terminology used—conclusion, proposition, thesis, and so forth—generalizations based upon particulars are opinions, nothing more and nothing less. The frequent statements concerning the capabilities of the equipment are intended to strengthen the argument by showing that high-quality equipment was used. The references to the scientists who worked on the Viking project are an appeal to authority.

Deduction

Deductive reasoning is another way to deal with evidence. While in inductive reasoning you move from the particular to the general, in **deductive reasoning** you move from the general to the particular. You start with some general principle, apply it to a fact, and draw a conclusion concerning the fact. Although you will seldom use the form of a syllogism in writing, we can best illustrate deductive reasoning with a **syllogism:**

1. All professional basketball players are good athletes.
2. Judy is a professonal basketball player.
3. Therefore, Judy is a good athlete.

Most often the general principle itself has been arrived at through inductive reasoning. For example, from long observation of lead, scientists have concluded that it melts at 327.4° C. They have arrived at this principle inductively from many observations of lead. Once they have inductively established a principle, scientists can use it deductively as in the following syllogism:

1. Lead melts at 327.4° C.
2. The substance in container A is lead.
3. Therefore, the substance in container A will melt at 327.4° C.

In expressing deductive reasoning, we seldom use the form of the syllogism. Rather, we present the relationship in abbreviated form. We may say, for instance, "Because Judy was a professional basketball player, I knew she was a good athlete." Or "The substance is lead. It will melt at 327.4° C."

Although induction is the more common organizing technique in argument, deduction is sometimes used, as in this example:

> Methanogens, like other primitive bacteria, are anaerobic: They live only in areas protected from oxygen. This makes sense, since there was virtually no oxygen in the atmosphere when bacteria first evolved. But once bacteria developed chlorophyll *a*, the pigment of green plants, they began to use carbon dioxide and water for photosynthesis and produced oxygen as a waste product. When massive colonies of these photosynthetic bacteria developed, they pumped large amounts of oxygen into the atmosphere. Oxygen is a powerful reactive gas, and most early bacteria were not equipped to survive with it. Some bacterial species that were adapted to the new gas, including the oxygen producers themselves, continued to thrive. Others presumably evolved special metabolisms to protect them from oxygen, found anaerobic environments, or disappeared.[2]

Presented formally, the syllogism in this paragraph would go something like this:

1. Methanogens cannot live in oxygen.
2. Oxygen was introduced into the methanogens' environment.
3. Therefore, methanogens either evolved special metabolisms to protect them from oxygen, found anaerobic environments, or disappeared.

Logical Fallacies

Many traps exist in induction and deduction for the unwary writer. When you fall into one of these traps, you have committed what logicians call a **fallacy.** Avoid a rush to either conclusion or judgment. Take your time. Don't draw inferences from insufficient evidence. Don't assume that just because one event follows another, the first caused the second—a fallacy that logicians call *post hoc, ergo propter hoc* (that is, *after this, therefore, because of this*). You need other evidence in addition to the time factor to establish a causal relationship.

For example, in the sixteenth century tobacco smoking was introduced into Europe. Since that time, the average European's life span has increased severalfold. It would be a fine example of the *post hoc* fallacy to infer that smoking has caused the increased life span, which in fact probably stems from improvements in housing, sanitation, nutrition, and medical care.

Another common error is applying a syllogism backwards. The following syllogism is valid:

1. All dogs are mammals.
2. Jock is a dog.
3. Therefore, Jock is a mammal.

But if you reverse statements (2) and (3) you have an invalid syllogism:

1. All dogs are mammals.
2. Jock is a mammal.
3. Therefore, Jock is a dog.

Jock, of course, could be a cat, a whale, a Scotsman, or any other member of the mammal family. You can often find flaws in your own reasoning or that of others if you break the thought process down into the three parts of a syllogism.

Comparison

In business and technical situations, you frequently have to choose between two or more alternatives. When such is the case, the method of investigating the alternatives will usually involve comparing the alternatives one to another. (Contrast is implied in comparison.) To be meaningful, the comparisons should be made by using standards or criteria. Perhaps you have bought a car recently. When you did, you had your choice of many alternatives. In reaching your decision, you undoubtedly compared cars using criteria such as price, comfort, appearance, gas mileage, and so forth. Perhaps you even went so far as to rank the criteria in order of importance, for example, giving price the highest priority and appearance the lowest. The more consciously you applied your criteria, the more successful your final choice may have been.

After you bought your car, no one asked you to make a report to justify your decision. However, in business it's common practice for someone to be given the task of choosing among alternatives. The completion of the task involves a report that makes and justifies the decisions or recommendations made. When such is the case, a comparison arrangement is a good choice. You can arrange comparison arguments by **alternatives** or by **criteria.**

Alternatives

Assume that you are comparing two alternative desalination processes: freezing and flash evaporation. Your criteria are cost, ease of maintenance, and purity of the water produced. After the necessary explanations of the processes and the criteria, you might organize your material this way:

- Freezing
    - Cost
    - Ease of maintenance
    - Purity
- Flash evaporation
    - Cost
    - Ease of maintenance
    - Purity

In this arrangement, you take one alternative at a time and run each through the criteria. This arrangement has the advantage of giving you the whole picture for each alternative as you discuss it. The emphasis is on the alternatives.

Criteria

In another possible arrangement, you organize by criteria:

- Cost
    - Freezing
    - Flash evaporation
- Ease of maintenance
    - Freezing
    - Flash evaporation
- Purity
    - Freezing
    - Flash evaporation

The arrangement by criteria has the advantage of sharper comparison. It also has an advantage for readers who read selectively. Not every reader will have equal interest in all parts of a report. For example, an executive reading this report might be most interested in cost, an engineer in ease of maintenance, a consumer in purity.

Recommendation reports based upon such comparisons are frequently cast in the form of a memorandum (see pages 414–424). Figure 9–2 shows a report based upon this comparison arrangement. Notice that because the report is for an executive the writer states his conclusions and recommendations before presenting his data.

MEMORANDUM

Date: 13 November 1991

To: Professor Milton Weller
Department of Entomology, Fisheries, and Wildlife

From: David M. Zellar DMZ

Subject: Choice of tape recorders

At your request I compared several available battery-powered tape recorders to determine which one would be most suitable for purchase by the Department of Entomology, Fisheries, and Wildlife (EFW). I compared four recorders: the Marantz Superscope CD-330, the RMF 740 AV Stereo Recorder, the AIWA TPR-945, and the Nakamichi 550. The criteria used for comparison, listed in priority order, are (1) frequency response, (2) cost, and (3) weight.

The Nakamichi 550 has the best frequency response and a low weight but has a high price tag. The AIWA TPR 945 meets all three criteria acceptably. Therefore, I recommend the AIWA TPR-945 at $350. However, if the Nakamichi would not be too expensive at $640, it could be a good alternative.

Findings

A summary of my findings follows.

Frequency response. EFW will purchase the tape recorder primarily to record bird vocalizations. The Nakamichi 550 has the best frequency response of 40-17,000 Hz. The AIWA TPR-945 is a close second at 50-15,000 Hz. The majority of bird vocalizations fall within the 100-12,000 Hz range, with only a rare few exceeding 15,000 Hz. Therefore, both the Nakamichi and the AIWA have very acceptable responses.

Cost. The Marantz Superscope CD-330 costs $249, the AIWA TPR-945, $350. The prices of both these machines are quite low. The other two machines cost considerably more: the RMF 740 AV Stereo Recorder, $579; the Nakamichi, $640.

Weight. None of the recorders is too heavy to carry on a field trip. The Marantz, RMF 740 AV, and the Nakamichi all weigh within a few ounces of 8 pounds. The AIWA weighs 12.5 pounds.

Figure 9–2 Comparison Report

Toulmin Logic

When you construct an argument by yourself, it's difficult at times to see the flaws in it. When you expose the same argument to your friends, even in casual conversation, they, being more objective about it, can often spot the flaws you have overlooked. **Toulmin logic** provides a way of checking your own arguments for those overlooked flaws.[3]

Discovering Flaws in Argument

Because using Toulmin logic is a way of raising those questions readers may ask, its use will make your arguments more reader oriented. Toulmin logic comprises five major components:

1. *Claim* — The major propositon or conclusion of the argument
2. *Grounds* — The evidence upon which the claim rests—facts, experimental research data, statements from authorities, and so forth
3. *Warrant* — That which justifies the grounds and makes them relevant to the claim
4. *Backing* — Further evidence for accepting the warrant
5. *Rebuttal* — Counter arguments, exceptions to the claim, warrant, or backing or reasons for not accepting them

We'll illustrate how Toulmin logic works, first with a simple example and then with one more realistically complex. In this first sample, we indicate the kinds of questions readers might ask that would lead to the next consideration.

Claim	We can't go on our picnic tomorrow.
Reader	How come?
Grounds	The weather will be too nasty. The National Weather Service predicts rain and low temperatures for tomorrow.
Reader	Can we trust the forecast?
Warrant	The Weather Service forecasts are accurate approximately 80% of the time.
Reader	How come the forecasts are so accurate?
Backing	Today's weather forecasting is based upon extensive observations, the application of scientifically sound principles, and the use of modern technology, such as radar.

Reader	What about the 20% of the time the forecasts are wrong?
Rebuttal	Of course, the forecasts are wrong about 20% of the time.

Claim, grounds, warrant, and backing, in this case, indicate that we have a strong argument, but the rebuttal demonstrates that it is not strong enough to say with absolute conviction that we can't go on the picnic. The situation calls for a *qualifier*. The claim would be better phrased as "We *probably* can't go on our picnic tomorrow." However, if a decision had to be made based upon this argument, calling off the picnic would seem to be justified.

What even this simple example shows is that arguments are rather complex chains of reasoning in which you have to argue not only for your claim but for the grounds upon which the claim is based. Toulmin logic helps you to construct the chain.

For a more complicated example, let's consider the greenhouse effect hypothesis.[4]

Claim	The accumulation of gases, particularly carbon dioxide (CO_2), emitted from the burning of fossil fuels will trap heat in the atmosphere, which will cause global warming, resulting in droughts, floods, and food shortages.
Grounds	In the past 100 years, CO_2 concentration in the atmosphere has risen from 270 parts per million (ppm) to 350 ppm. That this rise has been caused by the increased burning of fossil fuels seems indisputable. Researchers from Ohio State University report that central Asia has warmed by 1 to 3° C in the past century. Various computer models predict global temperatures rising by as much as 4° C in the next 50 years.
Warrant	Reputable scientists agree with this hypothesis. Mathematicians from AT&T Bell Labs report that "There is a 99.9% chance that the warming and the CO_2 rise are causally related." Climatologist James Hansen of NASA's Goddard Institute for Space Studies says that temperature data of the last 100 years show a worldwide rise of .4° C. He has further stated in Congressional testimony that the greenhouse effect has begun and will worsen.
Backing	Worldwide, environmentalists have called for stabilization of CO_2 by 2000. Prime Minister Margaret Thatcher of Great Britain, a noted conservative, asked the United Nations to effect treaties to restrict the emission of gases that can cause global warming.

Up to this point, the argument for the greenhouse hypothesis and its effects seems to be going well. But, if you dig further, you will find rebuttals.

Rebuttal	A study of ocean temperatures by scientists from the Massachusetts Institute of Technology shows no rise in ocean tempertures in the past 100 years.
Rebuttal	Scientists at the National Oceanographic and Atmosphere Institute reviewed U.S. climatic records and concluded, "There is no statistically significant evidence of an overall change in annual temperature or change in annual precipitation for the contiguous U.S. 1895–1987."
Rebuttal	Michael Schlesinger, a respected climatologist from the University of Illinois, says that Hansen's "statements have given people the feeling that the greenhouse effect has been detected with certitude. Our current understanding does not support that. Confidence in its detection is now down near zero."

And so on. Digging for evidence on the greenhouse effect shows a sharp division with reputable scientists coming down on both sides of the question. The claim has to be qualified, perhaps something like this: "Some scientific studies show a correlation between the rise of CO_2 in the atmosphere and global warming, but the evidence and methodology of such studies have not convinced all scientists of their validity." Applying Toulmin logic has resulted in a weaker claim, but it is a claim that can be supported with the existing evidence.

Arranging Argument for Readers

Toulmin logic can help you arrange your argument as well as discover it. Though you would not want to follow Toulmin logic in a mechanical way, thinking in terms of claim, grounds, warrant, backing, rebuttal, and qualifier can help you to be sure you have covered everything that needs to be covered. Obviously, claim and grounds must always be presented. In most business situations, as we have pointed out, the claim is likely to be presented first, particularly in executive reports. However, in a situation where the readers might be hostile to the claim, it might be preferable to reverse the order. If the grounds are strong enough, they may sway the readers to your side before they even see the claim. On the other hand, if a hostile audience sees the claim first they may not pay enough attention to the grounds to be convinced.

Rebuttals should always be considered and if serious be included in your presentation. You have an ethical responsibility to be honest with your readers. Furthermore, if your readers think of rebuttals that you do not deal with, it will damage your credibility. If you can counter the rebuttals successfully, perhaps by attacking their warrant or backing, your claim can stand. If you cannot counter them, you will have to qualify your claim.

How deeply you go into warrants and backing depends upon your readers. If your readers are not likely to realize what your warrant is (for example, "Respected scientists agree with this hypothesis."), then you had better include the warrant. If your readers will be likely to disagree with your warrant or discount its validity, then you had better include the backing. All in all, Toulmin logic can be a considerable help in discovering and arranging an argument. It is also extremely useful in analyzing the soundness of other people's arguments.

Planning and Revision Checklist

You will find the planning and revision checklists that follow Chapter 2, "Composing" (pages 33–35 and inside the front cover), and Chapter 4, "Writing for Your Readers" (pages 78–79), valuable in planning and revising any presentation of technical information. The following questions specifically apply to argument. They summarize the key points in this chapter and provide a checklist for planning and revising.

Planning

- What is your claim, that is, the major proposition or conclusion of your argument?
- What are your grounds? What is the evidence upon which your claim rests—facts, experimental research data, statements from authorities, and so forth?
- Do you need a warrant that justifies your grounds and makes them relevant?
- Do you need further backing for your grounds and warrant?
- Are there rebuttals—counterarguments, exceptions to the claim, warrant, or backing or reasons for not accepting them? Can you rebut the rebuttals? If not, should you qualify your claim?
- Are you choosing among alternatives? If so, what are they?
- What are the criteria for evaluating the alternatives?

Revision

- Is your claim clearly stated?
- Do you have sufficient grounds to support your claim?
- If needed, have you provided a warrant and backing for your grounds?
- Does any of your evidence cast doubt upon your claim? Have you considered all serious rebuttals?
- Have you dealt responsibly and ethically with any rebuttals?
- Have you remained fair and objective in your argument?
- Have you presented evidence for causality beyond the fact that one event follows another?
- If you have used deductive reasoning, can you state your argument in a syllogism? Does the syllogism demonstrate that you have reasoned in a valid way?

Planning

- Is your audience likely to be neutral, friendly, or hostile to your claim? If your audience is hostile, should you consider putting your claim last rather than first?

Revision

- Is your argument arranged so that it can be read selectively by readers with different interests?

Exercises

1. We have reprinted here an excerpt from the report of the Rogers Commission, which investigated the explosion of the space shuttle Challenger. The excerpt deals with the cause of the explosion. It presents a conclusion and defends it with induction and deduction. It demonstrates some of the components of Toulmin logic. Working in a group, analyze the argument's arrangement and presentation. Use this chapter's Planning and Revision Checklist to guide your analysis. Pay particular attention to how the argument's parts work together to defend its conclusion. Judge whether the argument is effective or ineffective. Write an analysis for your teacher that presents and supports your judgment.

The consensus of the commission and participating investigative agencies is that the loss of the space shuttle Challenger was caused by a failure in the joint between the lower segments of the right solid rocket motor. The specific failure was the destruction of the seals that are intended to prevent hot gases from leaking through the joint during the propellant burn of the rocket motor. The evidence assembled by the commission indicates that no other element of the space shuttle system contributed to this failure.

In arriving at this conclusion, the commission reviewed in detail all available data, reports and records; directed and supervised numerous tests, analyses and experiments by NASA, civilian contractors and various government agencies; and then developed specific failure scenarios and the range of most probable causative factors.

Findings

1. A combustion gas leak through the right solid rocket motor aft field joint initiated at or shortly after ignition eventually weakened and–or penetrated the external tank initiating vehicle structural breakup and loss of the space shuttle Challenger during STS Mission 51-L.
2. The evidence shows that no other STS 51-L shuttle element or the payload contributed to the causes of the right solid rocket motor aft field joint combustion gas leak. Sabotage was not a factor.
3. Evidence examined in the review of space shuttle material, manufacturing, assembly, quality control and processing of nonconformance reports found no flight hardware shipped to the launch site that fell outside the limits of shuttle design specifications.
4. Launch site activities, including assembly and preparation, from the receipt of the flight hardware to launch were generally in accord with established procedures and were not considered a factor in the accident.

5. Launch site records show that the right solid rocket motor segments were assembled using approved procedures. However, significant out-of-round conditons existed between the two segments joined at the right solid rocket motor aft field joint (the joint that failed).
6. The ambient temperature at time of launch was 36 degrees Fahrenheit, or 15 degrees lower than the next coldest previous launch.
7. The temperature at the 300 degree position on the right aft field joint circumference was estimated to be 28 degrees (plus or minus 5 degrees) Fahrenheit. This was the coldest point on the joint.
8. Experimental evidence indicates that due to several effects associated with the solid rocket booster's ignition and combustion pressures and associated vehicle motions, the gap between the tang and the clevis [of the joint] will open as much as .017 and .029 inches at the secondary and primary rings, respectively.
9. O-ring resiliency is directly related to its temperature.
   a. A warm O ring that has been compressed will return to its original shape much quicker than will a cold O ring when compression is relieved. Thus, a warm O ring will follow the opening of the tang-to-clevis gap. A cold O ring may not.
   b. A compressed O ring at 75 degrees Fahrenheit is five times more responsive in returning to its uncompressed shape than a cold O ring at 30 degrees Fahrenheit.
   c. As a result, it is probable that O rings in the right solid booster aft field joint were not following the opening of the gap between the tang and the clevis at time of ignition.
10. Of 21 launches with ambient temperatures of 61 degrees Fahrenheit or greater, only four showed signs of O-ring thermal distress; i.e., erosion or blow-by and soot. Each of the launches below 61 degrees Fahrenheit resulted in one or more O rings showing signs of thermal distress.
11. A series of puffs of smoke were observed emanating from the 51-L aft field joint area of the right solid rocket booster between 0.678 and 2.500 seconds after ignition of the shuttle solid rocket motors.

    The puffs appeared at a frequency of about three puffs per second. This roughly matches the natural structural frequency of the solids at liftoff and is reflected in slight cyclic changes of the tang-to-clevis opening.
12. This smoke from the aft field joint at shuttle liftoff was the first sign of the failure of the solid rocket booster O-ring seals on STS 51-L.
13. The leak was again clearly evident as a flame at approximately 58 seconds into the flight. It is possible that the leak was continuous but unobservable or non-existent in portions of the intervening period. It is possible in either case that thrust vectoring and normal vehicle response to wind shear as well as planned maneuvers reinitiated or magnified the leakage from a degraded seal in the period preceding the observed flames. The estimated position of the flame, centered at a point 307 degrees around the circumference of the aft field joint, was confirmed by [examination after] the recovery of two fragments of the right solid rocket booster.

Conclusion

In view of the findings, the commission concluded that the cause of the Challenger accident was the failure of the pressure seal in the aft field joint of the right solid rocket motor. The failure was due to a faulty design unacceptably sensitive to a number of factors. These factors were the effects of temperature, physical dimensions, the character of materials, the effects of reusability, processing and the reaction of the joint to dynamic loading.[5]

2. Rewrite the excerpt from the Rogers Commission report presented in Exercise 1 so that an average fifth-grade student could comprehend the argument and understand the cause of the accident.

3. Write a memo to an executive that recommends the purchase of some product or service the executive needs for the conduct of his or her business. Your memo should establish criteria and justify the choice of the product or service you recommend against other possible choices. See Chapter 13 for information on memo format.

4. Your new boss on your first job knows how important it is for the organization to stay aware of trends that may affect the organization. He or she asks you to explore such a trend. The possibilities are limitless, but you may be happier exploring some trend in your own field. For example, are you in computer science? Then you might be interested in the latest trends in artificial intelligence. Are you in forestry? Are there trends in the use and kinds of wood products?

Develop a claim about the trend, for example:

If trend *A* continues, surely *B* will result.
Trend *A* will have great significance for *X* industry.

Support your claim with a well-developed argument that demonstrates your ability to use induction, deduction, and Toulmin logic. Write your argument as a memorandum to your boss. See Chapter 13 for information on memo format.

5. You are a member of a consulting firm. Your firm has been called in to help a professional organization deal with a question of major importance to the members of the organization. For example, nurses have an interest in whether nurses should be allowed to prescribe medication and therapy. You will probably be most successful in this exercise if you deal with organizations and questions relevant to your major. Investigate the question and prepare a short report for the executive board of the organization. Your report should support some claim: for example, nurses should be allowed to prescribe medication and therapy. Use Toulmin logic in discovering and presenting your argument. That is, be aware of the need to provide grounds, warrants, backings, and qualifiers. Anticipate rebuttals and deal with them ethically and responsibly. Use Part III, "Document Design in Technical Writing" to help you format your report.

PART III

Document Design in Technical Writing

CHAPTER 10

Document Design

CREATING USEFUL FORMATS

HELPING READERS LOCATE INFORMATION

Desktop computers are now common in business. By the year 2000, "80% of the work force will be employed in jobs that involve either generating or transmitting information. In 1980, there was one electronic work station for every twenty-three white-collar employees; by 1989, one for every two."[1]

If you are planning a career in a scientific or technical field, you will almost certainly be using computers in your job. You may also be expected to write with word-processing software on a computer. Even if you don't have your own computer on your desk at work, the final copy of what you write will most likely be produced with software and printed on an electronic printer. You should know what you can do to make your reports look good and to make them easier to read. That's what this chapter is about.

With a computer, you have many options for writing and formatting your papers. You may be able to change the size and style of type. You may be able to integrate text and pictures on the same page. You may be able to turn data into charts with just a few keystrokes. Figures 10–1 and 10–2 show you some of the options that technical writers now have on microcomputers.

Computer technology has been changing rapidly and will continue to do so. In the last decade, major changes have taken place about every two-and-a-half years. That's four computer "generations" in 10 years.[2] By the time you've been at work for a decade, you'll see even more change.

Even if you don't have a computer, you have some options for formatting your papers. For example, on a typewriter, you control the margins and the spacing for your text. You control the placement and style of the section headings. You may have at least a few choices of type size.

How do you make good choices among the options that you now have? How can you best prepare yourself for the choices that you may have on your first job . . . and on a job 10 years from now? You'll find that the best preparation is to learn and practice the general principles of document design. In this chapter, we give you those principles and show you how to apply them with a variety of technologies. You'll find useful guidelines here whether you are using a typewriter, a stand-alone word processor, or a computer with both text and graphics.

Creating Useful Formats

Readers judge your professionalism from the presentation as well as from the content and style of your writing. In fact, a reader's first impression comes from the appearance of your work, not from what it says. A negative

USING TECHNOLOGY FOR WORD PROCESSING

You can put the title and author in a header on every page.

As an illustration of a physical object with fractional dimension, consider the coastline of an island.[1] Viewed on a map, the coastline appears fairly regular in shape, and the perimeter can, apparently, be well approximated by a smooth curve border. Viewed on a larger scale, bays and inlets become visible that had not been discernable earlier. Viewed by a person standing on the shore of the island, even smaller features of the coastline reveal themselves, each adding more and more length to the coastline. No matter on what scale a measurement is taken, at least down to the molecular level, a smaller measuring unit will reveal new structure and new length to the coastline.

Headings that are larger than the text stand out.

Other Real World Examples of Fractals

Many other examples of fractals can be found in the everyday world. In each case, as you look at the natural object more closely, new structures are revealed. Mandlebrot's book contains many interesting examples.[2] The following list gives four other examples that have been discussed in detail in the literature.

Boldface type lets you put a heading at the beginning of each example.

- **Noise in telephone lines.** Noise in telephone lines occurs in spurts and is never uniformly distributed over any time interval. In any time period in which there is excessive noise, there will be periods of quiet. The pattern is similar to that of the Cantor set, which we discuss in the next section.[3]

- **Blood vessels.** Blood vessels branch and divide in fractal patterns down to the level of individual blood cells.[4]

Listing the four examples makes them easier to remember.

- **Crumpled paper.** Rolled up paper balls are fractals. Their dimension has been measured to be approximately 2.51, depending on the surface density of the paper. This means that they are more space filling than a two-dimensional surface , but not as space filling as a three-dimensional volume.[5]

- **Dripping faucets.** When a faucet leaks slowly, the drips come at a steady pace. When it leaks faster, the pattern becomes chaotic and the frequencies form a fractal.[6]

The word processor numbers the footnotes automatically and adjusts the page for them.

It also lets you put the footnotes in smaller type.

[1] B. Mandlebrot, *Fractals: Form, Chance, and Dimension* (W. H. Freeman and Company, San Francisco, 1977) 30.

[2] *Ibid.*.

[3] J. Gleick, *Chaos* (R. R. Donnelley & Sons, Harrisonburg, VA, 1987) 92.

[4] *Ibid.*, p. 108.

[5] M. A. F. Gomes, Am. J. Phys. **55,** 649 (1987).

[6] R. Shaw, *The Dripping Faucet as a Model Chaotic System* (Aerial Publishing, Santa Cruz, CA, 1984).

You can set the word processor to number the pages as it prints them.

—7—

Figure 10–1 A technical paper with a header, list, and footnotes, prepared on a personal computer

Patricia Toth
Physics 273
Spring 1989

The Spectrum of Starlight as a Function of Time

Some word processors let you write even complex equations. This one, however, was created in a drawing program and then copied into the text.

The intensity of light that is received from a single star depends on its blackbody radiation spectrum, which is a function of temperature and frequency:

$$I_{BB}(f,T) = \frac{2hf^3/c^2}{e^{hf/kT} - 1}$$

Many word processors make it easy to use subscripts and superscripts.

where h = Planck's constant ~ 6.6×10^{-34} J-s and k = Boltzmann's constant ~ 1.38×10^{-23} J/°K.

Time on the main sequence

The length of time a star remains on the main sequence is estimated by the formula

$$t_{MS} = (t_* / t_{solar}) = (M_* / M_{solar})^{-2.3}$$

which is plotted in Figure 4 below:

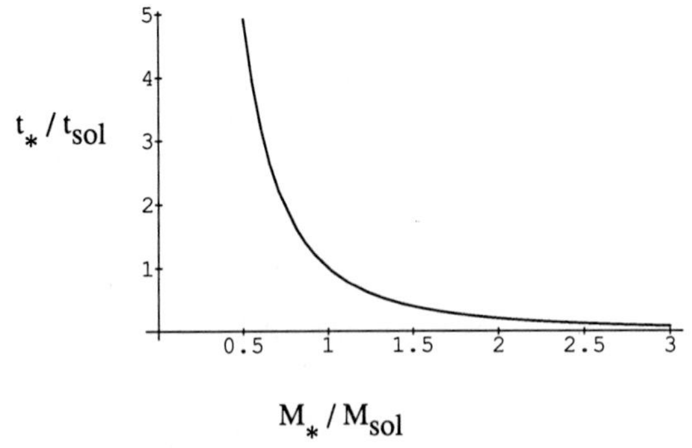

You can plot the figure in one program and then copy it into your report.

t_* / t_{sol}

5
4
3
2
1

0.5 1 1.5 2 2.5 3

M_* / M_{sol}

Fig. 4: The time a star spends on the main sequence (in units of the time the sun spends on the main sequence) as a function of the star's mass (in units of the sun's mass).

You can set up "footers" that include lines and text as well as page numbers. The footer appears on every page; only the page number changes.

A Model for the Evolution of Galactic Light
Page 6

Figure 10–2 A technical paper with text and graphics, prepared on a personal computer first impression may be difficult to overcome. A positive first impression may add to the persuasiveness of your position or convince your readers to put a little more effort into understanding your report.

Appearance is more than just aesthetics, however. The format of your document may actually help or hinder the reader. Look at the layouts of the pages in Figures 10–3 and 10–4. Which would you rather read? The page in Figure 10–3 provides no clues to the content. The page in Figure 10–4 helps the reader to see the structure of the writer's points. The headings, lists, and tables all help the reader.

Technical and business reports differ from literature in a very significant way. The writer of a short story or a novel hopes and expects that the reader will read every word from start to finish. The writers of technical or business reports know that many of their readers will not want or need the entire report. Rather, the readers will read selectively, scanning a report and looking for the sections that are pertinent to their needs and interests.

For instance, scientists reading through a professional journal will want to read thoroughly only those articles that relate closely to their specialities. They will want only summaries of the others. That's why most journals provide abstracts with each article to allow the scientific readers to extract key points without reading the entire article. Furthermore, the headings in the article, such as "Materials and Methods" or "Results and Discussion" allow the scientists to scan the article, looking for the sections that interest them.

Similarly, users working with a computer program are not likely to read the entire users' manual. They will go to the manual when they have a specific problem or need instructions for a specific task. They will look in the table of contents or index for words that match the task they are trying to do, such as "adding text to a graph" or "changing the colors in a pie chart."

In like manner, reports for executives should have formats that allow for selective reading. As we point out in Chapter 4, "Writing for Your Readers," not all readers of a report will be interested in the same parts of the report. (See pages 67–68.) By the way that you write the headings and combine them into the table of contents, you allow your executive readers to find the sections that they need. Executives also appreciate summaries and clearly labeled conclusions and recommendations that come early in the report.

Even if your readers are likely to read the entire document when they first get it, think about later uses, too. In many cases, readers are likely to come back to look for a specific section or for the answer to a specific question. Always keep in mind how your readers will be using your document. The choices that you make, such as where to put the headings, what size type to use, how wide to make the margins, all can make a difference in helping readers read selectively.

-- 3 --

With the substantial growth in computing in the College of Engineering during the past decade, the issue of linking the departments through a computer network has become critical. The network must satisfy a number of criteria to meet the needs of all of the engineering departments. We first state these criteria and then discuss them individually in detail.

To adequately serve both faculty and student needs in the present environment, the network must be able to handle the number of computers currently in use. In addition, the system must be able to expand and link in additional computers as the number of computers increases over the next few years. The different types of computers that the departments presently possess must all be linkable to the network, and the types of computers that are scheduled for purchase must also be able to be connected to the network. The network should permit the transfer of files in both text and binary form in order to facilitate student access to files and collaborative exchange among faculty and research associates. The network must also have adequate bandwidth in order to handle the expected traffic. Finally, the network must permit both students and faculty to link to the existing national networks.

The page with just text looks dense and uninviting.

Readers can't tell at a glance what the text is about.

Each department currently has both computer laboratories for students and computers that are associated with faculty research projects. The various departments possess different numbers of computers. The Aeronautical Engineering Department at present has 27 computers, while Civil Engineering has 12. The Electrical Engineering Department has the most in the College with 46. Mechanical has 22, and Nuclear Engineering, the smallest department in the College, presently has 7. This means that the entire College presently has 114 computers which will need to be networked.

In order to meet their different needs, each department has focused on the purchasing of computers with differing strengths. The computers provide for faculty and advanced students to program in a variety of languages including Pascal, C, and Fortran.

Figure 10–3 An example of poor formatting

With the substantial growth in computing in the College of Engineering during the past decade, the issue of linking the departments through a computer network has become critical. The network must satisfy a number of criteria to meet the needs of all of the engineering departments. We first list these criteria and then discuss them individually in detail.

Large headings make the topics and structure obvious.

What must the network do?

To serve both faculty and students, the network must be able to

A bulleted list makes the points more memorable.

Each item in the list becomes the heading for a subsection.

- handle the number of computers currently in use
- link different types of computers
- expand as the number of computers increases
- link to the national networks
- transfer and store both text and binary files

The network must also have adequate bandwidth in order to handle the expected traffic.

The subsection headings are also bold but smaller than the main section heading.

The shorter line length makes the text easier to read and makes the headings stand out.

Handling the number of computers currently in use

Each department has both computer laboratories for students and computers that are associated with faculty research projects. The following table shows the number of computers in each department at the end of the last fiscal year.

Aeronautical Engineering	27
Civil Engineering	12
Electrical Engineering	46
Mechanical Engineering	22
Nuclear Engineering	7
Total	114

The numbers are much clearer in a table.

Linking different types of computers

In order to meet their different needs, each department has focused its purchasing on machines with different strengths. The computers

The footer on every page reminds readers of the overall topic.

A Proposal to Install a Computer Network for the College of Engineering **page 3**

Figure 10–4 The same page reformatted

Planning Your Format

These six principles will help you plan your format:

1. Know what you'll have to work with.
2. Know what decisions you can make.
3. Choose a simple, functional format.
4. Plan for graphics.
5. Reveal your arrangement to your reader.
6. Keep the format consistent.

1. Know What You'll Have to Work With Plan the format for the machine you'll use to print your work. Do not plan to use features that are unavailable or unaffordable. For example, if you are developing a brochure or advertisement or manual that will be printed by a professional print shop, you may be able to have two or more colors in your work. If you will be producing your work on a typewriter or even a letter-quality desktop printer, color probably is not an option.

2. Know What Decisions You Can Make You may also be constrained by decisions that have already been made. Your company may have a standard format for reports or letters or proposals. Many journals have standard formats that people in the profession expect to see. Your professor may tell you to use specific formats for different types of writing.

Many software programs allow for style sheets, which are instructions for formatting that can be used repeatedly so that documents that are part of a series look alike. If someone has built style sheets for documents like the one you are writing, you may be expected to use them.

Don't change formats arbitrarily, just to be different. If your report is part of a series, making it look like the others in the series promotes the company's image and meets readers' expectations. If you think that the format that you are being asked to use won't work well for your readers, however, find out who makes decisions on format and present a case for the changes that you want.

3. Choose a Simple, Functional Format Provide no more complexity in the format than the situation requires. You'll impress readers most by providing just the information they need in a way that makes it easy for them to find and understand it. An overly elaborate design may give readers the impression that you didn't pay enough attention to the content.

Don't use more features than you need. If your report is short, say under five pages, a simple memorandum format will usually do. (See pages 414–424.) A memorandum may have headings, but it doesn't have a table of contents. If your report is longer, you'll need to provide elements such as a title page and a table of contents.

Weigh the value of additions, such as a glossary and appendixes, carefully. Add something only if you can justify it on the basis of functionality. For example, if your report contains only a few words that need to be defined, put the definitions in the text. On the other hand, if you use many terms that are unfamiliar to your readers, a glossary may be a good idea. Before including an appendix, ask yourself if your readers really need or want the extra information. You could instead tell them that it is available in your files and you would be happy to send it on request.

4. Plan for Graphics While planning your report, look for places where graphics can help explain the points. Instead of giving statistics in paragraphs of prose, turn them into simple tables and bar graphs. Many software programs now make it easy to integrate tables, charts, and graphs into your text. Visual ideas should be presented visually. Do not describe, for example, the keyboard of a computer. Show it in a photograph or drawing. Many programs now include "clip art," pictures that you can copy into your text.

5. Reveal Your Arrangement to Your Reader You see your report as you have arranged it. That is, in your mind, you see clearly all the divisions and subdivisions of your organizational plan. Your readers see your report page by page, and therefore, they may not grasp your arrangement or understand its significance. Give them some help. Even a two- or three-page memo will profit from a few well-placed headings that reveal your arrangement. Longer reports definitely need headings and a table of contents based on the headings.

6. Keep the Format Consistent Consistency helps readers. You should strive for consistency in both language and presentation. Figures 10–5 and 10–6 show two examples of consistency in formatting.

Setting Up a Good Format

To create a good format:

- Design the page for easy reading.
- Leave room in the margins.
- Use a medium line length.
- Set the spacing for easy reading.
- Use a ragged right margin (most of the time).
- Use white space to help people find information.
- Use lists appropriately.

Design the Page for Easy Reading In designing the overall page, you have to decide how many columns you will have and where you will put each of the elements of your format. Design with these goals in mind:

You can tell at a glance that this page shows procedures.

Each procedure is given in a consistent format.

Changing the Size of Your Card

You can change the size of your card, then change it back again.

The task is highlighted in color.

To shorten a card:

1. Move the cursor to a line between the last field line and the bottom of the card.
2. Press [Delete line] repeatedly until you have shortened the card as much as you want.

The steps are listed.

To lengthen a card:

1. Move the cursor to where you want to insert lines.
2. Press [Insert line] repeatedly until you have added all the lines you want.

Lengthening a card will cause a broken side border.

Each step begins with an imperative verb.

To fill in the breaks:

1. Touch [Redraw Border]

If you replace the card borders after you have adjusted the length of your card, you may see a double border at the bottom.

The numbers are in color.

To remove the extra border:

1. Move the cursor to a line below your last field.
2. Press [Clear display].

   The computer will remove everything below the cursor.
3. Touch [Redraw Border].

Figure 10–5 Consistent patterns in a software manual
Source: The User's Manual for Personal CardFile, © 1984, Hewlett-Packard Company. Used with permission.

- The page should look inviting.
- The page should look open—have space inside the text as well as in the margins.
- The design should help people find information quickly.
- The design should be appropriate for the type of document.

You can have a good design with one column of text if you keep wide margins. This format works well for letters and memos, as you can see in Figure 10–7 on page 228.

The heading announces the topic.

The Need for Continuing Evaluation

Recommendation

Quotes from members of the committee are in the margin in color.

". . . current practices and policies must change to emphasize the need of customers to understand the information provided by inserts."

Edwin Lee

Pacific Bell should regularly evaluate its inserts.

Action Steps

Each topic includes
—a recommendation
—action steps
—observations

1. The insert production team should meet after the bill inserts are completed to evaluate how the process worked.
2. As Pacific Bell begins to rewrite customer instruction material, especially bill inserts, focus groups should be used on a regular basis. Customer feedback is important in determining whether the inserts are understandable and useful. Focus groups should be used monthly during the first year and periodically after that to monitor the progress of Pacific Bell's plain language program.
3. Pacific Bell should reconvene Consumer Advisory Council IX in nine months so that we can comment on the company's progress.

Observations

The company's support of Consumer Advisory Council IX indicates that management recognizes that a problem exists with the documents that it sends to customers. It shows that the company is willing to change.

Figure 10–6 A consistent pattern in a report
Source: Plain Language and Pacific Bell's Inserts, A Report to Pacific Bell's Senior Management, © 1986 by Pacific Bell, San Francisco, CA. Used with permission.

For many situations, two columns of text or one narrower column of text with the headings in a wider left margin works best. Figure 10–8 (see page 228) is a fact sheet with a two-column format. Figure 10–9 (see page 229) is a page from the user's manual for a computer program.

Leave Room in the Margins The margins are the white space around the text. The white space makes the page look inviting. Too little space

(text continued on page 231.)

MEMORANDUM

To: All Department Managers

From: Mary Lenkowitz
Human Resources Manager

Date: June 4, 1990

Subj: Annual Performance Reviews

As you know, June is the month for our annual performance reviews. Last year, the process went very smoothly and everyone met the deadlines. Let's do it again this year.

Deadline for Return of Form HR-14 -- Change in Status

You must fill out and sign a Change in Status Form (HR-14) for every employee who is going to have a change in salary or employee level for the coming year. In order to process the changes for the July payroll, we must have the signed HR-14s in my office no later than **Monday, July 9.**

Schedule for the Performance Review Process

You may set your own internal department schedule for completing the process as long as you meet the July 9th deadline. Past experience, however, suggests that this schedule works well:

Week of	This happens
6/4	Human Resources Department distributes Personnel Review Forms to all employees
6/4 - 6/8	Employees fill out forms and give filled-out forms to supervisor
6/11 - 6/15	Supervisors review employees' forms, respond in writing, and fill out ratings
6/18 - 6/22	Supervisors meet with each employee
6/25 - 6/29	Managers review completed forms, meet with supervisor and employee if needed
7/2 - 7/6	Managers meet with all supervisors in the department

HR 90-323

Figure 10–7 One column works well for memos and letters

The large bold type and the dark line make the title stand out.

The title tells readers what the document is about.

The headings are easy to see because they are in bold type, larger than the text type, and separated from the text by white space.

If you are interested in only one question, you can find that one easily.

The text of each section is easy to read because the lines are short.

U.S. Department of Transportation
National Highway Traffic Safety Administration

Office of Public Affairs
Washington, D.C. 20590
(202) 426-9550

Testing How Well New Cars Perform In Crashes

April 1983

What is the New Car Assessment Program?

The Department of Transportation has an experimental program in which it tests cars to see how well they perform during a crash. NHTSA (the National Highway Traffic Safety Administration) is the agency within the Department of Transportation that conducts these tests. NHTSA publishes this fact sheet to give consumers information that can help them to compare the relative safety of cars they may be planning to buy.

Every new car sold in the United States must meet minimum safety standards that are set by the Federal government. NHTSA's crash test program goes beyond these minimum standards. It tests cars at 35 mph, a much stricter test than the 30 mph test that is required by the standards. A 35 mph crash is about 35 percent more violent than a 30 mph crash.

The graphs that come with this fact sheet show how different makes and models compare in the way they perform during these high speed crashes. In interpreting these test results, it is important to remember two points:

(1) **Drivers and passengers should always wear safety belts.** The human-like dummies used in the crashes are wearing safety belts. These tests measure how cars perform, not how people perform. The crashes are intended to illustrate potential injuries to people who are **properly seated and belted** in the car. Fifty percent of the deaths from road accidents could be avoided if drivers and passengers wore their safety belts.

(2) **Large cars usually offer more protection in a crash than small cars.** These test results are only useful for comparing the performance of cars in the **same size class**.

What do the graphs show?

The graphs show how badly a person's head could be injured in a head-on collision between two identical cars, if both were going at 35 mph.

In general, the lower the score on the head injury criteria, the **less** likely drivers and front-seat passengers will be to be seriously injured or killed in a frontal crash at 35 mph. If the dummy in a car model scores substantially higher than 1,000 on the head injury criteria, human drivers and front-seat passengers in that model will be **more** likely to suffer a serious head injury or be killed in a frontal crash at 35 mph.

How does NHTSA select the cars it will test?

NHTSA tests about 25 cars every year. The graphs show the test results for a variety of cars. Because NHTSA buys its test cars off the lot, just as you would, the tests cannot begin until after a new model year has started. Test results on new cars generally begin to be available in the late winter of each year.

NHTSA chooses the cars it will test to give useful information to as many consumers as possible. For example, a popular model is more likely to be chosen, because information on that model would be of interest to many consumers. For the same reason, very expensive cars are not tested as often.

What happens when model years change?

If a car that NHTSA has already tested remains essentially the same for the new model year, it will probably not be tested again. But a car that undergoes substantial changes in the new model year will be likely to be retested. In using any crash test data, you should always check the model year that was tested and whether NHTSA believes that the test results should be used to evaluate other model years.

Figure 10–8 Two equal columns is a good format for this fact sheet (note: reduced 59% from original).
Source: Testing How Well New Cars Perform in Crashes (Washington, DC: Natl. Highway Traffic Safety Administration, U.S. Dept. of Transportation, Apr. 1983).

This is an excellent layout when readers will be skimming a document.

1 Getting Ready to Use PCF

Starting PCF

To begin using PCF, you have to bring up the screen called Personal Applications Manager, or P.A.M., that has PCF as one of the choices. Here's how you do that.

The headings in the margin make it easy to scan the pages to find a particular set of instructions.

The text is easy to read because each line is short.

If You Have a Fixed Disc

If PCF is installed on your fixed disc, you can follow the four steps in this section. If it isn't, *Using Your HP Touchscreen Personal Computer* will tell you how to install an application. If you don't want to install PCF on your fixed disc, you can run the program from a flexible disc. Put the disc in a drive after you have done Step 1.

Follow these steps to start the program:

1. Make sure the disc drives are turned on.
2. Turn on the computer.

   You will see the P.A.M. Main menu with PCF as one of the choices.

Screen pictures fit into the text column.

The white space in the margin can also be used for printed notes, warnings, and small pictures, or for readers to write their own notes.

Callouts in the margin point to relevant parts of the picture.

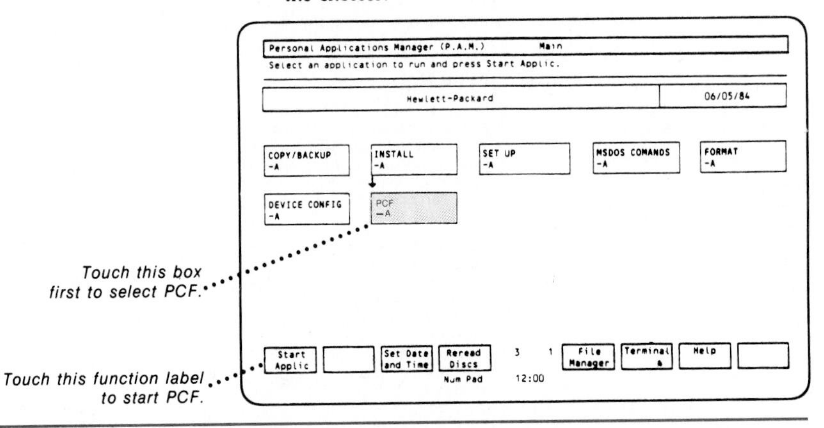

Figure 10–9 One column of text with headings in the margin works well for a computer manual

Source: From the tutorial for Personal CardFile, © 1984, Hewlett-Packard Company. Used with permission.

for margins makes the page look dense and hard to read. If you are putting your work in a binder, be sure to leave room for the binding. Don't punch holes through the writing. Think about whether a reader will want to punch holes in a copy later or put the work into a binder.

As you can see in Figure 10–10, you can use these guidelines for an 8½- by 11-inch page:

top margin	1 inch
bottom margin	1½ inches
left margin	1 inch, if not being bound 2 inches, if being bound
right margin	1 inch

If you are going to print or photocopy on both the front and back of the page, the binding will come at the left margin of odd-numbered pages, but at the right margin of even-numbered pages. Some word processing programs let you set the margins so that they alternate for right-hand (odd-numbered) pages and left-hand (even-numbered) pages. If you cannot set alternating margins, set both the right and left margin at about 1½ inches to allow for binding two-sided copies.

Figure 10–10 Page layouts showing margins for 8½- by 11-inch paper

You can use these margins if you know your paper will not be bound.

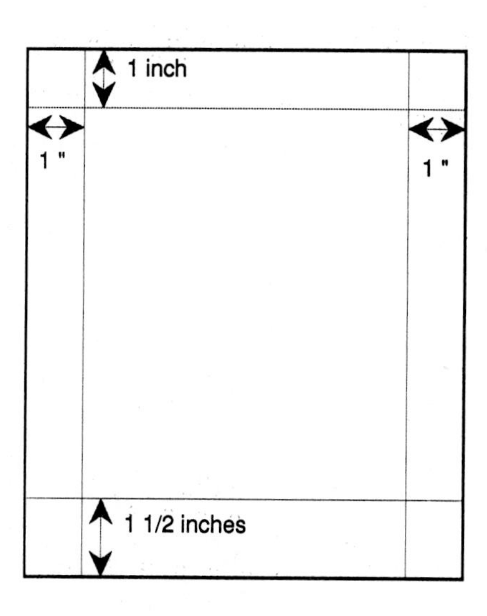

If your paper will be bound, leave extra room on the side that goes in the binding.

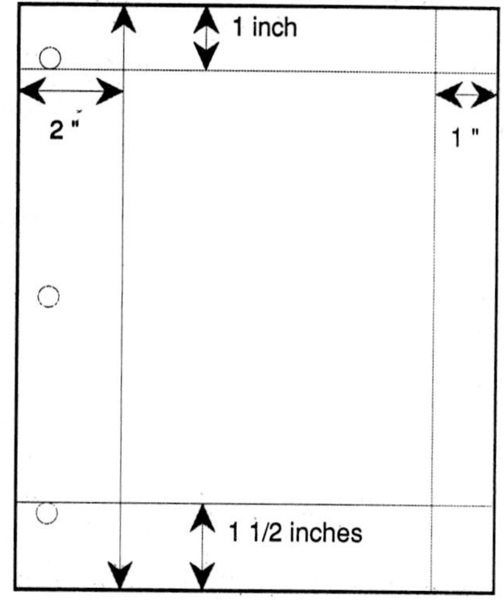

Use a Medium Line Length As you can see in the example, long lines of type are tiring to read. Moreover, readers are likely to lose their place in moving from the right margin of a long line back to the left margin of the next line.

Long lines of type are difficult for many people to read. Readers may find it difficult to get back to the correct place at the left margin. The smaller the type, the harder it is for most people to read long lines of type.

Very short lines are choppy to read. Readers have trouble keeping the sense of what they are reading. Very short lines also take up too much space.

Very short lines
take up too
much space
and make
comprehension
difficult.

Therefore, set the line length in a middle range.[3] The number of characters that you get in a certain amount of space depends in part on the size and style of the type that you are using. (See pages 239–247, later in this chapter, for more about choosing type sizes and styles.)

In general, keep the lines of type to about 50–70 characters, about 10–12 words. In a two-column format, keep each column to about 35 characters, about 5 words.

Set the Spacing for Easy Reading When should you use single-spacing? When should you use double-spacing? What other options are there?

It is common in the workplace to use single-spacing in the final copy of documents like letters and memos unless they are very short. When you use single-spacing, leave an extra space between paragraphs, as in this example:

Dear Mr. Linden:

The Testing Company is pleased to submit the enclosed proposal to conduct benchmark testing for Medical Company's new patient monitoring system. As you requested, our proposal includes costs for two rounds of testing.

Leave space between paragraphs

Our tests will give Medical Company the information that it needs to . . .

Drafts are usually typed in double-spacing because double-spacing is easier to read and gives editors and reviewers more room to write their corrections and notes. Your professor may ask you to use double-spacing even on your final copy for the same reasons. Sometimes, reports, proposals, manuals, and other iong documents in the workplace are kept in double-spacing so that people can scan them more easily.

When you use double-spacing, you have to show readers where a new paragraph starts. Your professor or your supervisor may tell you how to do this. If you aren't told which method to use, you may choose one of these methods:

- Indent the first line of each paragraph.
- Add one extra line between paragraphs.
- Add an extra double-space between paragraphs.

The last option leaves a lot of extra space, but in some word processing programs, it is very difficult to switch back and forth from double-spacing to single-spacing to get the extra single line between paragraphs.

On some typewriters and computers, you can adjust the spacing further. For example, you may have the option of putting one-and-a-half spaces between lines. Then, by pressing the carriage return or the enter key twice, you can leave three spaces between paragraphs. You may find that the result is a visually appealing and legible page that makes your document shorter than if you use double-spacing. Typesetting equipment and some desktop publishing systems allow even further adjustments. The key is to make the page look open and readable and to make the beginnings of each section and paragraph stand out.

Use a Ragged Right Margin Another choice that you have to make is how to line up the text vertically. By changing the alignment (justification) of the left and right margins, you can get four different styles, as in Figure 10–11.

This text is justified only on the left. This is the preferred style for most typewritten or word processed material.

This text is justifed on both the right and the left. This is typical of books and other printed materials.

This text is justified only on the right. This is highly unusual, except sometimes in headings.

This text has each line centered. Use this style only in headings.

Figure 10–11 The four ways that you can line up (justify) text

In technical writing, deciding what to do with the left margin is easy. Readers expect to see all the lines after the first one in a paragraph aligned neatly at the same place on the left. For the first line of a paragraph, you have two choices. Keeping the first line at the left margin is the typical format for letters, memos, and other short documents. Indenting the first line five spaces is the typical format for reports, proposals, and other long documents.

Block style

This is the typical paragraph style for short documents like memos and letters. All the lines in a paragraph, including the first, are at the left margin. This is called block style.

Indented style

This is the typical paragraph style for reports and other long documents. The first line of each paragraph is indented five spaces. In brochures and other documents with two or more columns, you can reduce the indent to three spaces.

For the right margin, the choice may not be as easy. Many word processing programs allow you to **justify** the type—that is, make all the lines end at exactly the same place on the page. Books are traditionally set with justified type, but now many documents, even typeset books, are being set with unjustified or **ragged** right margins

Be careful, however, in using justified type with your word processor. As the example below shows, justification can make text difficult to read.[4]

> To align the text along the right margin, the computer may work by inserting extra space between words. This can produce lines of text that are either unusually dense or unusually open. The unevenness inside the paragraph may be more difficult for readers than the unevenness at the end of the lines that you get with unjustified or ragged right type.

Even if your computer can microjustify—put the extra space evenly across the line so that you can't tell where it is—think about the purposes and audiences for your work. Justified type gives a document a formal tone; unjustified type gives a document a modern, friendly, more personal tone. There is some evidence that readers prefer unjustified text.[5] Poor readers may have difficulty reading justified text.[6]

Use White Space Don't think of the white space on the paper as just blank or empty. Think about how it can help your readers. To get your message across, you have to make your page look inviting. White space does that. The white space in the margins is important, but not enough. It's "passive" space; it just defines the block of the page that readers should look at. You also need to provide "active" white space inside the text.

You can create white space inside the text in several ways:

- Use headings frequently and surround them with white space. They make the structure of your document clear, and they make the sections easier to find. (See pages 248–258.)
- Use lists. They make your points or your instructions easier to find and follow. (See the next section.)
- Use pictures, tables, and other visuals. Visuals provide relief from the solid paragraphs of text. (See pages 330–332.)
- Separate paragraphs with an extra blank line or indent the first line of each paragraph.

The reference line in bold type summarizes the letter.

The white space makes the letter look inviting and easy to read.

Southern California Edison Company

September 19, 1990

Ms. Mary Smith
1234 Any Street
Culver City, CA 90230

RE: Account No. 62-42-421-3306-02

We are happy to provide you with a statement of your account.

Dear Ms. Smith:

The itemized statement of your account for electric service at 1234 Any Street, Culver City, that you requested is enclosed.

If you have any questions, please don't hesitate to call our Customer Service Specialist at (213) 402-2030.

Sincerely,

K. M. Brown

K. M. Brown
Customer Service Specialist

KMB:ep:BC-026
Enclosure

Figure 10–12 White space makes critical information stand out
Source: Used with permission of Southern California Edison Company.

Note how clearly the line that explains the subject of the letter stands out in Figure 10–12. The author used boldface and white space to make the one line of information easy to find.

Use Lists Listing is a very effective strategy for making information stand out on the page. Think about using lists when you are talking about a series of items or a set of conditions and when you are writing instructions.

Lists create active white space. They also break up the information so that it is both easier to remember and easier to find when the reader looks back at the page. These examples show the same instructions in paragraph form and in list form. Which set of instructions is easier to follow?

Poor style for instructions

Create a box by deciding where to put one corner of the box and moving the mouse to that position. Then press and hold the left mouse button. As you slide the mouse along the diagonal of the box that you are creating, you see the box appear on the screen. When the box is the size you want, release the mouse button.

Better style for instructions

To create a box:

1. Decide where to put one corner of the box.
2. Move the mouse to that position.
3. Press and hold the left mouse button.
4. Slide the mouse along the diagonal of the box. You see the box appear on the screen.
5. When the box is the size you want, release the mouse button.

When you are *writing* lists, keep these guidelines in mind:

1. *Write each item in the list so that it makes a grammatically correct sentence if read alone with the introduction.* In the following list, the writer has incorrectly written the sentence because the third item with the introduction reads: "To put together the bicycle, you will need **a two screwdrivers.**"

Poor

To put together the bicyle, you will need a

- wrench
- pliers
- two screwdrivers

The list would be grammatically correct if the writer had written it this way:

Better

To put together the bicycle, you will need

- a wrench
- pliers
- two screwdrivers

To check your lists, reread the introduction with each item in turn.

2. *Keep the items in the list parallel.* Use the same syntax for each item, unless you have a strong reason for not doing so. Which of the following lists can you read and remember more easily?

Poor

In case of fire:

- You should stay calm.
- The hotel operator must be called.
- A wet towel will help if it is stuffed under the door.

Better

In case of fire:

- Stay calm.
- Call the hotel operator.
- Wet a towel and stuff it under the door.

When you are *formatting* lists, keep these guidelines in mind:

1. *Indent the list.* Indenting sets the list off from the surrounding text.
2. *Use bullets for lists where the order is not critical.* This is a bullet •. You can create bullets on a typewriter by using a small letter o and filling it in later with ink. Some computer programs and printers support symbols, including filled-in circles or small squares. This book, for example, uses square bullets.

For lists inside of lists, use a dash, –, as in this example:

Good style

- two screwdrivers
  - a standard, slotted screwdriver
  - a Phillips-head screwdriver

3. *In reports, use numbered lists where you want people to remember the number of items.* In instructions, use numbered lists for the steps in each procedure. If you need a second level inside a numbered list, use small letters (a, b).
4. *Keep all the lines of each item in the list at the same indented margin.* Readers will more easily see where the list starts and ends if the numbers stand out together from the rest of the text, as in this example:

Let the numbers stand out

To save your graph:

1. Press Ctrl – S.
   A small window opens up with a place for you to name the graph.
2. Type a name for the graph. The name can be from one to eight characters long.
3. Press Enter.
   Grapher saves your file and then shows you the message: *File saved.*

Lists have an important place in technical writing, but don't overuse them. Think about your readers and how they will use the document. For example, readers may expect more lists in an instruction manual than in a report. In an instruction manual, a list is appropriate for a procedure even if the procedure only has two steps. In a report, a sentence with just two conditions might be more effective in connected prose than as a list.

Choosing Readable Type

Today's technology gives you many choices for showing text. Even many typewriters today allow you to change the type size and style—the way the letters look on the page.

On the job, you may or may not be responsible for choosing the size and style of type for your documents. Graphic artists and book designers, who are specialists in typography and design, may have those responsibilities. If you can, work with a graphic artist or book designer; the team approach is often the most effective way to produce a useful document. To make sure that your document meets your purposes for your readers, however, you have to be able to talk intelligently with these specialists. This section will help. It will also help you if you are responsible for all the aspects of your document.

To choose readable type:

- Choose a legible type size.
- Choose an appropriate typeface.
- Use highlighting effectively.
- Use lowercase letters.
- Use color carefully, if it is an option.
- Use special typefaces sparingly.
- Don't overdo it.

Choose a Legible Type Size Word processing programs usually measure type size with the graphic designer's term, "points." A **point** is 1/72nd of an inch. A letter in 36-point size is about ½ inch high. Many word processing programs offer a wide choice of point sizes, as you can see in Figure 10–13.

This is 8-point type.

This is 10-point type.

This is 12-point type.

This is 14-point type.

This is 18-point type.

This is 24-point type.

This is 36 point.

Figure 10–13 Type comes in different sizes

In general, use 9- to 12-point type for regular text.[7] Use larger sizes for titles and headings. (See pages 253–255.) If you use 12-point type for the regular text, you may want to use 10-point type for footnotes. Type that is smaller than about 8 points is difficult to read. Point size, by itself, isn't enough to ensure that you have legible type. The actual size of the type also depends on the typeface, the printer, and the way that the software and printer work together. With some typefaces on some computers, you may need to use 14-point type to make the material easy to read.

Point size is measured from the top of the tall letters like *h, k, l,* and *t* to the bottom of the long letters like *g, p,* and *y*, as in Figure 10–14.

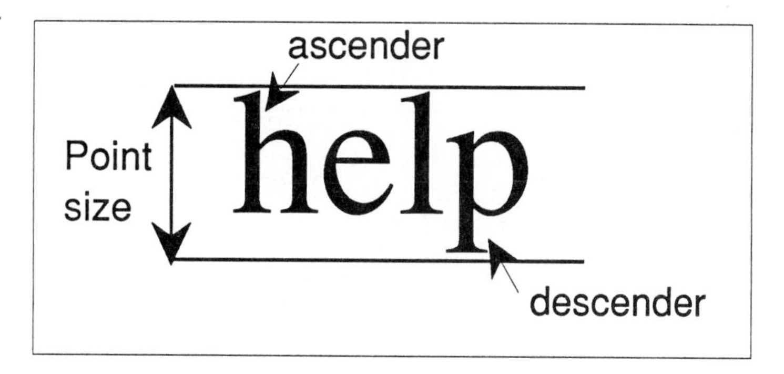

Figure 10–14 Type is measured from the top of ascenders to the bottom of descenders

You may have already noticed on your computer that two different typefaces in the same point size take up different amounts of space on a page. The way a typeface looks on the page also depends on the "x-height," the height of a lower case letter, like *x*. Look at Figure 10–15 and notice how much higher the top curve of the *h* is in the word on the left than it is in the word on the right. Both words are in the same point size, but they are in different typefaces.

With desktop publishing software, you can make other changes to type to move letters closer together or further apart.

Because of its smaller x-height, Times Roman gets more words on a line than Helvetica does.

64-point Helvetica

64-point Times Roman

Figure 10–15 Point size isn't enough to tell you how large a particular typeface will be on the page.

Choose an Appropriate Typeface On many computers, you can choose not only the size but also the style of type, called the typeface, or the **font.** There are hundreds of type fonts, but they break down into two main families, serif and sans serif. **Serif type** has extenders on the letters. *Sans* is the French word for *without.* **Sans-serif type** does not have the extenders. You can see the difference in Figure 10–16.

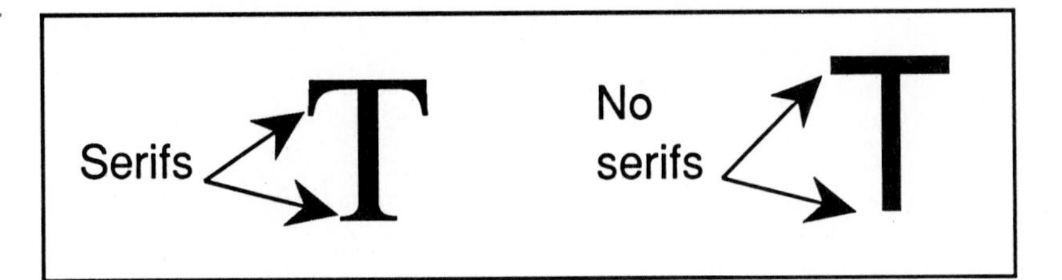

Figure 10–16 Type comes in two major styles: serif and sans serif (without serifs)

The serifs draw the readers' eyes across the page, so most books and other long documents are printed with serif type. However, sans-serif type can also be very readable, particularly if you use space-and-a-half or double-spacing.[8] Sans-serif type is often used in brochures and advertisements because it gives a document a more contemporary look. Sans-serif type also works well in the visuals for presentations. (See Chapter 19, "Oral Reports.") Figure 10–17 shows examples of some traditional typefaces and some of the newer ones that are available on computers like the Macintosh. Each name is printed in the typeface it names. Note which ones have serifs.

Figure 10–17 Some traditional and new typefaces (fonts) on the Macintosh computer

Serif	Sans-Serif
Times	Helvetica
Palatino	Geneva
New York	Chicago
Schoolbook	Helvetica Narrow
Courier	Monaco

Choose a typeface that is appropriate for your purposes and readers. Then stay with that typeface for the entire document. Use other features, such as type size, boldface, and italics to indicate differences. In selecting both typeface and point size, consider the system you are using. Type can be adjusted in many ways, so the output from a word processor or desktop publishing system depends on the way that the software and hardware are set up. Figure 10–18 shows the same sentence in the same typeface on three different computer systems. The best way to plan an important piece of work is to get a sample of the choices that are available to you and see just how they will look.

From WriteNow on the NeXT computer

This is 18-point bold Times Roman type.

From WordPerfect on an IBM computer and a Hewlett-Packard LaserJet II printer

This is 18-point bold Times Roman type.

From Microsoft Word on a Macintosh computer and an Apple Laser printer

This is 18-point bold Times Roman type.

Figure 10–18 The actual size and spacing of the type depend on the way your software and printer are set up

Use Highlighting Effectively There are many ways to draw the reader's attention to the points you want to emphasize. We have already seen two of them: using white space to set off elements and changing the size of the type.

With a typewriter, you can use:

underlining

ALL CAPITALS

Special characters before and after

"quotation marks"

A line above the type--for a heading

Lines above and below the sentence, for example, for a warning.

With a computer, you can use all of these, and also:

boldface *italics* SMALL CAPITALS

Boldface is a stronger highlighting technique than *italics*. Both are stronger than underlining. SMALL CAPITALS might have a place in an official document, if you need a typeface for specially defined words; but, otherwise, they are seldom used.

The most important guidelines to remember for using highlighting are these:[9]

1. *Don't overdo it.* Highlighting calls attention to special features. If you use too many different kinds of highlighting or use one kind too often, you dilute the forcefulness of the highlighting.
2. *Be consistent.* Highlighting helps readers find their way and understand the different kinds of information that you have in your text. To achieve these goals, however, you must use the same highlighting for the same feature throughout the text. If you use italics for specially defined words, use italics throughout. If you set warnings and cautions off with an indent and lines above and below, make sure that all warnings and cautions look like that.
3. *Match the highlighting to the importance of the information and to your readers' expectations.*

Large, bold headings stand out

Boldface works well for making a **word,** short phrase, or short sentence **stand out.**

Italics work well for emphasizing single words, such as *special.*

Italics are correct for book titles, such as *Reporting Technical Information.*

On a typewriter, use underlining or quotation marks for emphasizing special words, such as special or "special."

On a typewriter, use underlining for book titles, such as Reporting Technical Information.

4. *Don't use any of these highlighting techniques for more than a short sentence at a time.* Whole paragraphs in boldface, italics, underlining, or small capitals are difficult to read. Figure 10–19 shows a page that uses highlighting well.

The colored rule (line) over the heading marks a new section.

Creating regular text

Regular text goes inside fields that you've created. To create text, first you use the Browse tool to set the insertion point inside the field you want; and then you type.

Figure 5-12 Selecting the Browse tool

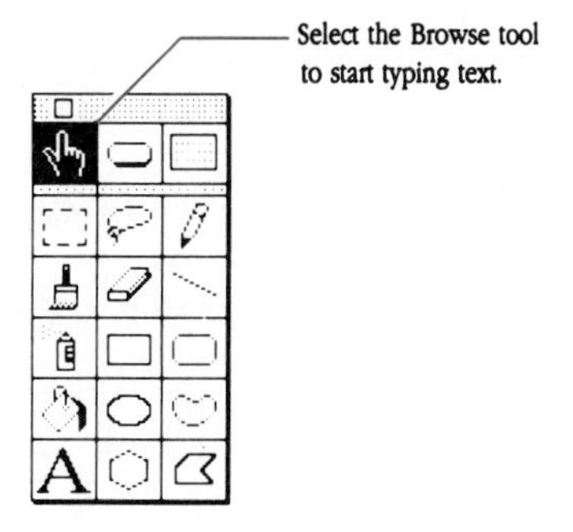
Select the Browse tool
to start typing text.

The triangle in the margin highlights important information.

△ **Important** Use the Browse tool, not the Paint Text tool, for entering text in fields. Regular text is always editable in an unlocked field; Paint text, however, is not editable once you have typed it and clicked the mouse button. △

Entering text

You use the keyboard to enter text in HyperCard. You must create a field before you can type text. (See "Creating a Field," earlier in this chapter.) The following steps explain how to enter text in a field:

Bold is used to make headings and instructions stand out.

1. **Select the Browse tool from the Tools menu.**
2. **Click in the field where you want to type.**

   (The insertion point appears at the left edge of the line you clicked.)

Figure 10–19 An example of useful highlighting in a software manual
Source: Macintosh® Hypercard® User's Guide, © 1989, Apple Computer, Inc. Used with permission.

Use Lowercase Letters Don't use all capitals for text. We use the shapes of the ascenders and descenders on lowercase letters to help us as we read.

Difficult to read

CAPITAL LETTERS GIVE US NO CLUES TO DISTINGUISH ONE LETTER FROM ANOTHER. THEREFORE, THE LETTERS BLUR INTO EACH OTHER VERY QUICKLY, AND WE WANT TO STOP READING. LOWERCASE LETTERS GIVE US CLUES TO THE SHAPES OF THE WORDS, AND WE USE THOSE SHAPES AS WE READ.

Easier to read

Capital letters give us no clues to distinguish one letter from another. Therefore, the letters blur into each other very quickly, and we want to stop reading. Lowercase letters give us clues to the shapes of the words, and we use those shapes as we read.

A sentence in all capitals takes about 13% more time to read than a sentence in the regular uppercase and lowercase letters that we expect. All capitals also take up about 30% more space on the page if you are using a typeface with proportional spacing.[10] Don't even use all caps for headings if you have other choices. Research shows that boldface works better than all caps for headings.[11]

Use Color Carefully, If It Is an Option Color won't be an option for most situations, but desktop printers and photocopiers that can produce different colors are becoming more common. Don't use color for text. High contrast between ink and paper is necessary for easy reading. Black ink on white paper provides the best contrast.

Color is not commonly used in technical reports, proposals, or feasibility studies. In annual reports, brochures, and advertisements, you will probably want full four-color art, which you should plan together with a designer. In manuals, one color in addition to the black ink that you use for the text can be very effective for titles, major headings, and specific features that you want to highlight.

Colored paper may be more of an option for you than colored ink. For example, in a long report, you may want to photocopy or print the appendixes on a different color paper to separate them more strongly from the body. Choose a light shade of the color to keep the contrast between ink and paper high. Strong colors may look garish. Dark yellow paper does not photocopy well.

Use Special Typefaces Sparingly Some computers offer unusual typefaces, as you can see in Figure 10–20. Use these sparingly, if at all, in reports and most other types of technical writing. They may be more appropriate in special short pieces, such as advertising, flyers, invitations, or brochures. Type carries messages; think about the message that your choice of type is sending.

THIS FONT LOOKS LIKE OLD-FASHIONED COMPUTER WRITING.

This might be appropriate for invitations.

Use outline fonts sparingly.

This looks like Ye Olde Font.

Figure 10–20 Unusual typefaces should be used with caution and only in appropriate situations
Source: Fonts from Arts & Letters, Computer Support Corporation, Dallas, TX, 75244.

Don't Overdo It The golden rule in document design is to make the page clean, clear, and consistent. Many writers, given the new options of word processing software and desktop printers, tend to put too many different features together. Keep it simple.

Printing Your Document

Setting up the format so that your document will be useful and readable is important. Paying attention to the printing is also important so that all the work you did in setting up the format looks as good as it can.

Make sure that you have a clean ribbon in your typewriter or dot-matrix printer or a clean cartridge in your laser printer. If the type is too light or not even across the page, change the ribbon or cartridge.

Use good paper so that the ink does not smudge. If you are typing the paper, take the time to retype a page if the erasures make it look messy. Remember that appearance is the first impression the reader has and that appearance conveys a message about how much you care for your readers.

Helping Readers Locate Information

Appearance and arrangement—design and organization—go hand in hand. To help your readers follow your document and find what they need quickly, you have to plan a useful structure for the document (organize it well) and show that structure to the readers (design it well).[12]

As we pointed out earlier in the chapter, most people read technical and business documents selectively. They may glance over the table of contents to see what the document is about and then pick and choose the sections to read by looking for headings that match their needs and interests. They may skim through the pages, stopping when a heading or example or table strikes them as relevant and important. They may need to go back to the document later to check specific facts—and they want to find the relevant pages quickly.

In the previous section of this chapter, we showed you how elements of page layout and typography can help people find information quickly. In this section, we show you how you can help your readers by giving them clues to the way that you have arranged your document. These clues include:

- Headings
- Headers and footers
- Page numbers
- Paragraph numbering, if you are required to use it

Writing and Designing Useful Headings

Headings are the short lines—the titles—that you put above each section and subsection. Even short documents, like memos, instructions, and letters, can benefit from headings. In longer documents, headings are essential to break the text into manageable pieces. Furthermore, in longer documents, the first few levels of headings become the table of contents.

The headings come from your outline. As you write, however, you may want to reorganize your material or rewrite the headings. You may want to divide the material with more headings than you have in the outline. You will probably also drop the numbers or letters from the outline and keep only the words in the headings. (See pages 260–261 about numbering systems in headings.)

The headings are the road map to your document. You have to think about both how you word them and how you incorporate them into your page design. Here are some guidelines both for writing useful headings and for designing them well.

Make the Headings Informative Generic headings, like "Part I" or "Section 2," give readers no clues to the content of your work. Make your headings carry weight; they can help to tell your story.

The best way to write headings is to put yourself in your reader's place. You can use questions, verb phrases, whole sentences, or key words or phrases as headings.

1. *Questions as headings.* In memos, brochures, fact sheets, policy and procedures manuals, and sometimes in letters, questions work very well as headings. Readers come to these documents with questions in mind; the document answers their questions. For example, a fact sheet or booklet might have headings like these:

Useful headings in a brochure

What does the gypsy moth look like?
How can we protect trees from gypsy moths?
How often should we spray?

or like these:

Useful headings in an information sheet

What is the student loan program?
Who is eligible for a loan?
How much money can I borrow?
How do I apply?

2. *Verb phrases as headings.* In instructions and in the action parts of proposals, verb phrases make excellent headings, as in this example:

Useful headings in instructions

Adding a graphic
Selecting the data
Selecting the type of graph to use
Labeling the axes
Adding a title
Adding a legend

3. *Whole sentences as headings.* If the thrust of your document is a set of guidelines or recommendations, making the guidelines or recommendations into the headings can be very effective.
4. *Key words as headings.* Long headings aren't always necessary. Single nouns or strings of nouns may work well if you are writing a standard document for readers who expect a certain set of headings in a specific order. A proposal for a student project might have this set of headings:

Useful headings if the audience expects them

Project Summary
Project Description
Rationale and Significance
Plan of Work
Facilities and Equipment
Personnel
Budget

A key word or phrase may be sufficient if the rest of a question or statement carries no real information or if all the headings in a section would repeat the same words.
Instead of

Poor

Selecting the paint to use
  Selecting the paint to use for interior jobs
  Selecting the paint to use for exterior jobs

you might have

Better

Selecting the paint to use
  For interior jobs
  For exterior jobs

Be Careful with Nouns as Headings Be careful when using single nouns or strings of nouns as headings, however. In many cases, the single noun or noun string does not make an informative heading. In a study of a government regulation, even readers who were the appropriate technical audience for a document had trouble predicting the information that would follow each key word heading. They also could not match the headings and the information in the sections when they were given a list of headings and the rest of the document without the headings. When the headings were rewritten as questions, the readers were much more successful at both tasks, predicting the text from the headings and matching the headings with the text.[13]

Headings that are single nouns or noun strings may be ambiguous, overly technical, or too general. Consider the headings in the following example. Are they clear to you?

Nouns make poor headings

INSURANCE POLICY
  Coverage
  Designation
  Liability
  Payment

The heading "Payment" is ambiguous. Does it mean payments from you to the insurance company or from the insurance company to you?

The heading "Designation" may be too technical. What information would you expect under "Designation"? You are part of the audience for insurance policies. If you don't understand the headings in an example like the one here, the headings aren't effective. "Designation" is an unnecessarily technical term for "naming someone to get the money if you die."

The heading "Coverage" may be too general. Where would you look for information on circumstances that are not covered? Wouldn't the policy be much easier to follow if it had separate headings for each major situation that is covered and that is not covered?

Keep the Headings Separate from the Text The heading is not part of the sentence following it. It stands outside the text. An example of poor writing would be

Poor

Point size. This is the way that type is measured.

The reader cannot understand the sentence by itself because "this" refers to the heading "point size." Repeat the key words somewhere in the first sentence that follows the heading so that the text is independent of the headings:

Better

Point size. Type is measured in point size.

Keep the Headings at Each Level Parallel Patterns help readers see relationships. Patterns also reduce the mental energy that readers have to expend to understand. The parallel headings in the second of these two examples are easier to grasp quickly.

Poor—not parallel

Graph Modifications
- Data selection updating
- To add or delete columns
- How to change colors or patterns
- Titles and legends can be included

Better—parallel and begins with action verbs

Modifying a graph
- Changing the data
- Adding or deleting columns
- Changing the colors or patterns
- Adding titles and legends

Parallelism is a very powerful tool in writing. (See pages 611–612 in "Part VI, Handbook.") A good way to check the parallelism of your headings is to write them out as a table of contents. When you have just the headings without any of the text, you can quickly see if your headings are parallel.

Indent each level as you would in an outline so that you can see the way you have actually done each level of heading. That way, you can also make sure that you have at least two headings at each level in each section. This rule holds for headings in a document just as it does for entries in an outline. (See pages 609–611.)

Some word processing programs allow you to mark headings so that the program collects and formats them as a table of contents. You set up the way that you want your table of contents to look. Then, you mark each heading with special tags to indicate that it is a heading and which level it belongs to. When you want to see the table of contents, you select the table of contents function. The program collects all the headings that you marked and formats them in the way that you specified. If your word processing program includes this function, use it with drafts to check the parallelism of your headings.

You can also collect the headings to check them and to create a table of contents by copying them one-by-one to a separate file or to the beginning or end of your text file. You can do this in most word processing programs by selecting (highlighting) the heading or by designating it as a "block" and then copying the highlighted "block." Copying in most word processing programs leaves the original just where it is and puts a second copy wherever you designate. "Block and copy" the headings one-by-one in order from the beginning to the end of your draft. Copy all the formatting features with the headings so that you can also check the consistency of your formatting.

Design the Headings to Show the Levels of Information The headings do more than outline your story. They also help people find specific parts quickly, and they show the relationships among the parts. To help people find specific parts quickly, the headings have to stand out from the text. To show the relationships among the parts, each level of heading has to be easily distinguished from the others. Design serves both these functions.

How Many Levels of Headings Should You Have? Student papers, and almost all business documents, rarely need more than four levels of headings. Readers may get lost and confused in a document that goes deeper than four levels of headings.

Memos and letters may have no headings at all. If they do, like brochures, fact sheets, and other short documents, they should not need more than one or two levels of headings. Reports, proposals, manuals, and other long documents should need no more than three or four levels of headings. If you find that you need more for a chapter of a long document, consider breaking the material into two chapters.

How Often Should You Have a Heading? The frequency of headings in a document depends on the purpose and audience for the document. In a short brochure, you might want a heading on every paragraph or every panel. In a report, you probably want a heading for every subsection; that might cover two or three paragraphs. In general, you want to have clues to the text's arrangement on every page.

What Features Can You Use to Show the Hierarchy of Headings? You can easily show four levels of headings. To show relationships, you can vary several features.

You can vary the size of the type:

Very Large Type

Large Type

Regular Size Type

You can vary the weight or style:

boldface

italic

underlining

color

You can vary the placement on the page:

Centered Heading

At the Left Margin

Indented and Above the Text
Text, text, text . . . .

Indented and run-in. Text, text, text . . .

You can vary the capitalization scheme:

ALL CAPITALS FOR ALL WORDS

Initial Capitals for Important Words

Sentence style, first word capitalized

You can vary the spacing of the letters:

REGULAR SPACING

E X P A N D E D C A P I T A L S

You can use other features, such as a line above or below the heading:

Main Heading

Second level heading

Note: Graphic designers call these lines, "rules."

Figure 10–21 shows how you can construct four levels of headings even with a typewriter. Figure 10–22 shows how you might do these same four levels of headings with a computer.

In planning your headings, keep these guidelines in mind:

1. *Use your readers' expectations about hierarchies.* We all think that bigger means more important. First-level headings should be larger than second-level and so on down to the level that is the same size as the text. Headings should never be smaller than the text. Similarly, most people see all-capital-letter words as being more important than initial capitals or sentence style. If you use all capitals, use it for the highest level or levels.
2. *Distinguish headings from the text in a uniform way.* Each level of heading has to be distinct from the others, but they should all say "heading" as opposed to "text." For example,

Level 1: centered, expanded capitals	C O N T R O L L I N G S O I L - B O R N E P A T H O G E N S I N T R E E N U R S E R I E S
Level 2: centered, non-expanded capitals	TYPES OF SOIL-BORNE PATHOGENS AND THEIR EFFECTS ON TREES Simply stated, the effects of soil-borne pathogens . . . .
Level 3: underlined, at the left margin, initial capitals	The Soil-borne Fungi At one time it was thought that soil-borne fungi . . . .
Level 4: underlined, run-in, first word capitalized	Basidiomycetes. The basidiomycetes are a class of fungi whose species . . . . Phycomycetes. The class phycomycetes is a very diversified type of fungi. It is the . . . . The Plant Parasitic Nematodes Nematodes are small, nonsegmented . . . . TREATMENTS AND CONTROLS FOR SOIL-BORNE PATHOGENS . . . .

Figure 10–21 Four levels of heading that you can create on a typewriter

putting all the headings in boldface is one way to distinguish the headings as a group from the text. Then you can use size, placement, and capitalization to distinguish one level of boldface heading from another. This book, for example, uses boldface and color to distinguish headings from text.

3. *Don't overdo it.* A heading in all capitals doesn't also need underlining. A run-in heading in color doesn't also need italics. Using too many cues in the same heading may confuse readers.[14]

Level 1: 18-point boldface, centered initial capitals

Controlling Soil-borne Pathogens in Tree Nurseries

Level 2: 14-point boldface, at the left margin, initial capitals

Types of Soil-borne Pathogens and Their Effects on Trees

Simply stated, the effects of soil-borne pathogens ....
....

Level 3: 11-point boldface, at the left margin, first word capitalized

The soil-borne fungi

At one time it was thought that the soil-borne fungi..................................
....

Level 4: 11-point (same as the text), boldface, run-in, first word capitalized

Basiodiomycetes. The basidiomycetes are a class of fungi whose species
....

Phycomycetes. The class phycomycetes is a very diversified typed of fungi. It is the ....

The plant parasitic nematodes

Nematodes are small, nonsegmented ....
....

Treatments and Controls for Soil-borne Pathogens

....
....

Figure 10–22 Four levels of heading that you can create on a word processor. This format might work well in a report. You might have a different format for a software manual.

4. *Put more space before the heading than after it.* You want readers to grasp the heading and the section that it goes with as a unit. If the spacing above the heading is the same or even less than the spacing below the heading, readers won't know how to relate the heading to the text. If you use a rule (line) as a feature of your page layout, consider putting the rule above the heading to emphasize that the heading goes with the text below it. This last guideline is most useful in manuals where people jump into the text to find a specific section quickly.
5. *Keep the heading with the section that it covers.* Don't leave a heading at the bottom of a page when the text appears on the next page. Make sure you have at least two lines of the first paragraph on the page with the heading. Some word processing programs let you protect the heading with the following text as a "block" that moves together. Highlight the heading and the first two lines of the following paragraph as a "block." Then choose the function that protects the block. If the page break comes in the middle of the block, the word processing program moves the entire block to the next page. Using this function is a better way to ensure good page breaks than adjusting each page in the draft yourself by adding extra lines or setting hard page breaks. If you adjust the page breaks yourself in one draft, you will have to check for extra spaces as well as poor page breaks every time you add or delete any text.

Numbering the Pages

Page numbers help readers keep track of where they are and provide easy reference points for talking about a document. Always number the pages of your drafts and final documents.

Brief manuscripts and reports that have little prefatory material almost always use Arabic numerals (1, 2) in order from the first page to the last. The commonly accepted convention is to center the page number below the text near the bottom of the page or to put it in the upper right-hand corner. Always leave at least one double-space between the text and the page number. Put the page number in the same place on each page. Page numbers that are centered near the bottom of the page often have a hyphen on each side, like this:

- 17 -

As reports grow longer and more complicated, the page-numbering system also may need to be more complex. If you have a preface or other material that comes before the main part of the report, it is customary to use small Roman numerals (i, ii, iii) for that material and then to change to Arabic numerals for the body of the report. The introduction may be part of the prefatory material or of the main body. The title page doesn't show the number but is counted as the first page. The following page is number 2 or ii.

When numbering the pages, you have to know if your document is going to be printed or photocopied single-sided or double-sided. If both sides of the paper will have printing on them, you may have to number some otherwise blank pages in your word processing files. New chapters usually start on a right-hand page, that is, on the front side of a page that is photocopied on both sides. The right-hand page always has an odd number. If the last page of your first chapter is page 9, for example, and your document will be photocopied double-sided, you have to include an otherwise blank page 10, so that the first page of your second chapter will be a right-hand page when the document is bound.

In reports, the body is usually paginated continuously from page 1 to the last page. For the appendixes, you may continue the same series of numbers or you may change to a letter-plus-number system. In that system, Appendix A is numbered A–1, A–2; Appendix B is numbered B–1, B–2. If your report is part of a series or if your company has a standard for the format of reports, you will need to make your page numbering match that of the series or standard. If you can choose how to paginate the report, you may want to use the letter-plus-number system for the appendixes.

Numbering appendixes with the letter-plus-number system has several advantages.

- It separates the appendixes from the body. Readers can tell how long the body of the report is as well as how long each appendix is.
- It clearly shows that a page is part of the appendixes and which appendix it belongs to. It makes pages in the appendixes easier to locate.
- It allows the appendixes to be printed separately from the body of the report. Sometimes, the appendixes are ready before the body of the report has been completed. Printing the appendixes first may help you meet a deadline.
- It allows changes to either the appendixes or the body without disturbing the other parts.

In large manuals, you will often see page numbering by chapter. Pages in the first chapter are numbered 1–1, 1–2; pages in the second chapter are numbered 2–1, 2–2. Numbering by chapter-plus-page has the same advantages as numbering the appendixes by the letter-plus-number system. Chapters can be printed as they are completed. You can change one chapter without having to change the entire document. If your table of contents and index do a good job of letting readers know which chapter they need, the page numbers that show the chapter number as well as the page number help readers locate information quickly.

If an appendix is a reprint of another document that already has page numbers, show both the original page number and the one for your document. Show the original without brackets and put the new page number in brackets, like this:

128

[A–26]

Including Headers or Footers

In multipage documents, it helps readers if you give them information about the document as well as the page number at the top or bottom of each page. If the information is at the top of the page, it is a **header.** If it is at the bottom of the page, it is a **footer.** Many word processing programs allow you to set up headers and footers so that they appear automatically in the same place on each page. You can include the page numbering and set it so that the computer numbers the pages in order as it prints. Designers sometimes call headers, "running heads," and call footers, "running feet."

With some programs, you can set up alternating headers or footers so that you get one on odd-numbered pages and another on even-numbered pages. If the page number is part of the header, it always goes nearest the edge of the page. For example, in this book, each left-hand page has a header with the page number followed by the title of the current part. Each right-hand page has a header with the chapter title followed by the page number.

A typical header for a *report* might look like this:

Jane Fernstein
Feasibility Study
June 1991

A typical header for a *letter* might look like this:

Dr. Thomas Menn -2- October 10, 1991

The header does not appear on the title page of the report or on the first page of a letter. Most word processing programs allow you to start the header on the second page. Either you put the header at the beginning of the file and then "suppress" it for the first page, or you put it in the file at the top of the second page.

Using Numbers with Your Headings

In many companies and agencies, the standard for organizing reports and manuals is to include a numbering system with the headings. The three systems in most common use are:

- The traditional outline system
- The century–decade–unit system (often called the Navy System)
- The multiple-decimal system

Figure 10–23 shows the three systems.

The rationale for these systems is that you can then refer to a section elsewhere in the report by its number. The numbering systems, however, have several disadvantages. In all these systems, if you want to add or remove a section, you have to renumber at least part of the report. Unless you have software that does this for you automatically, renumbering is tedious and highly susceptible to error. Many readers find it very difficult to follow these numbering systems, especially if you need more than three levels. The multiple-decimal system is particularly difficult for most people to use.

Moreover, better alternatives are available now. You can show the hierarchy of heading levels distinctively with changes in type size and placement instead of numbers. Then the numbering is redundant to the design and wording of the headings.

Unless you need extensive cross-references, readers should be able to find the section you are referring to if you give the section name and page number. Some software allows you to set up cross-references so that the computer finds and inserts the correct page number for each reference. If you have that feature in your word processing program, the computer will change the page numbers for the cross-references as they move with changes in the document. If you are using so many cross-references that finding them would be a chore without a numbering system, consider reorganizing your document.

If you are not required to use a numbering system, we suggest that you not institute one. However, many government agencies and many companies, especially those that prepare documents for the government, require one of these numbering systems. Therefore, it pays to be familiar with them.

If you use a numbering system with your headings, you must also put the numbers before the entries in your table of contents. You may be expected to use the numbers only with the discussion elements of your report. It is also acceptable, however, to use numbers with other major elements, such as introductions, conclusions, recommendations, and references.

Traditional outline system

```
TITLE
 I. FIRST-LEVEL HEADING
    A. Second-Level Heading
       1. Third-level heading
       2. Third-level heading
    B. Second-Level Heading

II. FIRST-LEVEL HEADING
    A. Second-Level Heading
       1. Third-level heading
       2. Third-level heading
    B. Second-Level Heading
```

Century-decade-unit system

```
TITLE
100 FIRST-LEVEL HEADING
    110 Second-Level Heading
        111 Third-level heading
        112 Third-level heading
    120 Second-Level Heading
200 FIRST-LEVEL HEADING
    210 Second-Level Heading
        211 Third-level heading
        212 Third-level heading
    220 Second-Level Heading
```

Multiple-decimal system

```
TITLE
1 FIRST-LEVEL HEADING
  1.1 Second-Level Heading
      1.1.1 Third-level heading
      1.1.2 Third-level heading
  1.2 Second-Level Heading
2 FIRST-LEVEL HEADING
  2.1 Second-Level Heading
      2.1.1 Third-level heading
      2.1.2 Third-level heading
  2.2 Second-Level Heading
```

Figure 10–23 Three types of numbering systems

Using Technology for Word Processing

A computer can be a very useful tool for technical writing, but it is still just a tool. It doesn't replace you as the writer. In this last section of the chapter, we give you some suggestions for making good use of the technology.

The Advantages of Using Word Processors

Word processors and word processing software on a computer offer many advantages.

You Can Edit and Revise Easily If you change your mind about what you want to say, you can type over what you have written. You can add to your text at any point. You can delete words, sentences, even whole paragraphs or pages. Some programs have an "undo" feature that allows you to get back the last piece (or the last several pieces) that you deleted if you change your mind again. One major advantage over using paper-and-pencil or a typewriter is that, no matter how many changes you make, your copy always looks clean.

You Can Move Text Around Easily With many word processing programs, you can "cut and paste" so that reorganizing is easy. You can move words, sentences, paragraphs, or larger blocks. You can see how a section reads with one organization, try it in a different order, and then go back to the first if that was better.

You Can Copy Material from One Place to Another If you can cut and paste, you can probably also copy material from one place in your paper to another. In many programs, when you copy material, you copy both the text and the way that the text is formatted. This can be a handy feature, for example, for making sure that you use the same format for all the headings at a certain level. If you have set up a way to do the heading in one place but have trouble remembering how you did it, copy the first heading to the place for the second one and then edit it.

You Can Save What You Have Typed and Work on It Again Later On a typewriter, if you stop in the middle of a page, it is difficult to put the page back into the typewriter later and get it lined up exactly as it was before. With a computer or word processor, *writing* and *printing* are two separate stages of preparing your paper.

You Can Save Different Drafts of the Same Text You can save several versions, even on the same diskette, as long as you give each draft a different name. For example, if you are not sure what approach to take in the early stages of writing a report, you might try a few different approaches. You can save them all. You might want to save the drafts of an assignment to compare the final version to an earlier version. If you are working on a collaborative writing assignment, each writer can prepare and save his or her part, and then the group can put the entire assignment together on the computer.

You Can Search and Replace Words and Phrases Another advantage that many word processing programs give you is the ability to search for specific words and phrases and even to replace them one at a time or all at once. For example, if you realize that the *Mr. Smith* you have been referring to actually spells his name *Smythe,* you can use the search and replace feature to change all the occurrences of *Smith* to *Smythe.*

You can use a search and replace feature in interesting ways. Here are a few:

1. *You can find out more about your own writing style.* Do you overuse "however" or "moreover" or "therefore"? You can search for the occurrences of each of these words and see how often you are using them.
2. *You can check on some specific common errors.* Do you sometimes type a possessive when you mean a plural? That is, do you sometimes type "student's" when you mean "students"? A program that checks for spelling errors (see pages 31–33) won't find this type of error because "student's" is an English word; it's just not the right one here. You can search for all occurrences of the apostrophe ' and check to be sure you mean each one.
3. *You may be able to search on format features.* For example, if you want to look at all of your headings and you used boldface for them, you can find the headings quickly by searching for boldface. The computer will move from heading to heading. (It will also show you any other places you have used boldface; and it will jump over a heading if you did not use boldface for that heading.)

You Can Set Up Shortcuts You can use the search and replace feature as a shorthand to save typing time and errors. If you are going to use the same long word or phrase many times in your report, you can choose a single letter to represent it while you are typing and then change that letter to the full name later.

For example, if you are writing about the Sudbury Farm Implement Company, you might choose to use the letter *s* to stand for the full name. In your draft, you would type "The *s* should consider . . ." Before you print your draft, you would use the search and replace feature to change each occurrence of *s* to the full name, Sudbury Farm Implement Company.

Be careful about the letter that you choose. The *s* works because the letter *s* by itself is not an English word. If you were writing about the Acme Company, you would have to choose a letter other than *a* because the computer won't be able to distinguish between the regular English word *a* and the letter you are using for your shorthand.

Also be careful when replacing your shorthand letter. You must search for "space-*s*-space." Otherwise, the computer will change every *s* in your draft to the full name even if the *s* is in another word.

With many word processing programs, you can set up a different type of shortcut for long words and phrases. You can set up the program so that every time you press a certain set of keys, the computer types out the word or phrase for you. Usually, you choose one of the extra keys, the one labeled *Ctrl* for "control" or the one labeled *Alt* for "alternate," along with a regular letter key. For example, you might choose to use the keys *Alt* and *s*. Every time you press those keys together, the computer types Sudbury Farm Implement Company into your draft.

Word processing programs call this "setting up a macro" or "assigning keystrokes." Look in the manual for your word processing program to see if you can do this and how to set it up in your program.

Cautions for Using Word Processors

Word processors may have many advantages over typewriters and even paper-and-pencil, but they won't by themselves make you a better technical writer. **You still need to think, plan, organize, write, and edit your work.**[15] The word processor, like the typewriter, is only a tool. It may be a fancier tool, but like the typewriter, it only puts on paper what you, the writer, type.

Learn to Type You'll never regret knowing how to type. Even if you plan to be a scientist, engineer, technical specialist, or someday a manager or executive, you'll find yourself using a keyboard. Even if you use a computer with a different type of device, like a mouse, or a touch screen, or a wand, you'll still need the keyboard for typing your text.

Be Aware of Your Writing Habits Many people find that when they first start using a computer, they get bogged down in revising too early. Because making changes is so easy, they spend too much time polishing single sentences instead of plowing ahead with the draft.

Think about the Overall Arrangement of Your Text With most computers, you can see only a small portion of your draft at one time. A typical computer screen shows you about one-third of a page of text. Sometimes, writers who compose with a word processor focus too much on the words that are in front of them. Their paragraphs are well structured, but the overall organization of the draft suffers.

Print Out Your Drafts Often To check on the overall organization of your draft, print out what you have written often. Then you can read through the entire draft, checking for coherence and consistency. If you've moved or deleted text, recheck all your cross-references.

Name Your Word Processing Files Logically On most computer systems, when you save what you have typed, you create a "file." Develop a naming system for your files that will be easy to remember. This is especially important if you keep multiple copies of the same assignment or if you are working on a collaborative assignment.

For a report, for example, you might use *chap1, chap2, chap3,* etc. If you have several assignments and want to save different versions with each assignment, you might name them by the assignment and the draft number, as in *resume, draft 1; resume, draft 2.* If your computer only allows you eight characters for naming files, you might shorten these to *resdra1, resdra2,* and so on.

A naming system like this can help you if you forget how you named the file. In most word processing programs, you can ask to see a list of your file names, and they come up on the screen in alphabetical order. If you have used names in order like *chap1, chap2,* or *resdra1, resdra2,* they will be next to each other in the list.

Start Assignments Early Leaving the writing to the last minute is just as disastrous with a word processor as it is with a typewriter or paper-and-pencil. The computer may make typing your assignment easier and may help you to prepare a clean, neat copy, but it doesn't shorten the composing process. You still need to go through all the steps in the process that we discussed in Chapter 2.

Writing Material to Be Read on the Computer

On the job, you may find yourself writing material that will be read on the computer but that may or may not ever be printed. For example, you may communicate with your colleagues by electronic mail. You may collaborate on a report with people in other groups by reviewing and revising everything on the computer. You may be given an assignment to write on-line documentation, such as help messages, for a computer program. What special considerations should you make for material that will be read directly from the computer screen?

Understanding the Differences between Screens and Paper Here are some of the differences between reading from paper and reading from the screen:

- *People read more slowly from the screen than from paper.* This situation may change as computer technology improves, but it is still a good idea to keep text on the screen at very low density. Put even fewer words on the screen than you would on paper. Put in even more white space. Use lists more. Write very short paragraphs.
- *You may have only a small amount of space in which to write on the screen.* Although some systems come with monitors (computer screens) that can show a full-page or even a two-page spread, many still have small screens that equal only about one-third of a piece of paper. Furthermore, you may be writing text that comes up in a window that occupies only part of the screen. Therefore, get to the point quickly. Be succinct. Put the most important information up front.

- *People make more mistakes editing and proofreading on the screen than on paper.* Therefore, make it easy for people to find and read information on the screen. Use words that your readers will recognize quickly. Do not use all capital letters. If you are giving examples, surround them with space so they are easy to see. If you are giving choices or codes, use a list or table format.

Working with Electronic Mail You may have several ways that you can communicate with your colleagues, your managers, and your clients. Your choices may include:

- *Face-to-face meeting*—talking directly in person to an individual or group.
- *Phone call*—talking directly, but not in person, to the individual or group.
- *Teleconferencing*—meeting with people who are located in different places. Teleconferencing usually means that at least some of the people are in a special room with video cameras so that each group can both hear and see the others. Sometimes the video connection is one-way and only the voice connection is two-way. Sometimes teleconferencing also includes exchanging electronic files.
- *Voice (audio) mail*—leaving a voice message, as on an answering machine. A voice-mail system may include many options that require the caller to listen to instructions and use the telephone keypad like a computer keyboard.
- *Electronic mail*—sending a written message that the person receives on his or her computer. Electronic mail messages usually are meant to be read on the screen although they also can be printed out and read from paper. You can send a message to a group of people at the same time.
- *Written mail*—sending a letter, memo, or report on paper.

Part of your planning must be to choose the appropriate medium. If you are giving sensitive or unpleasant news, for example, you would be more likely to choose a face-to-face meeting or a phone call than the "impersonal or cool electronic media."[16]

Electronic mail is somewhere between a telephone conversation and a written letter. It is often used informally like a telephone conversation, but you can't use tone of voice to convey emotions and you don't have the immediate feedback that tells you that the other person understands. Humor doesn't usually work in electronic mail; the recipient may take seriously what you meant as a joke.

Some people ramble more in electronic mail than they do on paper. Some people write more tersely in electronic mail than in regular paper correspondence, and they may come over as abrupt or dogmatic when they don't mean to be. Be wary of responding angrily to electronic mail and think carefully about how your readers are likely to respond to the style that you use in your electronic mail messages.

Electronic mail may also allow you to send messages to people with whom you might not regularly correspond. Know who is included in each electronic "group" that you can write to. If you can send electronic mail to "all," will the President of the Corporation be reading the note that you really meant for the colleagues at your level? Know how your system is set up. If you are writing confidential material, are you certain that the mail will only go to those who should see the message? Think about the political and organizational consequences of the choices you make in audience, content, and style. Even if the electronic mail system allows you to send a request directly to your manager's manager, are you being wise to bypass the regular channels?

Writing On-line Documentation **On-line documentation,** material that is meant to be read directly from the computer screen, hasn't yet removed all the paper from our offices. It is unlikely ever to do so. It is also unlikely that all documentation of the future will be read on the screen. However, the trend towards having more documents available on-line is growing, and technical writers should know about creating readable, useful on-line documents.[17]

The first point to realize is that you cannot simply take a printed book and make a usable on-line document by having it show up on the computer screen. The problems in paper documents are exacerbated on-line. If the document isn't arranged in very small pieces for selective reading, it will be difficult, if not impossible, to work with on-line. If the document is full of dense paragraphs, it will be even more difficult to read from most computer screens than from paper.

Readers have less tolerance for reading extended prose passages on the screen than on paper. They give up after about three screens of text. If the important information is at the bottom of the second page of a section of the paper document, it may be on the sixth screen on-line. Many readers won't get that far.

When you put documentation on-line, you may lose some of the features of book formatting. Some electronic systems don't show changes in type size or font on the screen. Some don't show italics. Some don't allow graphics. On the other hand, you may have more options on-line than you have on paper. Color, font changes, and graphics that were too expensive to put into the paper documentation may be possible to use at no extra cost on-line. **You have to plan, write, and design for each medium on its own terms.**

The screen is not a good place to expect people to use documents that require sustained reading. It is a good place for documents where people are going to be looking for specific pieces of information and where you are going to be able to give them that information quickly and succinctly.

The help system for a computer program works well on-line because users want the help while they are working, and they want information about specific choices or tasks. An employee benefits book might work well on-line because people go to it for answers to specific questions. A maintenance manual might work well on-line, especially in a system with both text and graphics, because readers are looking for specific information about a problem and how to fix it.

Considering Multimedia, Hypertext, and Hypermedia Technology is expanding our ideas about what a document is and how to present information to readers. In an electronic environment, we can combine not just text and static pictures but also voice, animation, and video. That is, we can have a **multimedia** environment. It's like having an overhead projector, a slide projector, a VCR (videocassette recorder), and a blackboard, as well as the textbook and the professor, all available to deliver the messages in a single document. The challenge is knowing when to use each medium and how to combine them effectively.

Technology also allows rapid access to large amounts of information. An entire encyclopedia can be stored on one CD-ROM (compact disk, read-only memory) along with the software needed to bring any part of the encyclopedia to the user's screen within a few seconds. However, just putting paper documents into electronic files on CDs won't help users. It takes technical writing skills to know how users are likely to search for information and how to present that information in screen-size chunks.

Hypertext is another feature that technology allows. Hypertext is nonlinear text, a way to let readers follow different paths through the information, depending on their interests and needs. In printed materials, the text is presented sequentially. Readers may not look at every page, but the pages are in a specific order. If the author has put in a long note, readers may have to turn several pages to get past the note. If the author has tables with technical details in an appendix, readers who are interested in the details have to turn to the appendix and then have the report open in two places at once. To get the definition of a word, readers have to find the glossary, find the word, and remember its meaning as they find their place back in the text.

With hypertext, the pieces of information don't have to be in any given order. They are stored as separate units in the computer and linked in a variety of ways. With hypertext, the notes and tables can be linked to the relevant text, but only readers who want to see the notes or tables need go to them—and then go directly back to the text or have both on the screen together. Words in the glossary can be linked directly to their definitions; if a reader asks for the definition, it can be shown in a window on the screen right next to the word.

Furthermore, we can combine the concepts of multimedia and hypertext. The combination of concepts is sometimes called **hypermedia.** We can have nonlinear chunks that are text, graphics, animation, video, voice, and various combinations of these media.

Even with paper documents, each reader interacts differently with the material, choosing what to read carefully, what to skim, what to skip. Multimedia and hypertext documents can give readers even more choice and opportunities for interaction. They also require even more planning and arranging than a linear printed text does. The writer has to visualize the paths that readers might want to take through the material. The writer has to divide the information into small pieces and yet help readers keep track of the "big picture" of the document at the same time. Creating useful hypertext and multimedia (or hypermedia) documents is not easy but may be an exciting challenge for technical writers.

Planning and Revision Checklists

The checklist on the following pages will help you plan and revise the design of your document. Use them in conjunction with the planning and revision checklists that follow Chapter 2, "Composing" (pages 33–35), and Chapter 4, "Writing for Your Readers" (pages 78–79). The questions summarize key points of this chapter and will help you plan and revise all your documents.

GENERAL QUESTIONS

Planning

- Have you planned the format?
- Have you considered your purposes, audiences, and how people will use your document?
- Have you checked on the software and hardware that you will be using to prepare both drafts and final copy? (Do you know what options are available to you?)
- Have you found out whether you are expected to follow a standard format or style sheet?
- Have you thought about all the features (headings, pictures, tables) that you will have, and have you planned a page design that works well with those features?
- Have you thought about how you will make the arrangement obvious to your readers? (What will you do to make it easy for people to read selectively in your document?)

Revision

- Is your document clean, neat, and attractive?
- Is the text easy to read?
- Will your readers be able to find a particular section easily?
- If your document is supposed to conform to a standard, does it?

QUESTIONS ABOUT SETTING UP A USEFUL FORMAT

Planning

- Have you set the margins so that there will be enough white space all around the page, including space for binding, if necessary?
- Have you set the line length and line spacing for easy reading?
- Have you decided how you are going to show where a new paragraph begins?
- Have you decided whether to use a justified or ragged right margin and set the software appropriately?
- Have you planned which features to surround with extra white space, such as lists, tables, graphics, and examples?

Revision

- Have you left adequate margins? Have you left room for binding?
- Are the lines about 50–70 characters long? If you are using two columns, are the lines about 35 characters long in each column?
- Is the spacing between lines and paragraphs consistent and appropriate?
- Can the reader tell easily where sections and paragraphs begin?
- Have you left the right margin ragged? If not, look over the paper to be sure that the justification has not made overly tight lines or left rivers of white.
- Have you used white space to help people find information on the page?
- Have you put white space around examples, warnings, pictures, and other special elements of the format?
- Have you used lists for steps in procedures, options, and conditions?

QUESTIONS ABOUT MAKING THE TEXT READABLE

Planning

- Have you selected a type size and typeface that will make the document easy to read?
- Have you planned for highlighting? Have you decided which elements need to be highlighted and what type of highlighting to use for each?
- Do you know if you can use color? If you can, have you planned what color to use and where to use it in the document?

Revision

- Is the text type large enough to be read easily?
- Have you been consistent in using one typeface?
- Have you used uppercase and lowercase letters for the text and for most levels of headings?
- Have you used highlighting functionally? Is the highlighting consistent? Does the highlighting make important elements stand out?

QUESTIONS ABOUT MAKING INFORMATION EASY TO LOCATE

Planning

- Have you planned your headings? Have you decided how many levels of headings you will need? Have you decided on the format for each level of heading?
- Will the format make it easy for readers to tell the difference between headings and text? Will the format make it easy for readers to tell one level of heading from another?
- Have you decided where to put the page numbers and what format to use?
- Have you decided on headers or footers (information at the top or bottom of each page)?
- Have you found out if you are expected to use a numbering system? If you are, have you found out what system to use and what parts of the document to include?

Revision

- Have you checked the headings? Are the headings informative? unambiguous? consistent? parallel?
- Will readers get an overall picture of the document by reading the headings?
- Is the hierarchy of headings obvious?
- Can readers tell at a glance what is a heading and what is text?
- If readers want to find a particular section quickly, will the size and placement of the headings help them?
- Have you checked the page breaks to be sure that you do not have a heading by itself at the bottom of a page?
- Are the pages numbered?
- Are there appropriate headers and footers?
- If you are using a numbering system, is it consistent and correct?

A QUESTION BEFORE YOU PRINT

- Does the printer have a clean ribbon or cartridge? Do you have enough of the correct size, color, and weight paper?

Exercises

1. Take an earlier assignment, and revise it with the design guidelines that you have learned in this chapter.
2. Have you recently seen a document that you found difficult to read or follow because it was not designed according to the guidelines in this chapter? Write a brief report to the company or agency that put out the document. Give specific recommendations for changing the design. Support your recommendations.
3. Redesign the document following your recommendations.

4. When college students receive a package of forms about financial aid, they also get a page of information about the financial aid program. Here are the headings from one state's version of this information page:

   THE FINANCIAL AID PROGRAM

   ELIGIBILITY

   DEFERMENTS

   TERMS AND CONDITIONS

   LOAN INSTITUTIONS

   ELIGIBLE SCHOOLS

   LIABILITY FOR REPAYMENT

   APPLICATION PROCESS

   Are these headings meaningful to you? Are they in the most logical order? Plan an information page that would be more useful to you. Write a set of headings that you would like to see on the information page. (Hints for this exercise: If you would ask questions, write the headings as questions. Don't just translate the nouns in this example into other words. Think about the content and arrangement that you, as the reader, would want to see.)
5. Prepare a tentative table of contents for your final report. Show the sentence style that you will use for each level of heading. Show the typography that you will use for each level of heading. Check your table of contents for parallelism.
6. Plan the page layout for your final report. Will you use a header or a footer? Where will you put the page numbers? Will you use a numbering system? If so, which one will you use? Prepare one or more sample pages using the word processor and printer that you will be using for the report. The pages do not have to have real text on them. For example, you can type, "Level-One Heading," and "This is what the text will look like." The pages should, however, indicate the margins, line length, spacing, type choices, and so on that you will use. Show all the levels of headings that you are planning to use. Also show how you will handle graphics with the text.

CHAPTER 11

Design Elements of Reports

PREFATORY ELEMENTS

Letter of Transmittal and Preface
Cover
Title Page
Table of Contents
List of Illustrations
Glossary and List of Symbols
Abstracts and Summaries

MAIN ELEMENTS

Introduction
Discussion
Ending

APPENDIXES

DOCUMENTATION

General Rules
Notes
Author–Date Documentation
MLA–APA Comparison
Copyright

Our approach to format is descriptive, not prescriptive; that is, we describe some of the more conventional practices found in technical reporting. We realize fully, and you should too, that many colleges, companies, and journals call for practices different from the ones we describe. Therefore, we do not recommend that you must follow at all times the practices in this chapter. If you are a student, however, your instructor may, in the interests of class uniformity, insist that you follow this chapter fairly closely.

Chapter 10, "Document Design," discusses strategies for achieving good design. In this chapter, we discuss the elements—the tools—you can use to carry out those strategies. We divide the elements into three groups: prefatory elements, main elements, and appendixes. Finally, we provide a section on documentation.

Prefatory Elements

The prefatory elements help your readers to get into your report. The letter of transmittal or preface may be the readers' first introduction to the report. The table of contents reveals the structure of your organization. In the glossary, readers will find the definitions of terms that may be strange to them. All the prefatory elements discussed in this section contribute to the success of your report.

Letter of Transmittal and Preface

We have placed the letter of transmittal and preface together because in content they are often quite similar. They usually differ in format and intended audience only. You will use the letter of transmittal when the audience is a single person or a single group. Many of your major reports in college will include a letter of transmittal to your professor, usually placed just before or after your title page. When on the job, you may handle the letter differently. Often, it is mailed before the report, as a notice that the report is forthcoming. Or it may be mailed at the same time as the report but under separate cover.

Generally, you will use the preface for a more general audience when you may not know specifically who will be reading your report. The preface or letter of transmittal introduces the reader to the report. It should be fairly brief. Always include the following basic elements:

- Statement of transmittal or submittal (included in the letter of transmittal only)
- Statement of authorization or occasion for report
- Statement of subject and purpose

Additionally, you may include some of the following elements:

- Acknowledgments
- Distribution list (list of those receiving the report—used in the letter of transmittal but not in the preface)
- Features of the report that may be of special interest or significance
- List of existing or future reports on the same subject
- Background material
- Summary of the report
- Special problems (including reasons for objectives not met)
- Financial implications
- Conclusions and recommendations

How many of the secondary elements you include depends upon the structure of your report. If, for example, your report's introduction or discussion includes background information, there may be no point in including such material in the preface or letter of transmittal. See Figures 11–1 and 11–2 for a sample letter of transmittal and a sample preface.

If the report is to remain within an organization, the letter of transmittal will become a memorandum of transmittal. This changes nothing but the format. See Chapter 13, "Correspondence," for letter and memorandum formats.

Cover

A report's cover serves three purposes. The first two are functional and the third aesthetic and psychological.

First, covers protect pages during handling and storage. Pages ruck up, become soiled and damaged, and may eventually be lost if they are not protected by covers. Second, because they are what readers first see as they pick up a report, covers are the appropriate place for prominent display of identifying information such as the report title, the company or agency by or for which the report was prepared, and security notices if the report contains proprietary or classified information. Incidentally, students should not print this sort of information directly on the cover. Rather, they should type the information onto gummed labels readily obtainable at the college bookstore, and then fasten the labels to the cover. A student label might look like the one in Figure 11–3.

Gatlin Hall
Weaver University
Briand, MA 02139

July 27, 1991

Dr. Ross Alm
Associate Professor of English
Weaver University
Briand, MA 02139

Dear Dr. Alm:

Statement of transmittal
Occasion for report

I submit the accompanying report entitled "Characteristics of Venus and Mercury" as the final project for English 430, Technical Writing.

Statement of subject and purpose

The report discusses the characteristics of both Venus and Mercury, covering size, mass, density, physical appearance, and atmosphere. I have made an effort to provide a base for understanding the significance of the space probes to Venus and Mercury carried on by NASA's Jet Propulsion Laboratory (JPL). Recent information about Mercury obtained by the most recent probe, Mariner 10, is incorporated into the report.

Acknowledgments

I am indebted to Ms. Mary Fran Buehler of JPL who has allowed me to quote extensively from her unpublished work on Mariner 10.

Sincerely,

Anne K. Chimato

Anne K. Chimato
English 430

Figure 11–1 Letter of Transmittal

Finally, covers bestow dignity, authority, and attractiveness. They bind a bundle of manuscript into a finished work that looks and feels like a report and has some of the characteristics of a printed and bound book.

Suitable covers need not be expensive and sometimes should not be. Students, particularly, should avoid being pretentious. All three purposes are frequently well served by covers of plastic or light cardboard, perhaps of 30- or 40-pound substance. Students can buy such covers in a variety of sizes, colors, and finishes.

While you are formatting your report, remember that when you fasten it into its cover about an inch of left margin will be lost. If you want an inch of margin, you must leave two inches on your paper. Readers grow

Occasion for report

Statement of subject and purpose

Feature of special interest

Existing reports on same subject

PREFACE

In recent years the National Aeronautics and Space Administration (NASA) has explored the inner planets of our solar system, Venus and Mercury, with unmanned space probes. This report, part of NASA's educational series for high school students, reports the information from the latest probe, Mariner 10. Characteristics of both Venus and Mercury—including size, density, physical appearance, and atmosphere—are discussed.

Of particular interest in this report is the surprising finding that Mercury, contrary to scientific expectation, has a magnetic field. This finding may cause present theories about the generation of magnetic fields within planets to be revised.

For lists of other reports on NASA's unmanned space probes write to NASA, Jet Propulsion Laboratory, California Institute of Technology, Pasadena, California 91125.

Figure 11–2 Preface

irritated when they must exert brute force to bend open the covers to see the full page of text.

Title Page

Like report covers, title pages perform several functions. They dignify the reports they preface, of course, but far more important, they provide identifying matter and help to orient the report users to their reading tasks.

CHARACTERISTICS OF VENUS AND MERCURY

by

Anne K. Chimato

English 430 27 July 1991

Figure 11–3 Student Label

To give dignity, a title page must be attractive and well designed. Symmetry and balance are important, as are neatness and freedom from clutter. The most important items should be boldly printed; items of lesser importance should be subordinated. These objectives are sometimes at war with the objective of giving the report users all the data they may want to see at once. Here, we have listed in fairly random order the items that sometimes appear on title pages. A student paper, of course, would not require all or even most of these items. The first four listed are usually sufficient for simple title pages.

- Name of the company (or student) preparing the report
- Name of the company (or instructor and course) for which the report was prepared
- Title and sometimes subtitle of the report
- Date of submission or publication of the report
- Code number of the report
- Contract numbers under which the work was done
- List of contributors to the report (minor authors)
- Name and signature of the authorizing officer
- Company or agency logo and other decorative matter
- Proprietary and security notices
- Abstract
- Library identification number
- Reproduction restrictions
- Distribution list (A list of those receiving the report. If the letter of transmittal does not contain this information, the title page should.)

Understandably, placing all of these items on an 8½- by 11-inch page would guarantee a cluttered appearance. Put down what you must, but no more.

Word processing now allows report writers to use different-size type and different type styles on a title page to indicate what is important and what is subordinate. Use this capability discreetly. Don't turn your title page (or any other part of your report) into a jumble of different typefaces. Generally, different type sizes and the use of boldface and plain style will suffice.

In Figure 11–4, we illustrate a title page in plain style. Figure 11–5 illustrates the use of word processing.

Pay particular attention to the wording of your title. Titles should be brief but descriptive and specific. The reader should know from the title what the report is about. A title such as "Effects of Incubation Temperatures on Sexual Differentiation in the Turtle, *Chelydra sepentina*" is illustrative. From it, you know specifically the research being reported.

CHARACTERISTICS OF VENUS AND MERCURY

Prepared for

Professor Ross Alm

English 430

Technical Writing

by

Anne K. Chimato

27 July 1991

Figure 11–4 Typed Title Page

To see how effectively this title works, leave portions of it out, and see how quickly your understanding of what the article contains changes. For example, "Sexual Differentiation in the Turtle" would suggest a much more comprehensive report than does the actual title. A title such as "Effects of Incubation Temperatures" could as well be about chickens as turtles. On the other hand, adding the words "An Investigation into" to the beginning of the title would add nothing really useful. The test of whether a title is too long or too short isn't in the number of words it

The Public Health Consequences of Disasters 1989

September 1989

U.S. Department of Health and Human Services
Public Health Service
Centers for Disease Control
Atlanta, Georgia 30333

Figure 11–5 Word Processed Title Page
Source: U.S. Department of Health and Human Services, Public Health Service, *The Public Health Consequences of Disasters 1989* (Atlanta: Centers for Disease Control, 1989) i.

contains but what happens if words are deleted or added. Keep your titles as brief as possible, but make sure they do the job.

Table of Contents

A table of contents (TOC) performs at least three major functions. Its most obvious function is to indicate by number the page on which discussion of each major topic begins; that is, it serves the reader as a locating device. Less obviously, a TOC forecasts the extent and nature of the topical coverage and suggests the logic of the arrangement and the relationship of the parts. Still earlier, in the prewriting stage, provisional drafts of the TOC enable the author to "think on paper"; that is, they act as outlines to guide the composition.

A system of numbers, letters, type styles, indentations, and other mechanical aids has to be selected so that the TOC will perform its intended functions. Figure 11–6 shows a TOC suitable for student reports. We have annotated the figure to draw your attention to a few key points. However, the annotations are suggestions only. There are many acceptable variations in TOCs. For example, the spaced leader dots in Figure 11–6 are used to carry the reader's eyes from the end of each title to the page number and at the same time tie the page together visually. But, this practice is by no means universal. Some people feel the leader dots clutter the page and therefore do not use them. If you use a numbering system in your report (see pages 260–261), the TOC should reflect that system.

In Figure 11–7, we reproduce a professionally done TOC that shows a good use of word processing capabilities. When you design your own TOC, as with the title page, avoid overcrowding. Seldom is there justification for listing parts subordinate to the subdivisions of sections. Very shortly a point is reached where users have almost as much trouble locating items in the TOC as they have in locating them by flipping through the pages of the report. Be sure that page numbers in your TOC match the page numbers in your final draft. Remember that the TOC entries and the headings on the text pages must be worded exactly the same. Every entry in the TOC must also be in the report. However, as we have already pointed out, every heading in the report need not be in the TOC.

List of Illustrations

If a report contains more than a few illustrations, say more than three or four, it is customary to list the illustrations either on a separate page or on the TOC page. Illustrations are of two major types: tables and figures. A **table** is any array of data, often numerical, arranged vertically in columns and horizontally in rows, together with the necessary headings and notes. **Figures** include photographs, maps, graphs, organization charts, and flow diagrams—literally anything that does not qualify as a table by the preceding definition. (For further details, see Chapter 12, "Graphical Elements of Reports.")

If the report contains both tables and figures, it is customary to use the page heading "Illustrations," a combining term, and to list all the figures first and then all the tables.

The titles of all illustrations should be as brief and yet as self-explanatory as can be. Avoid a cumbersome expression such as "A Figure

Make all major headings distinctive. All capitals or boldface are good choices.

Repeat any numbering system used internally in the document.

Indent five spaces.

In subheadings, capitalize first and last words and all principal words between. Do not capitalize articles, prepositions, and coordinating conjunctions unless they are a first or last word.

Line up all numbers, arabic and roman, on right-hand digit.

CONTENTS

ii

Figure 11–6 Typed Table of Contents

Showing Characteristic Thunderstorm Recording." Say, simply, "Characteristic Thunderstorm Recording." On the other hand, do not be overly economical and write just "Characteristic" or "A Comparison." At best, such generic titles are only vaguely suggestive.

Figure 11–8 shows a simple version that should satisfy most ordinary needs. Notice in the figure that Arabic numerals are used for figures and Roman numerals for tables. This practice is common but by no means

Contents

v

Figure 11–7 Word Processed Table of Contents
Source: U.S. Department of Health and Human Services, Public Health Service, *The Public Health Consequences of Disasters 1989* (Atlanta: Centers for Disease Control, 1989) v.

ILLUSTRATIONS

vi

Figure 11–8 List of Illustrations

standard. As with the TOC, be sure your page numbers are correct and that the titles listed accurately repeat the titles in the report.

Glossary and List of Symbols

Reports dealing with technical and specialized subject matter often include abbreviations, acronyms, symbols, and terms not known to the nonspecialist. Thus, a communication problem arises. Technically trained persons have an unfortunate habit of assuming that what is well-known to them is well enough known to others. This assumption is seldom justified. Terms, symbols, and abbreviations undergo changes in meaning with time and context. In one context, ASA may stand for American Standards Association; in another context, for Army Security Agency. The letter K may stand for Kelvin or some mathematical constant. The meaning given to Greek letters may change from one report to the next, even though both were done by the same person.

Furthermore, writers seldom have complete control over who will read their reports. A report intended by the author for an engineering audience may have to be read by members of management, the legal department, or sales. In doubtful instances, it is wise to play it safe by including a list of symbols or a glossary or both. Readers who do not need these aids can easily ignore them; those who do need them will be immeasurably grateful.

The list of symbols is normally a prefatory element. A glossary may be a prefatory element or placed as an appendix at the end of the report. When in your report you first use a symbol found in your list of symbols or a term found in the glossary, tell your reader where to find the list or the glossary. Figure 11–9 illustrates a list of symbols. Figure 11–10 illustrates a glossary.

Abstracts and Summaries

Abstracts and **summaries** are overviews of the facts, results, conclusions, and recommendations of a report. In many formats—for example, empirical research reports and feasibility reports—abstracts or summaries will be placed near the front of the report. In that position, they both summarize the report and allow busy readers to decide if they want to read further in the report. As we pointed out in Chapter 4, "Writing for Your Readers," both executives and experts expect the abstract or summary to come early in the report.

In more discursive reports, magazine articles, for instance, summaries most often come at the end of the report, where they serve to draw things together for the reader. Also, many reports, particularly empirical research reports, have an abstract at the beginning and a summary at the end. These facts raise the question of when is an overview an "abstract" and when is it a "summary." In general, these principles hold true:

- *Abstracts* are placed before technical reports, such as empirical research reports, meant for technical audiences.
- *Summaries* are placed before business and organizational reports, such as proposals and feasibility reports. When the audience is primarily an executive audience, the summary will be the type known as an *executive summary.*
- An overview placed at the end of a report will likely be called a "*summary.*"

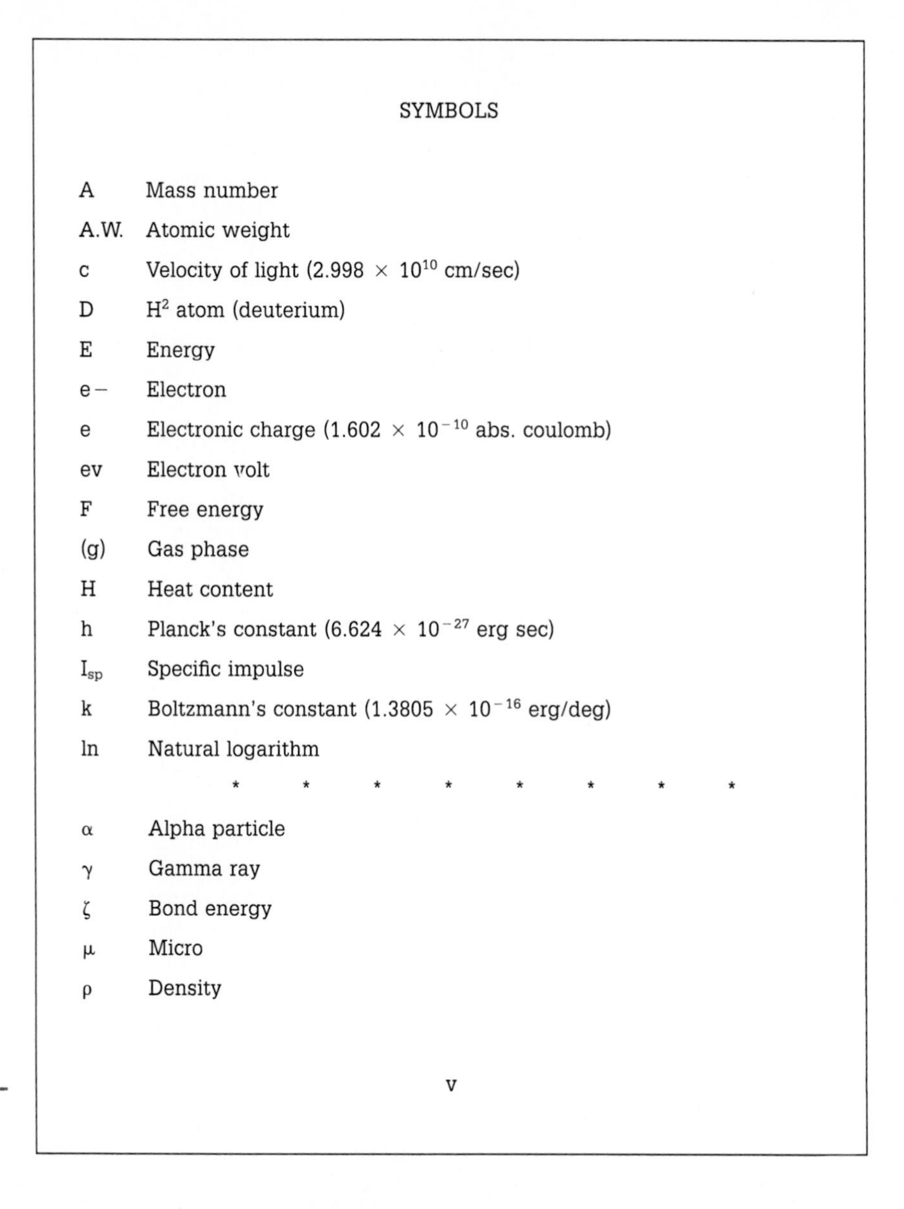

SYMBOLS

A	Mass number
A.W.	Atomic weight
c	Velocity of light (2.998×10^{10} cm/sec)
D	H^2 atom (deuterium)
E	Energy
e−	Electron
e	Electronic charge (1.602×10^{-10} abs. coulomb)
ev	Electron volt
F	Free energy
(g)	Gas phase
H	Heat content
h	Planck's constant (6.624×10^{-27} erg sec)
I_{sp}	Specific impulse
k	Boltzmann's constant (1.3805×10^{-16} erg/deg)
ln	Natural logarithm

* * * * * * * *

α	Alpha particle
γ	Gamma ray
ζ	Bond energy
μ	Micro
ρ	Density

v

Figure 11–9 List of Symbols

Because abstracts and executive summaries always appear as prefatory elements, we discuss them in this section. We discuss other types of summaries and conclusions and recommendations on pages 298–303, where we discuss how to end a report.

Abstracts Discussed here are abstract style and two kinds of abstracts: informative and descriptive.

Never use "I" statements in either kind of abstract. Report your infor-

Use parallel sentence fragments for glossary definitions. (See "Parallelism" in "Handbook")

Use complete sentences to add information to definition

GLOSSARY

Btu	the amount of heat required to raise the temperature of one pound of water one degree Fahrenheit.
degree day	a temperature standard around which temperature variations are measured.
design temperature	the maximum reasonable temperature expected during the heating or cooling season. Design calculations are based upon this number.
heat transmission coefficient	the quantity of heat in Btu transmitted per hour through one square foot of a building surface.
infiltration	the air leaking into a building from cracks around doors and windows.
sensible heat	heat that the human body can sense.
thermal conductivity	the quantity of heat in Btu transmitted by conduction per hour through one square foot of a homogeneous material for each degree Fahrenheit difference between the surfaces of the material.
thermal resistance	the reciprocal of thermal conductivity.

vii

Figure 11–10 Glossary

mation impersonally, as though it were written by someone else. The informative abstract in Figure 11–11 shows the style. This is not an arbitrary principle. If you were to publish your report, your abstract would likely be reprinted in an abstracting journal where the use of "I" would be inappropriate. Also, many companies, in the interest of good intracompany communication, publish the abstracts of all company research reports. The restriction on the use of "I," makes the use of passive voice common in abstracts. Because your full report contains complete documentation, you need not footnote or otherwise document the information in abstracts.

Research objectives

Methodology

Findings

ABSTRACT

This study investigated the role of signaling in helping good readers comprehend expository text. As the existing literature on signaling, reviewed in the last issue of the Journal, pointed to deficiencies in previous studies' methodologies, one goal of this study was to refine prose research methods. Two passages were designed in one of eight signaled versions each. The design was constructed to assess the individual and combined effect of headings, previews, and logical connectives. The study also assessed the effect of passage length, familiarity and difficulty. The results showed that signals do improve a reader's comprehension, particularly comprehension two weeks after the reading of a passage and comprehension of subordinate and superordinate inferential information. This study supports the hypothesis that signals can influence retention of text-based information, particularly with long, unfamiliar, or difficult passages.

Figure 11–11 Informative Abstract
Source: Jan H. Spyridakis, "Signaling Effect: Increased Content Retention and New Answers—Part II," *Journal of Technical Writing and Communication* 19 (1989): 395.

Informative Abstract Figure 11–11 illustrates an **informative abstract.** Informative abstracts are most often intended for an expert audience; therefore, their authors can use the technical language of the field freely. Like most informative abstracts, the one in Figure 11–11 summarizes three major elements of the full report:

- The objectives of the research or the report
- The methodology used in the research
- The findings of the report to include the results, conclusions and recommendations.

Most professional journals or societies publish stylebooks that include specifications about how to write an abstract. Many journals, because of high publication costs, will set arbitrary limits of under 200 words for abstracts. Because abstract writing uses many of the same techniques as summary writing, you might want to read what we say about that on pages 298–299 and 529–532.

Descriptive Abstract The main purpose of the **descriptive abstract** is to help busy readers decide if they need or want the information in the report enough to read it entirely. The descriptive abstract merely tells what the full report contains. Unlike the informative abstract, it cannot serve as a substitute for the report itself. Many reports contain descriptive abstracts, and many abstracting journals print them. The following example is typical of the content and style of a descriptive abstract:

> The management of the process by which technical documents are produced usually proceeds according to one of two models, the "division of labor" model or the "integrated team" model. This article reports on a survey that suggests the prevalence of each model and that gives insights into how the choice of a management model affects the practice of technical communication and the attitudes of technical communicators.[1]

The descriptive abstract discusses the *report,* not the subject. After reading this abstract, you know that this article "gives insights into how the choice of a management model affects the practice of technical communication and the attitudes of technical communicators," but you must read the article to gain the insights. Whether a report is 10 or 1,000 pages long, a descriptive abstract can cover the material in less than 10 lines.

Executive Summary Placed at the front of a report, the **executive summary** ensures that the points of that report important to an executive audience are immediately accessible to that audience. To that end, it is written in nontechnical language suited to an executive audience. Seldom more than one page long, typed doubled-spaced, it emphasizes the material that executives need in their decision-making process. It need not summarize all the sections of the report. For example, writing for a combined audience of scientists and executives, a writer might include a theory section in a report. The executive summary might skip this section altogether or treat it very briefly.

In their decision making, executives weigh things such as markets, risks, rewards, costs, and people. If your report recommends buying new equipment, they want to be assured that you have examined all reasonable alternatives and considered cost, productivity, efficiency, profits, and staffing. If you are reporting research, executives take your methodology for granted. They care very little for the physics, chemistry, or biology behind a development. What they want to know are your results and the implications of those results for the organization.

Figure 11–12 lists the questions that executives ask in different situations and that, therefore, an executive summary should answer. Before writing an executive summary, look at that figure and read pages 64–69, where we discuss the reading habits of executives. The annotated executive summary in Figure 11–13 illustrates the technique and the major parts of an executive summary.

Place an executive summary immediately before the introduction and label it "Summary" or "Executive Summary." In short reports and memorandum reports, the executive summary often replaces the introduction and is followed immediately by the major discussion.

Problems
What is it?
Why undertaken?
Magnitude and importance?
What is being done? By whom?
Approaches used?
Thorough and complete?
Suggested solution? Best? Consider others?
What now?
Who does it?
Time factors?

New Projects and Products
Potential?
Risks?
Scope of application?
Commercial implications?
Competition?
Importance to Company?
More work to be done? Any problems?
Required manpower, facilities, and equipment?
Relative importance to other projects or products?
Life of project or product line?
Effect on Westinghouse technical position?
Priorities required?
Proposed schedule?
Target date?

Tests and Experiments
What tested or investigated?
Why? How?
What did it show?
Better ways?
Conclusions? Recommendations?
Implications to Company?

Materials and Processes
Properties, characterisics, capabilities? Limitations?
Use requirements and environment?
Areas and scope of application?
Cost factors?
Availability and sources?
What else will do it?
Problems in using?
Significance of application to Company?

Field Troubles and Special Design Problems
Specific equipment involved?
What trouble developed? Any trouble history?
How much involved?
Responsibility? Others? Westinghouse?
What is needed?
Special requirements and environment?
Who does it? Time factors?
Most practical solution? Recommended action?
Suggested product design changes?

Figure 11–12 What Managers Want to Know
Source: James W. Souther, "What to Report," *IEEE Transactions on Professional Communication* PC-28 (1985): 6.

Main Elements

The body of a report contains its detailed information and interpretation. The body needs to be introduced and, normally, finished off with an ending of some sort that may include a summary, conclusions, recommendations, or simply a graceful exit from the report. We discuss all these elements in this section on main elements.

Introduction

A good introduction forecasts what is to follow in the rest of the report. It directs the reader's mind to what the subject and purpose are. It sets limits to the scope of the subject matter and reveals for the reader the plan of development of the report. Early in your paper, you also give any needed theoretical or historical background, and this is sometimes included as part of the introduction.

Problem definition

Alternate solutions considered

Recommendation

Effect of recommendation

SUMMARY

The University is steadily falling behind in the faculty and student use of computers. Our existing computer labs have insufficient numbers of computers, and those we do have are badly dated. We have faculty who are capable of designing computer programs for instructional use but who are reluctant to do so because their students do not have access to computers. Too many graduates are leaving the University as computer illiterates.

Information Resources has considered three solutions to the problem:

1. Require all freshmen to buy a microcomputer at an approximate cost of $1,500 each. At an interest rate of 8%, students could repay the University for their computers in 16 quarterly payments of $108.31 each.
2. Provide those students and faculty who want them with microcomputers through the University bookstore at deep discounts. Purchasers would arrange their own financing if needed.
3. Upgrade the University computer labs by providing $1 million over the next fiscal year to provide microcomputers, printers, software, and new furniture. Student computer lab fees of $25 per quarter will pay the cost of material and employees to run the labs.

We recommend both alternatives 2 and 3. We reject alternative 1 on the grounds that we are a public school and must not put educational costs out of reach for our students. Solutions 2 and 3 would make enough computers available for the immediate future to encourage their use by both students and faculty.

Figure 11–13 Executive Summary

Subject Never begin an introduction with a superfluous statement. The writer who is doing a paper on core memory in computers and begins with the statement "The study of computers is a vital and interesting one" has wasted the readers' time and probably annoyed them as well. Announce your specific subject loud and clear as early as possible in the introduction, preferably in the very first sentence. The sentence "This paper will discuss several of the more significant applications of the exploding wire phenomenon to modern science" may not be very subtle, but it gets the job done. The reader knows what the subject is. Often, in conjunction with the statement of your subject, you will also need to define some important terms that may be unfamiliar to your readers. For example, the student who wrote the foregoing sentence followed it with these two:

Defines subject

A study of the exploding wire phenomenon is a study of the body of knowledge and inquiry around the explosion of fine metal wires by a sudden and large pulse of current. The explosion is accompanied by physical manifestations in the form of a loud noise, shock waves, intense light for a short period, and high temperatures.

In three sentences the writer announced the subject and defined it. The paper is well under way.

Sometimes, particularly if you are writing for nonspecialists, you may introduce your subject with an interest-catching step. This step may be rather extended, as in this example:

Interest-catching introduction

Traveling in orbit around the earth at an altitude of some 270 miles, the Skylab astronauts rotated their spacecraft to establish a bearing on one of their principal check points: a cluster of several hundred round green spots, each half a mile in diameter, arrayed in an orderly pattern on the earth's surface below them. What they were viewing was a dense concentration of circular irrigated fields in north-central Nebraska, a pattern easily identified from space. Passengers on commercial jet airliners increasingly notice the same over many other areas of the continental U.S., including eastern Colorado, central Minnesota, the Texas Panhandle, the Pacific Northwest and northern Florida. Now the proliferating green circles can be seen even in the middle of the Sahara. What is being observed is perhaps the most significant mechanical innovation in agriculture since the replacement of draft animals by the tractor.

These circular green fields, most often found in arid or semiarid country, are being watered by the world's first successful irrigation machine.[2]

Or you may simply extract a particularly interesting fact from the main body of your paper. For example,

Interesting fact about subject

Last year, the Federal Aviation Administration attributed 16 aircraft accidents to clear air turbulence. What is known about this unseen menace that can cripple an aircraft, perhaps fatally, almost without warning?

In this example, the writer catches your interest by citing the accident rate caused by clear air turbulence and nails the subject down with a rhetorical question in the second sentence. Interest-catching introductions are used in brochures, advertisements, and magazine and newspaper articles. You will rarely see an interest-catching introduction in business reports or professional journals. If you do, it will usually be a short one.

Purpose Your statement of purpose tells the reader *why* you are writing about the subject you have announced. By so doing, you also, in effect, answer the reader's question "Should I read this paper or not?" For example, a writer who had word processors as her subject announced her purpose this way: "In this article I use illustrative narratives to argue that direct use of word processors can benefit any communicators in business, industry, or academe."[3] Readers who have no reason to be interested in such a discussion will know there is no purpose in their reading the report.

Another way to understand the purpose statement is to realize that it often deals with the *significance* of the subject. A writer who had the application of human engineering to technical writing as his subject announced his purpose this way:

Significance of subject

> Regardless of your writing specialty, however, you will be more effective as a technical writer if you become familiar with human engineering and learn to apply human-engineering principles.[4]

Scope The statement of scope further qualifies the subject. It announces how broad and, conversely, how limited the treatment of the subject will be. Often it indicates the level of competence expected in the reader for whom the paper is designed. For example, a student who wrote, "In this report I explain the application of superconductivity in electric power systems in a manner suitable for college undergraduates," declared her scope as well as her purpose. She is limiting her scope to superconductivity in electric power systems, and the audience she is designing her report for is not composed of high school students or graduate physicists but college undergraduates.

Plan of Development In a plan of development, you forecast your report's organization and content. The principle of psychological reinforcement is at work here. If you tell your readers what you are going to cover, they will be more ready to comprehend as they read along. The following, taken from the introduction to a paper on iron enrichment of flour, is a good example of a plan of development:

This study presents a basic introduction to three major areas of concern in regard to iron enrichment: (1) questions on which form of iron is best suited for enrichment use; (2) potential health risks from super enrichment—cardiovascular disease, hemochromatosis, and the masking of certain disorders; and (3) the inconsistencies in basic knowledge as they relate to the definitions, extent, and causes of iron deficiency.

You need not necessarily think of the announcement of subject, purpose, scope, and plan of development as four separate steps. Often, subject and purpose or scope and plan of development can be combined. In a short paper, perhaps two or three sentences might cover all four points, as in this example:

Subject

Purpose

Scope and plan of development

Concern has been expressed recently over the possible presence in our food supply of a class of chemicals known as "phthalates" (or phthalate esters). To help assess the true significance of this concern as to the safety and wholesomeness of food, this article discusses the reasons phthalates are used, summarizes their toxicological properties, and evaluates their use in food packaging materials in the context of their total use in the environment.[5]

Also, introductions to specialized reports may have peculiarities of their own. These will be discussed in Part IV, "Applications of Technical Writing."

Theoretical or Historical Background When necessary theoretical or historical background is not too lengthy, you can incorporate it into your introduction. In fact, when handled properly, such information may well catch the interest of the reader, as in the following example:

Climatologists attribute the warming trend to the furnaces of civilization which have been spewing forth increasing loads of carbon dioxide to the atmosphere. This colorless and odorless gas, exhaled by man and used by plants to make themselves green, restricts the escape into space of infrared radiation from the sun-warmed earth. Since increased CO_2 absorbs more of the infrared radiation than formerly, a larger amount of heat accumulates, causing a slight but significant increase in average global temperature. This impact of atmospheric CO_2 on climate—dubbed the "greenhouse effect"—has become more apparent in recent years because of the escalating rate at which power plants and industry throughout the world have burned coal, oil from shale, and synthetic oil and gas.[6]

If necessary background material is extensive, however, it more properly becomes part of the discussion.

Discussion

The discussion will be the longest section of your report. Your purpose and your content will largely determine the form of this section. Therefore, we can prescribe no set form for it. In presenting and discussing your information, you will use one or more of the techniques described in Part II, "The Techniques of Technical Writing," or the special techniques described in Part IV, "Applications of Technical Writing." In addition to your text, you will probably also use headings (see pages 248–257) and the various visual aids such as graphs, tables, and illustrations described in Chapter 12, "Graphical Element of Reports."

When thinking about your discussion, remember that almost every technical report answers a question or questions: What is the best method of desalination to create an emergency water supply at an overseas military base? How are substances created in a cell's cytoplasm carried through a cell's membranes? What is the nature of life on the ocean floor? How does a hydraulic pump work? Ask and answer the reporter's old standbys—Who? What? When? Where? Why? How? Use the always important "so-what?" to explore the significant implications of your information. However you approach your discussion, project yourself into the minds of your readers. What questions do they need answered to understand your discussion? What details do they need to follow your argument? You will find that you must walk a narrow line between too little detail and too much.

Too little detail is really not measured in bulk but in missing links in your chain of discussion. You must supply enough detail to lead the reader up to your level of competence. You are most likely to leave out crucial details at some basic point that, because of your familiarity with the subject, you assume to be common knowledge. If in doubt about the reader's competence at any point, take the time to define and explain.

Many reasons exist for too much detail, and almost all stem from writers' inability to edit their own work. When you realize that something is irrelevant to your discussion, discard it. It hurts, but the best writers will often throw away thousands of words, representing hours or even days of work.

You must always ask yourself questions like these: Does this information have significance, directly or indirectly, for the subject I am explaining or for the question I am answering? Does the information move the discussion forward? Does it enhance the credibility of the report? Does it support my conclusions? If you don't have a yes answer to one or more of these questions, the information has no place in the report, no matter how many hours of research it cost you.

Ending

Depending upon what sort of paper you have written, your ending can be (1) a summary, (2) a set of conclusions, (3) a set of recommendations, or (4) a graceful exit from the paper. Frequently, you'll need some combination of these. We'll look at the four endings and at some of the possible combinations.

It's also possible in reports written with executives as the primary audience that the "ending" may actually be placed at the front of the report. Remember the audience analysis research we discuss in Chapter 4, "Writing for Your Readers." It indicates that executives are more interested in summaries, conclusions, and recommendations than they are in the details of a report. This research leads many writers in business and government to move these elements to the front of their reports. They may be presented in separate sections labeled "Summary," "Conclusions," and "Recommendations," or combined into an executive summary. (See pages 291–292.) In either case, the body of the report may be labeled "Discussion" or even "Annexes."

Summary Many technical papers are not argumentative. They simply present a body of information that the reader needs or will find of interest. Frequently, such papers end with summaries. In a summary, you condense for your readers what you have just told them in the discussion. Good summaries are difficult to write. At one extreme, they may lack adequate information; at the other, they may be too detailed. In the summary you must pare down to material essential to your purpose. This can be a slippery business.

Suppose, for example, you are writing a report whose purpose is to explore the knowledge about the way the human digestive system absorbs iron from food. In your discussion you describe an experiment conducted with Venezuelan workers that followed isotopically labeled iron through their digestive systems. To enhance the credibility of the information presented, you include some details about the experiment. You report the conclusion that vegetarian diets decreased iron absorption.

How much of all this should you put in your summary? Given your purpose, the location and methodology of the experiment would not be suitable material for the summary. You would simply report that in one experiment vegetarian diets have been shown to decrease iron absorption.

In general, each major point of the discussion should be covered in the summary. Sometimes you may wish to number the points for clarity. The following, from a paper of about 2,500 words, is an excellent summary:

> The exploding wire is a simple-to-perform yet very complex scientific phenomenon. The course of any explosion depends not only on the material and shape of the wire but also on the electrical parameters of the circuit. An explosion consists primarily of three phases:
>
> **1.** The current builds up and the wire explodes.
> **2.** Current flows during the dwell period.
> **3.** "Post-dwell conduction" begins with the reignition caused by impact ionization.
>
> These phases may be run together by varying the circuit parameters.
>
> The exploding wire has found many uses: It is a tool in performing other research, a source of light and heat for practical scientific application, and a source of shock waves for industrial use.

Summaries should be concise, and they should introduce no material that has not been covered in the report. You construct a summary as you construct an informative abstract. You read your disucssion over, noting your main generalizations and your topic sentences. You smoothly blend these together into a paragraph or two. Sometimes you will represent a sentence from the discussion with a sentence in the summary. At other times you will shorten such sentences to phrases or clauses. The last sentence in the foregoing example represents a summary of four sentences from the writer's discussion. The four sentences, themselves, were the topic sentences from four separate paragraphs.

If you are working with word processing, you might do well to copy the material you are summarizing and then go through it, eliminating unwanted material to make your summary. Such a technique may be both easier and more accurate than retyping the material.

Conclusions Some technical papers work toward a conclusion. They ask a question, such as "Are nuclear power plants safe?" present a set of facts relevant to answering the question, and end by stating a conclusion: "Yes," "No," or sometimes, "Maybe." The entire paper aims squarely at the final conclusion. In such a paper, you argue inductively and deductively. You bring up opposing arguments and show their weak points. At the end of the paper, you must present your conclusions. **Conclusions** are the inferences drawn from the factual evidence of the report. They are the final link in your chain of reasoning. In simplest terms, the relationship of fact to conclusion goes something like this:

Facts	Conclusion
Car A averages 25 miles per gallon.	On the basis of miles per gallon, Car B is preferable.
Car B averages 40 miles per gallon.	

Because we presented a simple case, our conclusion was not difficult to arrive at. But even more complicated problems present the same relationship of fact to inference.

In working your way toward a major conclusion, you ordinarily have to work your way through a series of conclusions. In answering the question of nuclear power plant safety, you would have to answer a good many subquestions concerning such things as security of the radioactive materials used, adequate control of the nuclear reaction, and safe disposal of nuclear wastes. The answer to each subquestion is a conclusion. You may present these conclusions in the discussion, but it's usually a good idea to also draw them all together at the end of the report to prepare the way for the major conclusion.

Earlier we showed you the introduction to a report that questioned whether the class of chemicals known as phthalates endangered public health. Here are the conclusions to that report:

> Based on the observations made thus far, there is no evidence of toxicity in man due to phthalates, either from foods, beverages, or household products as ordinarily consumed or used.
>
> These observations, coupled with the limited use of phthalate-containing food packaging materials and the low rate of migration of the plasticizers from packaging material to food, support the belief that the present use of phthalates in food packaging represents no hazard to human health.[7]

All the conclusions presented are supported by evidence in the report.

In larger papers or when dealing with a controversial or complex subject, you would be wise to precede your conclusions with a summary of your facts. By doing so, you will reinforce in your reader's mind the strength and organization of your argument. For an example of such a combination, see Appendix A, pages 644–646. In any event, make sure your conclusions are based firmly upon evidence that has been presented in your report. Few readers of professional reports will take seriously conclusions based upon empty, airy arguments. Conclusions are frequently followed by recommendations.

Recommendations A conclusion is an inference. A **recommendation** is the statement that some action be taken or not taken. The recommendation is, of course, based upon the conclusions and is the last step in the process. You conclude that Brand X bread, for example, is cheaper per pound than Brand Y and just as nutritious and tasty. Your final conclusion, therefore, is that Brand X is a better buy. Your recommendation is "Buy Brand X."

Many reports such as feasibility reports, environmental impact statements, and research reports concerning the safety of certain foods or chemicals are decision reports that end with a recommendation. For example, we are all familiar with the government recommendations that have removed certain artificial sweeteners from the market and that have placed warnings on cigarette packages. These recommendations were all originally stated at the end of reports looking into these matters.

Recommendations are simply stated. They follow the conclusions, often in a separate section, and look something like this:

Based upon the conclusions reached, we recommend that our company

- Not increase the present level of iron enrichment in our flour.
- Support research into methods of curtailing rancidity in flour containing wheat germ.

Frequently, you may have a major recommendation followed by additional implementing recommendations, as in the following:

Major recommendation

We recommend that the Department of Transportation build a new bridge across the St. Croix River at a point approximately three miles north of the present bridge at Hastings.

Implementing recommendations

- The Department's location engineers should begin an immediate investigation to decide the exact bridge location.
- Once the location is pinpointed, the Department's right-of-way section should purchase the necessary land for the approaches to the bridge.

You need not support your recommendations when you state them. You should have already done that thoroughly in the report and in the conclusions leading up to the recommendations. It's likely, of course, that a full-scale report will contain, in sequence, a summary, conclusions, and recommendations. For more information about this subject and for more examples of summaries, conclusions, and recommendations, see pages 529–534 and 625–646.

Graceful Close A short, simple, nonargumentative paper often requires nothing more than a graceful exit. As you would not end a conversation by turning on your heel and stalking off without a "good-bye" or a "see you later" to cover your exit, you do not end a paper without some sort of close. In a short informational paper that has not reached a decision, the facts should be still clear in the readers' minds at the end, and they will not need a summary. One sentence, such as the following, which might end a short speculative paper on superconductivity, will probably suffice:

> Because superconductivity seems to have numerous uses, it cannot fail to receive increasing scientific attention in the years ahead.

Sometimes an appropriate quotation can be used to get you out of a report gracefully, as in the following:

Quotation to close

> As Roger Revelle and H. E. Seuss stated in 1957: Human beings are now carrying out a large-scale geophysical experiment of a kind that could not have happened in the past nor be repeated in the future. Within a few centuries we are returning to the atmosphere and oceans the concentrated organic carbon stored in the sedimentary rocks over hundreds of millions of years. This experiment, if adequately documented, may yield a far-reaching insight into the processes determining weather and climate.[8]

Combination Endings We have treated summaries, conclusions, and recommendations separately. And, indeed, it's likely that a full-scale report leading to a recommendation will contain in sequence separate sections labeled "Summary," "Conclusions," and "Recommendations." When such is the case, the summary will often be restricted to a condensation of the factual data offered in the body. The implications of the data will be presented in the conclusions, and the action to be taken in the recommendations. (See, for example, pages 529–534 and 644–646.)

However, in many reports, the major elements of factual summary, conclusions, and recommendations may be combined. A combination of summary, conclusions, and recommendations placed at the front of a report for a technical audience will likely be labeled an "Abstract." It will be, in fact, what we describe on page 290 as an informative abstract. The same combination located at the end of a report for any audience would likely be called a "Summary." A summary written specifically for an executive audience and located at the front of the report will be an executive summary. (See pages 291–292.)

It's unfortunate, perhaps, that there is a slight confusion of terms when these elements are used in different ways. Don't let the confusion in terminology confuse the essence of what is involved here. In all but the simplest reports, you must draw things together for your readers. You must condense and highlight your significant data and present any conclusions and recommendations you may have. Notice how this summary of a scientific research report smoothly combines all these elements:

Summary

Summary

In many turtles the hatchling's sex is determined by the incubation temperature of the egg, warm temperatures causing femaleness and cool temperatures maleness. Consequently, the population sex ratio depends upon the interaction of (i) environmental temperature, (ii) maternal choice of nest site, and (iii) embryonic control of sex determination. If environmental temperature differs between populations, then sex ratio selection is expected to adjust either maternal behavior or embryonic temperature-sensitivity to yield nearly the same sex ratio in the different populations.

Conclusion

Recommendation

To test this hypothesis in part, we have compared sex determining temperatures among embryos of emydid turtles in the northern and southern U.S. We predicted that embryos of southern populations should develop as male at higher temperatures than those of northern populations. The data offer no support for this prediction among the many possible comparisons between northern and southern species. The data actually refute the prediction in both of the North–South intraspecific comparisons. Further study is needed, in particular, of nest temperatures in the different populations.[9]

Appendixes

Appendixes, as the name implies, are materials appended to a report. They may be materials important as background information or needed to lend the report credibility. But they will not in most cases be necessary to meet the major purpose of the report or the major needs of the audience. For example, if you are describing research for an executive audience, they will likely be more interested in your results and conclusions than in your research methodology. If your audience consists totally of executives, you might include only a bare-bones discussion of your methodology in your report.

But suppose you had a primary audience of executives and a secondary audience of your fellow experts. You could satisfy both audiences by placing a detailed discussion of your methodology in an appendix—out of the executives' way but readily accessible for the experts. Like most decisions in technical writing, what goes into the body of a report, what goes into an appendix, and what is eliminated altogether are determined by your audience and purpose.

During the final stages of arranging your report, determine whether materials such as the following should be placed in appendixes:

- Case histories
- Supporting illustrations
- Detailed data
- Transcriptions of dialogue
- Intermediate steps in mathematical computation
- Copies of letters, announcements, and leaflets mentioned in the report
- Samples, exhibits, photographs, and supplementary tables and figures
- Extended analyses
- Lists of personnel
- Suggested collateral reading
- Anything else that is not essential to the sense of the main report

Before you place anything in an appendix, consider the effect on the report. Be certain that shifting an item to an appendix does not undermine your purpose or prevent the reader from understanding major points of the report.

Documentation

Different documentation systems are in use from college to college, journal to journal, company to company. Therefore, we cannot claim a universal application for the instructions that follow. Use them barring conflicting instructions from your instructor, college, employer, or the stylebook of the journal or magazine in which you hope to publish.

Before we go into the mechanics of documentation, it might be wise to discuss why and when you need to document.

First of all, documenting fulfills your moral obligation to give credit where credit is due. It lets your readers know who was the originator of an idea or expression and where his or her work is found. Second, systematic documentation makes it easy for your readers to research your subject further.

When do you document? Established practice calls for you to give credit when you borrow the following:

- Direct quotes
- Research data and theories
- Illustrations, such as tables, graphs, photographs, and so forth

You do not need to document general information or common knowledge. For example, even if you referred to a technical dictionary to find that creatinine's more formal name is methyglycocyamidine, you would not be obligated to show the source of this information. It is general information, readily found in many sources. If on the other hand you should include in your paper an opinion that the cosmos is laced with strands of highly concentrated mass energy called *strings,* you would need to document the source of this opinion.

In general, give credit where credit is due, but do not clutter your pages with references to information readily found in many sources. If in doubt as to whether to document or not, play it safe and document.

We explain two systems of documentation. The first involves using **notes,** the second, **author–date parenthetical references** keyed to a list of works cited. Notes, that is, footnotes and endnotes, are often used for documenting business and academic reports. Parenthetical documentation is in wide use in scientific papers and journals. Some general rules apply in both systems. We cover these rules first and then, in order, the note system and the author–date parenthetical system.

Teachers may specify one method or the other for class papers. In our explanation we follow *The MLA Style Manual,* a guide published by the Modern Language Association.[10] We include a brief comparison of the MLA author–date system to the author–date system of the American Psychological Association (APA).[11]

General Rules

The rules that follow apply whether you are using the note system or the author–date parenthetical system.

- You may use a short form of the publisher's name, for example, Wiley for John Wiley & Sons. Be consistent throughout your notes or citations, however.
- When page numbers are two digits, use both digits in the second number: 22–24 not 22–4.
- When page numbers are three or four digits, use only the last two digits in the second number: 112–15; 1034–39.

- When citing inclusive page numbers of an article, if the article is continued over to pages later in the publication, cite the first page of the article only, followed immediately by a plus sign, thus 22+.
- When a city of publication is not well known, include an abbreviated form for the state, province, or country in your note or citation. For states and provinces, use postal abbreviations, for example NY for New York and BC for British Columbia. (See page 384 for a list of state and province abbreviations.) For countries, use the abbreviations that can be found in most college dictionaries, for example, Arg. for Argentina.
- Use three-letter abbreviations for months with more than four letters in them: thus, Dec. and Apr. but May and June.
- When information on pagination, publisher, or date is missing in your source, at the point where you would put that information, put one of these abbreviations: n. pag. for no pagination, n.p. for no place of publication or no publisher, n.d. for no date.
- Illustrations are documented separately from the text. As we explain in Chapter 12, "Graphical Elements of Reports," they are documented directly on the table or figure. Because illustrations have complete documentation internally, their notes do not appear with either page footnotes or on a list of endnotes or works cited. The form of illustration notes, however, does follow the form of the notes that we describe here. For guidance, see Chapter 12 and the many notes we have provided on illustrations throughout this book, for example, those on pages 282 and 285.

Notes

In this section we first explain the note system of documentation and then provide a series of model notes that you can use as guides in constructing your notes.

The Note System Notes may be (1) displayed at the bottom of the page on which the documented material appears or (2) gathered together at the end of the report under the heading **Notes.** In the first method, the notes are called *footnotes* and in the second, *endnotes*. Footnotes are illustrated in Figure 11–14, endnotes in Figure 11–15. Endnotes are more common than footnotes in student and business reports and in journal articles. We use them in this book; see our "Chapter Notes" on pages 668–675.

Whether using footnotes or endnotes, number your notes in sequence

Current rhetorical theory indicates that this attempt, through analogy, to call on schemata for newspapers could affect readers' expectations about the writing in the newsletters, which in turn could influence the way these readers process the writing. Genre theory, for example, posits that generic patterns such as those in a newspaper, as part of our "cultural rationality,"[1] alert readers to ways of perceiving and interpreting documents.[2] In addition, theories of intertextuality, the concept that all texts contain explicit or implicit traces of other texts,[3] suggest that creating an analogy between newspapers and newsletters would affect readers' expectations, encouraging them to perceive and interpret material in a particular way.[4] We must ask, therefore, what readers' expectations about newspapers and hence, by analogy, about the newsletters, might be.

Place superscript note number after any punctuation at end of grammatical unit cited.

Use four spaces between text and notes.

Use one space between number and body of note.

Indent five spaces and repeat superscript number.

[1] C. R. Miller, "Genre as Social Action," Quarterly Journal of Speech 70 (1984): 165.

[2] Miller 159.

[3] J. E. Porter, "Intertextuality and the Discourse Community," Rhetoric Review 5 (1986): 34.

Single-space each note; double-space between notes.

[4] Porter 38.

Use no punctuation between name and page.

Figure 11–14 Footnotes on a Page
Source: Adapted with permission from Nancy Roundy Blyler, "Rhetorical Theory and Newsletter Writing," *Journal of Technical Writing and Communication* 20 (1990): 144.

through your paper. If your paper is divided into chapters, as is this book, number your notes in sequence through each chapter.

For both footnotes and endnotes, the note form is the same. In both cases the note number is indicated in the text by a superscript number, that is, a number placed above the line of type, as you can see in the many note numbers we give in this book and in Figure 11–14. Place the number in the text where it is relevant and where it disrupts the text the least. Generally, you should place the note at the end of the grammatical unit—for example, sentence or clause—that contains the material you are documenting.

Notes, whether footnotes or endnotes, appear in the order that the superscript note numbers occur in the paper. Indent the superscript number corresponding to the note number five spaces from the left margin. Begin the body of the note one space to the right of the number. If

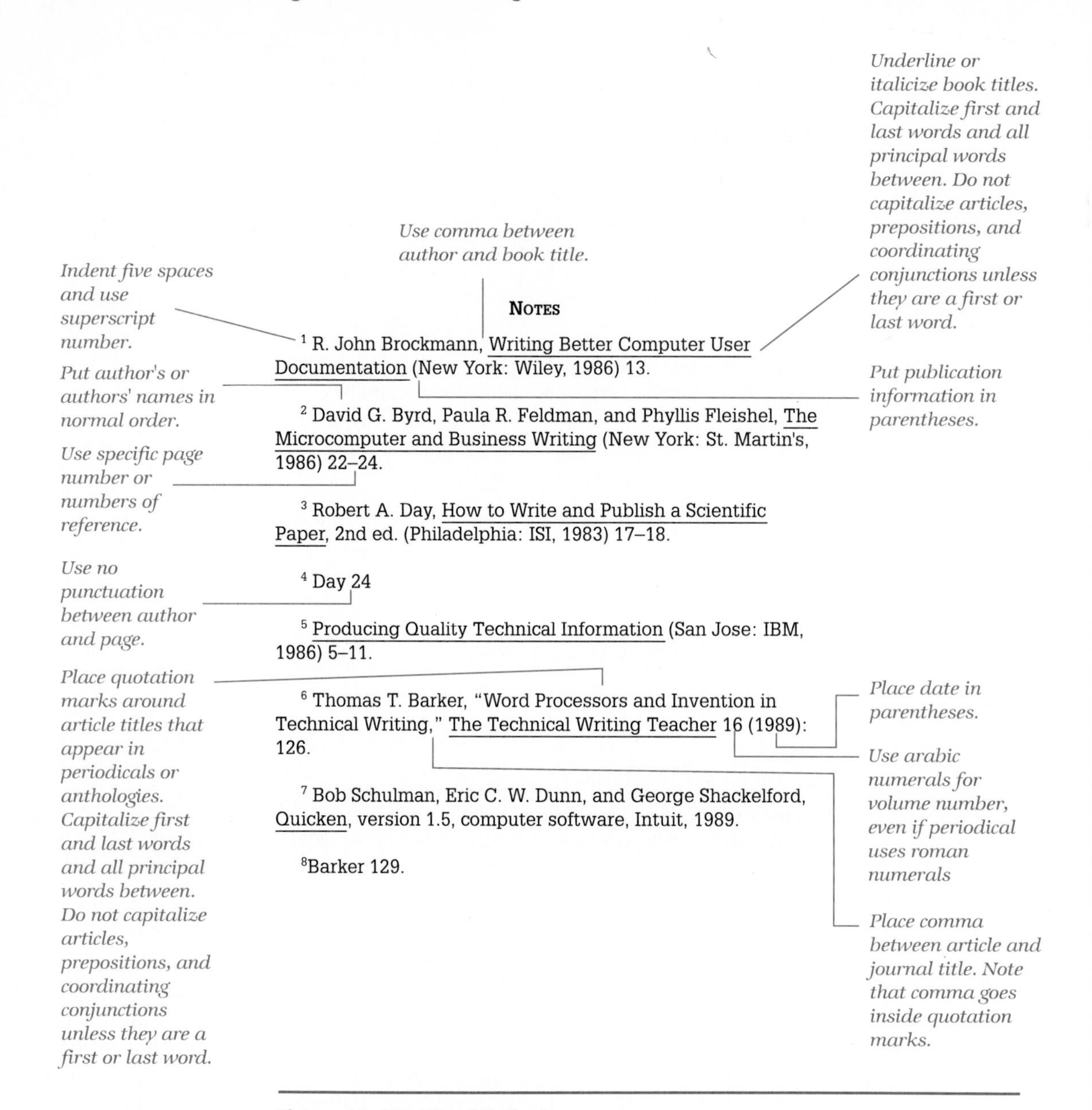

NOTES

[1] R. John Brockmann, Writing Better Computer User Documentation (New York: Wiley, 1986) 13.

[2] David G. Byrd, Paula R. Feldman, and Phyllis Fleishel, The Microcomputer and Business Writing (New York: St. Martin's, 1986) 22–24.

[3] Robert A. Day, How to Write and Publish a Scientific Paper, 2nd ed. (Philadelphia: ISI, 1983) 17–18.

[4] Day 24

[5] Producing Quality Technical Information (San Jose: IBM, 1986) 5–11.

[6] Thomas T. Barker, "Word Processors and Invention in Technical Writing," The Technical Writing Teacher 16 (1989): 126.

[7] Bob Schulman, Eric C. W. Dunn, and George Shackelford, Quicken, version 1.5, computer software, Intuit, 1989.

[8] Barker 129.

Use comma between author and book title.
Underline or italicize book titles. Capitalize first and last words and all principal words between. Do not capitalize articles, prepositions, and coordinating conjunctions unless they are a first or last word.
Indent five spaces and use superscript number.
Put author's or authors' names in normal order.
Put publication information in parentheses.
Use specific page number or numbers of reference.
Use no punctuation between author and page.
Place quotation marks around article titles that appear in periodicals or anthologies. Capitalize first and last words and all principal words between. Do not capitalize articles, prepositions, and coordinating conjunctions unless they are a first or last word.
Place date in parentheses.
Use arabic numerals for volume number, even if periodical uses roman numerals
Place comma between article and journal title. Note that comma goes inside quotation marks.

Figure 11–15 List of Endnotes

the note is longer than one line, begin subsequent lines at the left margin. In a student or company report, each line of an individual entry is single-spaced; double-spacing is used between separate entries. If you are preparing a report for publication, double-space all lines. Figures 11–14 and 11–15 illustrate these details.

Word processing can make placing and numbering footnotes and endnotes much easier than on a typewriter. Many programs will number the notes automatically. Write the note at the place in the text where it is relevant. If you have specified endnotes, the program saves the note with the other endnotes and puts them all after the text. If you have specified footnotes, the program puts the note on the bottom of the page, adjusting the rest of the text for it. If you move the relevant text to another place later, the note moves with it. When you add or delete a note, the program renumbers all the other notes. You can usually set up a different format for the notes than you have for the text. For example, you can double-space the text and single-space the notes.

Model Notes To construct your notes, consult the sample notes in Figures 11–14 and 11–15 and the examples that follow. We have annotated Figures 11–14 and 11–15 to draw your attention to certain distinctive note features, such as italics, underlining, punctuation, and spacing. (If you are typing your report, substitute underlining for italics. With word processing you may be able to use italics if you wish.) We categorize the model notes under the headings of **Books, Periodicals, Other,** and **Subsequent References** and provide an example of most of the notes you are likely to need in school or business. If you need a more complete set of notes than we have provided, consult *The MLA Style Manual* or the stylebook that guides the organization or journal where you intend to publish.

Books

The following examples illustrate various forms of book notes.

One Author

[1] R. John Brockmann, Writing Better Computer User Documentation (New York: Wiley, 1986) 13.

Two or More Authors

[2] David G. Byrd, Paula R. Feldman, and Phyllis Fleishel, The Microcomputer and Business Writing (New York: St. Martin's, 1986) 22-24.

An Anthology

[3] Paul V. Anderson, R. John Brockmann, and Carolyn R. Miller, eds., New Essays in Technical and Scientific Communication: Research, Theory, Practice (Farmingdale, NY: Baywood, 1983) 112-15.

Use the abbreviation *ed.* for one editor, *eds.*, for two or more. Use *trans.* for one or more translators.

An Essay in an Anthology

[4] Lee Odell, Dixie Goswami, Anne Herrington, and Doris Quick, "Studying Writing in Non-Academic Settings," New Essays in Technical and Scientific Communication: Research, Theory, Practice, eds. Paul V. Anderson, R. John Brockmann, and Carolyn R. Miller (Farmingdale, NY: Baywood, 1983) 27-28.

Second or Subsequent Edition

[5] Robert A. Day, How to Write and Publish a Scientific Paper, 2nd ed. (Philadelphia: ISI, 1983) 17-18.

Article in Reference Book

[6] "Petrochemical," The New Columbia Encyclopedia, 1975.

When a reference work is well known, you need cite only the name of the article, the name of the reference work, and the date of publication. If the work is arranged alphabetically, do not cite page numbers.

For less well known reference works, give complete information. Cite the author of the article if known; otherwise, begin with the name of the article.

[7] "Perpetual Calendar, 1775-2076," The New York Public Library Desk Reference, eds. Paul Fargis and Sheree Bykofsky (New York: Webster's New World, 1989) 10-13.

A Pamphlet

[8] Cataract: Clouding the Lens of Sight (San Francisco: American Academy of Ophthalmology, 1989) 1-2.

If the author of a pamphlet is known, give complete information in the usual manner.

Government or Corporate Publication

[9] U.S. National Aeronautics and Space Administration, Voyager at Neptune: 1989 (Washington, DC: GPO, 1989) 16.

Treat government and corporate publications much as you would any book, except that the "author" is often a government agency or a division within a company. GPO is the abbreviation for the U.S. Government Printing Office.

Anonymous book

[10] Producing Quality Technical Information (San Jose: IBM, 1986) 5-11.

When no human, government, or corporate author is listed, begin with the title of the book.

Proceedings

[11] Mary Fran Buehler, "Rules that Shape the Technical Message: Fidelity, Completeness, Preciseness," Proceedings 31st International Technical Communication Conference (Washington, DC: Society for Technical Communication, 1984) WE–9.

Unpublished Dissertation

[12] Penny Hutchinson, "Trauma in Emergency Room Surgery," diss., U of Chicago, 1990, 16.

Abbreviate "University" as "U"; A university press would be "UP," as in "Indiana UP."

Periodicals

The following examples illustrate various forms of periodical notes.

Journal with Continuous Pagination

[1] Thomas T. Barker, "Word Processors and Invention in Technical Writing," The Technical Writing Teacher 16 (1989): 126.

Journal That Pages Its Issues Separately

[2] David P. Gardner, "The Future of University/Industry Research," Perspectives in Computing 7.1 (1987): 5.

In this note, 7 is the volume number, 1 the issue number, and 5 the page number.

An alternative to the preceding is to treat such journals similarly to commercial magazines that page their issues separately.

[3] David P. Gardner, "The Future of University/Industry Research," Perspectives in Computing Spring 1987: 5.

Commercial Magazines and Newspapers

A weekly or biweekly magazine:

[4] Robert J. Samuelson, "End of the Third World," Newsweek 23 July 1990: 45.

Monthly or bimonthly magazine:

[5] Scott Beamer, "Why You Need a Charting Program," MacUser June 1990: 126.

Newspaper:

[6] Phillip M. Boffey, "Homeless Plight Angers Scientists," New York Times 20 Sep. 1988, national edition: 1+.

When the masthead of the paper specifies the edition, put that information in your note. Newspapers frequently change from edition to edition on the same day.

Anonymous article

[7] "Absolute," New Yorker 18 June 1990: 28.

When no author is given for an article, begin with the title of the article.

Other

Under **"Other,"** we show you model notes for computer software, information services, letters, and interviews.

Computer Software

[1] Bob Schulman, Eric C. W. Dunn, and George Shackelford, Quicken, version 1.5, computer software, Intuit, 1989.

Information Service

[2] R. Berdan and M. Garcia, Discourse-Sensitive Measurement of Language Development in Bilingual Children (Los Alamitos, CA: National Center for Bilingual Research, 1982) (ERIC ED 234 636).

Letters

[3] John S. Harris, letter to the author, 19 July 1990.

Interviews

[4] Herman Estrin, personal interview, 16 Mar. 1989.

Subsequent References

After the first complete note on an item, subsequent references need only briefly identfy the item. If the page reference is the same, use only the author's last name:

[5] Duin.

If the page numbers are different, include the new page numbers:

[6] Duin 188-90.

If you have two works by the same author, include a shortened version of the title to identify the correct work:

[7] Conniff, "Eye on the Storm" 21.

Author–Date Documentation

Author–date documentation combines parenthetical references in the text with an alphabetized list of all the works cited. The system is a common method of documentation in the sciences. Shortly, we show you how to make parenthetical references, but first we discuss how to construct the list of works cited.

Works Cited In a section headed "Works Cited," you will list all the works that you cite in your paper. Figure 11–16 illustrates how it is done. We have annotated Figure 11–16 to draw your attention to certain distinctive citation features, such as underlining, punctuation, and spacing.

The models that follow show you how to construct the citations you are likely to need in school and business. They look much like notes, but you will do some things quite differently.

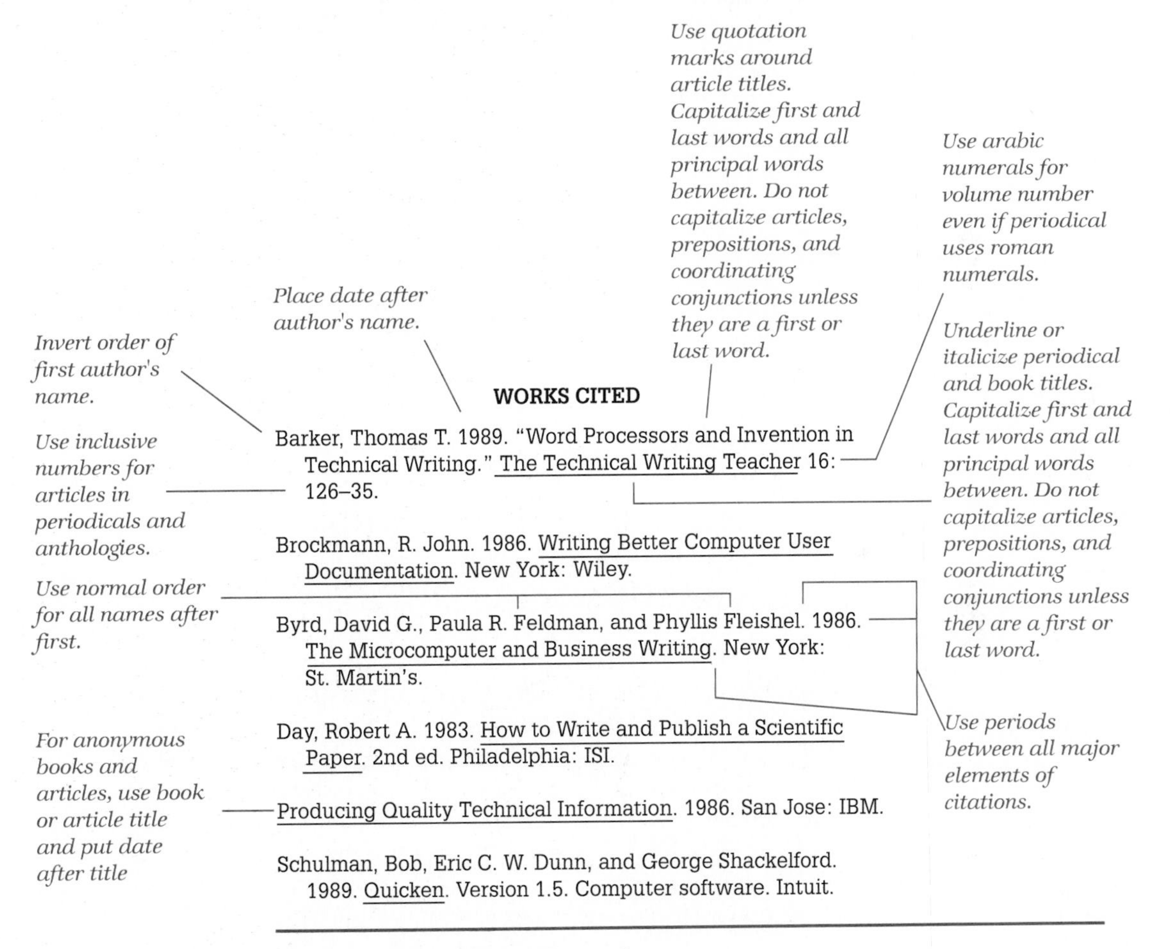

Use quotation marks around article titles. Capitalize first and last words and all principal words between. Do not capitalize articles, prepositions, and coordinating conjunctions unless they are a first or last word.
Use arabic numerals for volume number even if periodical uses roman numerals.
Place date after author's name.
Invert order of first author's name.
Underline or italicize periodical and book titles. Capitalize first and last words and all principal words between. Do not capitalize articles, prepositions, and coordinating conjunctions unless they are a first or last word.
Use inclusive numbers for articles in periodicals and anthologies.
Use normal order for all names after first.
Use periods between all major elements of citations.
For anonymous books and articles, use book or article title and put date after title
WORKS CITED
Barker, Thomas T. 1989. "Word Processors and Invention in Technical Writing." The Technical Writing Teacher 16: 126–35.
Brockmann, R. John. 1986. Writing Better Computer User Documentation. New York: Wiley.
Byrd, David G., Paula R. Feldman, and Phyllis Fleishel. 1986. The Microcomputer and Business Writing. New York: St. Martin's.
Day, Robert A. 1983. How to Write and Publish a Scientific Paper. 2nd ed. Philadelphia: ISI.
Producing Quality Technical Information. 1986. San Jose: IBM.
Schulman, Bob, Eric C. W. Dunn, and George Shackelford. 1989. Quicken. Version 1.5. Computer software. Intuit.

Figure 11–16 List of Works Cited

- Use periods between the basic components of the citation.
- Transpose the author's name: last, first, middle name or initial. If you have two or three authors, print the subsequent authors' names in normal order. If you have four or more authors, list the first author's name followed by "et al." for "and others."
- The year of publication follows the author's name and precedes the title. When you have no author's name, place the date after the title.
- Arrange the entries in alphabetical order. You need make each entry only once, doing away with the need for subsequent entry forms. Determine alphabetical order by the author's last name or, if no author is listed, by title (disregarding *the, a,* or *an*).
- Omit page numbers from whole-book entries. Give inclusive page numbers for articles in anthologies and periodicals.
- Do not number entries. Begin the first line at the left margin and indent the second and subsequent lines five spaces.

We have categorized the models under **"Books," "Periodicals,"** and **"Other."**

Books

The following examples illustrate various forms of book citations.

One Author

Brockmann, R. John. 1986. Writing Better Computer User Documentation. New York: Wiley.

Two or Three Authors

Byrd, David G., Paula R. Feldman, and Phyllis Fleishel. 1986. The Microcomputer and Business Writing. New York: St. Martin's.

Four or More Authors

Wilkinson, C. W., et al. 1986. Writing and Speaking in Business. 9th ed. Homewood, IL: Irwin.

Use the first author's name with et al. (and others) for four or more authors.

An Anthology

Anderson, Paul V., R. John Brockmann, and Carolyn R. Miller, eds. 1983. New Essays in Technical and Scientific Communication: Research, Theory, Practice. Farmingdale, NY: Baywood.

Use the abbreviation *ed.* for one editor, *eds.*, for two or more. Use *trans.* for one or more translators.

An Essay in an Anthology

Odell, Lee, et al. 1983. "Studying Writing in Non-Academic Settings." New Essays in Technical and Scientific Communication: Research, Theory, Practice. Eds. Paul V. Anderson, R. John Brockmann, and Carolyn R. Miller. Farmingdale, NY: Baywood. 17-40.

Second or Subsequent Edition

Day, Robert A. 1983. How to Write and Publish a Scientific Paper. 2nd ed. Philadelphia: ISI.

Article in Reference Book

"Petrochemical." 1975. The New Columbia Encyclopedia.

When a reference work is well-known, you need cite only the name of the article, the name of the reference work, and the date of publication. If the work is arranged alphabetically, do not cite page numbers.

For less well known reference works, give complete information. Cite the author of the article if known; otherwise, begin with the name of the article.

"Perpetual Calendar, 1775-2076." 1989. The New York Public Library Desk Reference. Eds. Paul Fargis and Sheree Bykofsky. New York: Webster's New World. 10-13.

A Pamphlet

Cataract: Clouding the Lens of Sight. 1989. San Francisco: American Academy of Ophthalmology.

If the author of a pamphlet is known, give complete information in the usual manner.

Government or Corporate Publication

U.S. National Aeronautics and Space Administration 1989. Voyager at Neptune: 1989. Washington, DC: GPO.

Treat government and corporate publications much as you would any book, except that the "author" is often a government agency or a division within a company. GPO is the abbreviation for the U.S. Government Printing Office.

Anonymous Book

Producing Quality Technical Information. 1986. San Jose: IBM.

When no human, government, or corporate author is listed, begin with the title of the book.

Proceedings

Buehler, Mary Fran. 1984. "Rules that Shape the Technical Message: Fidelity, Completeness, Preciseness." Proceedings 31st International Technical Communication Conference. Washington, DC: Society for Technical Communication. WE 9-12.

Unpublished Dissertation

Hutchinson, Penny. 1990. "Trauma in Emergency Room Surgery." Diss. U of Chicago.

Abbreviate "University" as "U"; a university press would be "UP," as in "Indiana UP."

Periodicals

The following examples illustrate various forms of periodical citations.

Journal with Continuous Pagination

Barker, Thomas T. 1989. "Word Processors and Invention in Technical Writing." The Technical Writing Teacher 16: 126-35.

Journal That Pages Its Issues Separately

Gardner, David P. 1987. "The Future of University/Industry Research." Perspectives in Computing 7.1: 4-10.

In this citation, 7 is the volume number, 1 is the issue number, and 4-10 are the inclusive pages of the article.

An alternative to the preceding is to treat such journals similarly to commercial magazines that page their issues separately.

Gardner, David P. 1987. "The Future of University/Industry Research." Perspectives in Computing Spring: 4-10.

Commercial Magazines and Newspapers

A weekly or biweekly magazine:

Samuelson, Robert J. 1990. "End of the Third World." Newsweek 23 July: 45.

Monthly or bimonthly magazine:

Beamer, Scott. 1990. "Why You Need a Charting Program." MacUser June: 126-38.

Newspaper:

Boffey, Phillip M. 1988. "Homeless Plight Angers Scientists." New York Times 20 Sep. National edition: 1+.

When the masthead of the paper specifies the edition, put that information in your note. Newspapers frequently change from edition to edition on the same day.

Anonymous article:

"Absolute." 1990. New Yorker 18 June: 28-29.

When no author is given for an article, begin with the title of the article.

Other

Under **"Other,"** we show you model citations for computer software, information services, letters, interviews, and two or more entries by the same author.

Computer Software

Schulman, Bob, Eric C. W. Dunn, and George Shackelford. 1989. Quicken. Version 1.5. Computer software. Intuit.

Information Service

Berdan, R., and M. Garcia. 1982. Discourse-Sensitive Measurement of Language Development in Bilingual Children. Los Alamitos, CA: National Center for Bilingual Research. ERIC ED 234 636.

Letter

Harris, John S. Letter to the author. 19 July 1990.

Interview

Estrin, Herman. Personal interview. 16 Mar. 1989.

Two or More Works by the Same Author

When you have two or more works by the same author or authors in your list of works cited, replace the author's name in the second citation with three unspaced hyphens and alphabetize by title.

Barker, Thomas. 1985. "Video Field Trip: Bringing the Real World into the Technical Writing Classroom." The Technical Writing Teacher 11: 175-79.

---. 1989. "Word Processors and Invention in Technical Writing." The Technical Writing Teacher 16: 126-35.

When you have two or more works in the same year by the same author or authors in your list of works cited, mark the years with lowercase letters, beginning with "a" and alphabetize by title.

Erowhurst, M. 1983a. Persuasive Writing at Grades 5, 7, and 11: A Cognitive-Development Perspective. Paper presented at the annual meeting of the American Educational Research Association, Montreal, Canada. ERIC ED 230 977.

--. 1983b. Revision Strategies of Students at Three Grade Levels. Final report. Educational Research Institute of British Columbia. ERIC ED 238 009.

Parenthetical Reference When you have completed your list of works cited, refer your reader to it through parenthetical references in your text. Figure 11–17 is a page using parenthetical references. We have annotated the figure to show you how the system works in the text. As Figure 11-17 illustrates, parenthetical references are placed before the mark of punctuation at the end of the clause or sentence that contains the material cited. Their purpose is to guide the reader to the corresponding entry in the list of works cited and, when appropriate, cite the specific pages of the reference. Some model references follow:

Author and Date

(Asher 1990)

Refers the reader to Asher's 1990 work in the list of works cited. Use this form when you are not citing a specific page.

Author, Date, and Page

(Asher 1990, 93)

Refers the reader to page 93 of Asher's 1990 work. Use this form when you are citing a specific page or pages.

Date and Page

(1990, 97)

Use this form when you have already mentioned the author's name in the passage leading up to the parenthetical reference, for example, "As Asher's research shows . . . ."

Pages Only

(324-27)

Use this form when you have mentioned both the author's name and the date in the passage leading up to the parenthetical reference, for example, "As Asher's research in 1990 shows . . . ."

Government or Corporate Author

(U.S. National Aeronautics and Space Administration 1989)

When you have a government agency or corporate division listed in the works cited as the author, you may use that in your parenthetical reference. If the name is long and clumsy, you might do better to work it into the passage leading up to the parenthetical reference. Alternatively, you might provide the reader with a shortened form, NASA, in this case.

Title of Work

(Producing Quality Technical Information 1986, 14.)

Use this form when you have no author's name and have listed the work by its title. As with the author–date reference, omit anything from the paragraph that you have mentioned in the passage leading up to the reference.

Two or Three Authors

(Berdan and Garcia 1982)

Name all the authors of a work by two or three authors.

Four or More Authors

(Odell et al. 1983, 28)

Use first author's name with et al. (and others) for four or more authors.

Two Works by the Same Author in Different Years

(Jarrett 1987, 18)
(Jarrett 1989)

When you have listed two works by the same author but written in different years, the dates will distinguish between them.

Two or More Works by the Same Author in the Same Year

(Jarrett 1987a, 18)
(Jarrett 1987b)

When you list two or more works by the same author in the same year, to distinguish them, mark them with lowercase letters, both in the parenthetical reference and the list of works cited.

MLA–APA Comparison

Becoming familiar with the MLA author–date system as described in this book will prepare you to use most of the author–date systems in use in documenting technical reports. Another prominent author–date system is that found in the *Publication Manual of the American Psychological Association* (APA). To alert you to the kinds of variations that exist from system to system, we provide a few samples to demonstrate the differences between the MLA and APA systems.

Both systems use parenthetical references in the text, but the APA system is slightly more complex:

MLA	(Asher 1990, 93)
APA	(Asher, 1990, p. 93)

Use specific page reference.

Place comma between date and page reference.

Current rhetorical theory indicates that this attempt, through analogy, to call on schemata for newspapers could affect readers' expectations about the writing in the newsletters, which in turn could influence the way these readers process the writing. Genre theory, for example, posits that generic patterns such as those in a newspaper, as part of our "cultural rationality" (Miller 1984, 165), alert readers to ways of perceiving and interpreting documents (Miller 1984, 159). In addition, theories of intertextuality, the concept that all texts contain explicit or implicit traces of other texts (Porter 1986, 34), suggest that creating an analogy between newspapers and newsletters would affect readers' expectations, encouraging them to perceive and interpret material in a particular way (Porter 1986, 38). We must ask, therefore, what readers' expectations about newspapers and hence, by analogy, about the newsletters, might be.

Place parenthetical notes before any punctuation that ends the material cited.

Figure 11–17 Parenthetical References on a Page
Source: Adapted with permission from Nancy Roundy Blyler, "Rhetorical Theory and Newsletter Writing," *Journal of Technical Writing and Communication* 20 (1990): 144.

The APA system places a comma between name and date and puts a "p" for "page" before the page number.

Both systems key their parenthetical references to a list of works cited. MLA calls the list "Works Cited." APA uses the title "References." Note the differences between typical book entries and periodical entries:

Book Entries

MLA

Title italicized; first, last, and principal words capitalized

Brockmann, R. John. 1986. Writing Better Computer User Documentation. New York: Wiley.

APA

Date in parentheses
First word only capitalized

Brockmann, R. J. (1986). Writing better computer user documentation. New York: Wiley.

Periodical Entries

MLA

Article title in quotes; first, last, and principal words capitalized

Barker, Thomas T. 1989 "Word Processors and Invention in Technical Writing." The Technical Writing Teacher 16: 126-35.

APA

Date in parentheses
Article title not in quotes; first word only capitalized

Barker, T. T. (1989). Word processors and invention in technical writing. The Technical Writing Teacher, 16, 126-35.

A full comparison of the two systems would reveal similar differences running throughout the various kinds of entries. A comparison with other systems would reveal still other differences. The moral of such comparisons is clear. You must obtain and use whatever style manual governs the publications or reports you write.

Copyright

Stringent copyright laws protect published work. When you are writing a student report that you do not intend to publish, you need not concern yourself with these laws. If, however, you intend to publish a report, you should become familiar with copyright law. Basically, copyright law requires that you get permission from the copyright holder for the use of illustrations and extended quotations. Look for information on the copyright holder on the title page of a publication or its reverse side. You can find a good summary of copyright law in *The MLA Style Manual,* published by the Modern Language Association.

Planning and Revision Checklist

You will find the planning and revision checklists that follow Chapter 2, "Composing" (pages 33–35 and inside the front cover), and Chapter 4, "Writing for Your Readers" (pages 78–79), valuable in planning and revising any presentation of technical information. The following questions specifically apply to the elements of reports. They summarize the key points in this chapter and provide a checklist for planning and revising.

Planning

- Which does your situation call for, a letter of transmittal or a preface?
- Will you bind your report? If so, remember to leave extra left-hand margin.
- Does your situation call for any information on your title page beyond the basic items: name of author, name of person or organization receiving the report, title of report, and date of submission?
- Have you used a system of headings that need to be repeated in your table of contents?
- Have you used a numbering system that needs to be repeated in your table of contents?
- Do you have enough illustrations to warrant a list of illustrations?
- Have you used symbols, abbreviations, acronyms, and terms that some of your readers will not know? Do you need a list of symbols or a glossary?
- Does your report require an abstract? Should you have an informative or descriptive abstract or both?
- Is your primary audience an executive one? Does the length of your report require an executive summary?

Revision

- Does your letter of transmittal contain clear statements of transmittal, the occasion for the report, and subject and purpose of the report? Do you need any other elements, such as a distribution list?
- Does your preface contain clear statements of the occasion for the report and the subject and purpose of the report? Do you need any other elements, such as acknowledgments?
- Is your cover suitable for the occasion of the report? Is it labeled with all the elements necessary to your situation?
- Is your title page well designed? If you have used word processing, have you kept your design simple? Does your title page contain all the elements required by your situation? Does your title describe your report adequately?
- Have you an effective design for your table of contents? Do the headings in your table of contents match their counterparts in the report exactly? Are the page numbers correct?
- Do you have a simple but clear numbering system for your illustrations? Do the titles in your list of illustrations match exactly the titles in the report? Are the page numbers correct?

Planning

- What will be the major questions in the executives' minds as they read your report? Plan to answer these questions in the executive summary.
- Are your subject, purpose, scope, and plan of development clear enough that you can state them in your introduction?
- Do you need an interest-catching step in your introduction?
- Do you need definitions or theoretical or historical background in your introduction?
- What information do your readers really need and want in your discussion? What questions will they have? What details do they want?
- What kind of ending do you need: summary, conclusion, recommendations, graceful exit, a combination of these?
- Do you have an executive audience? Should your "ending" come at the beginning of the report?
- Do you have material that would be better presented in an appendix rather than the discussion? Should you leave it out altogether?
- Do you have material you need to document: direct quotes, research data and theories, illustrations?
- Does your situation call for notes or author–date documentation? If notes, will you want footnotes or endnotes? Do you have word processing capability for your notes?
- Do you plan to publish your report? Does it contain material protected by copyright law? If so, begin seeking permission to use the material as early as possible.

Revision

- If needed, do you have a glossary and a list of symbols? Have you written your glossary definitions correctly?
- Are your abstracts written in the proper impersonal style? Does your informative abstract cover all the major points of your report? Have you avoided excessive detail? Does your abstract conform to the length requirements set for you?
- Does the executive summary answer the questions your executive audience will have? Have you included clear statements of your conclusions and recommendations? Have you held the length to one double-spaced page?
- Does the introduction clearly forecast your subject, purpose, scope, and plan of development? If they are needed, does the introduction contain definitions or theoretical or historical background?
- Does the discussion answer all the questions you set out to answer? Does it contain any material irrelevant to your subject and purpose?

Exercises

1. Reprinted in this exercise is the discussion section from a report titled *Mercury in Food.*[12] It was written for both food technologists and intelligent lay people. Its purpose is to examine the possible dangers of mercury poisoning in the food we eat. It reaches conclusions but does not offer recommendations. For the purpose of this exercise, pretend that you have written the report as a class assignment. Provide the following elements:

- Letter of transmittal to your writing teacher
- Title page that includes a descriptive abstract
- Introduction
- Summary
- Conclusions

Mercury and the Environment

During normal growth process, plants absorb mercury from the soil and air; in some instances, plants even concentrate it to small droplets of the metal. Some bacterial organisms, when exposed to inorganic mercury, can convert it to organic mercury compounds. These microscopic organisms and the material they produce, generally called alkyl mercury, may be consumed by fish or animals. Some animals and vegetables have the ability to convert organic forms of mercury back to inorganic compounds.

This constant cycling of mercury from one form to another has gone on for eons without any recognizable toxic effect on the food supply of the world. It is extremely doubtful that the concentration of mercury in the oceans has increased significantly as a result of man's use of the metal.

Awareness of Potential Dangerous Effects Increasing

Until the last two decades, man had only been vaguely aware of the problems arising from the misuse of mercury. A series of isolated incidents tragically demonstrated the potential dangers.

In 1953, a veritable epidemic hit the fishermen and their families in the villages on Minamata Bay, Japan. A number of people who were highly dependent on seafood showed signs of brain damage. Some of these cases were fatal. An intense investigation revealed that a local chemical plant was discharging a waste stream containing organic mercury into the bay. This was absorbed by the fish in the area and eventually passed on to the villagers.

After finding the cause of the problem, authorities were able to eliminate the source of pollution. Mercury levels in the bay returned to normal and once again the local fish was safe to eat.

Mercury showed its insidious effect in Sweden when naturalists noted that certain species of birds were diminishing. Here a study disclosed that these birds had been feeding on grain seeds that had been treated with mercury as a protection against fungus. A similar product was also a favorite pesticide. In 1966 the Swedish government banned or restricted these uses of mercury compounds. The bird life is reviving and the awareness of the potential problems is very much alive.

The practice of treating seeds with mercury compounds has also had its impact on humans. In 1968, a New Mexico farmer fed treated grain seed to his hogs and he and his family subsequently ate the meat from these animals. As a result, three of the farmer's children were crippled and a fourth child was born blind and retarded. Bags containing mercury treated seeds carry a label warning that the contents are poisonous to animals and humans, and such seeds are dyed a bright pink to differentiate them from untreated seeds. Unfortunately, however, these warnings were ignored or misunderstood by those involved. As a result of this, and similar incidents which occurred in other parts of the world, the United States Department of Agriculture in March of 1969 banned the practice of treating seeds with organic mercury compounds.

Maximum Allowable Exposure Levels Set

Urine specimens indicate that man and other animals absorb some mercury from their food and water. People who work in industries using mercury may show urine levels ten or twenty times higher than those who are not directly involved with the metal. The United States Department of Labor has established maximum allowable exposure levels to protect workers. Studies have shown that it is possible for a worker to absorb approximately one milligram of mercury per day from industrial exposure when working within the established allowable limits. Periodic physical examinations, which emphasize special methods of evaluation for possible neurological damage, indicate that the allowable levels are safe for the employees' health.

Guidelines Established for Levels in Food

The United States Food and Drug Administration has conducted routine analysis of foods for mercury content for several years. Practically all foods tested showed levels of mercury concentration well within the norms for the natural environmental content of the element. Only fish and fishery products showed concentrations greater than could be considered to be normal. As a result of their investigations and augmented by the Japanese and Swedish experiences, the Food and Drug Administration in 1969 established a 0.5-parts-per-million (ppm) guideline as the maximum safe limit for mercury in fish.

Only Fish Products Exceed Limits

Swordfish and tuna are the only commercially popular fish that have shown a mercury content exceeding 0.5 ppm. These two species of fish accumulate organic mercury compounds as they grow larger because they consume vast quantities of smaller fish.

Since both tuna and swordfish are caught at sea, far from any possible sources of industrial pollution, the mercury in their systems must come from natural sources. Most probably man, for many years, has eaten tuna and swordfish with concentrations of mercury higher than the established limit without signs of any harmful effect. Analysis of museum specimens of tuna caught during the period 1879 to 1909 reveals that they contain levels of mercury as high as those in fish being caught today. Scientists must, therefore, conclude that mercury levels in tuna, and very probably swordfish, have not appreciably changed in the past 90 years.

Why Man Hasn't Suffered from Eating Fish

Recent studies at the University of Wisconsin have discovered that some fish, including tuna, have a built-in mechanism for blocking and reducing the toxicity of mercury in their tissues. This research may answer the question of how man has safely eaten fish containing mercury concentrations higher than those allowed by the Food and Drug Administration. Most experts agree that 0.5 ppm level for fish has a considerable margin of safety built into it. It is heartening to recognize there have been no reported cases of mercury poisoning in the U.S.A. from eating fish.

2. This exercise uses the article on mercury in Exercise 1. Pretend you have written that article as a report for the executives and food scientists of a large food processing company. The company processes meat, fish, vegetables, and milk products such as cheese and butter. Write an executive summary for the mercury report that will answer the concerns and questions the executives would raise.
3. Compile a list of difficult and unusual terms you may have to introduce in the final report of your own selected project. Will the vocabulary problem be great enough to require you to include a glossary? If so, prepare a draft of the glossary.
4. Prepare at least a tentative table of contents for your proposed final report. Use at least two levels of headings.
5. Decide what material, if any, you should place in the appendix of your final report. If it is placed in an appendix, who would use the material and for what purpose? Why would you not either include this material somewhere in the discussion of your report or eliminate it altogether?
6. In a published report or textbook, locate as many of the elements discussed in this chapter as possible. What variations and departures do you find?
7. The following sources are given in the order in which they are cited in a report but without attention to proper format. List them in proper order and with proper format as they would appear in (a) a list of notes and (b) a list of works cited.

   David N. Dobrin. Writing and Technique. Urbana, Illinois. National Council of Teachers of English. 1989. Page 22 cited.

   Myra Kogen, editor. Writing in the Business Professions. Urbana, Illinois. National Council of Teachers of English. 1989. Page 46 cited.

   Jeanne W. Halpern. An Electronic Odyssey. Printed on pages 157 to 189 of Writing in Nonacademic Settings. Lee Odell and Dixie Goswami, editors. New York, New York. The Guilford Press. 1985. Pages 160 to 161 cited.

   Nancy Allen, Diane Atkinson, Meg Morgan, Theresa Moore, Craig Snow. Shared-Document Collaboration: A Definition and Description. Printed on pages 70 to 90 of Iowa State Journal of Business and Technical Communication. Volume 1. Issue number 2. September, 1987. Page 72 cited. Journal pages its issues separately.

   David H. Freedman. Common Sense and the Computer. Printed on pages 64 to 71 of Discover. August 1990. Page 67 cited.

David N. Dobrin. Writing and Technique. Urbana, Illinois. National Council of Teachers of English. 1989. Page 32 cited.

U.S. National Aeronautics and Space Administration. Search for Extra Terrestrial Intelligence. Washington, District of Columbia. U.S. Government Printing Office. No date. Page 18 cited.

David H. Freedman. Common Sense and the Computer. Printed on pages 64 to 71 of Discover. August 1990. Page 66 cited.

Richard C. Freed and David D. Roberts. The Nature, Classification, and Generic Structure of Proposals. Printed on pages 317 to 352 of Journal of Technical Writing and Communication. Volume 19. Issue number 4, 1989. Page 319 cited. Journal with continuous pagination.

Design and Typography. Printed on pages 561 to 584 of The Chicago Manual of Style. 13th Edition. Chicago, Illinois. The University of Chicago Press. 1982. Page 570 cited.

CHAPTER 12

Graphical Elements of Reports

OBJECTS

- Photographs
- Drawings

CONCEPTS AND PROCESSES

TRENDS, RELATIONSHIPS, AND SUMMARIES

Tables
Graphs

COMPUTER GRAPHICS

Prose informs its readers in a sequential way; a **graphic** informs in an instantaneous way. We can illustrate this concept easily enough with the cartoon in Figure 12–1. Look at the cartoon for a moment, then render into words what you have seen. Your rendering will be a series of statements that pick out features of the cartoon and describe them in a sequential way. The nature of language makes any other approach impossible. Because many technical concepts are more easily understood when the reader can see them in the instantaneous, overall way that you see the cartoon, graphics are often indispensable in technical writing.

Graphics in technical writing are not decoration. Graphics simplify and clarify concepts. You can use them to reenforce points and to summarize masses of statistical data. Graphics can show trends and relationships in an instant that might take many words to explain. The planning and invention of graphics should proceed at the same time and pace as the planning and invention of your prose. The same factors of situation, audience, and purpose that guide your writing choices will guide your graphics choices.

In technical writing you will often combine the use of words and graphics, as in the flowchart in Figure 12–2. The combination of words and graphics in that figure illustrates the process described there far more swiftly and efficiently than either words or graphics could alone. The

Figure 12–1 Cartoon
Source: Public Health Service, *Noninsulin-Dependent Diabetes* (Washington, DC: U.S. Department of Health and Human Services, 1987) 7.

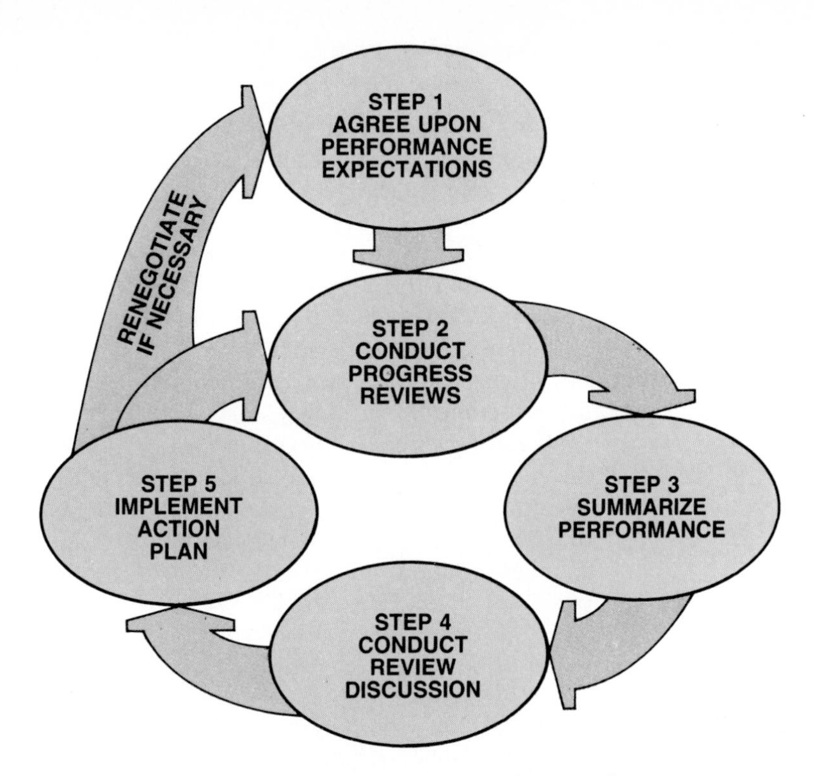

The Performance Appraisal and Review Process

The Performance Appraisal and Review Process consists of five interrelated steps illustrated in Figure 1. The goal of the process is increased employee performance improvement and job satisfaction. To attain this goal, the process has been designed to clarify performance expectations and to provide a way to evaluate performance and improvement levels throughout the entire appraisal period. The five steps to the process are:

Step 1—Agree upon Performance Expectations

Step 2—Conduct Progress Reviews

Step 3—Summarize Performance

Step 4—Conduct Review Discussion

Step 5—Implement Action Plan

Figure 12–2 Flowchart: The combination of words and graphics illustrates the process better than either could alone.
Source: A Guide for Honeywell Avionics Division Supervisors (Minneapolis: Honeywell, 1982) 4. Reprinted by permission.

amount of prose explanation you provide with a graphic will depend on the importance of the graphic to your exposition, the complexity of the graphic, and the technical expertise of your audience. Generally speaking, the more important and complex the graphic, and the less expert your audience, the more prose explanation you need. In any event, always at least mention the graphic to your reader. And mention it early in the discussion. Do not lead readers through a complicated prose explanation, and then refer to the graphic that simplifies the whole explanation. Send them to the graphic immediately, and they can cut back and forth between the prose and the graphic as necessary.

When a graphic is important to your explanation, locate it as close to the explanation as you can. If it is small, put it right on the text page. If the graphic itself takes up a page, place it facing the prose explanation or at least on an adjoining page. When graphics are less important to your explanation, they may be located further away, perhaps in an appendix.

In this chapter we provide you with some basic information about the theory and practice of graphics. We tell you and demonstrate for you how and when to use graphics to show objects, illustrate concepts and processes, and show trends and relationships.[1] We conclude with a section on the use of computer graphics.

Objects

In showing objects, you have a choice of photographs or drawings.

Photographs

Use photographs when you want to add realism to reports and sometimes a bit of drama. As Figure 12–3 illustrates, photographs are as close as you can come to showing the object as it really is without showing the object itself. The photograph of the gyro in Figure 12–4 illustrates a simple way to show scale. Be sure your photographs are as high a quality as technology and your budget allow. When using photomicrographs, for which the quality may be lower than in standard photographs, be sure to tell your readers what they are seeing and what the size or magnification is, as illustrated in Figure 12–5.

Drawings

Despite the realism of photographs, drawings have several advantages over them. For one thing, the artist can include as much or as little detail as purpose and audience demand. The features of the underwater re-

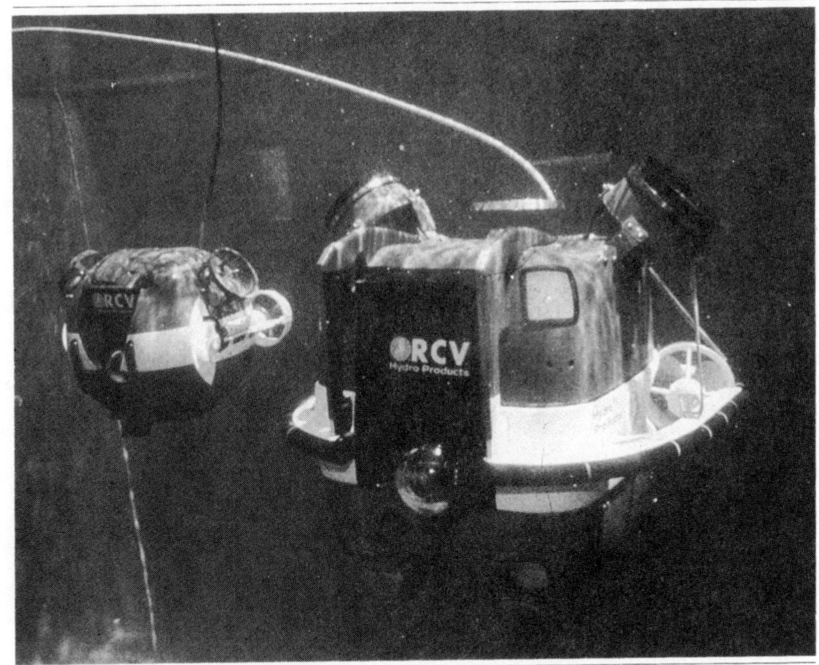

Remotely Controlled Underwater Vehicles

Figure 12–3 Photograph: Photographs lend themselves well to realistic portrayal.
Source: Scientific Honeyweller June 1984: cover. Courtesy of *Scientific Honeyweller,* all rights reserved.

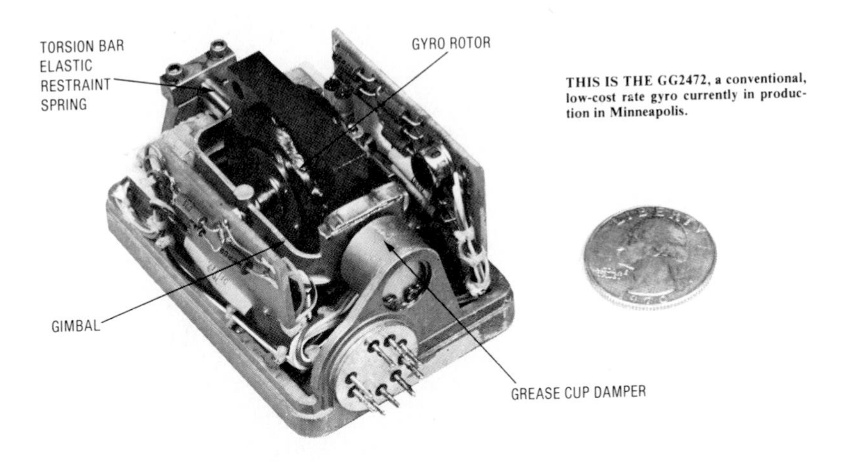

Figure 12–4 Annotated Photograph: Notice use of the coin to show scale.
Source: Jack Lower and Hank Dinter, "A History of Honeyweller Gyros and Accelerometers," *Scientific Honeyweller* September 1982: 9. Courtesy of *Scientific Honeyweller,* all rights reserved.

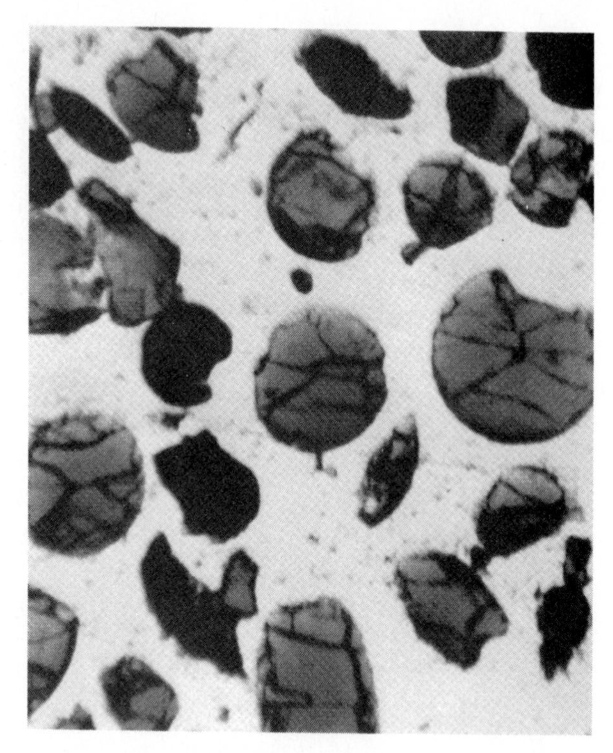

Photomicrograph of the Apollo 17 orange soil showing that it is composed of glass spherules measuring about 100 microns (four-thousandths of an inch) in diameter.

Figure 12–5 Photomicrograph of Lunar Soil: In a micrograph, always indicate size of object or degree of magnification.
Source: Maria Zuber et al., eds., *Planetary Geosciences*—1988 (Washington, DC: NASA, 1989) 58.

motely controlled vehicle stand out more clearly in the drawing in Figure 12–6 than they do in the photograph in Figure 12–3.

In drawings the artist can "cut away" barriers to sight and allow the reader to see inside the object, as in Figures 12–7 and 12–8.

The artist can also "explode" the object to show clearly the relationships among the parts, as in Figure 12–9.

Frequently, photographs and drawings are used together, as was the case with Figures 12–3 and 12–6. The photographs provide realism and drama, the drawings simplicity and accurate reference.

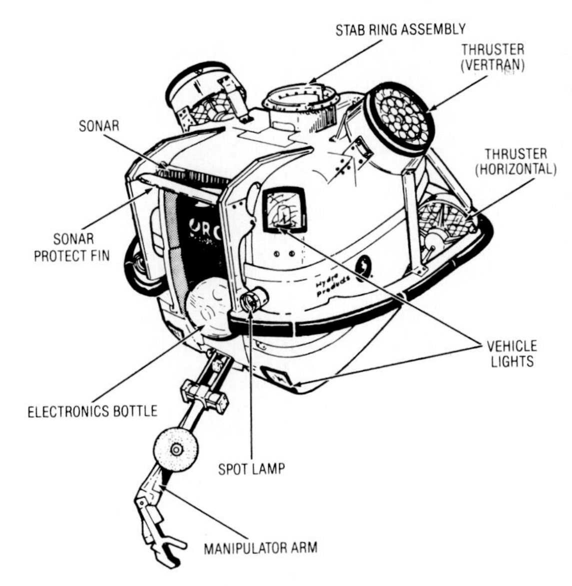

RCV-150 HAS FOUR THRUSTERS: two horizontal thrusters and two vertran (vertical and tranverse) thrusters. The selective powering of the thrusters allows the vehicle to move forward at 2.5 knots, backward at 2 knots, to port or starboard at 1.5 knots, up or down at 1 knot and to rotate at a rate of 50 degrees per second. The vehicle ascends into its launcher from below and is secured by the stab-ring assembly. The manipulator arm, shown extended, is normally stowed within the lower framework of the vehicle.

Figure 12–6 Drawing: Drawings allow better selection of detail than photographs. Generally annotation is clearer on drawings than photographs. Compare with Figure 12–3.
Source: Arthur A. Billet, "Underwater Remotely Controlled Vehicles," *Scientific Honeywеller* June 1984: 8. Courtesy of *Scientific Honeywеller,*

Figure 12–7 Cutaway drawing of Sealab III: Cutaways as in this figure and Figure 12–8 allow readers to see inside objects.

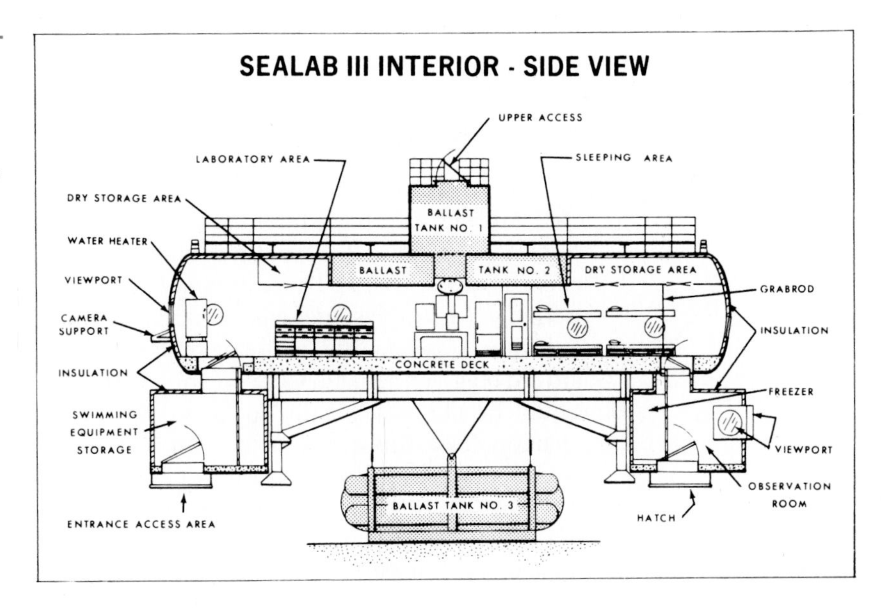

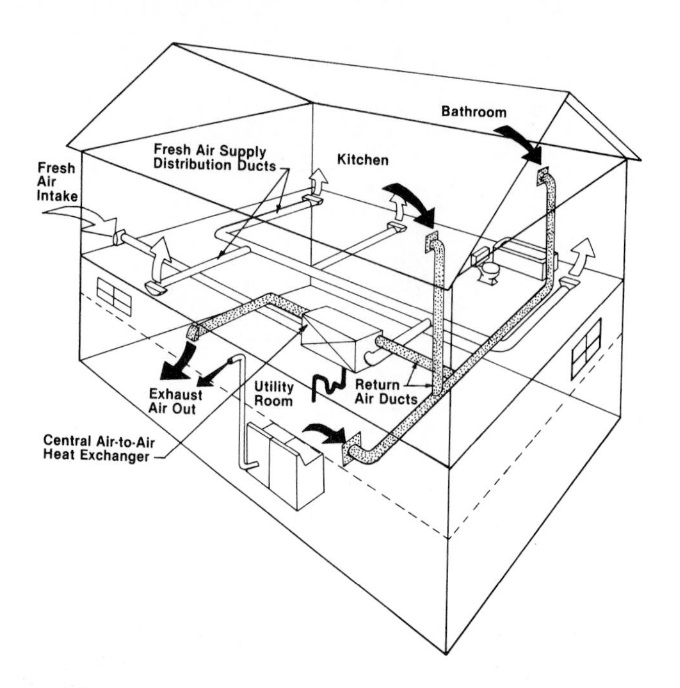

A fully ducted central air-to-air heat exchanger has the capability of distributing fresh air to and stale air from all areas of the house.

Figure 12–8 Cutaway drawing of central air-to-air heat exchanger: Notice use of annotations.
Source: U.S. Department of Energy, *Heat Recovery Ventilation for Housing* (Washington, DC: GPO, 1984) 9.

Annotations are of major importance in graphics, and you will want to do them properly. Look at the annotations in Figures 12–4 and 12–6 through 12–10. Notice that they can be done in capital letters (Figures 12–6 and 12–7) and a combination of uppercase and lowercase letters (Figure 12–8). The combination of uppercase and lowercase is probably the simplest and easiest to read. Annotations must be horizontal to the page, as they are in all our figures, rather than parallel to the lines and arrows connecting them to the object. Lines and arrows, if used, should be obvious but not so thick and heavy as to divert the reader's attention from the annotations. Use of a second color for the lines, if within your budget, aids clarity.

Figure 12–10 illustrates how you can bracket several lines of text, if that is what you are annotating. It also demonstrates how lining up your annotations simplifies the reading of them.

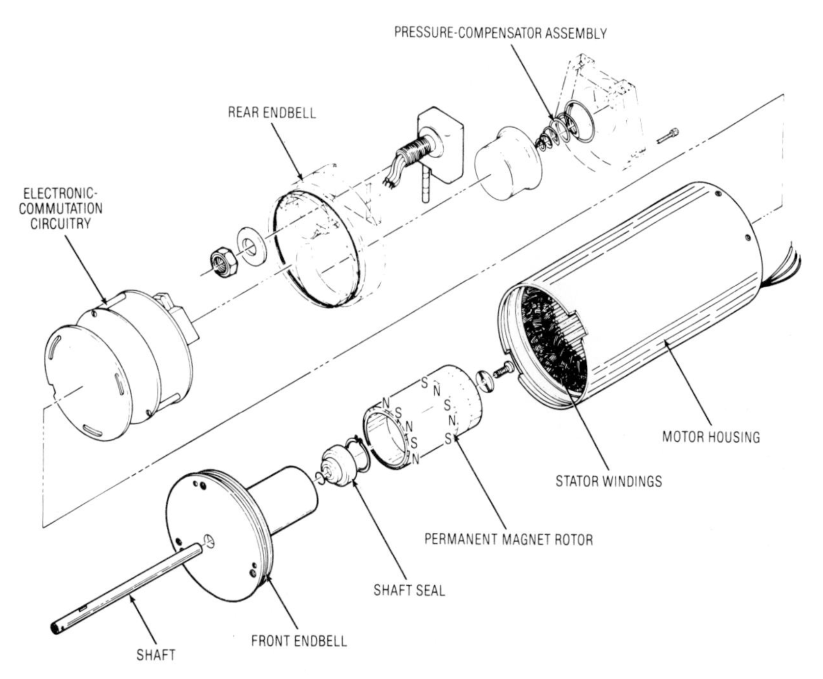

STRONG PERMANENT MAGNETS AND ELECTRONIC COMMUTATION made it possible to achieve higher thrust without increasing the size or weight of the motors in the RCV-225. Hall-effect devices that detect the position of samarium-cobalt magnets control the switching of solid-state power transistors that pass current to stator windings. The selective switching of the transistors provides the rotating magnetic field that drives the rotor. The motor is immersed in oil so that it can be pressure compensated and the weight of a pressure-protective housing can be avoided.

Figure 12–9 Exploded Drawing: Exploded drawings clarify relationships among parts.

Source: Arthur A. Billet, "Underwater Remotely Controlled Vehicles," *Scientific Honeyweller* June 1984: 6. Courtesy of Scientific Honeywell, all rights reserved.

Concepts and Processes

You can express concepts and processes in graphics. You can express **concepts**—that is, general ideas or understandings—in graphics for all levels of audience. Figures 12–11 to 12–13 illustrate graphics suitable for audiences with little technical knowledge. In Figure 12–11, the artist has superimposed a greenhouse over the earth to illustrate the concept of the greenhouse effect. In Figure 12–12, the artist with a simple drawing and a few arrows conceptualizes one way in which radon is driven from a house.

Eastman Kodak Company wanted to dramatize the concept that photographers would get better results if they used matched equipment in

Explanations are lined up for fast reading.

Arrows are unobtrusive. They lead the eye into information accurately and quickly and don't distract from it.

Color helps because there is little space between the lines in the listing, and colored lines are more visible. To accomplish the same effect, if you cannot use color, use more space and a slightly heavier line.

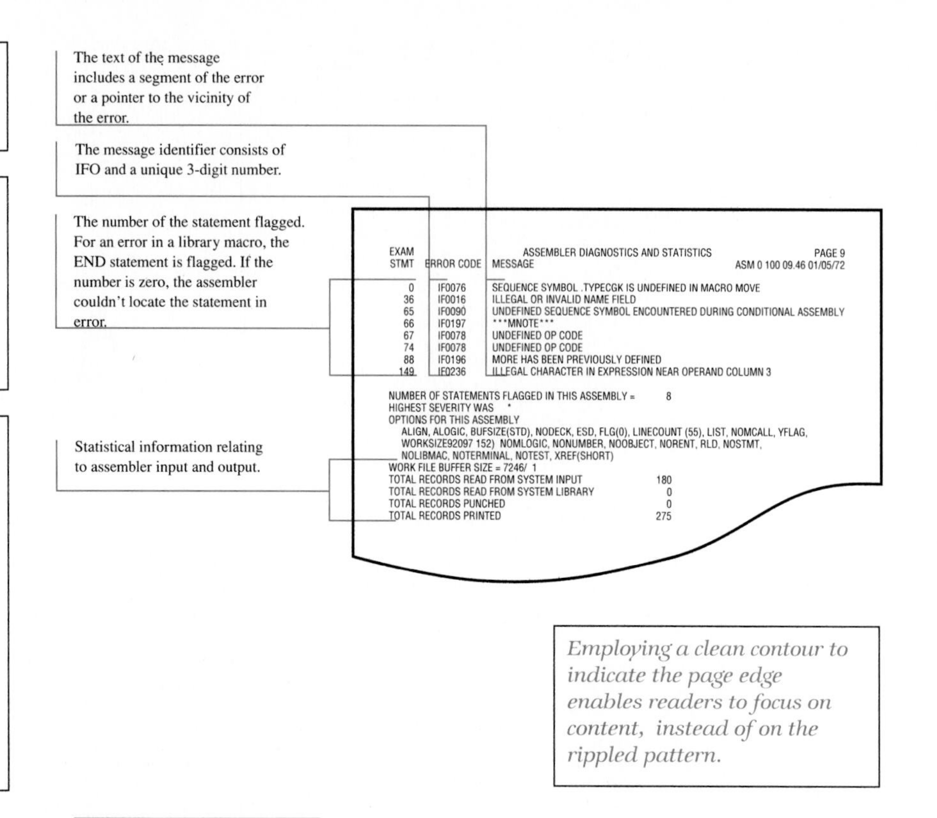

Figure 12–10 Annotated Text
Source: Information Development, Santa Teresa Laboratory, *Producing Quality Technical Information* (San Jose, CA: IBM, 1986) 28.

their work. To do so, as shown in Figure 12–13, they presented in a fresh, visual way the old saying that a chain is no stronger than its weakest link.

Often, words and graphics will be combined to explain or define a concept. In Figure 12–14, writing for a lay audience, the author supports the verbal definition of a cataract with two graphics. One shows light rays in a normal eye; the second, light rays in an eye with a cataract.

You can illustrate concepts visually for expert and technical audiences as well as for lay and executive audiences. Figure 12–15 demonstrates how words and drawings can work together to illustrate for such audiences the concept of a continuously variable transmission. Figure 12–15 also illustrates well how graphics can be placed adjacent to the verbal text, allowing the reader to move easily back and forth between the two.

Like concepts, **processes**—that is, operations or actions—can be well

Figure 12–11 Greenhouse Effect Concept: Placing stylized greenhouse over globe graphically illustrates greenhouse effect.
Source: Agricultural Research March 1988: cover.

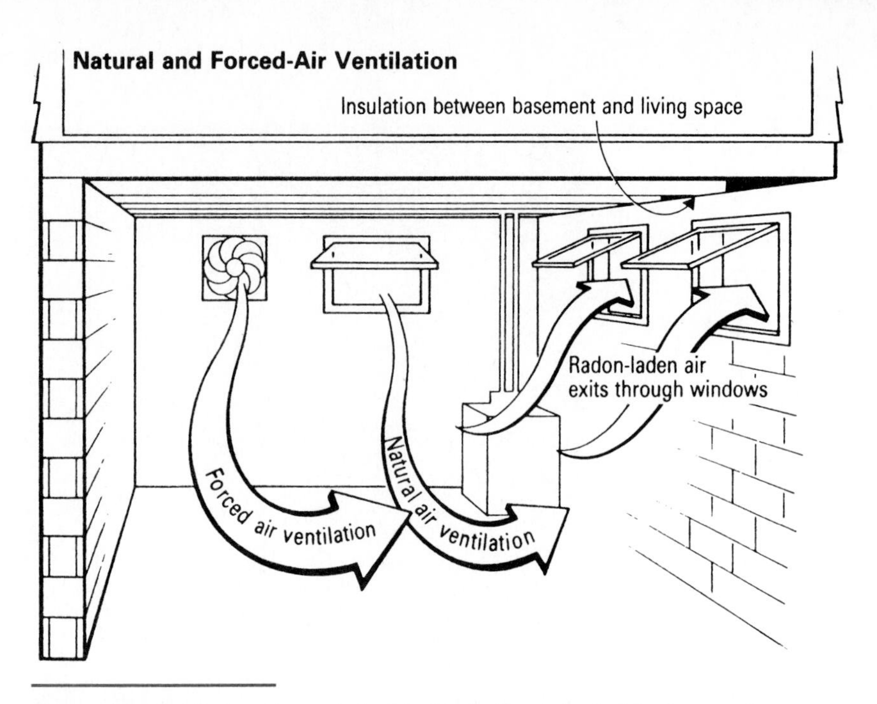

Figure 12–12 Radon Concept: In this illustration intended for lay people, arrows show the effect of ventilation.
Source: U.S. Environmental Protection Agency, *Radon Reduction Methods* (Washington, DC: GPO, 1989) 18.

Figure 12–13 A Chain Is No Stronger than Its Weakest Link.
Source: Copyright © Eastman Kodak Company, 1986. Courtesy of Eastman Kodak Company, all rights reserved.

Figure 12–14 What Is a Cataract? Graphics and words are combined to define a cataract for the lay reader. *Source: Cataract: Clouding the Lens of Sight* (San Francisco: American Academy of Ophthalmology, 1989) 1–2. Copyright © American Academy of Ophthalmology, 1989, all rights reserved.

What Is A Cataract?

A cataract is a clouding of the normally clear and transparent lens of the eye. It is not a tumor or a new growth of skin or tissue over the eye, but a fogging of the lens itself. When a cataract develops, the lens becomes cloudy like a frosted window and may cause a painless blurring of vision.

The lens, located behind the pupil, focuses light on the retina at the back of the eye to produce a sharp image. When a cataract forms, the lens can become so opaque and unclear that light cannot easily be transmitted to the retina. Often, however, a cataract covers only a small part of the lens and if sight is not greatly impaired, there is no need to remove the cataract. If a large portion of the lens becomes cloudy, sight can be partially or completely lost until the cataract is removed.

There are many misconceptions about cataracts. For instance, cataracts do not spread from eye to eye, though they may develop in both eyes at the same time. A cataract is not a film visible on the outside of the eye. Nor is it caused from overuse of the eyes or made worse by use of the eye. Cataracts rarely develop in a matter of months. They usually develop gradually over many years. Finally, cataracts are not related to cancer. Nor does having a cataract mean a person will be permanently blind.

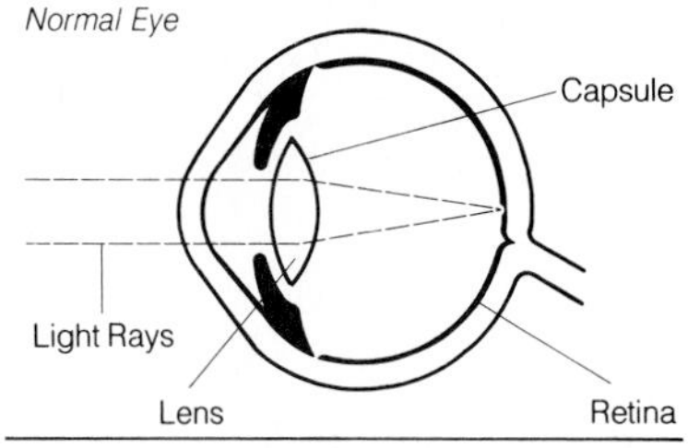

The lens focuses light on the retina.

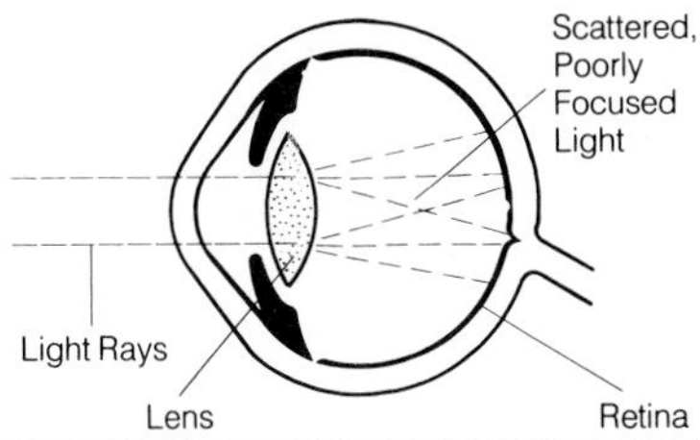

As a cataract forms, the lens becomes opaque and light cannot easily be transmitted to the retina.

illustrated for any audience by graphics or a combination of graphics and prose. In Figure 12–16, the author demonstrates for a lay audience the process by which a heat exchanger warms a house. For the sake of simplicity, the structure of the heat exchanger is reduced to a **schematic,** that is, a diagram of the actual mechanism. The verbal text, in the form of annotations, supports the visual presentation.

In Figure 12–17, writing for a technically educated audience, the writer uses several schematics to clarify key points in the explanation.

Diagrams that illustrate the flow of a process are called, not unnaturally, **flowcharts.** The flowchart in Figure 12–2 on page 331 demonstrates how a business process can be illustrated for an executive audience. The flowchart in Figure 12–18 displays a physical process for a technical and expert audience. Also for a technical and expert audience, the flowchart in Figure 12–19 shows the use of symbolic artwork, in this case, telephones and computer terminals. Such artwork is now available

Continuously Variable Transmission

A chain slides along two cones, in a novel transmission concept.

Langley Research Center, Hampton, Virginia

A transmission proposed at Langley Research Center includes a chain drive between two splined shafts. The chain sprockets follow the surfaces of two cones. As one chain sprocket moves toward the smaller diameter, the other chain sprocket moves toward the larger diameter, thereby changing the "gear" ratio. The movement is initiated by tension applied to the chain by a planetary gear mechanism.

The speed ratio between drive and driven shafts is a function of the position of the chain with respect to the conical surfaces. For example, if the plane of rotation of the chain is at the end of the axes of the cones, as shown in Figure 1, then the speed ratio (output speed/input speed) is at its minimum value. However, if the chain is positioned at the opposite ends of the cone axes, the speed ratio between the shafts is at its maximum.

Positioning the chain along the surfaces of the cones is accomplished by the movement of the hubs along the splined inner shafts of the cones. This movement of the hubs is transmitted to the chain through gear trains that move up or down cogs that radiate from the hubs. Movement of the gear assembly up and down the cogs, as shown in Figure 2, causes a simultaneous rotation of the pinion on the rack. This rotation of the pinion produces a rotation of the worm, which in turn causes a rotation of the worm wheel and chain sprocket. Thus, the rotation of the chain sprocket is synchronized with the change in circumference of the chain so that a constant tension is maintained in the chain as the chain and gear assembly move parallel to the cone surface.

Previous devices such as spur gears, belt and chain-link drives, and conventional hydraulic and mechanical automobile transmissions do not provide a simple, compact, or efficient method for continuously varying the speed ratio between driving and driven components. This device is positive, simple, and efficient over a wide range of speed ratios.

This work was done by David C. Grana of **Langley Research Center.** *No further documentation is available.*
LAR-12844

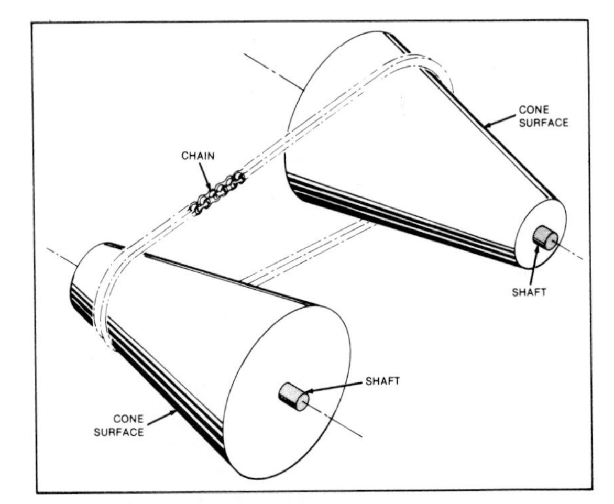

Figure 1. **The Speed Ratio** between driving and driven shafts is a function of the position of the chain with respect to the conical surfaces.

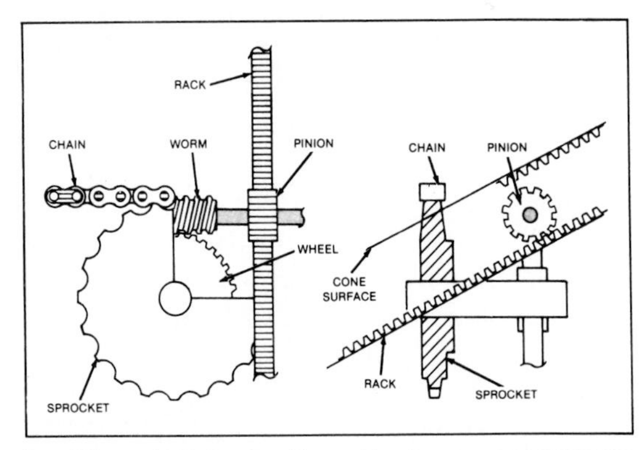

Figure 2. **Movement of the Gear Assembly** up and down the cogs causes a simultaneous rotation of the pinion on the rack.

Figure 12–15 Words and drawings working together: Technical people frequently combine words and illustrations to make their point. *Source: NASA Tech Briefs* Spring 1985: 128.

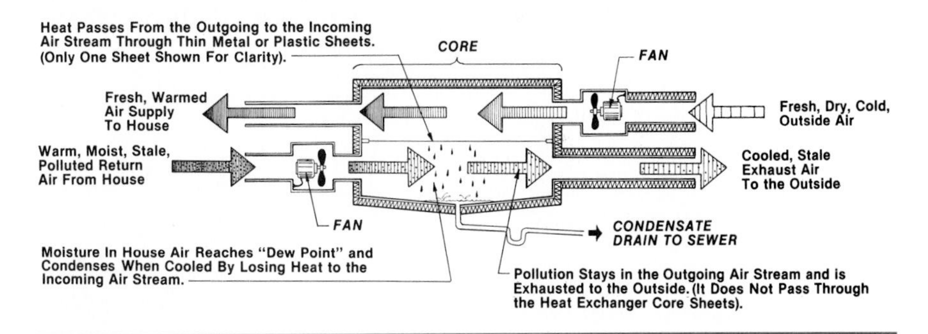

Figure 12–16 Heat exchanger: The combination of schematics and annotations illustrates the process.
Source: U.S. Department of Energy, *Heat Recovery Ventilation for Housing* (Washington, DC: GPO, 1984) 5.

through the "clip art" files of computer graphics programs, increasing the ease with which interesting and useful charts can be produced.

For ease of reading, the words in flowcharts, like the words in annotations, are printed horizontal to the page. You will see exceptions to this, for example in Figure 12–2, but you should violate this principle only with good reason.

Algorithms—processes with decision points in them—can be presented in specialized flowcharts that display how the flow through a process can be rerouted or stopped, depending on certain variables. They are particularly useful for technical and expert audiences, as illustrated in Figure 12–20. However, they also are useful for leading nontechnical audiences through complicated instructions, as demonstrated in Figure 12–21. You can also use informal tables or lists to display algorithms, as in Figure 12–22.

Trends, Relationships, and Summaries

A major use of graphics is displaying trends and relationships, for example, showing employment trends over time or comparing the volume of electricity used from one region to another. A closely related use, one that is often combined with showing trends and comparisons, is summarizing statistical data. For all three uses, tables and graphs are your primary tools.

(Text continued on page 348.)

Controlling the Focus in Electron-Beam Welders

A set of parallel whirling wires samples the beam to determine its focus.

Marshall Space Flight Center, Alabama

A detector using two whirling wires measures the focus of the beam in an electron-beam welder. Conventional beam-focus detectors employ only a single whirling tungsten wire.

In the single-whirling-wire detector, the potential difference between the wire and ground is sampled, and the time the wire spends in the beam is determined. An oscilloscope displays the beam focus as a variable-width peak, which is monitored as the focus control is adjusted. However, the intense beam at the sharp focus sometimes melts the wire.

In contrast, the multiple-wire beam-sampling method provides for a simple null-meter focus indication that is easily controlled by the operator. The detector not only operates at high beam currents but also eliminates the need for an oscilloscope.

The principle of operation of the beam-focus detector is shown in the figure. Two or more wires, located at two different heights and separated by insulated spacer rings, sample the beam. The whirling wires sample the beam as before; but in this case the outputs from the wires are compared by an electronic circuit, and the resulting signal is fed to a null-meter. The operator merely adjusts the focus control to obtain a null reading on the meter. At that time the beam focus will be centered between the two parallel layers of wires.

This work was done by Douglas I. Macfarlane and Kirk W. Spiegel of Rockwell International Corp. for **Marshall Space Flight Center.** *No further documentation is available.*
MFS-19814

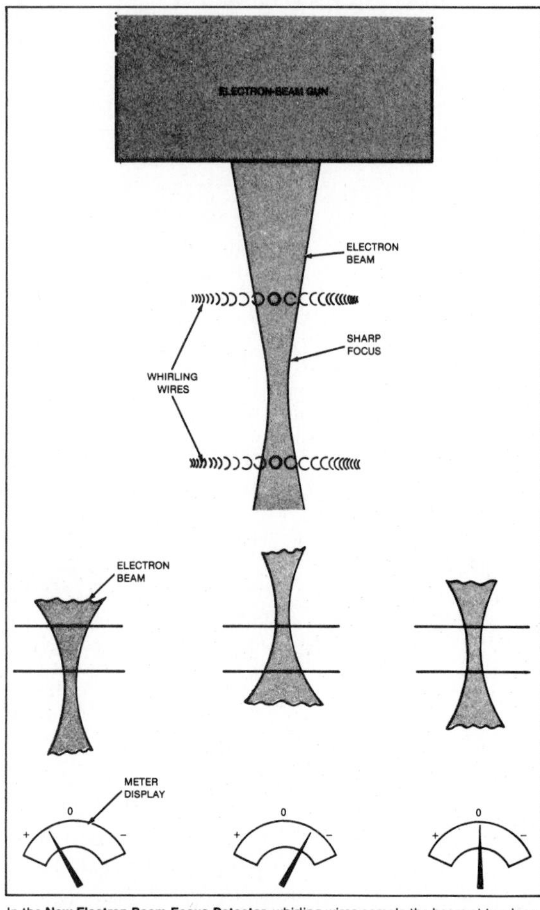

In the **New Electron-Beam Focus Detector,** whirling wires sample the beam at two locations. The new detector can be used at high levels of beam power because the wires are not located at the sharp focus of the beam.

Figure 12–17 Process illustration: Schematics, text, and annotations are combined in a process description.
Source: NASA Tech Briefs Winter 1983: 259.

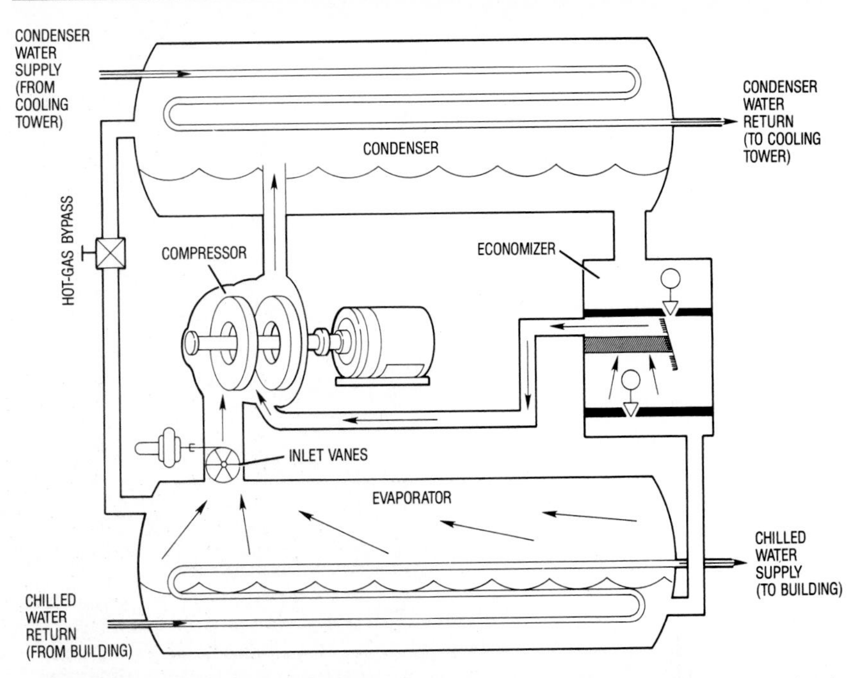

CENTRIFUGAL-CHILLER SYSTEMS extract heat from the air inside a building and transfer it to the outside. Within the building, chilled water circulates in cooling coils. Inside air is blown through the cooling coils, cooling the air and warming the water. The warmed water is recooled in the evaporator, where its heat is transferred to a refrigerant. The temperature differential between water entering the evaporator and water leaving it is about 10 degrees Fahrenheit. The evaporated refrigerant passes into the compressor through inlet vanes. The rotation of the compressor's impeller increases the refrigerant's pressure enough to condense it. When this phase transition occurs, a substantial amount of heat is released. The heat is transferred from the condenser to the water in the condenser coils. The condenser water is then sent to a cooling tower, where the heat is released to the outside air by evaporation. The hot-gas bypass helps keep the chiller operating properly at low loads. The economizer pre-cools the refrigerant before it enters the evaporator.

Figure 12–18 Physical process flowchart: A flowchart can clarify complicated physical processes.
Source: Charles H. Culp, "Mentor and AI Technology Transfer," *Scientific Honeywell* Summer 1988: 30. Courtesy of *Scientific Honeywell*, all rights reserved.

Figure 12–19 Symbolic artwork: Symbols make graphs interesting and useful.
Source: Cornell Drentea, "Local Area Networks," *Scientific Honeywell* December 1984: 42. Courtesy of *Scientific Honeywell*, all rights reserved.

HYBRID LOCAL-AREA NETWORK, the product of Ztel Inc. in Wilmington, Massachusetts, combines local-area networks and private branch exchanges (the system processing units). It consists of two networks, a token-passing packet ring *(hatched ring)* over which data is transmitted at high speed and a circuit-switched ring *(solid ring)* over which pulse-code-modulated speech and data are transmitted at lower speed. As the connections at the bottom of the illustration show, a data call from a microcomputer to a computer on a local-area network would be sent over the token-passing ring, whereas a telephone call from one location to another would be routed on the circuit-switched ring. Both data and telephone calls can also be switched directly between an incoming and an outgoing line. The rings, however, greatly reduce the amount of wiring necessary to connect many nodes, substituting multiplexing on one medium for many connections.

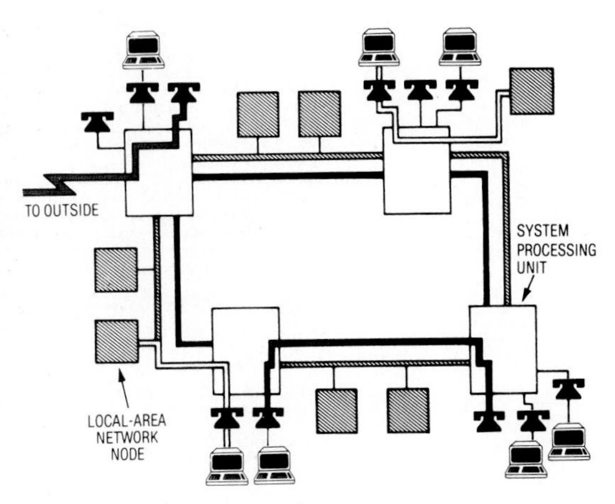

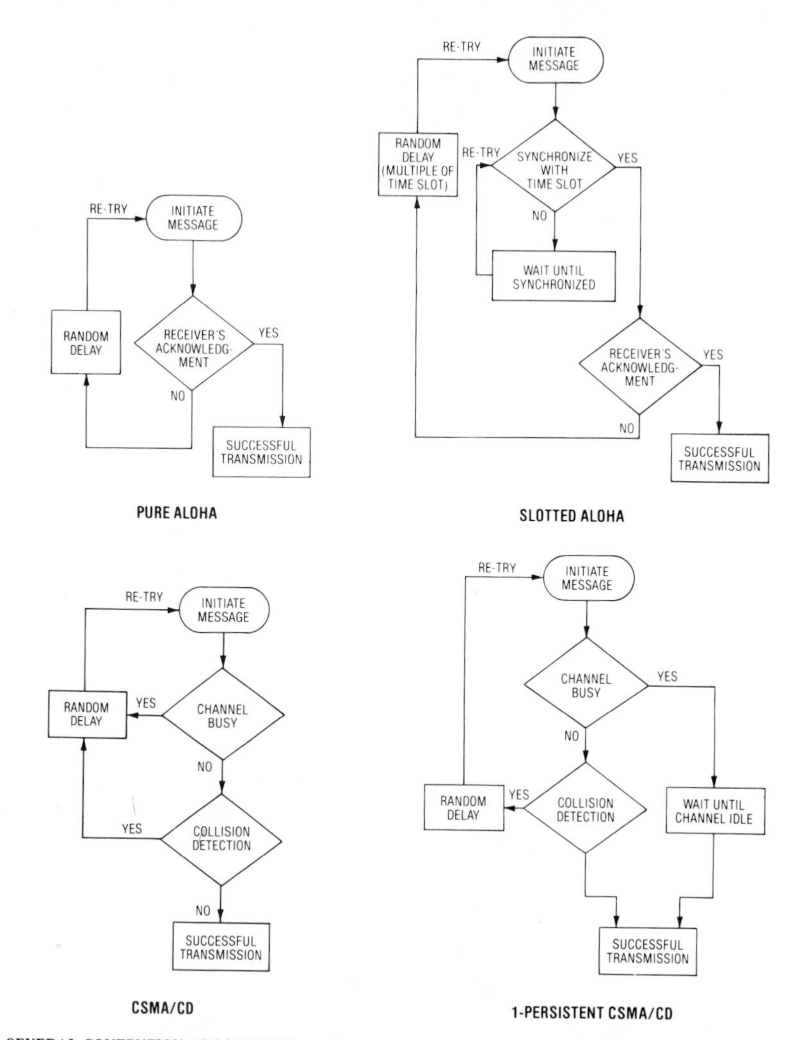

SEVERAL CONTENTION ALGORITHMS are illustrated by these flow charts. In pure ALOHA no attempt is made to prevent collisions from occurring; each node with data to transmit simply does so. In slotted ALOHA, the node must wait for the arrival of a clock signal that divides the transmission time into intervals longer than the packet transmission time. Under this scheme if a node is the only one ready at the beginning of an interval, its transmission will not be interrupted by nodes that become ready during it. In CSMA/CD networks the nodes sense the carrier before transmitting so that collisions can occur only if two nodes become ready almost simultaneously. The nodes also listen while transmitting so that collisions are detected immediately and time is not wasted transmitting the rest of a message that has already been garbled. The last control strategy is a variation of CSMA/CD in which the nodes sense the carrier continually rather than at random intervals and so transmit the moment it becomes idle. With each refinement the maximum possible throughput of the network increases. All of these strategies, however, may be vulnerable to heavy traffic. The heavier the traffic, the more collisions occur and the more bandwidth is consumed by colliding packets and by retransmissions. Intuitively it seems likely that under sufficiently heavy loads messages would encounter infinite delays and the network would be disabled.

Figure 12–20 Algorithm Flowcharts: Flowcharts clearly show the decision points in algorithms.

Source: Cornell Drentea, "Local Area Networks," *Scientific Honeyweller* December 1984: 39.

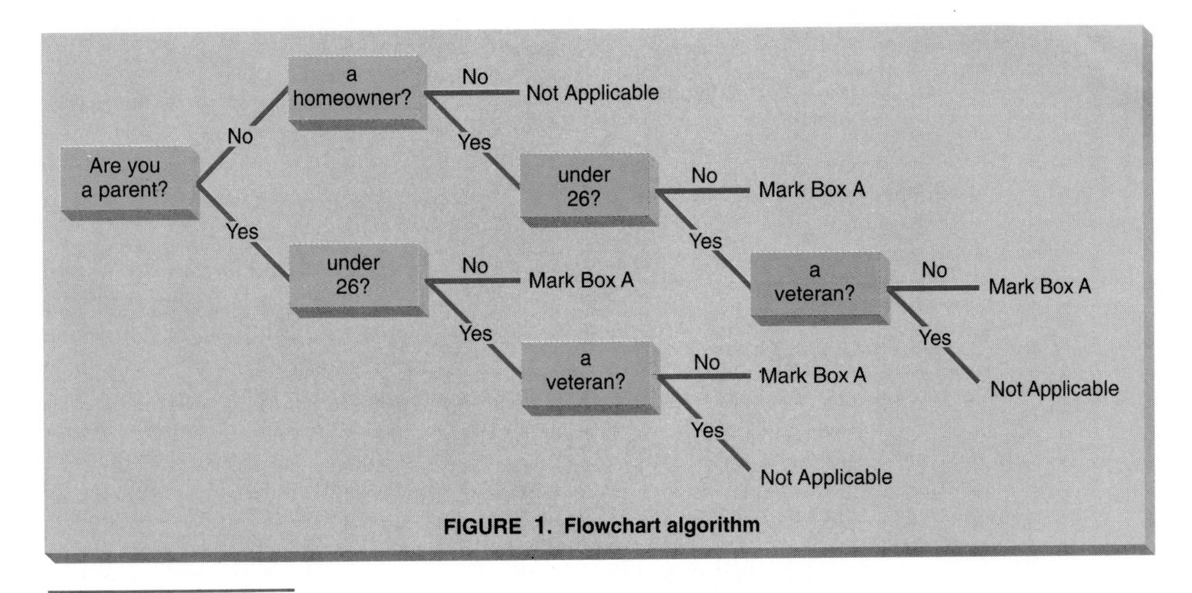

FIGURE 1. Flowchart algorithm

Figure 12–21 Algorithm Flowcharts for Nontechnical Audience: Simple flowchart intended for nontechnical audiences.
Source: Simply Stated January 1982: 3.

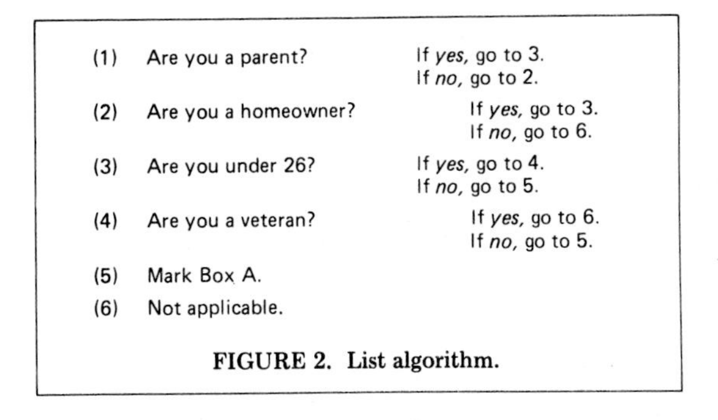

(1)	Are you a parent?	If *yes*, go to 3. If *no*, go to 2.
(2)	Are you a homeowner?	If *yes*, go to 3. If *no*, go to 6.
(3)	Are you under 26?	If *yes*, go to 4. If *no*, go to 5.
(4)	Are you a veteran?	If *yes*, go to 6. If *no*, go to 5.
(5)	Mark Box A.	
(6)	Not applicable.	

FIGURE 2. List algorithm.

Figure 12–22 List Algorithm: Well-constructed lists illustrate algorithms well.
Source: Simply Stated January 1982: 3.

Tables

A **table** is any arrangement of data set up in vertical columns and horizontal rows, classified as either formal or informal.

Informal Tables We "read" an informal table much as we would read normal prose conveying essentially the same information. Just as we usually read a paragraph only once, so do we scan through an informal table only once. It follows that informal tables must be fairly simple and immediately clear. As you may guess, the writer who inserts an informal table is actually exercising a choice between using prose and using the tabular display.

Informal tables have certain earmarks to distinguish them from the formal variety:

- They are brief and simple.
- They are not identified by table number and often have no title.
- So that they will be seen as a continuation of the text, they have no ruled frame around them, and they have only the necessary minimum of internal ruled lines.
- Lacking titles and table numbers, informal tables cannot be listed in the list of tables at the front of the report or in the index.

The example in Figure 12–23 shows a common variety of informal table. Portions of the surrounding text have been reproduced to illustrate how the sense of the table ties in with that of the prose preceding and following it.

Informal tables have certain virtues. Because they use minimal verbal machinery, they are space-savers, often conveying factual information that would require prose occupying several times the same space. They give readers relief by breaking up pages of solid prose. They signal readers that something different or special is being said. Readers can slow down and be sure of picking up the sense.

Formal Tables Formal tables play a major role in reporting technical information. For clarity of discussion here, we first identify in Figure 12–24 the standard parts of such tables by name, location, and function.

You should notice several points in our skeleton table. First, the number and title of the table are traditionally placed above the table, either centered or at the left margin. Usually, tables are numbered consecutively throughout a report with either Arabic or Roman numerals or occasionally capital letters. If a report is divided into chapters, the tables may be numered consecutively throughout each chapter: I–A, I–B, II–A, II–B; or 1–1, 1–2, 2–1, 2–2, and so on.

Since data on attainment were first collected in the 1940 census, the percentage of people 25 years or older who have finished high school and college has steadily risen:

Four years of high school or more:

1988	**76%**
1940	**24%**

Four or more years of college:

1988	**20%**
1940	**4%**

While the proportions of men and women who finish high school are not significantly different, the proportions who are college graduates are significantly different. In 1988, 24% of men and 17% of women had completed 4 or more years of college.

Figure 12–23 Informal Table: Informal tables save words and provide visual relief for readers.
Source: Adapted from U.S. Bureau of Census, *Population Profile of the United States: 1989* (Washington, DC: GPO, 1989) 22.

Every column and subcolumn and every line or group of lines must have a heading that clearly identifies the data. All headings should have parallel grammatical form (see pages 611–612). Substantive headings consisting of nouns or noun phrases are usually most appropriate. The stub heading classifies the line headings only; it should not introduce the column headings. In other words, stub and column headings have vertical

TABLE NUMBER
TITLE

Stub Heading[a]	Column Heading[b]	Column Heading	
		Subheading	Subheading
Line Heading			
Subheading	Individual "cells" for tabulated data		
Subheading			
Line Heading			

[a]Footnote
[b]Footnote

Figure 12–24 Typical Table Format: Use this table to identify the standard parts of a table.

reference, whereas line headings have horizontal reference. Violations of these principles are shown in the very faulty Table A in Figure 12–25.

Note how a few simple revisions make Table B in Figure 12–25 more logical and easier to read. However, even Table B's arrangement is not the best. Assume that the main intent of the table is to compare the different models. In such a case, people clearly prefer Table C, in which the data directly compared are arranged in columns rather than rows.

Notice that in these sample tables we have eliminated the vertical lines of the skeleton table. Eliminate as much clutter from tables as possible. In many instances, you can separate vertical columns with spacing rather than lines. Also, the horizontal double line between the headings and data is frequently eliminated and a single line substituted. Let clarity for the reader guide you in deciding how many and what kinds of internal lines you want in your table. Be careful with your spacing. Too little results in crowded, hard-to-read tables. Too much results in readers' having difficulty seeing the relationships you want them to see.

TABLE A

Model Number	100	101	102	103
Year of origin	1980	1984	1988	1992
Lens f-number	5.6	3.5	2.8	1.4
Shutter speed (max)	1/300	1/500	1/800	1/1000
Retail cost new	$120	$135	$200	$225

TABLE B

	Model Numbers			
Descriptive Data	100	101	102	103
Year of origin	1980	1984	1988	1992
Lens f-number	5.6	3.5	2.8	1.4
Shutter speed (maximum)	1/300	1/500	1/800	1/1000
Retail cost new (dollars)	120	135	200	225

TABLE C

Model Number	Year of Origin	Lens f-number	Shutter Speed (maximum in seconds)	Retail Cost New ($)
100	1980	5.6	1/300	120
101	1984	3.5	1/500	135
102	1988	2.8	1/800	200
103	1992	1.4	1/1000	225

Figure 12–25 Evolution of a Table: Table C is more logical and easier to read than either Table A or B.

Guidelines for Constructing Tables

- Make every table as simple, clear, and logical as it can be made.
- Mention every table in the text, preferably before it appears.
- Make every reasonable effort to insert a table into the text where the reader is expected to refer to it.
- If it is necessary to refer to an earlier or later table, give its page number as well as table number.
- Keep titles clear and succinct. Avoid phrases like "A Summary of" and "Presentation of Data Concerning" as in "A Presentation of Data Concerning Energy Production and Consumption: 1970–1990." Simply say, "Energy Production and Consumption: 1970–1990."
- Arrange column headings and line headings in some rational order: alphabetical, geographical, temporal, or quantitative.
- Where applicable, include units of measure, such as miles, degrees, or percentages, in column and line headings.
- In columns, align whole numbers on the right-hand digits, fractional numbers on the decimal points.
- Acknowledge the source of data appearing in a table either in your text or in a note on the table. For example, see Figures 12–26 and 12–27.
- As indicated in our skeleton table in Figure 12–14, footnotes are internal to the table. Designate them by superscript lowercase letters or by symbols (* † ‡). Many stylebooks recommend against numbers as footnotes for fear they will be mistaken for data. However, you will see numbers used, notably in the *Statistical Abstract of the United States.*

Figures 12–26 and 12–27 illustrate typical tables. Key parts of the table in Figure 12–26 are annotated. Notice the unit indicator and headnote directly beneath the title that give information needed for interpreting the table. Notice also the source line at the bottom. When, as in Figure 12–26, all the information in a table comes from one source, such a line is commonly used. When the information comes from various sources, as in Figure 12–27, footnotes are used.

Graphs

The basic kinds of graphs you can construct to illustrate your reports are bar graphs, circle graphs, and line graphs. Before your read the ground rules of graphing that follow, look at Figure 12–28 to get a general idea of what distinguishes each kind of graph from the others.

Table Number and title · Unit indicator · Footnote indicator · Parallel rule · Headnote · Spanner · Column heads · Stub · Field · Footnotes · Heavy rule

No. 57. Householder and Marital Status of Population, 15 Years Old and Over: 1987

[**In thousands, except percent.** As of **March.** See headnote, table 51]

HOUSEHOLDER AND MARITAL STATUS	Total, 15 yrs. and over [1]	MALE					FEMALE				
		Total [1]	20–24 years	25–44 years	45–64 years	65 yr. and over	Total [1]	20–24 years	25–44 years	45–64 years	65 yr. and over
Total persons	**186,688**	**89,368**	**9,499**	**37,671**	**21,428**	**11,578**	**97,320**	**9,859**	**38,597**	**23,472**	**16,397**
Householder	89,479	61,735	2,884	28,861	19,304	10,487	27,744	1,915	10,346	6,774	8,511
Single	12,071	5,985	1,224	3,514	734	391	6,086	1,356	3,365	601	603
Married, spouse present	51,537	48,573	1,523	22,405	16,371	8,197	2,964	239	1,630	817	260
Married, spouse absent	4,206	1,569	59	815	506	188	2,638	152	1,535	709	227
Widowed	11,291	1,721	–	41	363	1,317	9,570	5	362	2,385	6,819
Divorced	10,374	3,887	77	2,086	1,331	393	6,487	164	3,455	2,261	603
Not householder	97,209	27,634	6,615	8,809	2,124	1,091	69,575	7,944	28,252	16,698	7,887
Single	37,114	20,801	6,155	5,058	514	135	16,313	4,634	2,622	411	295
Married, spouse present	53,035	3,713	277	1,966	960	479	49,322	2,999	24,203	15,566	6,165
Married, spouse absent	1,847	1,021	106	639	183	67	825	156	423	141	62
Widowed	1,952	399	11	25	59	305	1,553	3	59	239	1,252
Divorced	3,261	1,699	66	1,121	406	104	1,562	151	944	342	111
PERCENT DISTRIBUTION											
Total persons	**100.0**	**100.0**	**100.0**	**100.0**	**100.0**	**100.0**	**100.0**	**100.0**	**100.0**	**100.0**	**100.0**
Householder	47.9	69.1	30.4	76.6	90.1	90.6	28.5	19.4	26.8	28.9	51.9
Single	6.5	6.7	12.9	9.3	3.4	3.4	6.3	13.8	8.7	2.6	3.7
Married, spouse present	27.6	54.4	16.0	59.5	76.4	70.8	3.0	2.4	4.2	3.5	1.6
Married, spouse absent	2.3	1.8	.6	2.2	2.4	1.6	2.7	1.5	4.0	3.0	1.4
Widowed	6.0	1.9	–	.1	1.7	11.4	9.8	.1	.9	10.2	41.6
Divorced	5.6	4.3	.8	5.5	6.2	3.4	6.7	1.7	9.0	9.6	3.7
Not householder	52.1	30.9	69.6	23.4	9.9	9.4	71.5	80.6	73.2	71.1	48.1
Single	19.9	23.3	64.8	13.4	2.4	1.2	16.8	47.0	6.8	1.8	1.8
Married, spouse present	28.4	4.2	2.0	5.2	4.5	4.1	50.7	30.4	62.7	66.3	37.6
Married, spouse absent	1.0	1.1	1.1	1.7	.9	.6	.8	1.6	1.1	.6	.4
Widowed	1.0	.4	.1	.1	.3	2.6	1.6	–	.2	1.0	7.6
Divorced	1.7	1.9	.7	3.0	1.9	.9	1.6	1.5	2.4	1.5	.7

– Represents or rounds to zero. [1] Includes 15–19 year olds.
Source: U.S. Bureau of the Census, *Current Population Reports*. series P-20, No. 423, and earlier reports.

Figure 12–26 Annotated Table: Key parts of the table are annotated. Notice particularly the unit indicator and the footnotes.

Source: U.S. Bureau of the Census. *Statistical Abstract of the United States: 1989,* 109th ed. (Washington, DC: GPO, 1989) xiii.

Guidelines for Constructing Graphs

- Keep the graph simple. Provide no more detail than you absolutely need, no more ideas than the reader can easily grasp, only essential facts. Keep in mind the purpose and audience of the graph.
- As with tables, keep titles clear and succint. In order to balance your page you may vary title placement, but be as consistent as possible.
- Make the graph big enough to be legible. The answer to what is "big enough" is subjective but also rather obvious. No squinting should be necessary.
- Keep the graph orderly. In making comparisons, present the quantities in some such order as from large to small or small to large or chronologically, according to your purpose.

Amino acid composition of processed wild rice, rice, oat groats, and wheat

Amino Acid	Wild rice[a] Minn.	Wild rice[a] Wisc.	Brown rice[b]	Polished rice[b]	Oat groats[b]	Whole wheat[b]
	g per 100 g recovered amino acids					
Alanine	6.2	5.9	6.3	5.4	5.0	3.5
(Ammonia)	–	2.0	2.4	2.8	2.7	3.9
Arginine	8.2	7.3	9.3	8.1	6.9	4.7
Asparatic acid	10.6	10.1	10.6	9.2	8.9	5.1
Cysteine	0.2	1.4	1.2	2.7	1.6	2.3
Glutamic acid	19.5	19.2	22.9	17.6	23.9	30.8
Glycine	4.9	5.0	5.2	4.5	4.9	4.0
Histidine	2.8	2.6	2.7	2.2	2.2	2.3
Isoleucine	4.4	4.4	3.8	4.6	3.9	3.9
Leucine	7.5	7.4	8.8	8.1	7.4	6.8
Lysine	4.5	4.4	4.0	3.4	4.2	2.6
Methionine	3.1	2.6	1.9	2.7	2.5	1.7
Phenylalanine	5.1	5.2	5.6	5.3	5.3	4.6
Proline	4.0	4.3	4.6	4.5	4.7	9.5
Serine	5.5	5.6	5.9	5.0	4.2	4.9
Threonine	3.3	3.6	3.8	3.5	3.3	3.0
Tyrosine	3.9	3.1	3.5	4.5	3.1	3.1
Valine	6.2	5.9	5.4	6.5	5.3	4.8
SLTM[c]	10.9	10.6	9.7	9.6	10.0	7.3
	percent					
Total protein	14.2	13.5	12.8	7.3	17.1	12.0

[a]Data from University of Minnesota, Agricultural Experiment Station and University of Wisconsin–Madison, Agricultural Experiment Station.

[b]Data from table in "1975 Report on Wild Rice Processors' Conference" by University of Wisconsin–Madison.

[c]Sum of lysine, threonine, and methionine.

Figure 12–27 Sample Table with Footnotes: Notice use of letters for footnotes. *Source:* Ervin A. Oelke et al., *Wild Rice Production in Minnesota* (St. Paul: U of Minnesota Agricultural Extension Service, 1982)

Figure 12–28 Graph Types

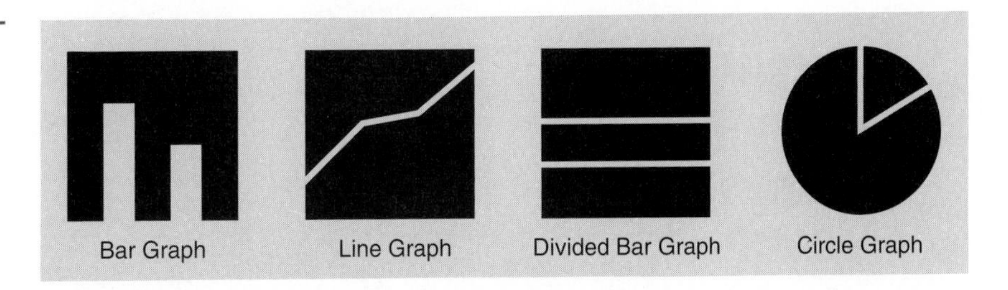

- Make the illustration attractive. A graph should be well balanced and blend into the page design.
- Use any device possible to help the reader interpret your graph. Color, arrows, heavier lines, and annotation may be used to this end. Including a summary of the graph material, either on the face of the graph or in a caption below the title, will aid the reader's understanding.
- For most readers, avoid three-dimensional presentation of data, as illustrated in Figure 12–29. Such graphs may be appropriate on occasion, but graphs that attempt to depict volume are very difficult to interpret correctly. Many computer programs, which we'll discuss shortly, can make beautiful three-dimensional graphs. We advise you not to be tempted. Don't sacrifice clarity for beauty.
- Make the units agree. Design your graphs so that the reader cannot possibly compare dollars with tons, or miles with hours. Make sure that the graph will be read so as to compare dollars with dollars, tons with tons, and so on.
- Acknowledge your sources for graphs as you do for tables, either in your text or in a note on the graph. For example, see Figure 12–31.

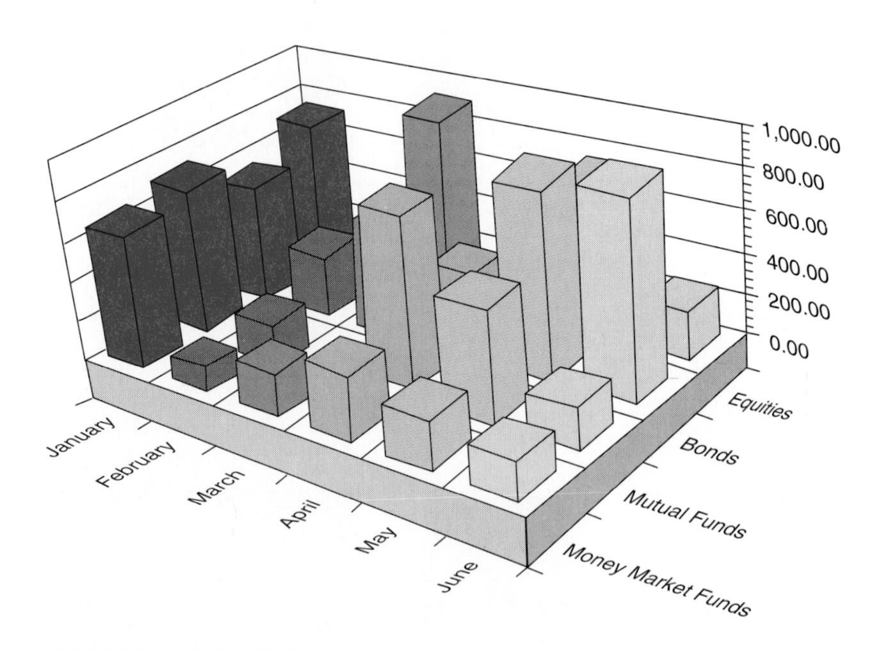

Figure 12–29 Three-Dimensional Graphs: Most readers find three-dimensional graphs hard to interpret.

Bar Graphs With bar graphs you can show relationships among data, as demonstrated by the multiple bar graph in Figure 12–30 and the divided bar graph in Figure 12–31. Bar graphs are used in technical reports for every audience from lay person to expert. However, because bar graphs are easily and accurately interpreted, they are excellent choices for lay people and executives.

In general, it makes no difference whether the bars run vertically or

Largest Central City as a Percentage of Total Metropolitan Area Population, for the 25 Largest Metropolitan Areas: 1986

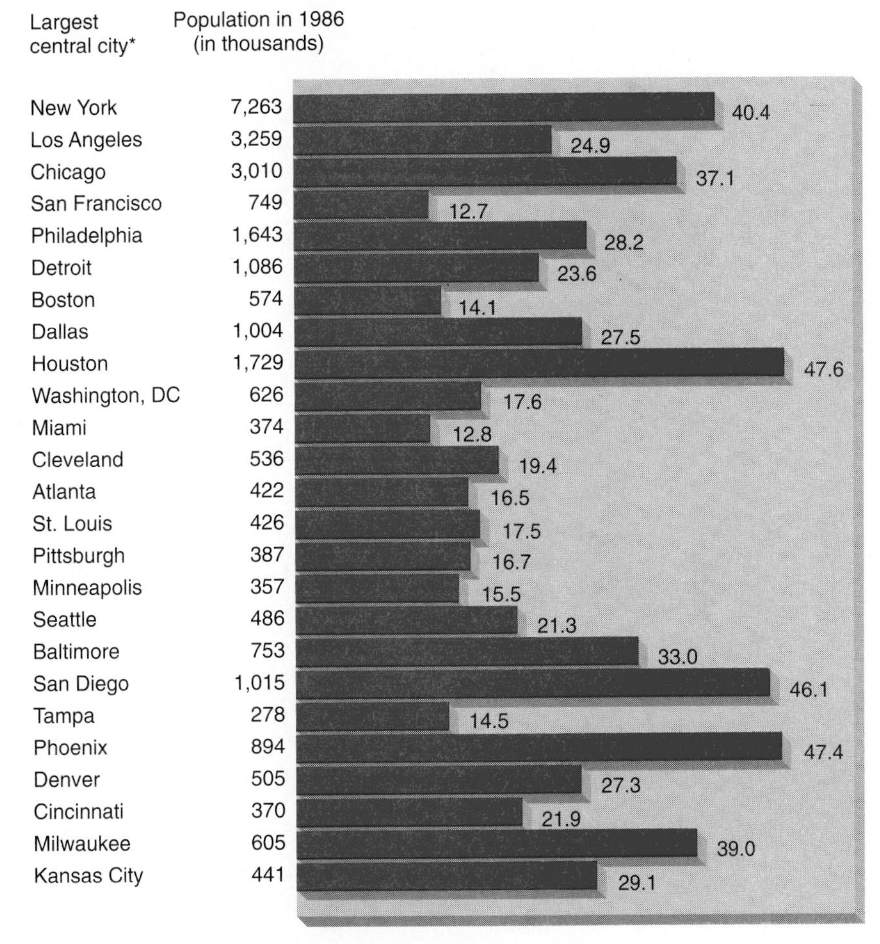

* Shown in order of metropolitan area population in 1986.

Figure 12–30 Multiple Bar Graph: Bar graphs are an excellent choice for every level of audience.
Source: U.S. Bureau of the Census, *Population Profile of the United States: 1989* (Washington, DC: GPO, 1989) 15.

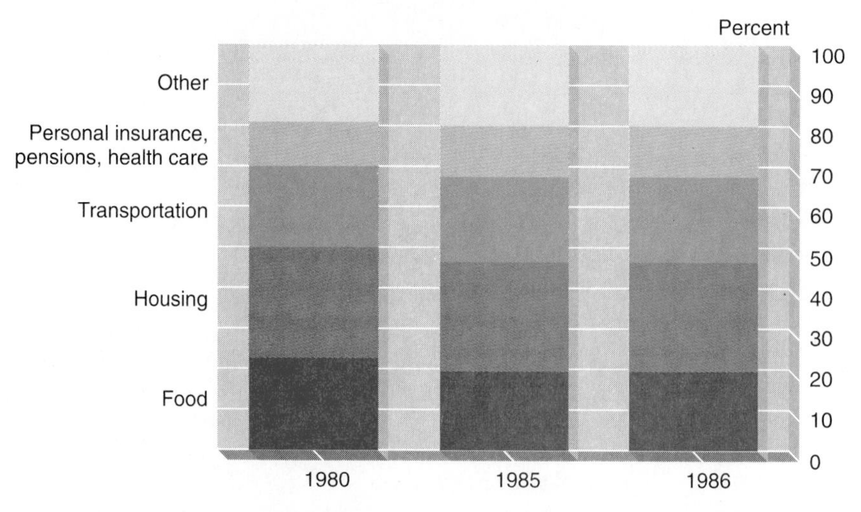

Figure 12–31 Divided Bar Graph
Source: U.S. Bureau of the Census, *Statistical Abstract of the United States: 1989,* 109th ed. (Washington, DC: GPO, 1989) 439.

horizontally, but there are some exceptions. One factor to be considered is page layout. Another is the nature of the data. People expect certain kinds of comparisons to be read up and down—altitudes for example—and others, like distances, to be read from side to side. The degree of reading precision you desire will determine whether you use a scale and, if you use it, how many lines to include on it.

Avoid, when possible, using keys to graphs. Put information on the bar itself or right next to it, as in Figures 12–30 and 12–31. When space or other limitations require the use of keys, keep them simple, as in Figure 12–32.

Circle Graphs The circle or pie graph makes for ease and accuracy of reading. It shows proportions, the relative size of related quantities within a whole. The pie graph is commonly found in business and financial reports and is an excellent graph for executives. Figure 12–33 presents two typical pie graphs. In both graphs, note the orderly arrangement of the parts. Notice that where space allows the percentage figures and identifications are placed inside the wedge. Whether they are inside or outside, percentage figures and identifications should be horizontal to the page and not to the lines separating the wedges.

Many graphic artists consider it good practice to begin the largest segment at twelve o'clock and then proceed clockwise in descending order of magnitude. However, this practice can be changed when you have good reason to do so. For example, the graphs shown in Figure 12–33 do not follow the conventional practice because to do so would lead to an inconsistency in the order in which the information is presented. In

College Graduates, by Age and Sex: 1988
(in percent)

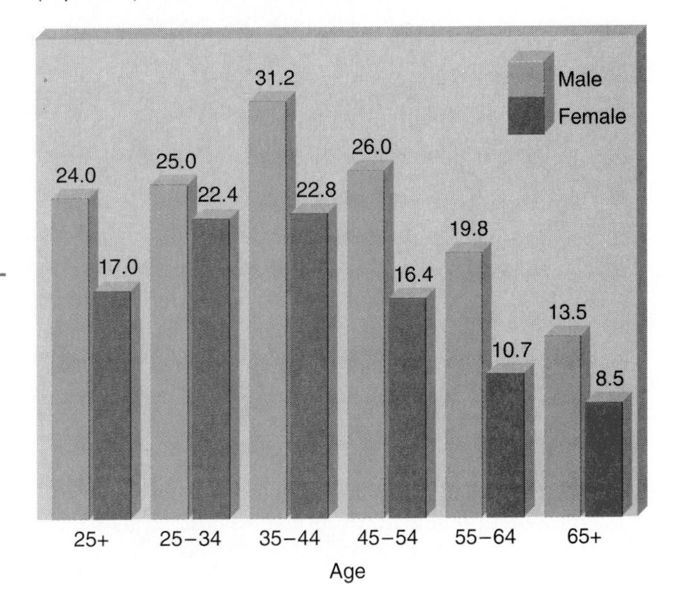

Figure 12–32 Bar Graph with Key: When possible avoid keys by labeling graph directly. When this is not practical, keep keys simple.
Source: U.S. Bureau of the Census, *Population Profile of the United States: 1989* (Washington, DC: GPO, 1989) 22.

Electric Energy Production, by Source of Energy

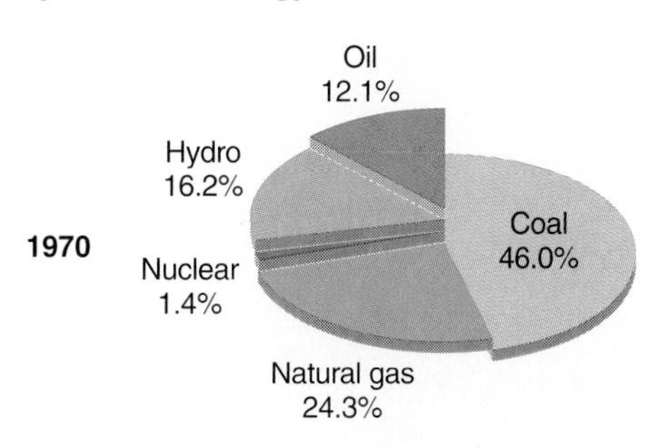

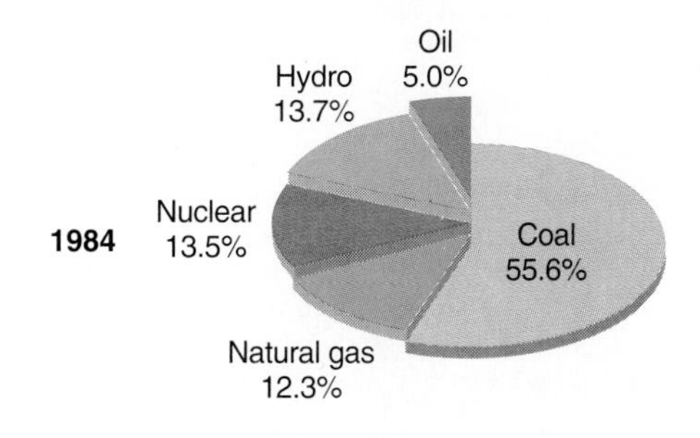

Figure 12–33 Circle Graphs: Circle graphs are excellent graphs for executives.
Source: U.S. Bureau of the Census, *Statistical Abstract of the United States: 1986*, 106th ed. (Washington, DC: GPO, 1986) xxviii.

Figure 12–34, the convention is set aside to allow the "exploded" segment to balance properly on the page.

In general, you should follow the conventions established by experienced graphic artists. Such conventions have come into being because they lead to ease and accuracy in interpreting graphs. But when you have good reasons to not follow such conventions, don't be fanatical about observing them. Bend the rules if by so doing you can help the reader.

Line Graphs Of all graphs used by scientists and engineers, the line graph is by far the most popular and useful. Certainly, it lends itself the best to purely technical information. One use for the line graph is to plot the behavior of two variables: the independent variable and the dependent variable. The independent variable is normally plotted horizontally, that is, along the *abscissa.* The dependent variable is normally plotted vertically; that is, along the *ordinate.* The dependent variable is affected by changes in the independent variable.

To avoid a cluttered appearance, values needed are often shown in hash marks on the abscissa and ordinate, as in Figure 12–35, rather than on a complete grid. Figure 12–35 also illustrates the use of the "suppressed

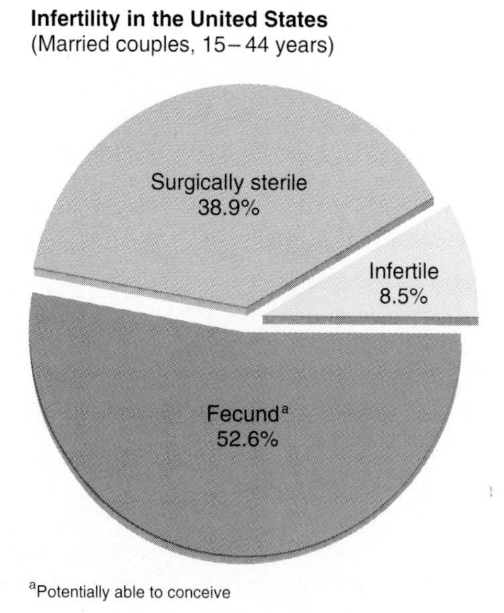

Figure 12–34 Exploded Circle Graph: The exploded segment emphasizes the key information.
Source: U.S. Congress, *Infertility: Medical and Social Choices* (Washington, DC: Office of Technology Assessment, 1988) 2.

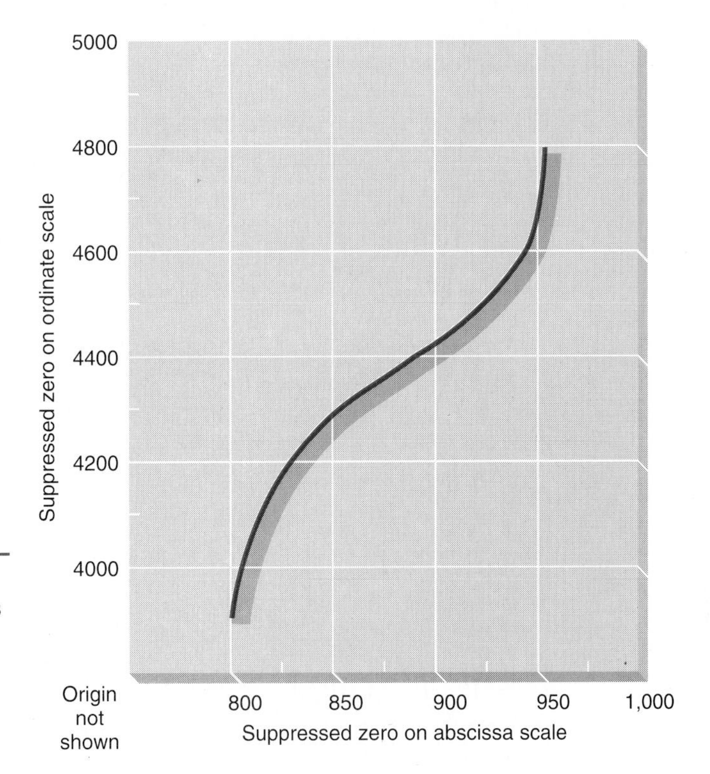

Figure 12–35 Line Graph with Hash Marks and Suppressed Zero: Line graphs display information well for technical and expert audiences.

zero." When the values plotted are so high that it would be impractical to start with zero, a scale starting at a point above zero is used. Whenever you follow this practice, be sure your reader understands clearly that the zero has been suppressed. One way to do this is to separate the scale, as in Figure 12–35.

When accuracy is of paramount importance, use a complete grid. Figure 12–36 illustrates a line graph plotted on graph paper with a fine grid that permits accuracy of construction and interpretation.

Line graphs are also excellent for showing and comparing trends. Figure 12–37 is a good example of such use. As in Figure 12–37, various devices, such as lines, dots, and dashes can be used to keep the lines distinct. When color is appropriate, you may be able to use it as well. However, if the graph will need to be photoreproduced, you will likely have to stay with black and white. Also, most journals will not print color because of the expense.

Pictograms A pictogram presents data pictorially. For example, the bars of a bar graph may be drawn to represent men or women or factories or whatever. In drawing a pictogram the problem is to make sure that the reader interprets the drawing as a pictogram, for example, in Figure 12–38, as a pile of coins rather than as the volume or area covered. If the

Model Hybrid Rocket Engine
(Thrust vs. chamber pressure)

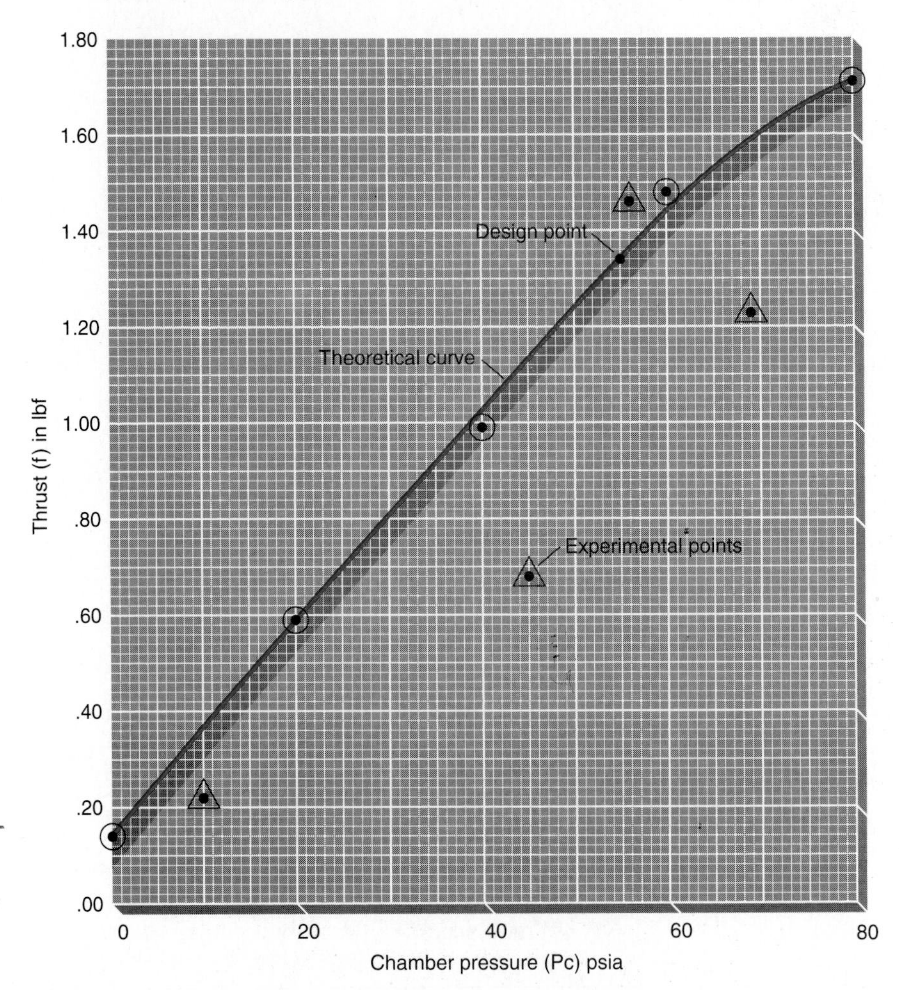

Figure 12–36 Line Graph Requiring Fine Grid: When accuracy of interpretation is vital, use a complete grid.

design is drawn so that it is seen in terms of the desired units, it can be read quite accurately. A pictogram has the attention-getting value of novelty and is sometimes useful for a lay audience.

Table–Graph Relationship Obviously, often you could present the same data in a table, in a bar graph, or in a line graph. Which method or combination of methods you choose will depend on your intention and your audience.

In the table in Figure 12–39, you see the number of hours and minutes that an automobile water pump survived testing at various revolutions per minute (rpm). The table summarizes the data and shows clearly how the life of the water pump decreases as its speed of revolution is increased.

A pie chart is an excellent way to show percentages of the whole.

Pie charts work best with two to six wedges.

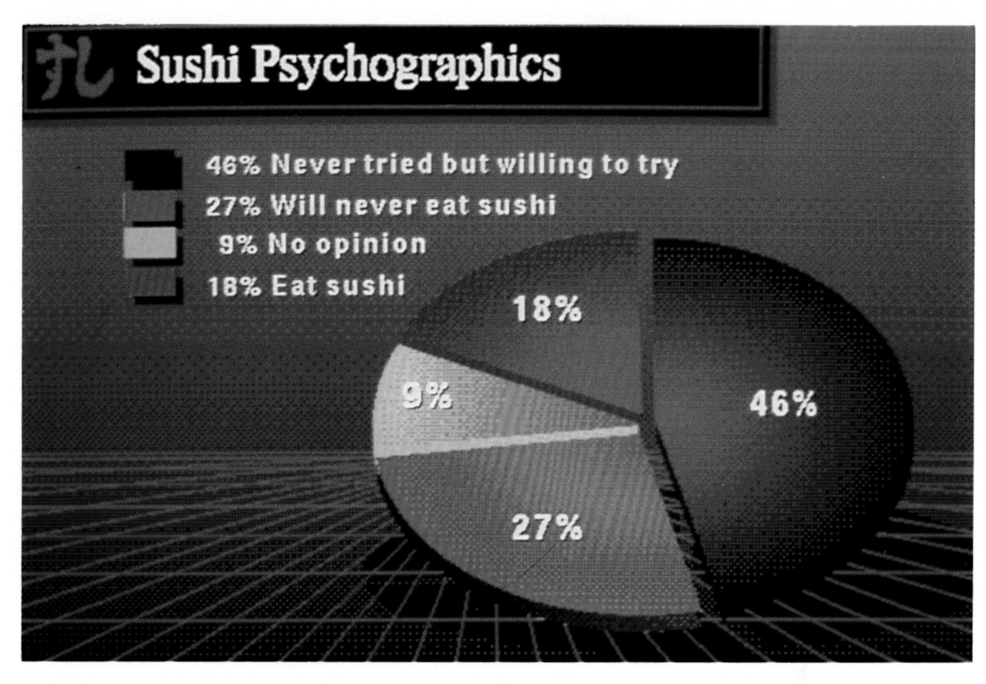

Source: © Aldus Corporation 1988–1989. Used with the express permission of Aldus Corporation. Aldus®, PageMaker®, Aldus Freehand®, and Persuasion® are registered trademarks of Aldus Corporation. All rights reserved.

Note how clearly this picture tells you what it is about.

The wedges of a pie must always add up to 100%.

This stacked area chart is a useful way to show three sets of information together:
- trends over time;
- relative size of the different groups;
- total size of the group.

The legend, which tells you what each color represents, is critical.

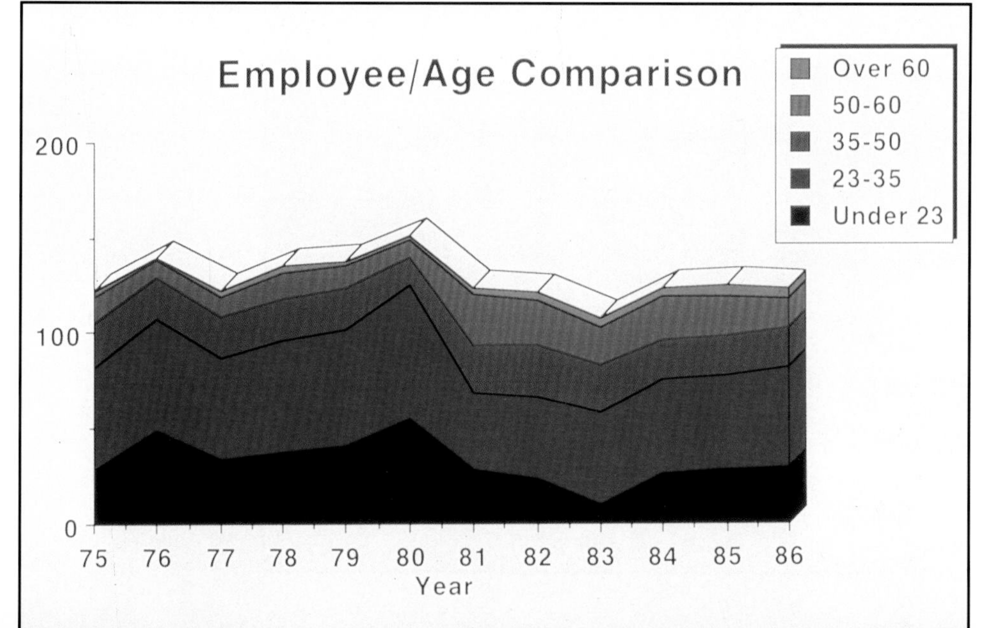

The author has stacked the areas in logical order, by age, not by the size of the segment.

Source: CA-Cricket Graph, Computer Associates International, Inc. Used with permission.

Bar charts are useful for making comparisons across time or among different groups.

Be sure to put a title on each bar chart.

Be sure to label both the *x* axis and the *y* axis.

Created with Harvard® Graphics 2.3, Software Publishing Corporation, Mountain View, California

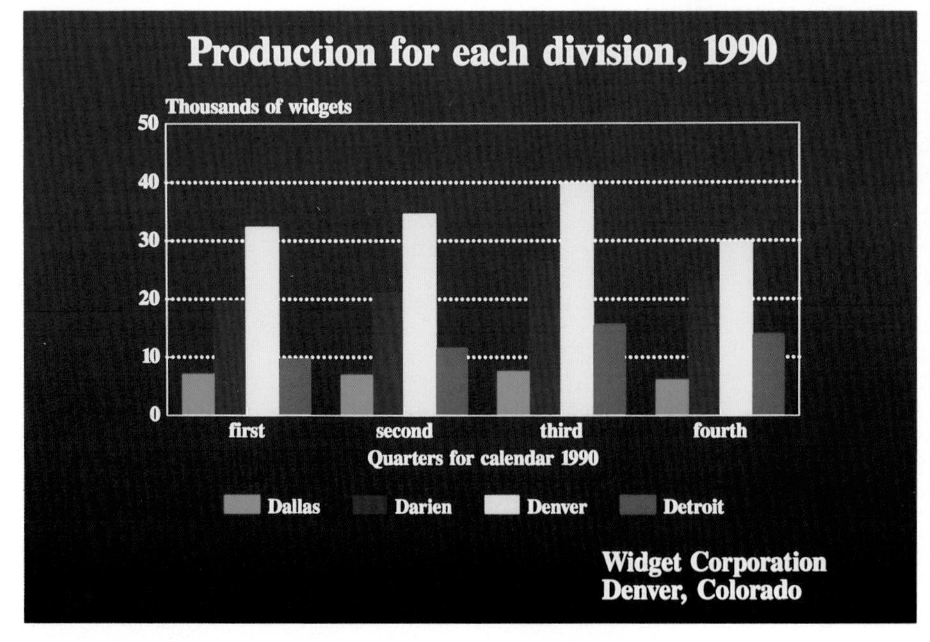

Each color represents one division, and the legend matches the color to the division name.

Keep charts simple. Three-dimensional versions of bar charts may hide important information if smaller numbers come behind larger numbers.

Combining stored graphics with data lets you show information visually.

Every chart should have a title.

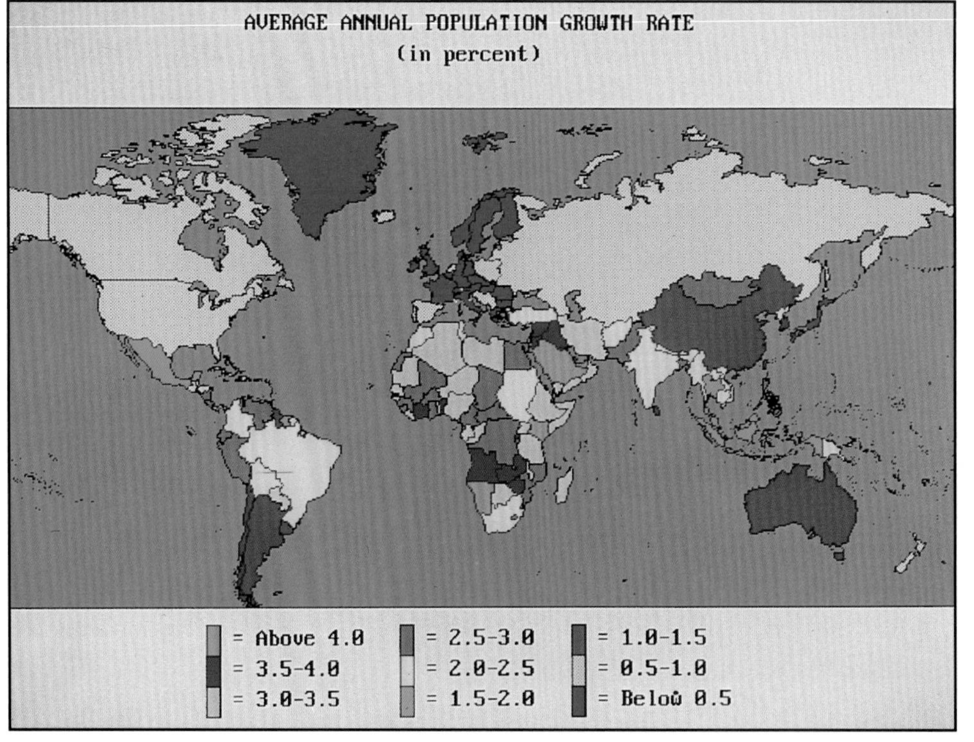

You can see at a glance that population growth is highest in Africa and lowest in Europe.

The legend is neatly stacked in columns at the bottom. It doesn't interfere with the picture.

The colors follow a logical order with cooler colors (blues and greens) for low rates and hotter colors (pinks and reds) for high rates.

***Source:** Copyright 1990, PC Globe, Inc. Used with permission.*

Attractive art adds value to good writing.

The text is in two columns. You can read the paragraph to the left of the sheep without crossing over the picture.

The piece is well written. Don't use art to try to mask poor writing.

Source: *Full Write software, Courtesy of Ashton-Tate Corporation*

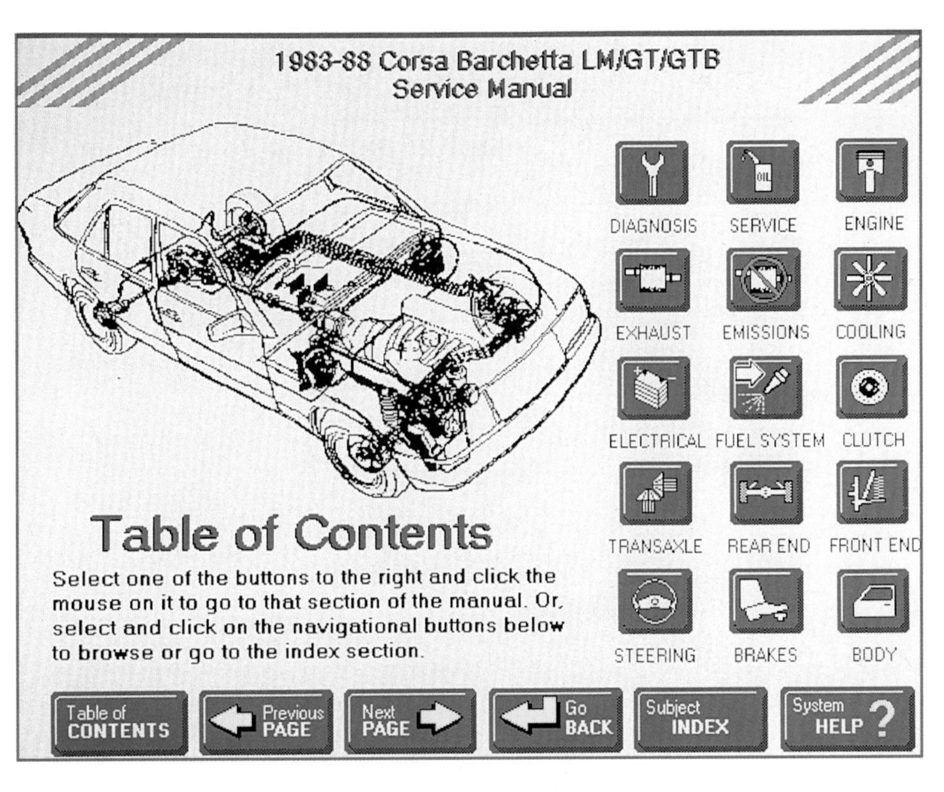

You can develop technical documents to be used "on-line," that is, directly on the computer and not on paper.

Developing a hypertext manual for use on-line requires special skills. You can't just take a paper manual and expect to use it on the computer.

With a hypertext program, users can go directly to the section that they need.

With these navigation tools, users can move around in the manual quickly and flexibly.

Source: *Guide 3.0, copyright 1990, Owl International, Inc. Used with permission.*

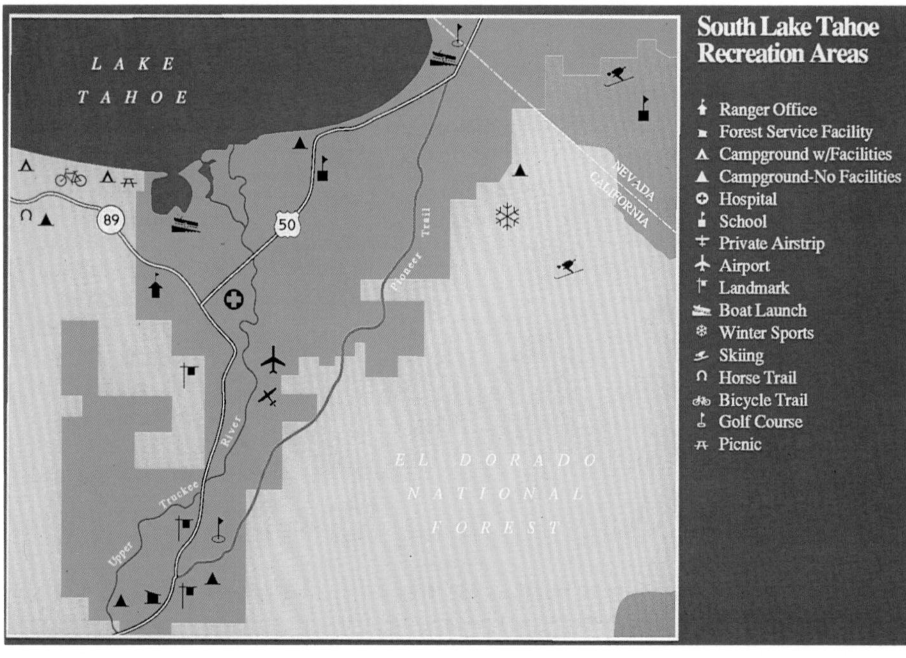

Computers can help you construct technical drawings, like maps, to be part of your reports or other documents.

The map itself is not cluttered with words. Only the lake is named on the map.

The symbols are explained in the legend, which is listed neatly down the side.

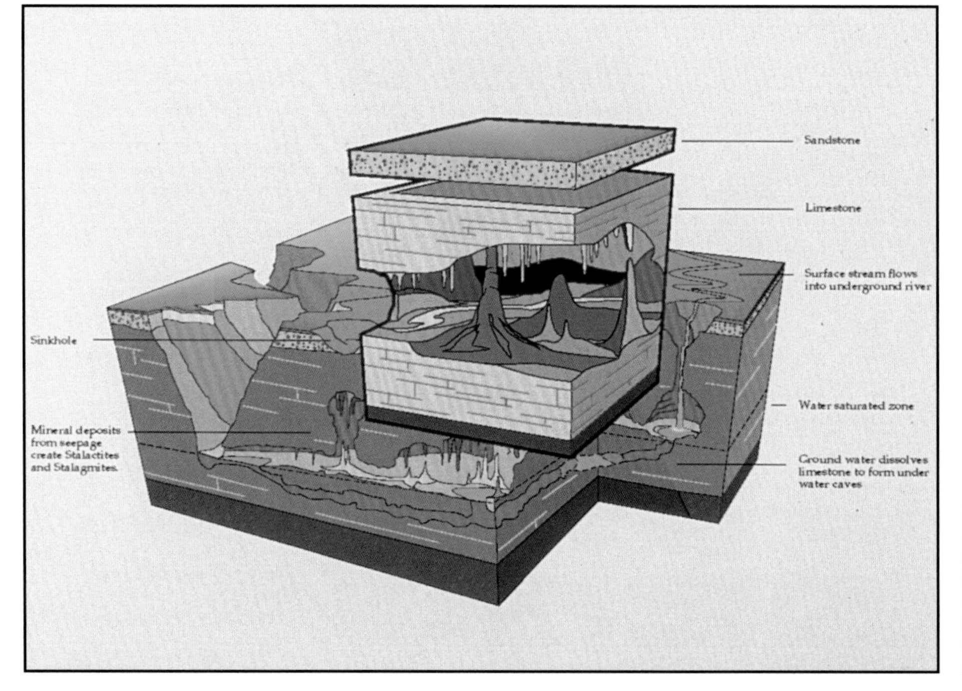

A geologist's picture of the effects of underground water.

Callouts in the picture label the different parts.

You can pull one section out and "explode" it to show more details.

The function sin*(xy)*. The colors represent ranges in the values of the function.

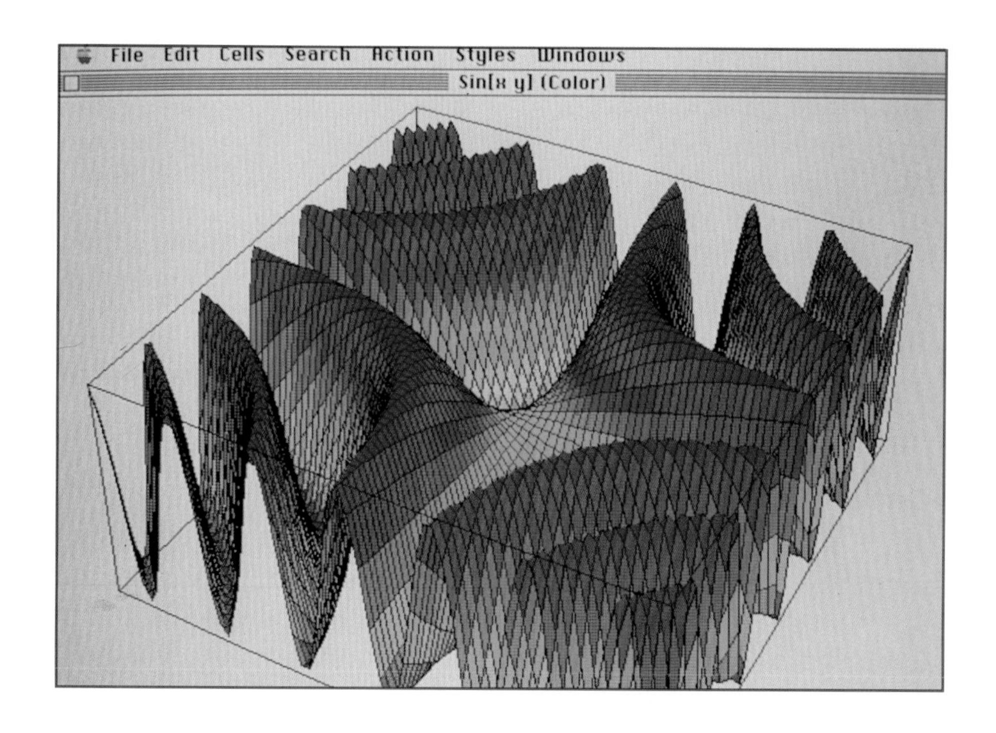

You can plot mathematical functions into spectacularly colored graphics.

Source: *Computer graphics generated in* Mathematica. *Wolfram Research, Inc. Used with permission.*

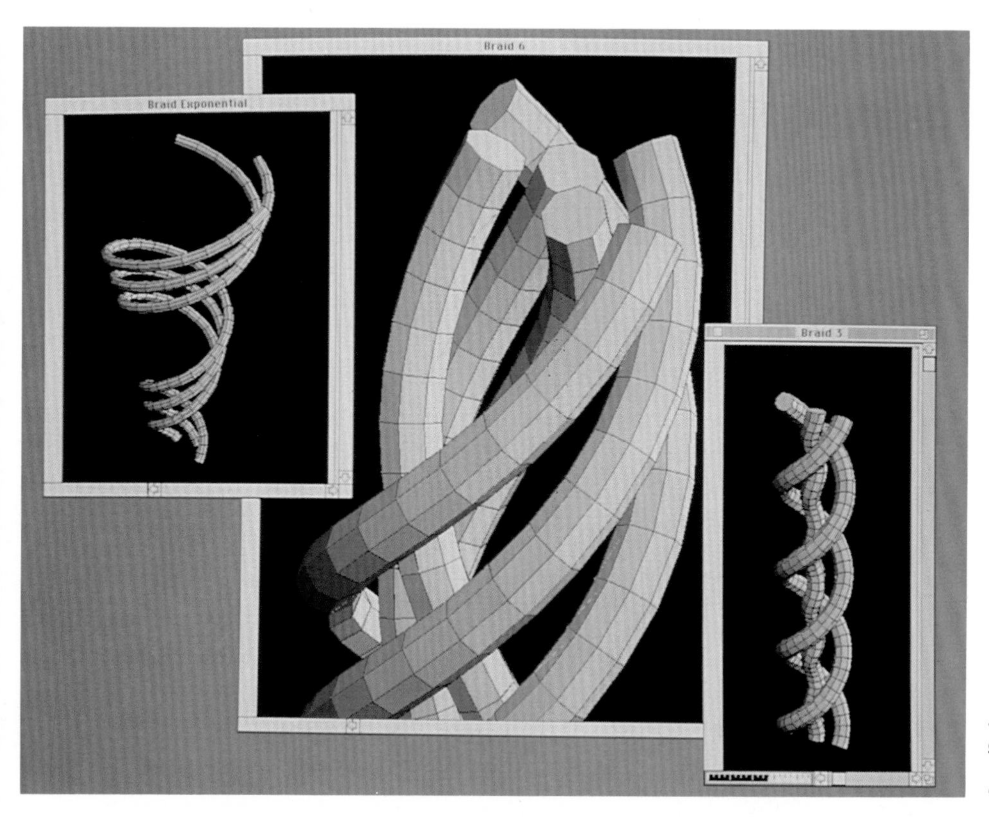

Three examples of braids. Braid theory, a branch of knot theory, is used in fields as diverse as molecular biology and physics.

You can include several pictures in one illustration.

Source: *Computer graphics generated in* Mathematica. *Wolfram Research, Inc. Used with permission.*

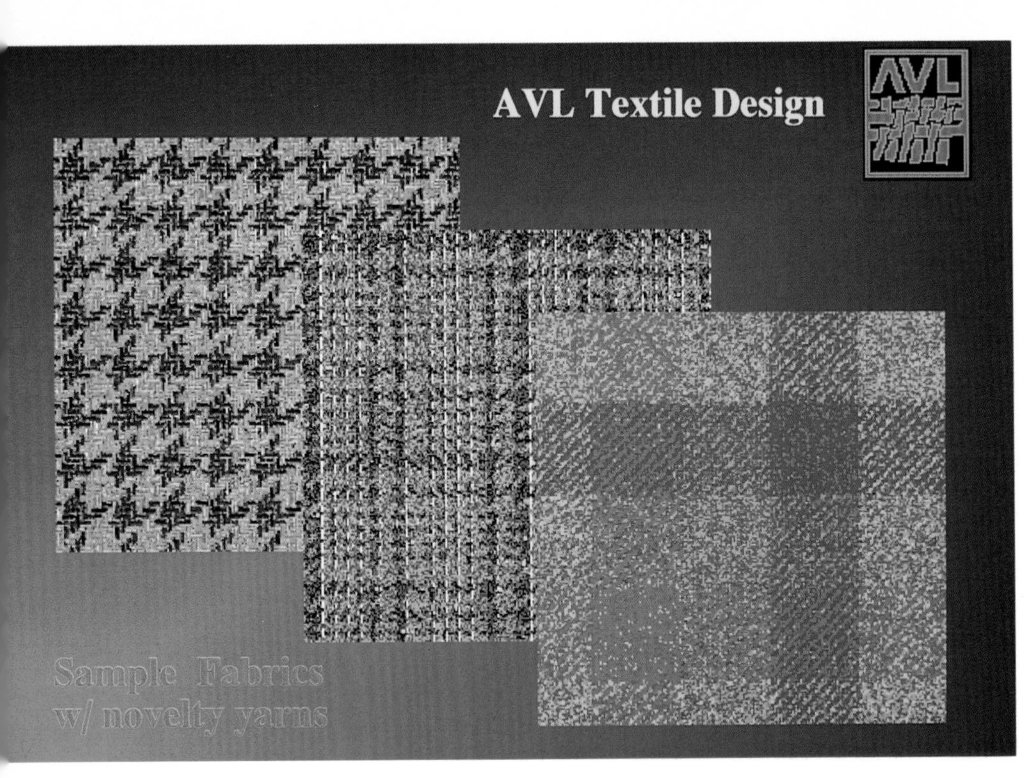

With a product like Design and Weave, you can show very realistically what a finished textile will look like.

Source: *AVL Looms, Inc. Used with permission.*

With another program, MultiColor, you can plan and show the design of printed textiles.

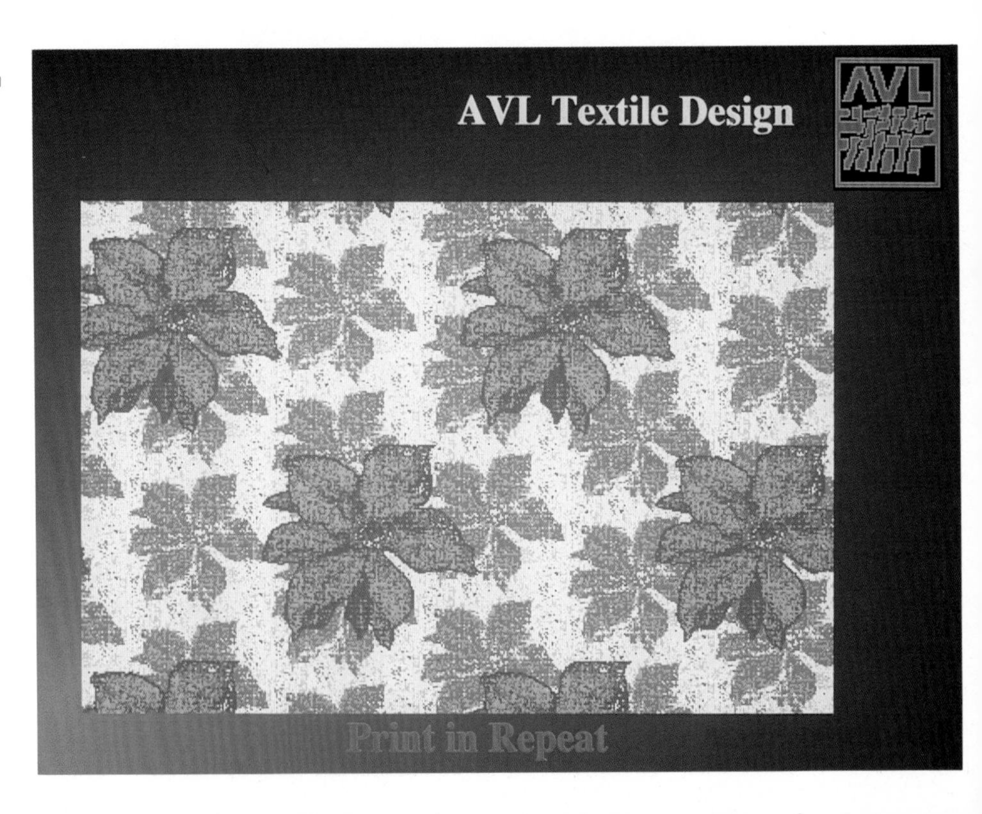

Source: *AVL Looms, Inc. Used with permission.*

You can plan and show architectural drawings.

Source: *© Spencer & Maxwell Architects/Architrion II by Gimeor, Inc. Used with permission.*

You can illustrate interiors from different perspectives.

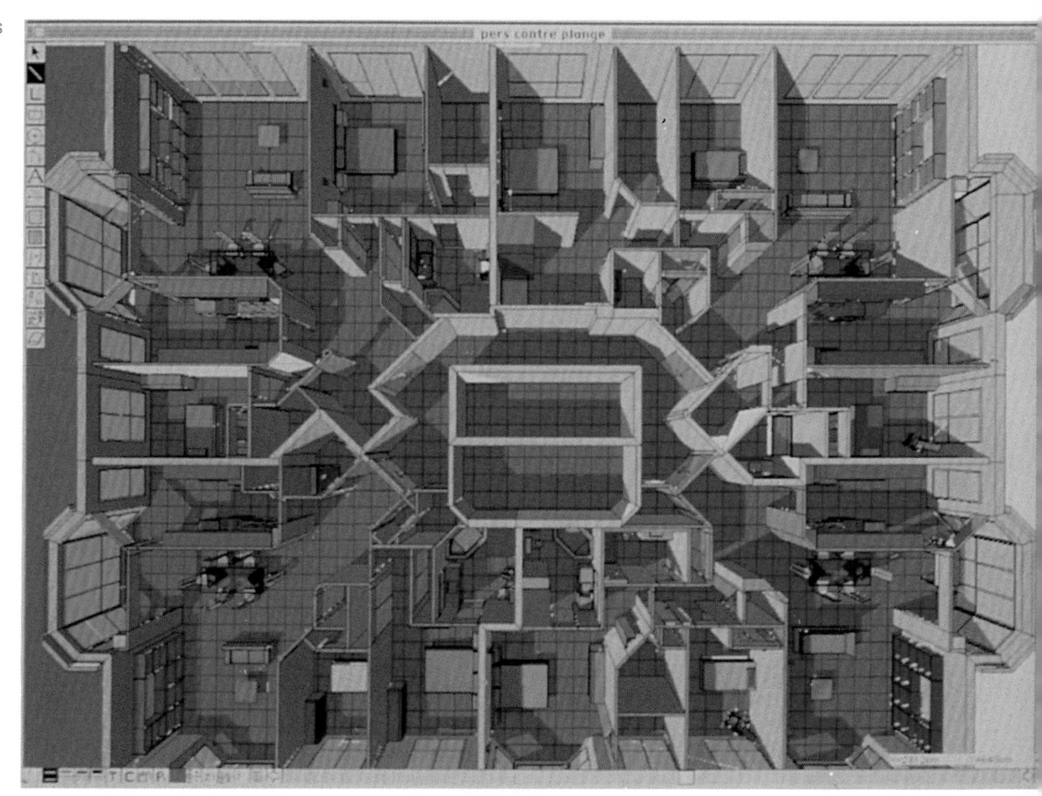

Source: *© B.A.G.H./ Architrion II by Gimeor, Inc. Used with permission.*

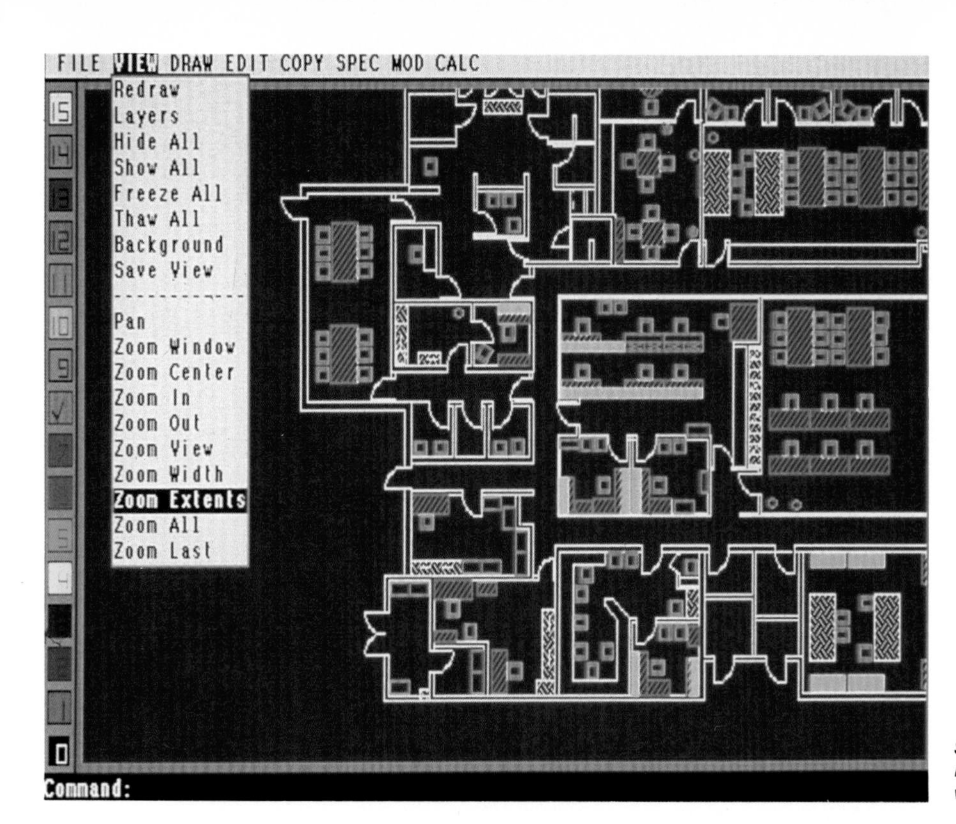

With prints from a computer-aided design program, you can show a finished design or stages in the development of the design.

Source: *EasyCAD 2 by Evolution Computing. Used with permission.*

Computer-aided design programs can be useful to students and professionals in many fields.

Source: *EasyCAD 2 by Evolution Computing. Used with permission.*

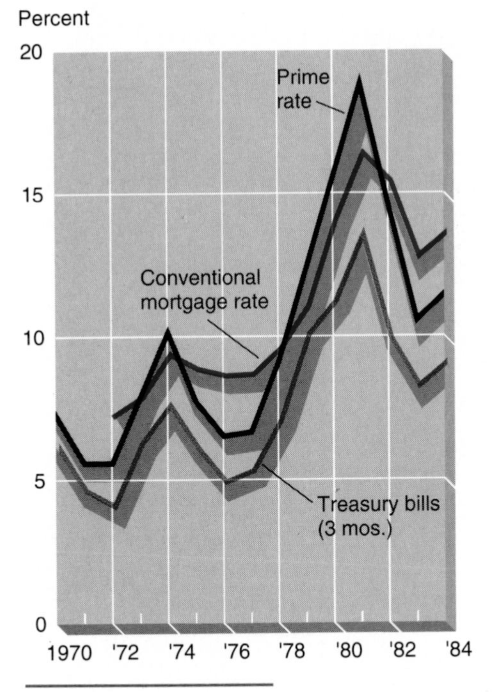

Figure 12–37 Line Graph: Line graphs illustrate trends well.
Source: U.S. Bureau of the Census, *Statistical Abstract of the United States: 1986,* 106th ed. (Washington, DC: GPO, 1986) xxv.

But the table does not show very clearly the shape of the relationship of rpm to time before failure. If you were writing this paper for a predominantly lay audience, you might choose to show this shape with a simple bar graph. The bar graph in Figure 12–39 shows clearly the relationship between rpm and the time before failure. Because it is customary to show time on the horizontal axis, the graph ignores the convention that the dependent variable is plotted on the ordinate.

An audience primarily composed of technically minded people would probably prefer the line graph in Figure 12–39. They would recognize, as nontechnical readers might not, that the increases in rpm actually constitute a continuous variable. For them, the slope of the curve would show the continuous relationship of rpm and time to failure. The line graph also reverts to the conventional and plots the dependent variable along the vertical. The zero is suppressed on the abscissa, but the numbers on the scale make this fact quite clear.

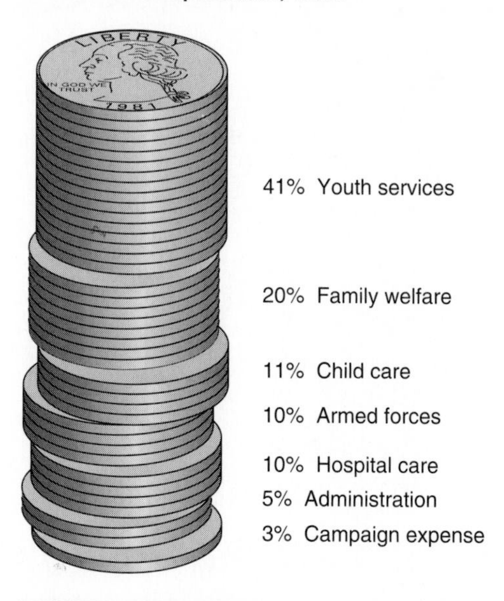

Figure 12–38 Pictogram: A pictogram has the advantage of novelty and is well suited to lay audiences.

Sometimes a writer will give the same information in a graph and in a table. In Figure 12–40, the line graph shows the trend of median age at first marriage. The table in Figure 12–41, taken from the same article, allows the reader to get exact data without interpreting the graph.

Computer Graphics

Computer graphics have greatly increased the ease with which graphics can be produced. Many available programs allow you to enter numbers into a personal computer, push a few keys, and wait for your choice of graphs—bar, line, or circle—to come out. Some sophisticated programs allow you to draw your own graphs by manipulating on screen such elements as color, lines, arcs, rectangles, and circles. You can add words and symbols to your graphs from an electronic library of typefaces and "clip art." **Clip art libraries** may include dozens of symbols such as cars, phones, people, animals, factories, clocks, keys, flowers, and so forth. Maps and templates for various kinds of graphs are available in many programs. To help you use such programs well, we have illustrated and interpreted several computer-produced graphics for you in a special insert beginning facing page 360.

Some programs, referred to collectively as "desktop publishing," allow you to produce reports, memos, and brochures in which graphics and

Table IV: Testing Time Before Failure

Test no.	RPM	Hours–minutes
1	4,000	4:08
2	5,000	4:02
3	6,000	3:55
4	7,000	3:45
5	8,000	3:26
6	9,000	2:45
7	10,000	1:15

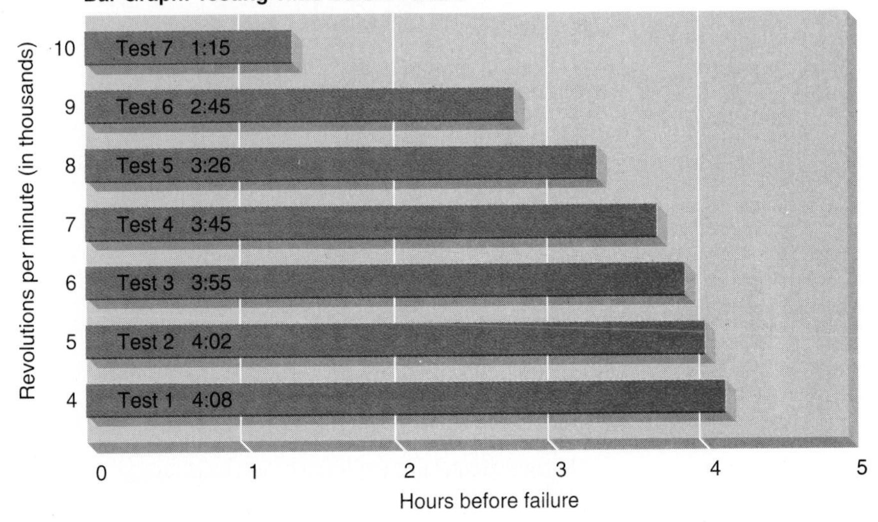

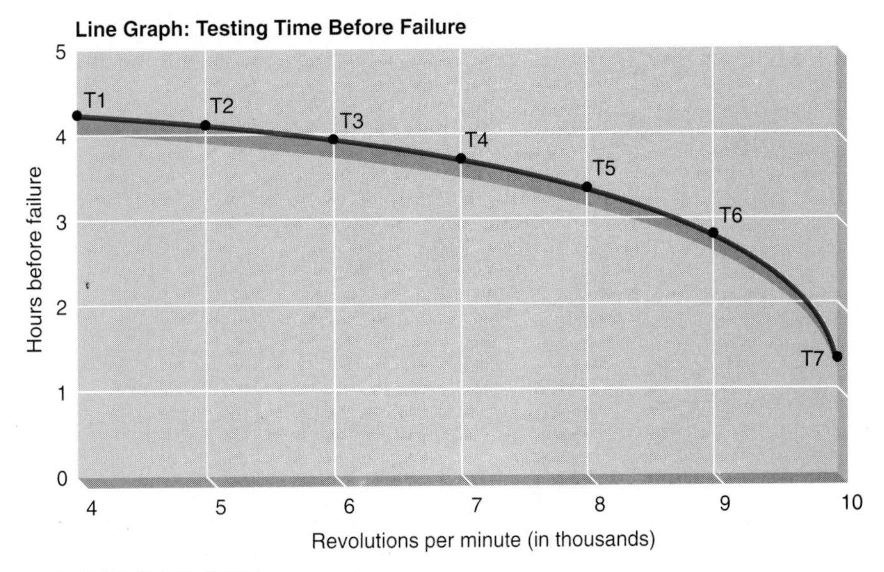

Figure 12–39 Table–graph Relationship: The two graphs show clearly the relationship between rpm and time before failure. The line graph shows the relationship as a continuous variable and is well suited to technical and expert audiences.

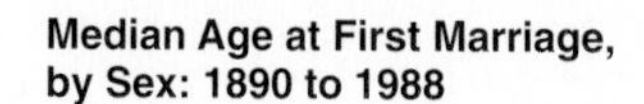

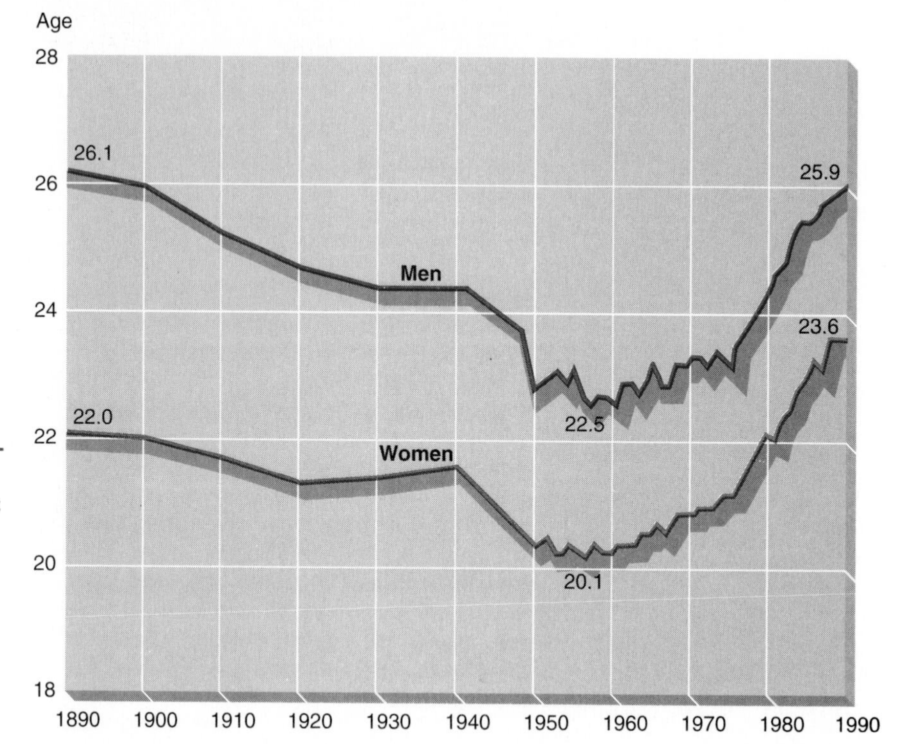

Figure 12–40 Line Graph of Marriage Ages: The line graph shows the shape of a trend. *Source:* Arlene F. Saluter, "Singleness in America," *Studies in Marriage and the Family* (Washington, DC: U.S. Department of Commerce, 1989) 1.

various typefaces are integrated without the old bother of cut and paste followed by photographic production. Clearly, as such programs become more common, the need for even casual report writers to know something about layout and design will grow. We provide you with some basic information on these subjects in Chapter 10, "Document Design."

In this chapter we have presented you with an introduction to graphics.

Figure 12–41 Table of Marriage Ages *Source:* Arlene F. Saluter, "Singleness in America," *Studies in Marriage and the Family* (Washington, DC: U.S. Department of Commerce, 1989) 2.

Table B. Median Age at First Marriage, by Sex: 1890 to 1988

Year	Men	Women	Year	Men	Women
1988	25.9	23.6	1955	22.6	20.2
			1950	22.8	20.3
1985	25.5	23.3	1940	24.3	21.5
1980	24.7	22.0	1930	24.3	21.3
1975	23.5	21.1	1920	24.6	21.2
1970	23.2	20.8	1910	25.1	21.6
1965	22.8	20.6	1900	25.9	21.9
1960	22.8	20.3	1890	26.1	22.0

Note: A standard error of 0.1 years is appropriate to measure sampling variability for any of the above median ages at first marriage, based on Current Population Survey data.

For more information, we recommend you to the books on the subject in our bibliography. Also, if you intend to publish in a journal, read the stylebook that governs that journal for special instructions in the preparation of graphics. For example, the *Council of Biology Editors Style Manual* devotes 12 pages to the subject.

Planning and Revision Checklist

You will find the planning and revision checklists that follow Chapter 2, "Composing" (pages 33–35 and inside the front cover), and Chapter 4, "Writing for Your Readers" (pages 78–79), valuable in planning and revising any presentation of technical information. The following questions specifically apply to graphics. They summarize the key points in this chapter and provide a checklist for planning and revising.

Planning

- How important are graphics to your presentation?
- How complex are your graphics likely to be?
- How expert is your audience in reading graphics?
- Based on your answers to the previous questions, how much prose explanation of your graphics are you likely to need?
- Do you have objects to portray?
- What do you want to illustrate about the objects?
- Do you want to draw attention to certain aspects of the objects and not to others?
- Will exploded or cutaway drawings of the objects serve your purpose?
- Would photographs of the objects add realism or drama to your report?
- Based on your answers to the previous questions about objects, what kinds of photos and drawings are you likely to need?
- Will any of your photos or drawings need a scale reference?
- Will any of your photos or drawings need annotation?
- Are you working with any concepts that can be best presented visually or in a combination of words and graphics?
- Do you have any definitions that should be presented visually in whole or in part?

Revising

- Are your graphics suited to your purpose and audience?
- Are your graphics well located and easy to find?
- Have you shown scale on your photos when necessary?
- Are your annotations horizontal to the page? Are they easy to find and read?
- Will your readers grasp the concepts you have shown visually? Do your verbal and graphic elements complement each other?
- Will your readers easily follow any processes you have shown graphically?
- Do you have any blocks of data that should be converted to informal or formal tables?
- Are your tables and graphs properly titled and properly numbered?
- Are your tables and graphs simple, clear, and logical?
- Have you referred to your tables and graphs in your text?
- Have you when necessary included units of measure in your tables and graphs?
- Are the numbers in your tables aligned correctly? Whole numbers on right-hand digits? Fractional numbers on the decimal points?
- Have you acknowledged the sources for your tables and graphs?
- Are your graphs legible?

- Do you have processes or algorithms that should be depicted visually?
- Would a flowchart of any of your processes or algorithms aid your readers?
- Will you be presenting information on trends or relationships? Should some of the information be presented in tables and graphs?
- Do you have masses of statistics that should be summarized in tables?
- Would some informal tables help you present your data?
- Which are your readers most likely to comprehend, bar and circle graphs or line graphs?
- For each graph you plan, ask this question: Is this graph intended to give the reader a general idea, such as the shape of a trend, or should the reader be able to extract precise information from the graph?
- Would a pictogram add interest to your report?
- Are your graphs attractive?
- When necessary, have you helped your readers to interpret your graphs with commentary or annotations?
- Do your graphs need a grid or hash marks for more accurate interpretation?
- Have you avoided the use of keys? If not, have you kept them simple?
- Have you plotted your graphs according to the conventions—independent variable horizontally, dependent variable vertically? If not, do you have a good reason for your variation?
- If you have used a suppressed zero, will it be obvious to your reader?
- Do your tables and graphs complement each other?

Exercises

1. Graphs are used to depict objects, concepts, processes, trends, and relationships. Looking through such sources as newspapers, magazines, textbooks, and professional journals, find an example of each such use. Photocopy your examples (or cut them out if the newspaper or magazine belongs to you) and bring them to class. Be prepared to discuss the strategy involved in each graphic and how purpose, situation, and audience shaped the graphic.
2. You are a graduate assistant for a professor of political science. This morning the professor came to you with the table we have printed as Figure 12–42. She informed you that she is writing an article on world population growth that will be published in the *New York Times*. One emphasis in the article is the danger to political stability of underdeveloped countries caused by a rapid growth of population. She asked that you make her a black and white graph that will help the readers comprehend the key information in the table and to understand how important that information is. The table contains more information than you would want to put on one graph. Therefore, you must select what information you choose to portray. As in writing, base your selection on purpose, situation, and audience. Your completed graph should have a title and any prose

No. **1402.** World Population and Vital Statistics, by Continent and Region, 1970 to 1988, and Projections, 1990 and 2000

[Crude birth rate: Number of births during 1 year per 1,000 persons (based on midyear population). Crude death rate: Number of deaths during 1 year per 1,000 persons (based on midyear population). Total fertility rate: Average number of children that would be born per woman if all women lived to the end of their childbearing years and, at each year of age, they experienced the birth rates occurring in the specified year. For explanation of "more developed" and "less developed" categories, see text, section 31]

CONTINENT AND REGION	MIDYEAR POPULATION (millions)					ANNUAL RATE OF GROWTH[1] (percent)			CRUDE BIRTH RATE		CRUDE DEATH RATE		TOTAL FERTILITY RATE (number)		LIFE EXPECTANCY (years)	
	1970	1980	1988	1990	2000	1970–1980	1980–1990	1990–2000	1988	2000	1988	2000	1988	2000	1988	2000
World total	**3,721**	**4,476**	**5,143**	**5,320**	**6,241**	**1.8**	**1.7**	**1.6**	**26.8**	**23.2**	**9.7**	**8.4**	**3.49**	**3.04**	**61.3**	**65.0**
More developed regions	1,049	1,137	1,198	1,211	1,269	.8	.6	.5	14.9	13.3	9.7	9.7	1.95	1.91	73.3	75.6
Less developed regions	2,672	3,339	3,945	4,109	4,971	2.2	2.1	1.9	30.4	25.8	9.7	8.1	3.96	3.33	59.5	63.6
Percent of world	72	75	77	77	80	(X)	(X)	(X)	(X)	(X)	(X)	(X)	(X)	(X)	(X)	(X)
Africa[2]	376	493	622	660	887	2.7	2.9	3.0	43.9	39.4	14.1	10.7	6.26	5.48	52.7	57.7
Asia	2,111	2,594	3,004	3,111	3,655	2.1	1.8	1.6	27.1	22.3	9.1	7.8	3.44	2.81	61.0	65.3
More developed regions	104	117	123	124	129	1.1	.6	.4	11.9	12.6	7.1	9.0	1.79	1.78	77.8	78.5
Less developed regions	2,007	2,477	2,882	2,988	3,526	2.1	1.9	1.7	27.7	22.7	9.2	7.7	3.51	2.85	60.7	65.0
East Asia[3]	991	1,181	1,304	1,334	1,481	1.8	1.2	1.0	18.9	15.2	6.7	6.7	2.20	1.79	69.4	73.1
South Asia[3]	1,120	1,412	1,700	1,777	2,174	2.3	2.3	2.0	33.4	27.2	10.9	8.5	4.39	3.51	57.3	62.4
Latin America[2]	286	365	436	455	551	2.4	2.2	1.9	29.1	23.9	7.1	6.1	3.62	2.92	66.8	70.4
Middle America[2 3]	70	93	112	116	141	2.9	2.2	2.0	29.8	24.3	6.3	5.3	3.65	2.75	67.8	71.4
Caribbean[2 3]	25	29	33	34	39	1.6	1.4	1.4	25.3	21.1	8.2	7.2	2.98	2.60	65.8	69.0
South America[2 3]	191	242	291	304	370	2.4	2.3	2.0	29.3	24.0	7.2	6.2	3.68	3.02	66.5	70.2
Northern America[3 4]	226	252	272	277	296	1.1	.9	.7	15.2	12.5	8.6	8.9	1.84	1.84	75.5	76.9
Europe[4]	460	484	497	499	510	.5	.3	.2	13.2	12.1	10.6	10.6	1.76	1.76	74.3	75.9
Soviet Union[4]	243	266	286	291	312	.9	.9	.7	18.8	16.2	10.6	9.4	2.45	2.27	68.9	73.2
Oceania	19	23	26	26	30	1.6	1.5	1.3	19.8	17.7	8.1	8.3	2.58	2.37	69.2	71.2
Australia and New Zealand[4]	15	18	20	20	22	1.4	1.2	1.0	15.6	14.0	7.7	8.3	1.92	1.92	75.9	76.9
Less developed regions	4	5	6	6	8	2.5	2.4	2.2	34.0	29.4	9.5	8.2	4.81	3.79	59.0	62.6

X Not applicable. [1] Computed by the exponential method. For explanation of average annual percent change, see Guide to Tabular Presentation. [2] Less developed regions. [3] For component countries, see table 1405. [4] More developed region.

Source: U.S. Bureau of the Census, *World Population Profile, 1989,* forthcoming.

Figure 12–42 World Population Table
Source: U.S. Bureau of the Census, *Statistical Abstract of the United States: 1989,* 109th ed. (Washington, DC: GPO, 1989) 812.

needed to help the readers understand what the graph depicts. Acknowledge the source of the information in the graph. Also, accompany your graph with a paragraph for your instructor that explains how purpose, situation, and audience guided you in constructing your graph.

3. Construct a flowchart for some purpose and audience of your own choosing. For example, you could chart the flow of some technical process in your field to show your parents what you are learning in school. Or you could chart some algorithm to instruct others in how to follow some process. There are, indeed, many processes you are learning about that could be charted to aid some reader to understand them or do them. Your completed chart should have a title and accompanying prose that helps the reader to understand what the chart depicts. Also, accompany your chart with a paragraph for your instructor that explains how purpose, situation, and audience guided you in constructing your chart.
4. Prepare a list of the graphics you intend to include in your project report. Rough out the design for several of them. Be prepared to discuss your graphics with your instructor.

PART IV

Applications of Technical Writing

Correspondence

Instructions

Proposals

Progress Reports

Feasibility Reports

Empirical Research Reports

CHAPTER 13

Correspondence

COMPOSING LETTERS AND MEMORANDUMS

Topic and Purpose
Audience

STYLE AND TONE

Clear Style
Human Tone

FORMAT

Heading
Date Line
Inside Address
Attention Line
Subject Line
Salutation
Body
Complimentary Close
Signature Block
End Notations
Continuation Page

LETTERS OF INQUIRY

Identification
Inquiry
Need
Choice of Recipient
Courteous Close
Sample Letter

REPLIES TO LETTERS OF INQUIRY

LETTERS OF COMPLAINT AND ADJUSTMENT

Letter of Complaint
Letter of Adjustment

THE CORRESPONDENCE OF THE JOB HUNT

Preparation
Letter of Application
The Resume
Follow-up Letters

LETTER AND MEMORANDUM REPORTS

Introduction
Summary
Discussion
Action

Even in this day of instant communication through telephones, fax machines, and electronic mail, typed—or, more likely, word processed—correspondence still plays a major role in getting an organization's work done. For example, executives may reach a decision in a telephone conversation or a teleconference, but that decision has to be documented. People may forget, or worse, misremember, what the decision was. A letter or memo records the decision for everyone.

In fact, as we point out in Chapter 10, "Document Design," it's a myth that electronic technology is creating the paperless office. American business and government turn out more paper in the electronic age than ever before. Much of that paper consists of correspondence, and, on the job, you will often be responsible for a share of your organization's letters and memorandums.

We do not attempt in this chapter to give you a course in business correspondence. There are many fine books on the subject, some of them listed in our bibliography. However, we do cover those aspects of correspondence that will be most useful to you. We discuss the composition of business letters and memorandums, their style and tone, and their format. Then we instruct you in how to write letters of inquiry, replies to letters of inquiry, letters of complaint and adjustment, the correspondence of the job hunt, and memorandum and letter reports.

Composing Letters and Memorandums

Composing letters and memorandums—memos, for short—is little different from composing any piece of writing as we describe the process in Chapter 2, "Composing." But a few points about topic, purpose, and audience might be worth reemphasizing because of their importance to good correspondence.

Topic and Purpose

People on the job use letters and memos to inform, to instruct, to analyze and evaluate, to argue—the whole range of activities, in fact, that occupy people at work. Letters go to people outside the company or organization; memos go to people within. Any of the following would be a typical topic and purpose:

- Instructing the office secretaries in how to complete the new travel payment requests

- Convincing the boss that the office needs two new microcomputers
- Answering a complaint

Whatever your topic and purpose may be, be sure to announce them immediately. Nothing irritates a reader more than to read through several paragraphs of details with no idea of where the letter or memo is heading. Be directive. Tell your reader what your topic and purpose are with an opening like this one:

> Additional report writing responsibility in our office has increased our need for greater word processing capability. This memo describes the problems caused by the increased work load and the benefits of purchasing two microcomputers and a laser printer for the use of our writers.

The only exception to such a clear statement of topic and purpose may be when the recipient is likely to be hostile to your conclusions. We discuss this possibility and how to handle it when we discuss bad-news letters on pages 394–397.

Audience

In composing your letter or memo, consider your readers: Who are they and what do they know already? What is their purpose in reading your letter or memo?

Who Are the Readers? Letters and memos may be addressed to one person or to many. Even when you address a letter or memo to one person, you may send copies of it to many others. When such is the case, you have to think about the knowledge and experience of those receiving copies as well as those receiving the original. If, for example, those receiving copies lack background in the topic under discussion, you may have to take time to fill them in.

In another common situation, you may be explaining a technical problem and its solution to a colleague with technical knowledge equal to your own. However, you may be sending a copy to your boss, who lacks that technical knowledge. When such is the case, you would be wise to lead into your discussion with what amounts to an executive summary. (See pages 291–292.) Fill the boss in quickly on the key points, and tell him or her the implications of what you are saying.

What Is the Reader's Purpose? You have a purpose in writing your letter or memo, but your reader has a purpose as well. Be sure to match your purpose to your reader's. For instance, your reader may be reading to evaluate your recommendation and accept or reject it. Your purpose must be to provide enough information to make that evaluation and decision possible.

In another situation, your purpose may be to explain why a project is running late. Your reader, somewhat skeptically no doubt, will be reading to determine if your reasons are valid and acceptable. Provide enough information to justify your position. If you don't match your purpose to your reader's, you may not reach the result you intended.

Style and Tone

In your correspondence, work for a clear style and a human tone.

Clear Style

Everything we say about clear and readable style in Chapter 6 goes doubly when you are writing letters and memos. Letters must be clear, so clear that the reader cannot possibly misunderstand them. Use short paragraphs, lists, clear sentence structure, and specific words. Above all, avoid pomposity and the cold formality of the passive voice.

Do not make your letters and memos a repository for all the clichés that writers before you have used. Avoid expressions like these:

Poor

- We beg to advise you that. . . .
- We are in receipt of your letter that. . . .
- It is requested that you send a copy of the requested document to our office.

There are literally hundreds of such expressions that weigh down business correspondence. To protect yourself from such prose pachyderms, remember our closing advice from Chapter 6: Ask yourself if you would or could say in conversation what you have written. If you know you would strangle on the expression, don't write it. Restate it in simpler language, like this:

Better

- We'd like you to know that. . . .
- We received your letter that. . . .
- Please send us a copy of your latest tax return.

Human Tone

To get a human tone into your correspondence, focus on the human being reading your letter or memo. Develop what has been labeled the *you-attitude.*

To some extent, we suppose, the you-attitude means that your letters and memos contain a higher percentage of *you*'s than *I*'s. But it goes beyond this mechanical usage of certain pronouns. With the you-attitude, you see things from your reader's point of view. You think about what the letter or memo will mean to the reader, not just what it means to you. We can illustrate simply. Suppose you have an interview scheduled with a prospective employer and, unavoidably, you must change dates. You could write as follows:

I-attitude

Dear Ms. Moody:

A change in final examination schedules here at school makes it impossible for me to keep our appointment on June 10.

I am really disappointed. I was looking forward to coming to Los Angeles. I feel inconvenienced, but I hope we can work out an appointment. Please let me know on what date we can arrange a new meeting.

Sincerely yours,

Now this letter is clear enough, and it may even get a new appointment for its writer. But it has the I-attitude, not the you-attitude. The persona projected is of a person who thinks only of himself. The reader may be vaguely or even greatly annoyed. This next version will please a reader far more:

You-attitude

Dear Ms. Moody:

A change in final examination schedules here at school makes it impossible for me to keep our appointment on June 10. When I should have been talking to you, I'll be taking an exam in chemistry.

I hope this change will not seriously inconvenience you. Please accept my regrets.

Will you be able to work me in at a new time? Final exam week ends on June 12. Please choose any date after that convenient for you.

Sincerely yours,

Mechanically, the second letter contains more *you*'s, but more to the point, it considers the inconvenience caused the reader, not the sender. And it makes it easy for the reader to set up a new date.

Notice also that the second letter is a bit more detailed than the first. Many people have stressed the need for brevity in letters and memos, and certainly it is a good thing to be concise. But do not get carried away with the notion. Letters and memos are not telegrams. When they are too brief, they give an impression of brusqueness, even rudeness. Often, a longer letter or memo gives a better impression. Particularly avoid brevity when you must refuse people something or disappoint their expectations. People appreciate your taking time to explain in such a situation.

With the you-attitude, you consider both your reader's feelings and needs. Here is the way a camera company politely corrected an amateur photographer's film-loading procedure:

Dear Mr. Bayless:

Courteous and complete

We are sending you a roll of film to replace the one overexposed when you used your new Film-X camera for the first time. And we'll certainly replace the camera if it proves to be defective.

Trouble such as you have experienced is often caused by the photographer inadvertently forgetting to set the proper film speed while loading the camera. The ASA dial on the front of the camera must be set to match the ASA film speed number of the film you are working with. Failure to do so will result in either under- or overexposed film such as you experienced.

If you'll check Instruction 5 on page 3 of your Film-X Manual, you'll find complete information on how to set the ASA dial.

We hope that your problem was nothing more than an improperly set ASA. If not, however, follow the warranty instructions that came with the camera, and we'll replace your Film-X.

Sincerely yours,

With this letter, the writer begins by offering a gift. Next he puts the reader's mind at ease by telling him that the company will replace the camera if need be. Then he attempts to correct the reader's film-loading procedure, the real reason, in the writer's opinion, for the trouble.

Suppose the letter writer had taken an attitude like the one expressed in this letter:

Dear Mr. Bayless:

Concise but discourteous

The type of film failure you wrote to us about stems from faulty loading of the Film-X. Your method and not the camera is defective.

Review the instructions that came with the camera. Pay particular attention to Instruction 5.

Sincerely yours,

The second letter would surely irritate the reader. Do people really write foolish, rude letters like this one? Unfortunately, perhaps because they are pressed for time, or because of a sense of superiority, they do—and worse. When people make a mistake, correct the consequences without making them feel like fools. We all make mistakes, and we appreciate gentle correction, not the lash of sarcasm.

Our advice about style and tone can be summed up with what we call the four **C**'s of correspondence: Correspondence should be **clear, concise, complete,** and **courteous.** Sometimes, **concise** and **complete** may be in conflict. If in doubt, opt for completeness. Always remember the importance of being **courteous.** Taking time for the you-attitude is the good and human way to act. Luckily for us all, it's also very good business.

Format

Almost any organization you join will have rules about its letter and memo formats. You will either have a secretary to do your correspondence, or you will have to learn the rules for yourself. In this section we give you only enough rules and illustrations so that you can turn out a good-looking, correct, and acceptable business letter or memo on your own. If for no other reason, you will find this a necessary skill when you go job hunting.

Figures 13–1 and 13–2 illustrate the block and semiblock styles on

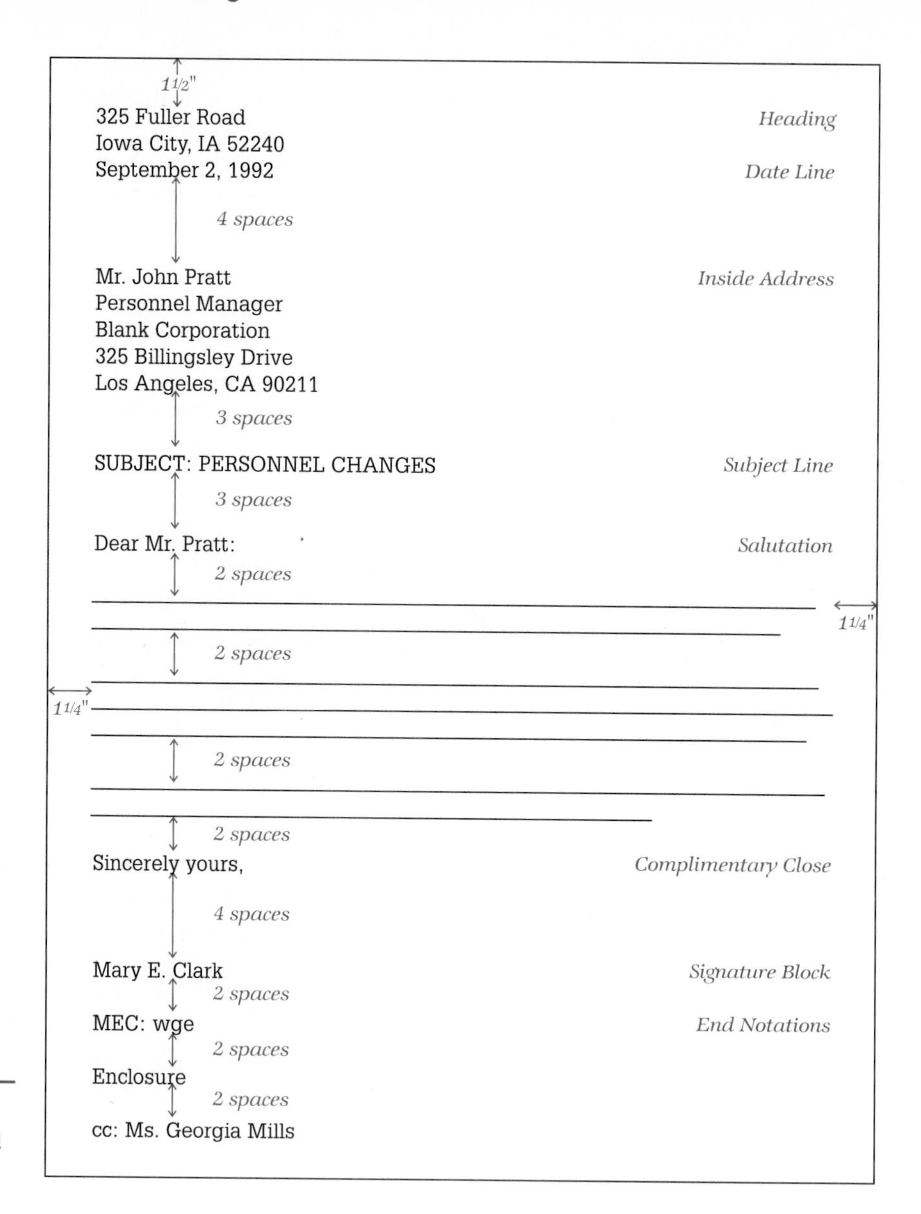
325 Fuller Road
Iowa City, IA 52240
September 2, 1992

Mr. John Pratt
Personnel Manager
Blank Corporation
325 Billingsley Drive
Los Angeles, CA 90211

SUBJECT: PERSONNEL CHANGES

Dear Mr. Pratt:

Sincerely yours,

Mary E. Clark

MEC: wge

Enclosure

cc: Ms. Georgia Mills

Figure 13–1 Block Letter on Nonletterhead Stationery

nonletterhead stationery. Figures 13–3 and 13–4 illustrate the block style and the simplified style on letterhead stationery.

The chief difference between memos and letters is format. Figure 13–5 illustrates a typical memo format. Figure 13–6 illustrates the heading used for the continuation pages of either a letter or a memo. We have indicated in these samples the spacing, margins, and punctuation you should use. In the text that follows, we discuss briefly the different styles and then

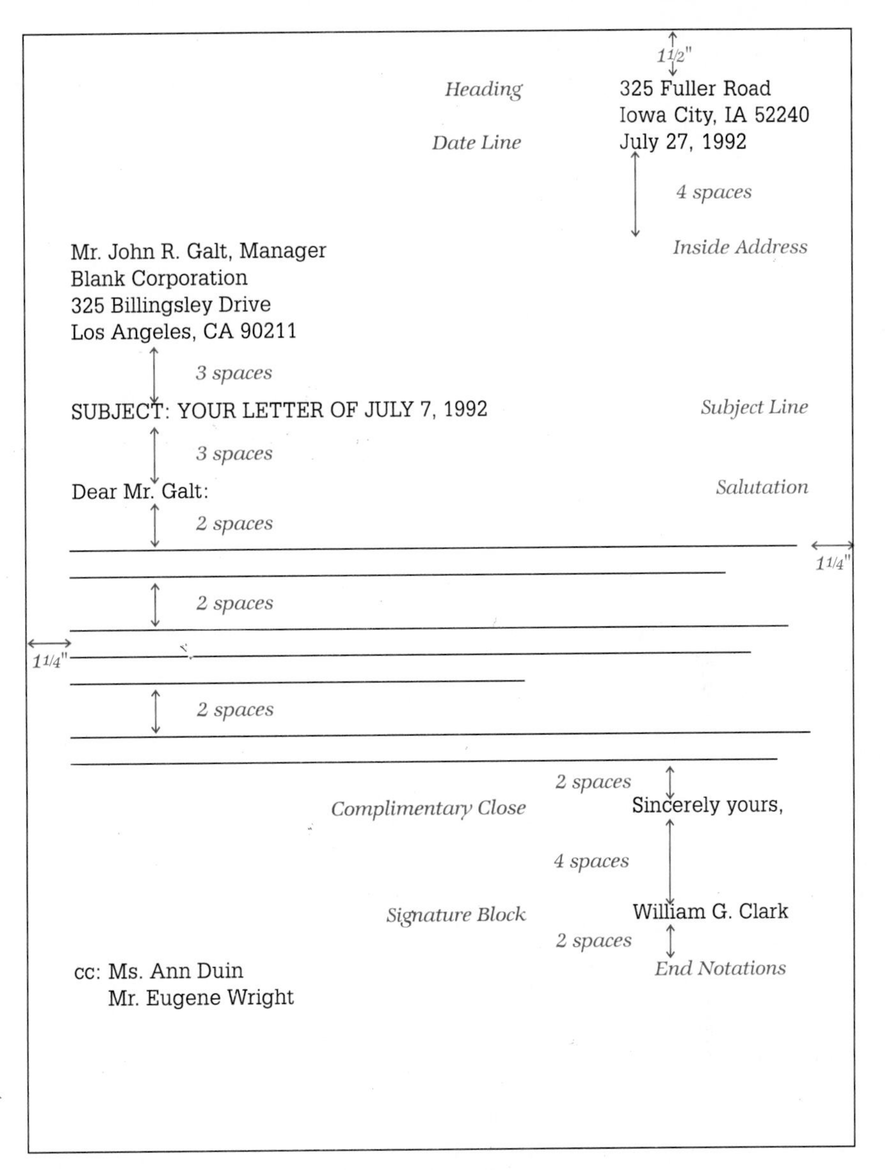

Figure 13–2 Semiblock Letter on Nonletterhead Stationery

give you some of the basic rules you should know about the parts of a letter or memo. Before continuing with the text, look at Figures 13–1 through 13–6.

For most business letters you may have to write, any of the styles shown would be acceptable. In letters of inquiry or complaint, where you probably do not have anyone specific in a company to address, we suggest the simplified style. For letters of application we suggest the block or

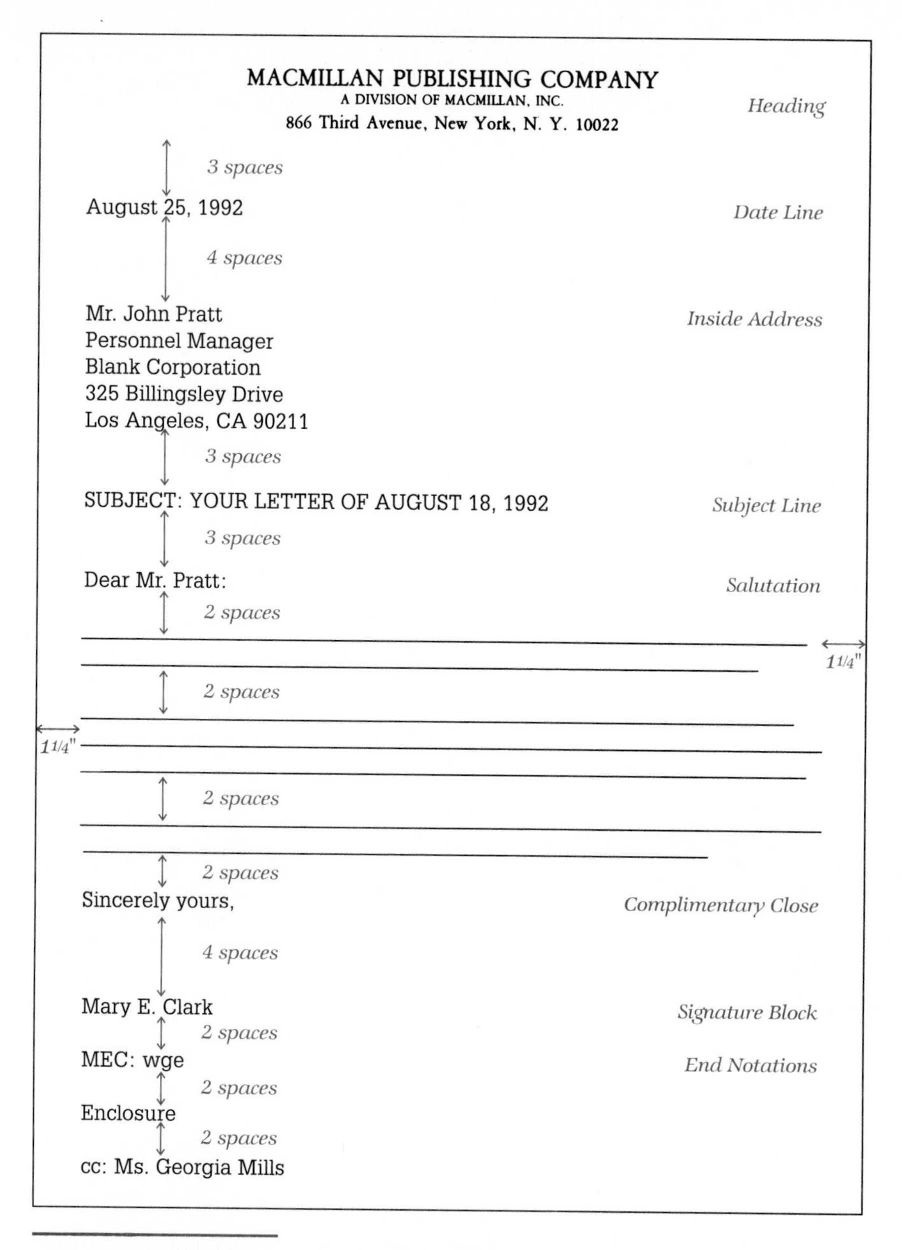

MACMILLAN PUBLISHING COMPANY
A DIVISION OF MACMILLAN, INC.
866 Third Avenue, New York, N. Y. 10022

Heading

3 spaces

August 25, 1992

Date Line

4 spaces

Mr. John Pratt
Personnel Manager
Blank Corporation
325 Billingsley Drive
Los Angeles, CA 90211

Inside Address

3 spaces

SUBJECT: YOUR LETTER OF AUGUST 18, 1992

Subject Line

3 spaces

Dear Mr. Pratt:

Salutation

2 spaces

1 1/4"

2 spaces

1 1/4"

2 spaces

2 spaces

Sincerely yours,

Complimentary Close

4 spaces

Mary E. Clark

Signature Block

2 spaces

MEC: wge

End Notations

2 spaces

Enclosure

2 spaces

cc: Ms. Georgia Mills

Figure 13–3 Block Style on Letterhead Stationery

semiblock style without the use of a subject line. Some people still find the simplified letter without the conventional salutation and complimentary close a bit too brusque. Unless you know for certain that the company you are applying to prefers the simplified form, do not take a chance with it.

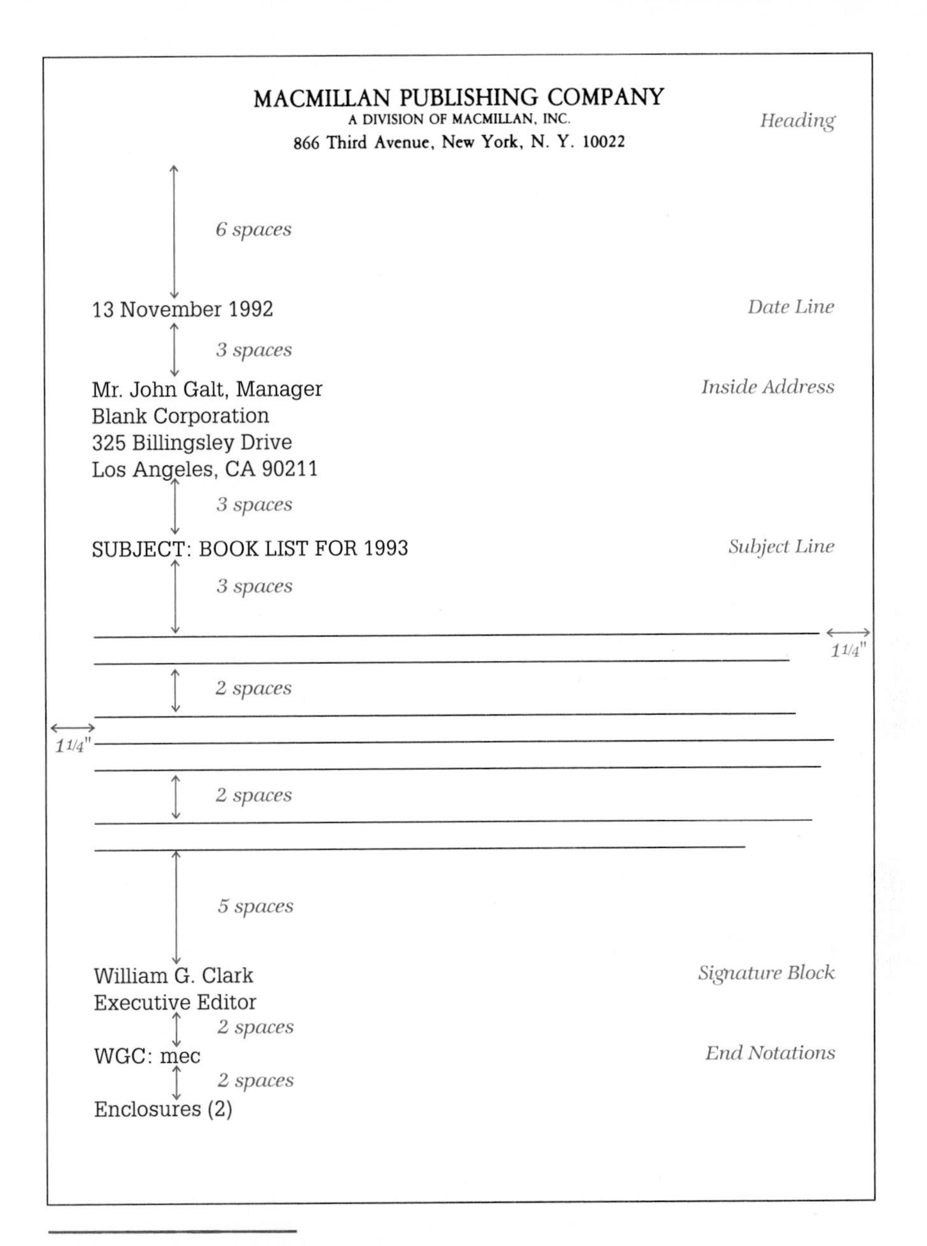
MACMILLAN PUBLISHING COMPANY
A DIVISION OF MACMILLAN, INC.
866 Third Avenue, New York, N. Y. 10022

13 November 1992

Mr. John Galt, Manager
Blank Corporation
325 Billingsley Drive
Los Angeles, CA 90211

SUBJECT: BOOK LIST FOR 1993

William G. Clark
Executive Editor

WGC: mec

Enclosures (2)

Figure 13–4 Simplified Style on Letterhead Stationery

If you are an amateur typist doing your own typing or word processing, we suggest the block style as the best all-around style. All the conventional parts of a letter are included, but everything is lined up along the left-hand margin. You do not have to bother with tab settings and other complications. Some people feel that a block letter looks a bit lopsided,

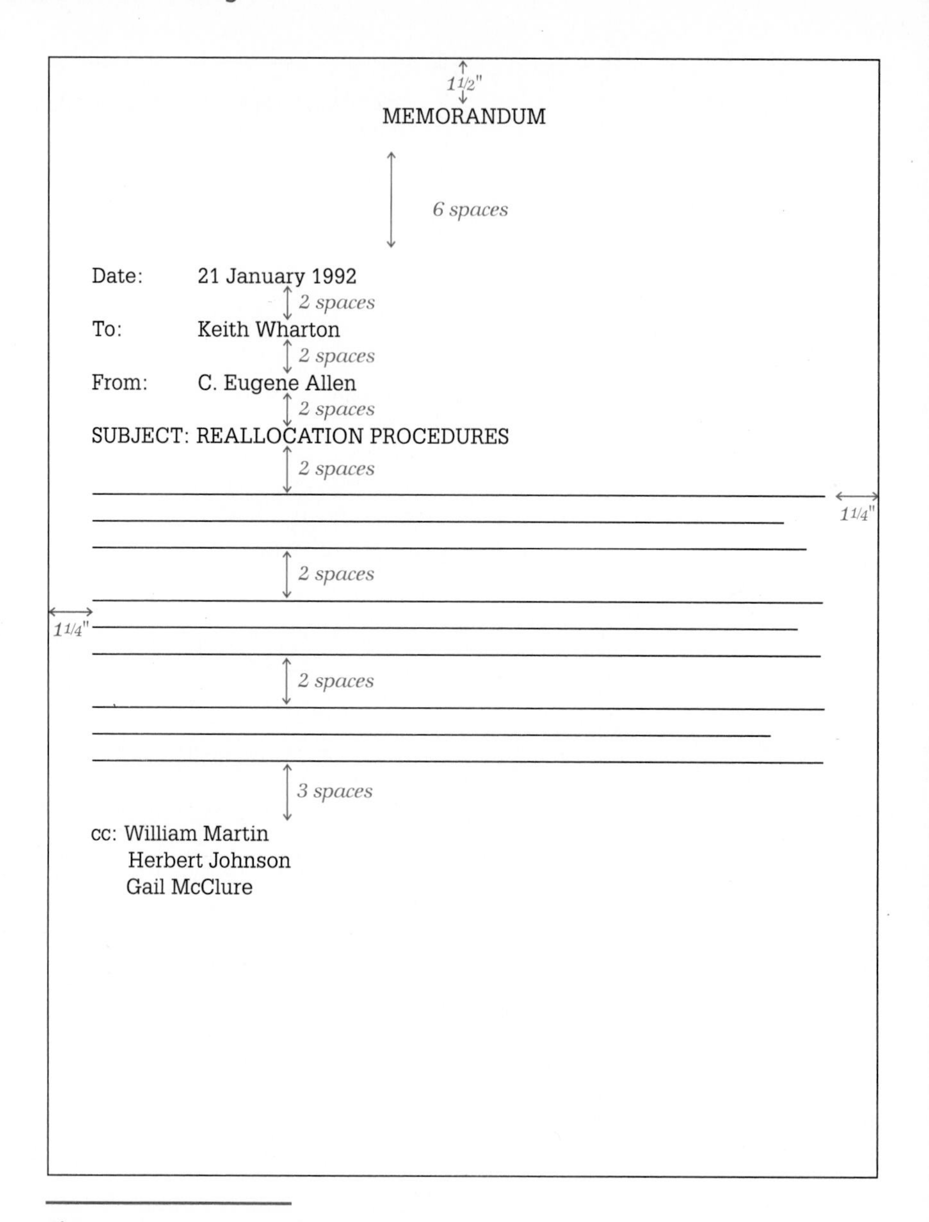
MEMORANDUM

Date: 21 January 1992

To: Keith Wharton

From: C. Eugene Allen

SUBJECT: REALLOCATION PROCEDURES

cc: William Martin
Herbert Johnson
Gail McClure

Figure 13–5 Memorandum Format

but it is a common style that no one will object to. No matter which style you choose, leave generous margins, from an inch to two inches all around, and balance the first page of the letter vertically on the page. Because letters should look good with lots of white space, you should seldom allow paragraphs to run more than seven or eight lines.

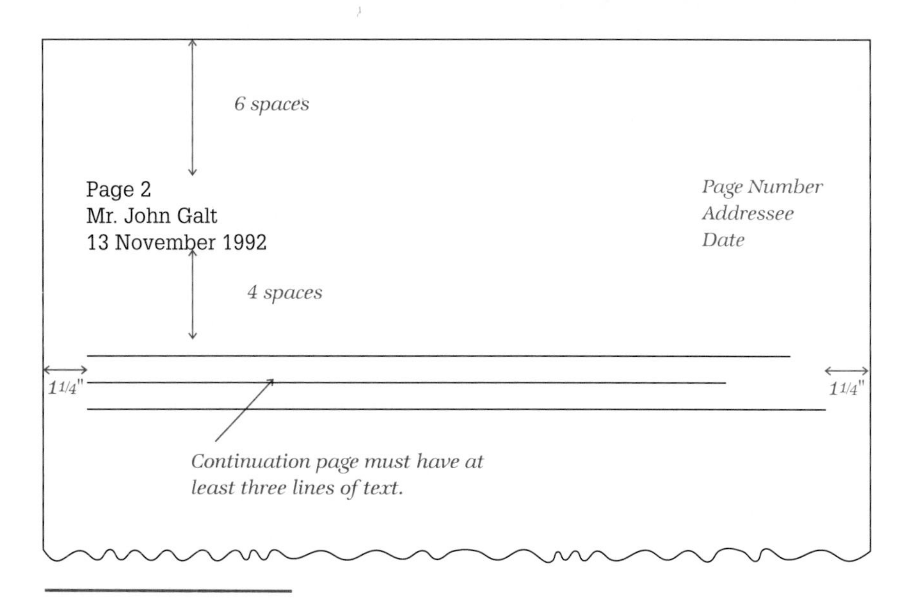

Figure 13–6 Continuation Page

Heading

When you do not have letterhead stationery, you will have to type your heading. In the semiblock style, the heading is approximately flush right. In the other formats shown, the heading is flush left. Do not abbreviate words like *street* or *road*. Write them out in full. You may abbreviate the names of states and provinces. Figure 13–7 gives you the state abbreviations for the United States and province abbreviations for Canada. If you do have business letterheads made up, have them printed on good-quality white bond in a simple style. With word processing and a laser printer, it's now possible to print your own letterheads. We have two cautions if you choose to do so. Only the most expensive laser printers will turn out printing as crisp and sharp as commercial printing will. Second, be careful not to design a too-elaborate letterhead. Keep it simple in a standard design.

Date Line

In a letter without a printed letterhead, the date line is part of your heading in the block, semiblock, and simplified styles. When you do have a printed letterhead, the date line is flush left in the block and simplified styles and approximately flush right in semiblock. Place it three to six spaces below a printed letterhead in a manner to help balance the letter vertically on the page. Write the date out fully either as June 3, 19XX or

United States				Canada	
Alabama	AL*	Missouri	MO	Alberta	AB
Alaska	AK	Montana	MT	British Columbia	BC
American Samoa	AS	Nebraska	NE	Labrador	LB
Arizona	AZ	Nevada	NV	Manitoba	MB
Arkansas	AR	New Hampshire	NH	New Brunswick	NB
California	CA	New Jersey	NJ	Newfoundland	NF
Colorado	CO	New Mexico	NM	Nova Scotia	NS
Connecticut	CT	New York	NY	Northwest Territories	NT
Delaware	DE	North Carolina	NC	Ontario	ON
District of Columbia	DC	North Dakota	ND	Prince Edward Island	PE
		Ohio	OH	Quebec (Province de Quebec)	PQ
Florida	FL	Oklahoma	OK	Saskatchewan	SK
Georgia	GA	Oregon	OR	Yukon Territory	YT
Guam	GU	Pennsylvania	PA		
Hawaii	HI	Puerto Rico	PR		
Idaho	ID	Rhode Island	RI		
Illinois	IL	South Carolina	SC		
Indiana	IN	South Dakota	SD		
Iowa	IA	Tennessee	TN		
Kansas	KS	Texas	TX		
Kentucky	KY	Utah	UT		
Louisiana	LA	Vermont	VT		
Maine	ME	Virginia	VA		
Maryland	MD	Virgin Islands	VI		
Massachusetts	MA	Washington	WA		
Michigan	MI	West Virginia	WV		
Minnesota	MN	Wisconsin	WI		
Mississippi	MS	Wyoming	WY		

*Notice that both letters of the abbreviation are capitalized and that no period is used.

Figure 13–7 Geographic Abbreviations for the United States and Canada

3 June 19XX. Do not abbreviate the month or add *st, nd,* etc. (e.g., 1st, 2nd), to the number of the day.

Inside Address

The inside address is placed flush left in all the formats shown. Make sure the inside address is complete. Follow exactly the form used by the person or company you are writing to. If your correspondent abbreviates *Company* as *Co.,* you should also. Use *S. Edward Smith* rather than *Samuel E. Smith* if that is the way Smith wants it. Do use courtesy titles such as *Mr., Dr.,* and *Colonel* before the name. The usual abbreviations used are *Mr., Ms.,* and *Dr. Miss* and *Mrs.* are still in use, but *Ms.* has become standard usage in the workplace. Place one-word titles such as

Manager or *Superintendent* immediately after the the name. When a title is longer than one word, place it on the next line by itself. Do not put a title after the name that means the same thing as a courtesy title. For example, don't write *Dr. Samuel E. Smith, Ph.D.*

Attention Line

On occasion you may wish to write to an organizational address but also draw your letter to the attention of some individual. It's a way of saying, in effect, "Anyone there can answer this letter, but if Mr. Smith is there he is the best person to handle the matter." When you use an attention line, type it flush left two spaces below the inside address. Capitalize only the "A." You can use a colon or not between the "Attention" and the name:

Attention Mr. Frank Rookard

or

Attention: Mr. Frank Rookard

Subject Line

Place the subject line flush left in all the styles. In the block and semiblock styles, it is usually preceded by the heading "SUBJECT" though sometimes you will see it with nothing before it. Type both the heading and the subject line itself in capital letters. In the simplified style omit the heading, and type the line in all-capital letters. In this line you can give the date of the letter being answered or a substantive heading such as "complaint adjustment."

Salutation

Place the salutation flush left. Convention still calls for the use of *Dear.* Always use a name in the salutation when one is available to use. When you use a name, be sure it is in the inside address as well. Also, use the same courtesy title as in the inside address, such as *Dear Dr. Sibley* or *Dear Ms. McCarthy.* You may use a first-name salutation, *Dear Sarah,* when you are on friendly terms with the recipient. Follow the salutation with a colon.

What do you do when you are writing a company blindly and have no specific name to address? Some people use a *Dear* followed by the name of the department being written to, such as *Dear Customer Relations Department.* Still others substitute a *Hello* or *Good Day* for the traditional salutation. One solution, perhaps the best, is to choose the simplified style, where no salutation is used.

In any case, do **not** begin letters with *Dear Person* or *Dear People,* which are distasteful to many, *Gentlemen* or *Dear Sir,* which are sexist, or *To Whom It May Concern,* which is old-fashioned.

Body

In typing the body of an average-length letter or memo, single-space between the lines and double-space between the paragraphs. In a particularly short letter, double-space throughout the body and use five-space indentations to mark the first lines of paragraphs. Avoid splitting words between lines. Never split a date or a person's name between two lines.

Complimentary Close

In the block style, place the complimentary close flush left. In the semiblock style, align the close with the heading (or with the date line in a letterhead letter). Settle for a simple close, such as *Sincerely yours* or *Very truly yours*. Capitalize only the first letter of the close and place a comma after the close.

Signature Block

Type your name four spaces below the complimentary close in the block and semiblock styles, five spaces below the last line in the simplified style. Use your first name, middle initial, and last name or, if you prefer, your first initial and middle name, as in *M. Lillian Smith*. We don't recommend the use of initials only, as in *M. L. Smith*, because this form puts your correspondents at a disadvantage. People who don't know you will not know whether to address you as *Dear Mr.* or *Dear Ms.*

If you have a title, type it below your name. Sign your name immediately above your typed name. Sign it legibly without fancy flourishes. Your signature and your typed name should agree. In memos it's customary to initial next to your typed name in the "From" line.

End Notations

Various end notations may be placed at the bottom of a letter or memo, always flush left. The most common ones indicate identification, enclosure, and carbon copy.

Identification The notation for identification is composed of the writer's initials in capital letters and the typist's initials in lowercase:

DHC:lnh

Enclosure The enclosure line indicates that additional material has been enclosed with the basic letter or memo. You may use several forms:

Enclosure
Enclosures (2)
Encl: Employment application blank

Copy The copy line informs the recipient of a letter or memo when you have sent a copy of the letter to someone else. In form, the copy notation looks like this:

cc: Ms. Elaine Mills

See Figures 13–1 to 13–5 for proper spacing and sequence of these three notations.

Continuation Page

Use a continuation page or pages when you can't fit your letter or memo onto one page. Do not use letterhead stationery for a continuation page. Use plain bond of the same quality and color as the first page.

As shown in Figure 13–6, the continuation page is headed by three items: page number, name of addressee, and date.

When you have a continuation page, the last paragraph on the preceding page should contain at least two lines. The last continuation page must have at least three lines of text to accompany the complimentary close and the signature block.

Letters of Inquiry

As a student, business person, or simply a private person, you will often have occasion to write letters of inquiry. Students often overlook the rich sources of information for reports that they can tap with a few well-placed, courteous letters. Such letters can bring brochures, photographs, samples, and even very quotable answers from experts in the field the report deals with.

Sometimes companies solicit inquiries about their products through their advertisements and catalogs. In such cases your letter of inquiry can be short and to the point:

Your advertisement on page 89 of the January 1992 *Scientific American* invited inquiries about your new Film-X developing process.

I am a college student and president of a 20-member campus photography club. The members of the club and I would appreciate any information about this new process that you can send us.

We are specifically interested in modernizing the film-developing facilities of the club.

As in the letter just quoted, you should include three important steps:

- Identify the advertisement that solicited your inquiry.
- Identify yourself and establish your need for the information.
- Request the information. Specify the precise area in which you are interested. Obviously, in this step you also identify the area in which the company may expect to make a sale to you. You thus, in a subtle way, point out to the company why it is in its best interest to answer your inquiry promptly and fully: a good example of the you-attitude in action.

An unsolicited letter of inquiry cannot be quite so short. After all, in an unsolicited letter you are asking a favor, and you must avoid the risk of appearing brusque or discourteous. In an unsolicited letter you include five steps:

- Identify yourself
- State clearly and specifically the information or materials that you want.
- Establish your need for the requested information or materials.
- Tell the recipient why you have chosen him or her as a source for this information or materials.
- Close courteously, but do **not** say "thank you in advance."

The first four steps may be presented in various combinations rather than as distinct steps, or in different order, but none of the steps should be overlooked.

Identification

In an unsolicited letter, identify yourself. We mean more here than merely using a title in your signature block. Rather, you should identify yourself in terms of the information sought. That is, you are not merely a student, you are a student in a dietetics class seeking information for a paper about iron enrichment of flour. Or you are a member of a committee investigating child abuse in your town. Or you are an engineer for a state department of transportation seeking information for a study of noise walls on city freeways. Certainly, the more prestigious your identification of yourself, the more likely you are to get the information you want. But do not misrepresent yourself in any way. Misrepresentation will usually lead to embarrassment.

Some years ago, one of the authors of this text naively wrote for information on bookshelves using his college letterhead instead of his private stationery. He was deluged with brochures that were followed up by a long-distance telephone call from a sales manager asking how many shelves the college was going to need: "Is the college expanding its library facilities?"

"No," the lame answer came, "I just wanted to put up a few shelves in my living room."

Do not misrepresent yourself. But do honestly represent yourself in the best light you can. Most companies are quite good about answering student inquiries. They recognize in students the buyers and the employees of the future and are eager to court their goodwill.

Inquiry

State clearly what you want. Be very specific about what you want, and do not ask for too much. Avoid the shotgun approach of asking for "all available information" in some wide area.

Particularly, do not expect other people to do your work for you. Do not, for example, write to someone asking for references to articles on some subject. A little time spent by you in the library will produce the same information. On the other hand, it would be quite appropriate to write to an engineer or scientist asking for a clarification or amplification of some point in a recent article that he or she had written. Science thrives on the latter kind of correspondence.

If you do need involved information, put your questions in an easy-to-answer questionnaire form. If you have only a few questions, include them in your letter. Indent the questions and number them. Be sure to keep a copy of your letter. The reply you receive may answer your questions by number without restating them.

For a larger group of questions, or when you are sending the questionnaire to many people, make the questionnaire a separate attachment. Be sure in your letter to refer the reader to the attachment. Questionnaires are tricky. Unless they are presented properly and carefully made up, they will probably be ignored. Do not ask too many questions. If possible, phrase the questions for yes or no answers or provide multiple-choice options. Sometimes, meaningful questions just cannot be asked in this objective way. In such cases, do ask questions that require the respondent to write an answer. But try to phrase your questions so that answers to them can be short. Provide sufficient space for the answer expected. If you are asking for confidential information, stress that you will keep it confidential. The questionnaire should be typed, printed, or photocopied. For further information about putting together a questionnaire, see pages 93–99.

Need

Tell the recipient why you need the information. Perhaps you are writing a report, conducting a survey, buying a camera, or simply satisfying a healthy, scientific curiosity. Whatever your reason, do not be complicated or devious here. Simply state your need clearly and honestly. Often this step can best be combined with the identification step. If there is some deadline by which you must have an answer, say what it is.

Choice of Recipient

Tell the recipient of the letter why you have chosen him or her or the recipient's company as a source for the information. Perhaps you are writing a paper about stereo equipment and you consider this company to be one of the foremost manufacturers of FM tuners. Perhaps you have read the recipient's recent article on space medicine. You specifically want an amplification of some point. Obviously, this section is a good place to pay the recipient a sincere compliment, but avoid flattery or phoniness.

Point out any benefit the recipient may gain by answering your request. For example, if you are conducting a survey among many companies, promise the recipient a tabulation and analysis of results when the survey is completed.

Courteous Close

Close by expressing your appreciation for any help the receiver may give you. But do not use the tired, old formula, "Thanking you in advance, I remain. . . ." Later, do write a thank-you note even if all you get is a refusal. Who knows? The second letter may cause a change of heart.

Sample Letter

A complete letter of inquiry might look like this one:

Dear Mr. Hanson:

Identification

Need

I am a second-year student at Florida Technological Institute. For a course in technical writing that I am currently taking, I am writing a paper on the proper way to educate Americans in the use of the metric system. The paper is due on 3 March.

Choice of source

In the journals where I have been researching the subject, you are frequently mentioned as a major authority in the field. Would you be kind enough to give me your opinion about how metrics should be taught?

Specific information requested

The specific question that concerns me is whether metric measurements should be taught in relation to present standard measurements, such as the foot and pound, or whether they should be taught independently of other measurements. I see both methods in use.

Courteous close

Any help you can give me will be greatly appreciated, and I will, of course, cite you in my paper.

Sincerely yours,

Replies to Letters of Inquiry

In replying to a letter of inquiry, be as complete as you can. Probably you will be trying to avoid a second exchange of letters of questions and answers. If you can, answer the questions in the order in which they appear in the inquiry. If the inquirer represents an organization and has written on letterhead stationery, you may safely assume that a file copy of the original letter has been retained. In that case, you don't have to repeat the original questions—just answer them. If the inquirer writes as a private person, you cannot assume that a copy of the letter has been kept. Therefore, you will need to repeat enough of the original question or questions to remind the inquirer of what has been asked.

You might answer the earlier letter about metric education in the following fashion:

Dear Ms. Montez:

Repetition of question

The question of how to educate people about the metric system concerns a great many educators. I suppose it's inevitable that those familiar with the present system will be tempted to convert metric measurements to ones they already know. For example, people will say, "A kilogram, that's about two pounds."

Answer to question

In my opinion, however, such conversion is not the best way to teach metrics. Rather, people should be taught to think in terms of what the metric measurement really measures, to associate it with familiar things. Here are some examples:

--A paper clip is about a centimeter wide.
--A dollar bill weighs about a gram.
--A comfortable room is 20 degrees C.
--Water freezes at 0 degrees C.
--At a normal walking pace, we can go about five kilometers an hour.

These are the kinds of associations we have made all our lives with the present system. We need to do the same for metric measurements. As in learning a foreign language, we really learn it only when we stop translating in our heads and begin to think in the new language.

Source of additional information

I hope this answers your question. You can get a valuable little book about metrics by writing to the Office of Public Information, Federal Reserve Bank of Minneapolis, Minneapolis, Minnesota 55480. Ask for The United States and the Metric System.

Sincerely yours,

Because the writer of the letter of inquiry wrote as a private person, she is, in this answer, reminded of the original question. The tone of the letter is friendly. The question is answered succinctly but completely. An additional source of information is mentioned, an excellent idea that should be followed when possible.

When you can, include as enclosures to your letter previously prepared materials that provide adequate answers to the questions asked. If you do so, provide whatever explanation of these materials is needed.

Letters of Complaint and Adjustment

Mistakes and failures happen in business as they do everywhere else. Promised deliveries don't arrive on time or at all. Expensive equipment fails at a critical moment. If you're on the receiving end of such problems, you'll probably want to register a complaint with someone. Chances are good that you'll want an adjustment—some compensation—for your loss or inconvenience. To seek such adjustment, you'll write a letter of complaint or, as it is also called, a claim letter.

Letter of Complaint

Your attitude in a letter of complaint should be firm but fair. There is no reason for you to be discourteous. You'll do better in most instances if you write the letter with the attitude that the offending company will want to make a proper adjustment once they know the problem. Don't threaten to withdraw your business on a first offense. Of course, after repeated offenses you will seek another firm to deal with, and that should be made clear at the appropriate time.

Be very specific about what is wrong and detail any inconvenience

26 Shady Woods Road
White Bear Lake MN 55101
28 October 1991

Customer Relations Department
Chapter Products, Inc.
1925 Jerome Street
Brooklyn, NY 11205

SUBJECT: BROKEN SANDING BELTS

Product information

This past September, I purchased two Chapman sanding belts, medium grade, #85610, at the Fitler Lumber Company in St. Paul, Minnesota. I paid a premium price for the Chapman belts because of your reputation. However, the belts have proved to be unsatisfactory, and I am returning them to you in a separate package (at a cost to me of $2.36).

Problem

The belt I have labeled #1 was used only 10 minutes before it broke. The belt labeled #2 was used only 5 minutes when the glue failed and it broke.

Inconvenience caused

I attempted to return the belts to Fitler's for a refund. The manager refused me a refund and said I'd have to write to you.

I am disturbed on two counts. First, I paid a premium price for your belts. I did not expect an inferior product. Second, does the retailer have no responsibility for the Chapman products he sells? Do I have to write to you every time I have a problem with one of your products?

Motivation for fair adjustment

I'm sure that Chapman is proud of its reputation and will want to adjust this matter fairly.

John Griffin
John Griffin

Figure 13–8 Letter of Complaint

caused you. Be sure to give any necessary product identification such as serial numbers. At the end of your letter, motivate the receiver to make a fair adjustment. If you know exactly what adjustment you want, spell it out. If not, allow the suggested adjustment to come from the company you are dealing with. Figure 13–8 shows you what a complete letter of complaint looks like. Note that it is in simplified format and addressed to the Customer Relations Department. You can safely assume that most companies have an office or an employee specifically responsible for complaints. In any case, if the letter is addressed in this manner, it will reach the appropriate person much more rapidly.

Letter of Adjustment

What happens at the other end of the line? You have received a letter of complaint and must write the adjustment letter. What should be your attitude? Oddly enough, most organizations welcome letters of complaint. At least they prefer customers who complain rather than customers who think "never again" when a mistake occurs and take their business elsewhere. Most organizations will go out of their way to satisfy complaints that seem at all fair. A skillful writer will attempt to use the adjustment letter as a means of promoting future business.

Letters of adjustment fall into two categories—granting of the adjustment requested or a refusal. The first is good news and easy to write. The second is more difficult, as indicated by its common name—the bad-news letter.

The Good-News Letter When you are granting an adjustment, be cheerful about it. Remember, your main goal is to build goodwill and future business. Follow these three steps:

- Begin by expressing regret about the problem or stating that you are pleased to hear from the customer—or both. Our earlier comments about the you-attitude should be much in your mind while writing an adjustment letter.
- Explain the circumstances that caused the problem. State specifically what the adjustment will be.
- Handle any special problems that may have accompanied the complaint and close the letter.

Figure 13–9 shows you such a letter.

The Bad-News Letter A letter refusing an adjustment is obviously more difficult to write. You want, if possible, to keep the customer's goodwill. You want at the very least to forestall future complaints. In stating your refusal, you must exercise great tact. Bad-news letters usually consist of five steps:

- Begin with a friendly opener. Try to find some common ground with the complainant. Express regret about the situation. Even though you may think the complaint is totally unfair, don't be discourteous. Incidentally, not everyone writes a letter as courteous as the one in Figure 13–8. Sometimes, in fact, people are downright abusive. If so, attempt to shrug it off. Just as you

CHAPMAN PRODUCTS, INC.
1925 Jerome Street
Brooklyn, NY 11205

November 11, 1991

Mr. John Griffin
26 Shady Woods Road
White Bear Lake, MN 55101

Dear Mr. Griffin:

Expression of regret

Thank you for your letter of October 28. We're sorry that you had a problem with a Chapman product. But we are happy that you wrote to us about your dissatisfaction. We need to hear from our customers if we are to provide them satisfactory products.

Explanation of circumstances

The numbers on the belts you returned indicate that they were manufactured in 1981. Sanding belts, like many other products, have a "shelf life," and the belts you purchased had exceeded theirs. Age, heat, and humidity had weakened them.

Statement of adjustment

Mr. Griffin, we stand behind our products. Although we are sure that the belts you purchased were not defective when we shipped them, we wish to replace them for you. You are being shipped a box of 10 belts, medium grade. We're sure that these belts will live up to the Chapman name.

Handling of special problems

We also suggest that you look in the yellow pages of your telephone directory under "Hardware-Retail" for authorized Chapman dealers. We can only suggest to independent dealers how they should shelve and sell our products. We can exert more quality control with our own dealers. We know that you can find a Chapman dealer who will give you excellent service.

Sincerely,

Theresa R. Brummer

Theresa R. Brummer
Customer Service

TB/ay

cc: Fitler Lumber Company

Figure 13–9 Good-News Letter of Adjustment

would not pour gasoline on a fire, don't answer abuse with abuse. Pour on some cooling words instead.

- Second, explain the reason for the refusal. Be very specific here and answer at some length. The very length of your reply will help to convince the reader that you have considered the problem seriously.
- Third, at the end of your explanation, state your refusal in as inoffensive a way as possible.
- Fourth, if you can, offer a partial or substitute adjustment.
- Finally, close your letter in a friendly way.

Companies selling products are not the only organizations that receive letters of complaint. Public service organizations do also. The letter that follows illustrates a refusal from such an organization—a state department of transportation. In this case, a citizen had written stating that a curved section of highway near her home was dangerous. She requested that the curve be rebuilt and straightened. The reply uses the strategy that we have outlined.

In the case of this letter, the goal is not to keep a paying customer but to keep a taxpayer friendly. In either case, the strategy is to offer an honest, detailed, factual explanation in a cheerful way.

Dear Mrs. Ferguson:

Friendly opener

Thank you for your letter concerning the section of Trunk Highway (TH) 50 near your home. The Department of Transportation shares your concern about the safety of TH 50, particularly the section between Prestonburg and Pikeville, near which your home is located.

Reasons for refusal

TH 50, as you mentioned, has a goodly number of hills and curves and, because of the terrain, some steep embankments. However, its accident rate--3.58 accidents per million vehicle miles--is far from the worst in the state. In fact, there are 64 other highways with worse safety records.

We do have studies under way that will result ultimately in the relocation of TH 50 to terrain that will allow safer construction. These things take time, as I'm sure you know. We have to coordinate our plans with county and town authorities and the federal government. Money is assigned to these projects by priority, and judged by its accident rate, TH 50 does not have top priority.

Accidents along TH 50 are not concentrated at any one curve. They are spread out over the entire highway. Reconstruction at any one location would cause little change in the overall accident record.

Statement of refusal

For all the above reasons we do not contemplate rebuilding any curves on TH 50. However, we are currently evaluating the need

Offer of substitution

for guardrails along the entire length of the highway. Within a year we will likely construct guardrails at a number of locations. Most certainly, we will place a guardrail at the curve that concerns you. This should correct the situation to some extent.

Friendly close

We appreciate your concern. Please write to us again if we can be of further help.

Sincerely yours,

The five-step strategy of the bad-news letter can be useful on many occasions. Any time you have to disappoint someone's expectations or you expect a hostile audience, consider using the bad-news approach.

The Correspondence of the Job Hunt

In many cases, the first knowledge prospective employers will have of you is the **letter of application** and **resume** that you send to them. A good letter of application, sometimes called a *cover letter*, and resume will not guarantee that you get a job, but bad ones will probably guarantee that you do not. In this section we'll describe the preparation for writing the letter of application and the resume and then discuss them and several follow-up letters needed during the job hunt.

Preparation

The good letter of application begins before you ever sit down to write your letter of application and resume. As in all writing you must first discover and select your material. In this case, your material is found in information gathering about potential employers and in analyzing your own interests and skills.

Information Gathering Begin by finding out all you can about the company you are applying to. Talk to friends who may know something about the company. Read the company's advertisements. If they have a company magazine, get it and read it. Using a letter of inquiry, send away for copies of their brochures and annual reports. Investigate the company in sources such as *Standard Statistics, Standard Corporation Reports,* and Dun and Bradstreet's *Middle Market Directory, Million Dollar Directory,* and *Reference Book of Manufacturers.*

Consult magazines and newspapers that regularly carry business news such as *Forbes,* the *Wall Street Journal,* and *Business Week.* To see what has recently appeared in the business press about the company, see the *F & S Index of Corporations and Industries* and the *Business Periodicals Index.* For general coverage see the *New York Times Index* and the *Reader's Guide to Periodical Literature.* Analyze the company to see what achievements it is most proud of and to determine its goals.

A major source of information is the *CPC Annual,* published each year by the College Placement Council and found in all college placement offices. The *CPC Annual* covers most major companies and many governmental organizations, providing information about their products, services, locations, job opportunities, and even their management philosophy. For example, a recent entry for the Jet Propulsion Laboratory in Pasadena, California, states the following:

> The emphasis here is on a creative engineering and science environment with a free exchange of innovative ideas and opinions. Management philosophy is to develop staff members to their highest level of maturity in an atmosphere of involvement. The work is exciting, and creative expression is encouraged.

The *CPC Annual* provides the necessary addresses you'll need to contact the organizations listed. Also included in each *Annual* is advice on interviewing and the correspondence of the job hunt.

For information about the occupation you plan to enter, look into the *Occupational Outlook Handbook,* published by the U.S. Bureau of Labor Statistics, or the *Career Index.* For information about jobs with the federal government and how to get them, look into *Working for the USA* and the *Federal Career Directory.* The latter is published yearly. For jobs at the state level, ask the state department of human resources for information.

We have listed only some of the most basic sources where you can find information to prepare for the job hunt. Your librarian and college placement officer can help you find many more, some perhaps quite specifically oriented to the occupation you are most interested in. We cannot emphasize enough how important this preparation is. Listen to what one interviewer had to say on the subject:

> It's really impressive to a recruiter when a job candidate knows about the company. If you're a national recruiter, and you've been on the road for days and days, you have no idea how pleasant it is to have a student say, "I know your company is doing such and such, and has plants here and here, and I'd

> like to work on this particular project." Otherwise I have to go into my standard spiel, and God knows I've certainly heard myself give that often enough.[1]

Self-Analysis Analyze yourself as well as potential employers. What are your strengths? What are your weaknesses? How well have you performed in past jobs? Have you shown initiative? Have you improved procedures? Have you accepted responsibility? Have you been promoted or given a merit raise? How can you present yourself most attractively? What skills do you possess that relate directly to what the employer seems to need? Are you really qualified for the job you want? When you have finished your information gathering and your self-analysis, you are ready to begin your letter of application and your resume.

Letter of Application

Plan the mechanics of your **letter of application** carefully. Buy the best quality white bond paper. This is not time to skimp. Plan to type or word process your letter, of course, or have it done. Use a standard typeface. Do not use italics. Make sure your letter is mechanically perfect, free from erasures and grammatical errors. Be brief but not telegraphic. Keep the letter to one page. Don't send a letter duplicated in any way. Accompany each letter, however, with a duplicated resume. We have more to say about this feature later.

Pay attention to the style of the letter and the resume that accompanies it. The tone you want in your letter is one of self-confidence. You must avoid both arrogance and humility. You must sound interested and somewhat eager, but not fawning. Do not give the impression that you must have the job. Nor do you want to seem uncaring about whether you do get the job.

When describing your accomplishments in the letter and resume, use action verbs. They help to give your writing brevity, specificity, and force. For example, don't just say that you worked as a salesclerk. Rather, tell how you maintained inventories, sold merchandise, prepared displays, implemented new procedures, and supervised and trained new clerks. Here's a sampling of such words:

administer	edit	oversee
analyze	evaluate	plan
conduct	exhibit	produce
create	expand	reduce costs
cut	improve	reorganize
design	manage	support
develop	operate	was promoted
direct	organize	write

You cannot avoid the use of *I* in a letter of application. But take the you-attitude as much as you can. Think about what you can do for the prospective employer. The letter of application is not the place to be worried about salary and pension plans. Above all, be mature and dignified. Forget about tricky and flashy approaches. Write a well-organized, informative letter that highlights those skills your analysis of the company shows it desires most. We will discuss the application letter in terms of a beginning, a body, and an ending.

The Beginning Beginnings are tough. Do not begin aggressively or cutely. Beginnings such as "WANTED: an alert, aggressive employer who will recognize an alert, aggressive young forester" will usually send your letter wastebasket-bound. If it is available to you, a bit of legitimate name dropping is a good beginning. Use this beginning only if you have permission and if the name dropped will mean something to the prospective employer. If you qualify on both counts begin with an opener like this:

> Dear Ms. Marchand:
>
> Professor John J. Jones of State University's Food Science faculty has suggested that I apply for the post of food supervisor that you have open. In June I will receive my Bachelor of Science degree in Food Service from State University. Also, I have spent the last two summers working in food preparation for Memorial Hospital in Melbourne.

Remember that you are trying to arouse immediate interest about yourself in the potential employer. Another way to do this is to refer to something about the company that interests you. Such a reference establishes that you have done your homework. Then try to show how some preparation on your part ties you into this special interest. See Figure 13–10 for an example of such an opener.

Sometimes the best approach is a simple statement about the job you seek accompanied by something in your experience that fits you for the job, as in this example:

> Your opening for a food supervisor has come to my attention. In June of this year, I will graduate from State University with a Bachelor of Science in Food Service. I have spent the last two

summers working in food preparation for Memorial Hospital in Melbourne. I believe that both my education and work experience qualify me to be a food supervisor on your staff.

Be specific about the job you want. As the vice-president of one firm told us, "We have all the people we need who can do anything. We want people who can do something." Quite often, if the job you want is not open, the employer may offer you an alternative one. But employers are not impressed with vague statements such as, "I'm willing and able to work at any job you may have open in research, production, or sales."

The Body In the body of your letter, you select items from your education and experience that show your qualifications for the job you seek. Remember always, you are trying to show the employer how well you will fit into the job and the organization.

In selecting your items, it pays to know what things employers value the most. One thorough piece of research[2] shows that for recent college graduates employers give priority as follows:

First Priority	Major field of study Academic performance Work experience Plant or home-office interview Campus interview
Second Priority	Extracurricular activities Recommendations of former employer Academic activities and awards
Third Priority	Type of college or university attended Recommendations from faculty or school official
Fourth Priority	Standardized test scores In-house test scores Military rank

Try to include information from the areas that employers seem to value the most, but emphasize those areas in which you come off best. If your grades are good, mention them prominently. If you stand low in your class—in the lowest quarter, perhaps—maintain a discreet silence. Speak to the employer's interests, and at the same time highlight your own accomplishments. Show how it would be to the employer's advantage to

635 Shuflin Road
Watertown, CA 90233
March 23, 1991

Mr. Morrell R. Solem
Director of Research
Price Industries, Inc.
2163 Airport Drive
St. Louis, MO 63136

Dear Mr. Solem:

Opener showing knowledge of company

I read in the January issue of Metal Age that Dr. Charles E. Gore of your company is conducting extensive research into the application of X-ray diffraction to the solutions of problems in physical metallurgy. I have conducted experiments at Watertown Polytechnic Institute in the same area under the guidance of Professor John J. O'Brien. I would like to become a research assistant with your firm and, if possible, work for Dr. Gore.

Specific job mentioned

Highlights of education

In June, I will graduate from WPI with a Bachelor of Science Degree in Metallurgical Engineering. At present, I am in the upper 25 percent of my class. In addition to my work with Professor O'Brien, I have taken as many courses relating to metal inspection problems as I could.

Highlights of work experience

For the past two summers, I have worked for Watertown Concrete Test Services where I have qualified as a laboratory technician for hardened concrete testing. I know how to find and apply the specifications of the American Society for Testing and Materials. This experience has taught me a good deal about modern inspection techniques. Because this practical experience supplements the theory learned at school, I could fit into a research laboratory with a minimum of training.

Reference to resume

You will find more detailed information about my education and work experience in the resume enclosed with this letter. I can supply job descriptions concerning past employment and the report of my X-ray diffraction research.

Request for interview

In April, I will attend the annual meeting of the American Institute of Metallurgical Engineers in Detroit. Would it be possible for me to talk with some member of Price Industries at that time?

Sincerely yours,

Jane E. Lucas

Jane E. Lucas

Enclosure

Figure 13–10 Letter of Application

hire you. The following paragraph, an excellent example of the you-attitude in action, does all these things:

> I understand that the research team of which I might be a part works as a single unit in the measurement and collection of data. Because of this, team members need a general knowledge and ability in fishery techniques as well as a specific job skill. Therefore, I would like to point out that last summer I worked for the Department of Natural Resources on a fish population study. On that job I gained electro-fishing and seining experience and also learned how to collect and identify aquatic invertebrates.

By being specific about your accomplishments, you avoid the appearance of bragging. It is much better to say, "I was president of my senior class," than to say, "I am a natural leader."

One tip about job experience: The best experience is that which relates to the job you seek, but mention job experience even if it does not relate to the job you seek. Employers feel a student who has worked is more apt to be mature than one who has not.

Do not forget hobbies that relate to the job. You are trying to establish that you are interested in, as well as qualified for, the job.

Do not mention salary unless you are answering an advertisement that specifically requests you to. Keep the you-attitude. Do not worry about pension plans, vacations, and coffee breaks at this stage of the game. Keep the prospective employer's interests in the foreground. Your self-interest is taken for granted.

If you are already working and not a student, you construct the body in much the same fashion. The significant difference is that you will emphasize work experience more than college experience. Do not complain about your present employer. Such complaints will lead the prospective employer to mistrust you.

In the last paragraph of the body, refer the employer to your enclosed resume. Mention your willingness to supply additional information such as references, letters concerning your work, research reports, and college transcripts.

The Ending The goal of the letter of application is an interview with the prospective employer. In your ending you request this interview. Your request should be neither humble nor overaggressive. Simply indicate that you are available for an interview at the employer's convenience, and give any special instructions needed for reaching you. If the prospective employer is in a distant city, indicate (if you can) some convenient time and place where you might meet a representative of the company, such as the convention of a professional society. If the employer is really interested, you may be invited to visit the company at its expense.

The Complete Letter Figure 13–10 shows the complete letter of application. Take a minute to read it now. The beginning of the letter shows that the writer has been interested enough in the company to investigate it. The desired job is specifically mentioned. The middle portion highlights the course work and work experience that relate directly to the job sought. The close makes an interview convenient for the employer to arrange.

The word processor is a great convenience when you are doing application letters. It allows you to store basic paragraphs of your letter that you can easily modify to meet the needs and interests of the organization you are writing to. Such modification for each organization is truly necessary. Personnel officers read your letter and its accompanying resume in about 30 seconds or so. If you have not grabbed their interest in that time, you are probably finished with that organization.

The Resume

With your letter of application, you enclose a **resume** that provides your prospective employer with a convenient summary of your education and experience. To whom should you send letters and resumes? When answering an advertisement, you should follow whatever instructions are given there. When operating on your own, send them, if at all possible, to the person within the organization for whom you would be working; that is, the person who directly supervises the position. This person normally has the power to hire for the position. Your research into the company may turn up the name you need. If not, don't be afraid of calling the company switchboard and asking directly for the name and title you need. Write to human resources directors only as a last resort. Whatever you do, write to *someone* by name. Don't send "To Whom It May Concern" letters on your job hunt. It's wasted effort. Sometimes, of course, you may gain an interview without having sent a letter of application—for example, when recruiters come to your campus. When you do, bring a resume with you and give it to the interviewer at the start of the interview. He or she will appreciate this help tremendously.

Research indicates that, as in the letter of application, good grammar, correct spelling, neatness and brevity—ideally only one page—are of major importance in your resume.[3] If you have wide experience to report, you will find that the smaller type available with word processing or at the printers will help you fit it on one page. But don't do any smaller than 10-point type or your resume will be too hard to read and too cramped looking.

You have three alternatives for producing your resume. You can type it and have it photoreproduced. You can have it printed by a commercial printer. You can word process it, providing you have access to a letter-quality printer. Word processing is probably the best of the three alternatives because it allows you the flexibility of modifying your resume for different employers.

Use good paper in a standard color such as white or off-white. It's best if your letters and resumes are on matching paper. Make the resume good-looking. Leave generous margins and white space. Use distinctive headings and subheadings. The samples in Figures 13–11, 13–12, and 13–13 provide you with good models of what a resume should look like. Take a look at them now before you read the following comments.

The resumes shown are in the three most used formats: chronological, functional, and targeted. All have advantages and disadvantages.

Chronological Format The advantages of a chronological resume, Figure 13–11, are that it's traditional and acceptable. If your education and experience show a steady progression toward the career you seek, the chronological resume portrays that progression well. Its major disadvantage is that your special capabilities or accomplishments may sometimes get lost in the chronological detail. Also, if you have holes in your employment or educational history, they show up clearly.

Address In a chronological resume, put your address at the top. Give your phone number and don't forget the area code.

Career Objective A career objective entry specifies the kind of work you want to do and sometimes the industry or service area where you want to do it, something like this:

> Work in food service management.

or like this:

> Work in food service management in a metropolitan hospital.

If you are going to have only one resume printed or photocopied, you might be wise to omit the career objective. It may not be appropriate for every potential employer you send your resume to. You can always state your career objective in your letter of application. On the other hand, if you are using word processing for your resume and can modify it for different employers, including a career objective is a good idea. Place the career objective entry immediately after the address and align it with the rest of the entries, as shown in Figure 13–13.

Education For most students, educational information should be placed before business experience. People with extensive work experience, however, may choose to put that first. List the colleges or universities you have attended in reverse chronological order; in other words, list the

(Text continued on page 409.)

RESUME OF JANE E. LUCAS

65 Shuflin Road
Watertown, California 90233
Phone: (213) 596-4236

Education

1990–1992 WATERTOWN POLYTECHNIC INSTITUTE
WATERTOWN, CALIFORNIA

Fragmentary sentences used

Highlights of education

Special activities

Candidate for Bachelor of Science degree in Metallurgical Engineering in June 1992. In upper 25% of class with GPA of 3.2 on 4.0 scale. Have been yearbook photographer for two years. Member of Outing Club, elected president in senior year. Elected to Student Intermediary Board in senior year. Oversaw promoting and allocating funds for student activities. Wrote a report on peer advising that resulted in a change in college policy. Earned 75% of college expenses.

1988–1990 SAN DIEGO COMMUNITY COLLEGE
SAN DIEGO, CALIFORNIA

Honors and activities

Received Associate of Arts degree in General Studies in June 1990. Made Dean's List three of four semesters. Member of debate team. Participated in dramatics and intramural athletics.

Business Experience

1990, 1991 summers WATERTOWN CONCRETE TEST SERVICES
WATERTOWN, CALIFORNIA

Special work skills

Qualified as laboratory technician for hardened concrete testing under specification E329 of American Society for Testing and Materials (ASTM). Conducted following ASTM tests: Load Test in Core Samples (ASTM C39), Penetration Probe (ASTM C803-75T), and the Transverse Resonant Frequency Determination (ASTM C666-73). Implemented new reporting system for laboratory results.

1990–1992 WATERTOWN ICE SKATING ARENA
WATERTOWN, CALIFORNIA

During academic year, work 15 hours a week as ice monitor. Supervise skating and administer first aid.

Summary of early employment

1986–1990 Summer and part-time jobs included newspaper carrier, supermarket stock clerk, and salesperson for large department store.

Personal Background Grew up in San Diego, California. Travels include Mexico and the Eastern United States. Can converse in Spanish. Interests include reading, backpacking, photography, and sports (tennis, skiing, and running). Willing to relocate.

References Personal references available upon request.

February 1992

Figure 13–11 Chronological Resume

Academic, work, and extracurricular activities categorized by capabilities

Summary of work experience in reverse chronological order

RESUME OF JANE E. LUCAS

65 Shuflin Road
Watertown, California 90233
Phone: (213) 596-4236

Education 1992	Candidate for degree in Metallurgical Engineering from Watertown Polytechnic Institute in June 1992.
Technical	• Qualified as laboratory technician for hardened concrete testing under specification E329 of American Society for Testing and Materials (ASTM). • Conducted following ASTM tests: Load Test in Core Samples (ASTM C39), Penetration Probe (ASTM C803-75T), and the Transverse Resonant Frequency Determination (ASTM C666-73). • Will graduate in upper 25% of class with a GPA of 3.2 on a 4.0 scale.
People	• Elected President of Outing Club. • Elected to Student Intermediary Board. • Oversaw promoting and allocating funds for student activities. • Participated in dramatics and intramural athletics.
Communication	• Wrote a report on peer advising that resulted in a change in Institute policy. • Participated in intercollegiate debate. • Worked two years as yearbook photographer. • Completed courses in advanced speaking, small group discussion, and technical writing.
Work Experience	
1990, 1991 summers	• Watertown Concrete Test Services, Watertown, California: laboratory technician.
1990-1992	• Watertown Ice Skating Arena, Watertown, California: Ice monitor.
1986-1990	• Summer and part-time jobs included newspaper carrier, supermarket stock clerk, and salesperson for large department store. • Earned 75% of college expenses.
References	References available upon request.

February 1992

Figure 13–12 Functional Resume

Capabilities listed separately

Achievements that support capabilities listed

Summary of work experience in reverse chronological order

RESUME OF JANE E. LUCAS

65 Shuflin Road
Watertown, California 90233
Phone: (213) 596-4236

Job objective — Research assistant in a testing or research laboratory.

Education
1992 — Candidate for degree in Metallurgical Engineering from Watertown Polytechnic Institute in June 1992.

Capabilities

- Find and apply specifications of the American Society for Testing and Materials (ASTM).
- Conduct X-ray diffraction tests.
- Work individually or as a team member in a laboratory setting.
- Take responsibility and think about a task in terms of objectives and time to complete.
- Report research results in both written and oral form.
- Communicate persuasively with nontechnical audiences.

Achievements

- Will graduate in upper 25% of class with a GPA of 3.2 on a 4.0 scale.
- Earned 75% of college expenses.
- Qualified as laboratory technician for hardened concrete testing under Specification E329 of ASTM.
- Conducted major ASTM tests and implemented new reporting system for laboratory results.
- Completed courses in advanced speaking, small group discussion, and technical writing.
- Wrote a report on peer advising that resulted in a change in Institute policy.

Work Experience

1990, 1991 summers
- Watertown Concrete Test Services, Watertown, California: laboratory technician.

1990-1992
- Watertown Ice Skating Arena, Watertown, California: Ice monitor.

1986-1990
- Summer and part-time jobs included newspaper carrier, supermarket stock clerk, and salesperson for large department store.

References — References available upon request.

February 1992

Figure 13–13 Targeted Resume

school you attended most recently first, the one before second, and so on. Do not list your high school.

Give your major and date or expected date of graduation. Do not list courses, but list anything that is out of the ordinary, such as honors, special projects, and emphases in addition to the major. Extracurricular activities also go here.

Business Experience As you did with your educational experience, put your business experience in reverse chronological order. To save space and to avoid the repetition of *I* throughout the resume, use phrases rather than complete sentences. The style of the sample resumes makes this technique clear. As we advise you to do in the letter of application, emphasize the experiences that show you in the best light for the kinds of jobs you seek. Use active verbs in your descriptions. Do not neglect less important jobs of the sort you may have had in high school, but use even more of a summary approach for them. You would probably put college internships and work–study programs here, though you might rather choose to put them under education. If you have military experience, put it here. Give highest rank held, list service schools attended, and describe duties. Make a special effort to show how military experience relates to the civilian work you seek.

Personal Background You may wish to provide personal information about yourself. Personal information can in a subtle way point out desirable qualities you possess. Recent travels indicate a broadening of knowledge and probably a willingness to travel. Hobbies listed may relate to the work sought. Participation in sports, drama, or community activities indicates a liking for working with people. Cultural activities indicate you are not a person of narrow interests.

If you indicate you are married, you might want to say that you are willing to relocate. Don't say anything about health unless you can describe it as excellent. Because of state and federal laws concerning fair employment practices, certain information should not appear in your resume. Do not mention handicaps and do not include a photograph.

References You have a choice with references. You can list several references with addresses and phone numbers or simply put in a line that says "References available on request." Both methods have an advantage and a disadvantage. If you provide information on references, a potential employer can contact them immediately, but you use up precious space that might be better used for more information about yourself. Conversely, if you don't provide the reference information, you save the space but put an additional step between potential employers and information they may want. It's a judgment call, but, on balance, we favor saving space by omitting the reference information. Your first goal is to interest the potential employer about yourself. If that happens then it will not be difficult to provide the reference information at a later time.

In any case, do have at least three references available. Choose from among your college teachers and past employers people who know you well and are likely to say positive things about you. Get their permission, of course. Also, it's a smart idea to send them a copy of your resume. If you can't call on them personally, send them a letter that requests permission to use them as a reference, reminds them of the association with you, and sets a time for their reply, like this:

Dear Ms. Pickford:

Request for reference

In June of this year, I'll graduate from Watertown Polytechnic Institute with a B.S. in metallurgical engineering. I'm getting ready to look for work. May I have permission to use you as a reference?

Past association

During the summers of 1990 and 1991, I worked as a laboratory technician in your testing lab at Watertown. They were good summers for me, and I qualified, with your help, to carry out several ASTM tests.

Time for reply

I will send my resume out to some companies by 15 March and will need your reply by that time. I enclose a copy of my resume for you, so that you can see what I've been doing.

Thanks for all your help in the past.

Best regards,

Dateline At the bottom of the resume, place the date—month and year—in which you completed the resume.

Functional Format A main advantage of the functional resume (Figure 13–12) is that it allows you to highlight those experiences that show you to your best advantage. Extracurricular experiences show up particularly well in a functional resume. Its major disadvantages are that you don't show as clearly a steady progression of work and education. Also, the functional resume is a newer format than the chronological, and some employers may not find it as acceptable as the older format.

The address portion of the functional resume is the same as the chronological. After the address, you may provide a job objective line if you like.

For education, simply give the school from which you received your degree, your major, and your date of graduation.

The body of the resume is essentially a classification. You sort your experiences—educational, business, extracurricular—into categories that reveal capabilities related to the jobs you seek. Remember that in addition to professional skills employers want good communication skills and good people skills. Possible categories are *technical, professional, team building, communication, research, sales, production, administration,* and *consulting.*

The best way to prepare a functional resume is to brainstorm it. Begin by listing some categories that you think might display your experiences well. Brainstorm further by listing your experiences in those categories. When you have good listings, select the categories and experiences that show you in the best light. Remember, you don't have to display everything you've ever done, just those things that might strike a potential employer as valuable.

Finish off the functional resume with a brief reverse chronological work history and a dateline, as in the chronological resume.

Targeted Format The main advantage of the targeted resume is also its main disadvantage. You zero in on one goal. If you can achieve that goal, fine, but the narrowness of the approach may block you out from other possibilities. The targeted resume displays your capabilities and achievements well, but like the functional resume it's a newer format that may not have complete acceptance with all employers.

The address and education portions of the targeted resumes are the same as the functional. The whole point of a targeted resume is that you are aiming at a specific job objective. Therefore, you express your job objective as precisely as you can.

Next, you list your capabilities that match the job objective. Obviously, you have to understand the job sought to make the proper match. Capabilities are things you could do if called upon to do so. To be credible, they must be supported by the achievements or accomplishments that are listed next in your resume. You finish off the targeted resume with a reverse chronological work history and a date line, as in the functional resume.

As with the functional resume, brainstorming is a good way to discover the material you need for your targeted resume. Under the headings *capabilities* and *achievements,* make as many statements about yourself as you can. When done, select those statements that best relate to the job sought.

General Advice Look again at the three resumes in Figures 13–11, 13–12, and 13–13. Notice that all three use informative headings and allow enough white space to be inviting. Figure 13–11 is done with standard typewriter print. Figures 13–12 and 13–13 show the possibilities if you have a word processor with access to different typefaces or if you have your resume set by a printer. The addition of the boldface type in the headings provides a more professional polish. We strongly recommend such printing if you can manage it.

Follow-up Letters

Write follow-up letters (1) when you have had no answer to your letter of application in two weeks; (2) after an interview; (3) when a company refuses you a job; and (4) to accept or refuse a job.

No Answer When a company has not answered your original letter of application, write again. Be gracious, not complaining. Say something like this:

Dear Mr. Souther:

Recall of original letter

On 12 April I applied for a position with your company. I have not heard from you, so perhaps my original letter and resume have been misplaced. I enclose copies of them.

Request for decision

If you have already reached some decision concerning my application, I would appreciate your letting me know.

I look forward to hearing from you.

Sincerely yours,

After an Interview Within a week's time, follow up your interview with a letter. Such a letter draws favorable attention to yourself as someone who understands business courtesy and good communication practice. Express appreciation for the interview. Draw attention to any of your qualifications that seemed to be important to the interviewer. Express your willingness to live with any special conditions of employment such as relocation. Make clear your hope for the next stage in the process, perhaps a further interview. If you include a specific question in your letter, it may hasten a reply. Your letter might look like this one:

Dear Ms. Marchand:

Expression of appreciation

Thank you for speaking with me last Tuesday about the food supervisor position you have open.

Reminder of special qualifications

Working in a hospital food service relates well to my experience and interests. A feasibility study I am currently writing as a senior project deals with a hospital food service's ability to provide more varied diets to people with restricted dietary requirements. May I send you a copy next week when it is completed?

Agreement with special conditions

I understand that my work with you would include alternating weekly night shifts with weekly day shifts. This requirement presents no difficulty for me.

Times for interview

Tuesdays and Thursdays are best for me for any future interviews you may wish, but I can arrange a time at your convenience.

Sincerely yours,

After a Job Refusal When a company refuses you a job, good tactics dictate that you acknowledge the refusal. Express thanks for the time spent with you, and state your regret that no opening exists at the present time. If you like, express the hope that they may consider you in the future. You never know; they might.

Accepting or Refusing a Job Writing an acceptance letter presents few problems. Be brief. Thank the employer for the job offer and accept the job. Settle when you will report for work and express pleasure at the prospect of working for the organization. A good letter of acceptance might read as follows:

Dear Mr. Solem:

Acceptance of job

Thank you for offering me a job as research assistant with your firm. I happily accept. I can easily be at work by 1 July as you have requested.

Expression of pleasure

I look forward to working with Price Industries and particularly to the opportunity of doing research with Dr. Gore.

Sincerely yours,

Writing a letter of refusal can be difficult. Be as gracious as possible. Be brief but not so brief as to suggest rudeness or indifference. Make it clear that you appreciate the offer. If you can, give a reason for your refusal. The employer who has spent time and money in interviewing you and corresponding with you deserves these courtesies. And, of course, your own self-interest is involved. Some day you may wish to reapply to an organization that for the moment you must turn down. A good letter of refusal might look like this one:

Dear Ms. White:

Expression of appreciation

I enjoyed my visit to the research department of your company. I would very much have liked to work with the people I met there. I thank you for offering me the opportunity to do so.

Refusal and reason for it

However, after much serious thought, I have decided that the research opportunities offered me in another job are closer to the interests I developed at the University. Therefore, I have accepted the other job and regret that I cannot accept yours.

Courteous close

I appreciate the courtesy and thoughtfulness that you and your associates have extended me.

Sincerely yours,

Letter and Memorandum Reports

Many business reports run from two to five pages. They are too short to need the elements of more formal reports, such as title pages and tables of content. Usually, such a short report will be written as a letter or memo. All of the reports that we discuss in Chapters 14 through 18 can be and often are written as memos or letters. All of the strategies discussed in Chapters 7, 8, and 9, used singly or in combination, are found in letter and memorandum reports. A common plan for either a letter report or a memo report calls for an introduction, a summary, a discussion, and an action step. Figures 13–14 and 13–15 illustrate how such a report can be presented as either a memo that stays within the organization or a letter that goes outside.

Introduction

Begin by telling the reader the subject and purpose of the report. Perhaps you are reporting on an inspection tour or summarizing the agreements reached in a consultation between you and the recipient. You may be reporting the results of a research project, or the beginning of one. Or you may be writing a progress report on a project that is under way but not completed. Whatever the subject and purpose may be, state them clearly. If someone has requested the report, name the requester. Typically, the introduction in a letter or memo report will not have a heading.

Southern Wire Company
Memorandum

Date: 15 January 1991

To: Louise Carson, President

From: Thomas Kehoe
Chief Engineer JE

Subject: Research in High-Temperature Superconductivity

Combination executive summary and introduction

Since technical breakthroughs in high-temperature superconductivity (HTS) in 1987, the federal government has been supporting research in HTS at an increasing rate, spending $145 million in fiscal year 1988. In the next 10 years, the government will fund a number of joint ventures in HTS between government laboratories and private industry to prepare for a market in HTS technology it estimates to be worth $12 billion by 2000.

Wire fabrication will be major research area and a major piece of the HTS market. If we are to afford significant research and to maintain a market share in HTS, we must act quickly to seek a joint venture with a government laboratory. To that end, this report, after a brief discussion of HTS technology and government policy, recommends that management set up a small task force to explore and report on the funding possibilities and to produce a model proposal for management's consideration.

Discussion

In ordinary conductivity, electrons meet resistance in passing through a conductor, such as copper wire, and, therefore, lose efficiency. In superconductivity, electrons meet little resistance in passing through the conductor and efficiency increases. Low-temperature superconductivity has existed for about 30 years, but at extremely low temperatures of 0 Kelvin (K), about −464° Fahrenheit. Expensive liquid helium is required for such temperatures. High-temperature superconductivity operates at about 125K (−234° F). Widely available liquid nitrogen will produce such a temperature and provide an inexpensive and reliable refrigeration system. The cost savings are shown in the following table that compares the life-cycle savings of HTS with those of low-temperature and conventional systems in several applications.

Figure 13–14 Memorandum Report
Source: Adapted from material in Solar Energy Research Institute, *Superconductivity for Electric Power Systems* (Washington, DC: U.S. Department of Energy, 1989).

Continued

Advantages of tables:
- *Provide white space*
- *Make comparisons easy to see*

Page 2
Louise Carson, President
15 January 1991

Life-Cycle Cost Savings (%) with High-Temperature Superconductors

	Compared with Low-Temperature System	Compared with Conventional System
Generator (300 megawatt)	27	63
Transformer (1,000 megavolt-ampere)	36	60
Transmission Line (10,000 megavolt-ampere, 230-kilovolt)	23	43
Motors	11	21
Magnetic Separators	15	20

The lower cost and higher efficiency of HTS will have a tremendous effect on virtually all applications of electrical power, such as motors, generators, heat pumps, transmission lines, magnetic storage, and computing. The Department of Energy (DOE) estimates the market to be $12 billion by 2000.

Research in HTS is beginning around the world. The following table shows the amount spent by the major players in the field in fiscal year 1988:

	FY 1988 Funding ($ millions)
United States	145
Japan	135
United Kingdom	25
France	20
West Germany	15

Figure 13–14
(continued)

Page 3
Louise Carson, President
15 January 1991

The federal government is funding research in all aspects of HTS at the rate of about $200 million a year. DOE has established three pilot programs to make the resources of the national laboratories at Argonne, Los Alamos, and Oak Ridge available to private industry and is seeking joint ventures in research and development. Furthermore, companies acting in such joint ventures can obtain the patents and licenses needed to go commercial with HTS.

The major "so-what"

As a wire fabricator, we clearly have a major stake in HTS. If we do not move swiftly and surely to develop wire suitable for HTS, we run the risk of losing a share of what will be a major market.

Action

Recommendations

I recommend that management establish a small task force composed of myself and one representative each from marketing, finance, and the legal department and charge it with three tasks:

- Contact DOE to obtain information on federal funding for joint ventures.
- Report on the funding and research possibilities available to us.
- Develop a model proposal for a joint venture between us and DOE.

Copies for those directly concerned

cc: Joseph Roberts
Vice President, Marketing

Emmie Ingram
Chief Financial Officer

Thomas Scanlan
Director, Legal Department

Figure 13–14
(continued)

ENERGY CONSULTING

2221 K Street NW
Washington, DC 20007

15 January 1991

Ms. Louise Carson, President
Southern Wire Company
651 River Street
Savannah, GA 31455

SUBJECT: REPORT ON HIGH-TEMPERATURE SUPERCONDUCTIVITY RESEARCH

Dear Ms. Carson:

Reason for report

Combination executive summary and introduction

At your request Energy Consulting has looked into high-temperature superconductivity (HTS) and the funding possibilities for Southern Wire Company in HTS. We can report favorable news. Since technical breakthroughs in HTS in 1987, the federal government has been supporting research in HTS at an increasing rate, spending $145 million in fiscal year 1988. In the next 10 years, the government will fund a number of joint ventures in HTS between government laboratories and private industry to prepare for a market in HTS technology it estimates to be worth $12 billion by 2000.

Wire fabrication will be a major research area and a major piece of the HTS market. If Southern Wire is to conduct significant research and maintain a market share in HTS, you should act quickly to seek a joint venture with a government laboratory. To that end, this report, after a brief discussion of HTS technology and government policy, recommends that Southern Wire management set up a small task force to explore and report on the funding of possibilities and to produce a model proposal for management's consideration.

Figure 13–15 Letter Report
Source: Adapted from material in Solar Energy Research Institute, *Superconductivity for Electric Power Systems* (Washington, DC: U.S. Department of Energy, 1989).

Page 2
Louise Carson, President
15 January 1991

Discussion

In ordinary conductivity, electrons meet resistance in passing through a conductor, such as copper wire, and, therefore, lose efficiency. In superconductivity, electrons meet little resistance in passing through the conductor and efficiency increases. Low-temperature superconductivity has existed for about 30 years, but at extremely low temperatures of 0 Kelvin (K), about −464° Fahrenheit. Expensive liquid helium is required for such temperatures. High-temperature superconductivity operates at about 125K (−234° F). Widely available liquid nitrogen will produce such a temperature and provide an inexpensive and reliable refrigeration system. The cost savings are shown in the following table that compares the life-cycle savings of HTS with those of low-temperature and conventional systems in several applications.

Advantages of tables:
- *Reduce wordage*
- *Provide relief for reader*

Life-Cycle Cost Savings (%) with High-Temperature Superconductors

	Compared with Low-Temperature System	Compared with Conventional System
Generator (300 megawatt)	27	63
Transformer (1,000 megavolt-ampere)	36	60
Transmission Line (10,000 megavolt-ampere, 230-kilovolt)	23	43
Motors	11	21
Magnetic Separators	15	20

The lower cost and higher efficiency of HTS will have a tremendous effect on virtually all applications of electrical power, such as motors, generators, heat pumps, transmission lines, magnetic storage, and computing. The Department of Energy (DOE) estimates the market to be $12 billion by 2000.

Figure 13–15
(continued)

Continued

Page 3
Louise Carson, President
15 January 1991

Research in HTS is beginning around the world. The following table shows the amount spent by the major players in the field in fiscal year 1988:

	FY 1988 Funding ($ millions)
United States	145
Japan	135
United Kingdom	25
France	20
West Germany	15

The federal government is funding research in all aspects of HTS at the rate of about $200 million a year. DOE has established three pilot programs to make the resources of the national laboratories at Argonne, Los Alamos, and Oak Ridge available to private industry and is seeking joint ventures in research and development. Furthermore, companies acting in such joint ventures can obtain the patents and licenses needed to go commercial with HTS.

As a wire fabricator, Southern Wire has a large stake in HTS. If you do not move swiftly and surely to develop wire suitable for HTS, you run the risk of losing a share of what will be a major market.

Action

I recommend that management establish a small internal task force composed of one representative each from engineering, marketing, finance, and the legal department and charge it with three tasks:

Figure 13–15
(continued)

List format for recommendations

Page 4
Louise Carson, President
15 January 1991

- Contact DOE to obtain information on federal funding for joint ventures.
- Report on the funding and research possibilities available to Southern Wire.
- Develop a model proposal for a joint venture between Southern Wire and DOE.

If you would like Energy Consulting to work with the task force or to help Southern Wire in working with DOE, we would be happy to do so.

Sincerely yours,

Walter Mazura

Walter Mazura
Vice President and Director

WM: ekg

Figure 13–15
(continued)

How to write a business letter

Some thoughts from Malcolm Forbes

President and Editor-in-Chief of Forbes Magazine

International Paper asked Malcolm Forbes to share some things he's learned about writing a good business letter. One rule, "Be crystal clear."

A good business letter can get you a job interview.

Get you off the hook.

Or get you money.

It's totally asinine to blow your chances of getting *whatever* you want—with a business letter that turns people off instead of turning them on.

The best place to learn to write is in school. If you're still there, pick your teachers' brains.

If not, big deal. I learned to ride a motorcycle at 50 and fly balloons at 52. It's never too late to learn.

Over 10,000 business letters come across my desk every year. They seem to fall into three categories: stultifying if not stupid, mundane (most of them), and first rate (rare). Here's the approach I've found that separates the winners from the losers (most of it's just good common sense)—it starts *before* you write your letter:

Know what you want

If you don't, write it down—in one sentence. "I want to get an interview within the next two weeks." That simple. List the major points you want to get across—it'll keep you on course.

If you're *answering* a letter, check the points that need answering and keep the letter in front of you while you write. This way you won't forget anything—*that* would cause another round of letters.

And for goodness' sake, answer promptly if you're going to answer at all. Don't sit on a letter—*that* invites the person on the other end to sit on whatever you want from *him*.

Plunge right in

Call him by name—not "Dear Sir, Madam, or Ms." "Dear Mr. Chrisanthopoulos"—and be sure to spell it right. That'll get him (thus, you) off to a good start.

(Usually, you can get his name just by phoning his company—or from a business directory in your nearest library.)

Tell what your letter is about in the first paragraph. One or two sentences. Don't keep your reader guessing or he might file your letter away—even before he finishes it.

In the round file.

If you're answering a letter, refer to the date it was written. So the reader won't waste time hunting for it.

People who read business letters are as human as thee and me. Reading a letter shouldn't be a chore—*reward* the reader for the time he gives you.

Write so he'll enjoy it

Write the entire letter from his point of view—what's in it for *him*? Beat him to the draw—surprise him by answering the questions and objections he might have.

Be positive—he'll be more receptive to what you have to say.

Be nice. Contrary to the cliché, genuinely nice guys most often finish first or very near it. I admit it's not easy when you've got a gripe. To be agreeable while disagreeing—that's an art.

Be natural—write the way you talk. Imagine him sitting in front of you—what would you *say* to him?

Business jargon too often is cold, stiff, unnatural.

Suppose I came up to you and said, "I acknowledge receipt of your letter and I beg to thank you." You'd think, "Huh? You're putting me on."

The acid test—read your letter *out loud* when you're done. You might get a shock—but you'll know for sure if it sounds natural.

"Be natural. Imagine him sitting in front of you—what would you say to him?"

Don't be cute or flippant. The reader won't take you seriously. This doesn't mean you've got to be dull. You prefer your letter to knock 'em dead rather than bore 'em to death.

Three points to remember:

Have a sense of humor. That's refreshing *anywhere*—a nice surprise

Courtesy International Paper Company.

Figure 13–16 How to Write a Business Letter. Malcolm Forbes's thoughts about successful business letter writing agree so well with ours that we reproduce them here. They illustrate that business people take good communication seriously.
Source: Courtesy International Paper Company.

in a business letter.

Be specific. If I tell you there's a new fuel that could save gasoline, you might not believe me. But suppose I tell you this:

> "Gasohol" – 10% alcohol, 90% gasoline – works as well as straight gasoline. Since you can make alcohol from grain or corn stalks, wood or wood waste, coal – even garbage, it's worth some real follow-through.

Now you've got something to sink your teeth into.

Lean heavier on nouns and verbs, lighter on adjectives. Use the active voice instead of the passive. Your writing will have more guts.

Which of these is stronger? Active voice: "I kicked out my money manager." Or, passive voice: "My money manager was kicked out by me." (By the way, neither is true. My son, Malcolm Jr., manages most Forbes money – he's a brilliant moneyman.)

"I learned to ride a motorcycle at 50 and fly balloons at 52. It's never too late to learn anything."

Give it the best you've got

When you don't want something enough to make *the* effort, making *an* effort is a waste.

Make your letter look appetizing – or you'll strike out before you even get to bat. Type it – on good-quality 8½" x 11" stationery. Keep it neat. And use paragraphing that makes it easier to read.

Keep your letter short – to one page, if possible. Keep your paragraphs short. After all, who's going to benefit if your letter is quick and easy to read?

You.

For emphasis, underline important words. And sometimes indent sentences as well as paragraphs.

> Like this. See how well it works? (But save it for something special.)

Make it perfect. No typos, no misspellings, no factual errors. If you're sloppy and let mistakes slip by, the person reading your letter will think you don't know better or don't care. Do you?

Be crystal clear. You won't get what you're after if your reader doesn't get the message.

Use good English. If you're still in school, take all the English and writing courses you can. The way you write and speak can really help – or *hurt*.

If you're not in school (even if you are), get the little 71-page gem by Strunk & White, *Elements of Style*. It's in paperback. It's fun to read and loaded with tips on good English and good writing.

Don't put on airs. Pretense invariably impresses only the pretender.

Don't exaggerate. Even once. Your reader will suspect everything else you write.

Distinguish opinions from facts. Your opinions may be the best in the world. But they're not gospel. You owe it to your reader to let him know which is which. He'll appreciate it and he'll admire you. The dumbest people I know are those who Know It All.

Be honest. It'll get you further in the long run. If you're not, you won't rest easy until you're found out. (The latter, not speaking from experience.)

Edit ruthlessly. Somebody ~~has~~ said that words are ~~a lot~~ like inflated money – the more ~~of them that~~ you use, the less each one ~~of them~~ is worth. ~~Right on.~~ Go through your entire letter ~~just~~ as many times as it takes. ~~Search out and~~ Annihilate all unnecessary words, ~~and~~ sentences – even ~~entire~~ *paragraphs*.

"Don't exaggerate. Even once. Your reader will suspect everything else you write."

Sum it up and get out

The last paragraph should tell the reader exactly what you want *him* to do – or what *you're* going to do. Short and sweet. "May I have an appointment? Next Monday, the 16th, I'll call your secretary to see when it'll be most convenient for you."

Close with something simple like, "Sincerely." And for heaven's sake sign legibly. The biggest ego trip I know is a completely illegible signature.

Good luck.

I hope you get what you're after.

Sincerely,

Malcolm S. Forbes

Years ago, International Paper sponsored a series of advertisements, "Send me a man who reads," to help make Americans more aware of the value of reading.

Today, the printed word is more vital than ever. Now there is more need than ever before for all of us to *read* better, *write* better, and *communicate* better.

International Paper offers this *new* series in the hope that, even in a small way, we can help.

For reprints of this advertisement, write: "Power of the Printed Word," International Paper Co., Dept. 1-A, P.O. Box 900, Elmsford, New York 10523.

Summary

In most cases the summary will be an executive summary and will have a heading labeling it as a summary. (See pages 291–292.) It will emphasize the things in the report important to an executive's decision-making process and state clearly any conclusions, recommendations, and decisions that have been reached. Often, the functions of the introduction and the executive summary are combined. When this is the case, the combination usually will not have a heading.

Discussion

Give your discussion a heading. You might label it simply as "Discussion," or perhaps call it "Findings," "Results and Discussion," or the like. Develop your discussion using the same techniques and rhetorical principles that you would use in a longer report. Remember to consider your audience precisely as you do in longer reports. For example, if you are writing to an executive, do not fill your letter or memo with jargon and technical terms.

If you must report a mass of statistics, try to round them off. If absolute accuracy is necessary, perhaps you can give the figures in an informal table. (See pages 348–349.) Some word processing equipment now makes it possible to incorporate graphs into memos and letters. If you have such a capability, take advantage of it by following the principles developed in Chapter 12, "Graphical Elements of Reports." You may also find listing a useful technique in letters and memos. (See pages 116–117 and 237–239.) Use subheadings just as you would in a longer report. (See pages 248–257.)

Action

If your letter or memo report recommends action on someone's part, include an action section with an "Action" or "Recommendation" heading. In this section state—or sometimes, more diplomatically, suggest—what that action should be.

Planning and Revision Checklists

You will find the planning and revision checklists that follow Chapter 2, "Composing" (pages 33–35 and inside the front cover), and Chapter 4, "Writing for Your Readers" (pages 78–79), valuable in planning and revising any presentation of technical information. The following questions specifi-

cally apply to correspondence. They summarize the key points in this chapter and provide a checklist for planning and revising.

GENERAL

Planning

- What is your topic and purpose?
- Who are your primary readers? Secondary readers? Do they have different needs? How can you satisfy all your readers?
- Why will your readers read your letter or memo?
- What format will best suit the situation?
- Do you have all the information you need to address your letter or memo? Names, titles, addresses, and so forth?

Revision

- Have you stated your topic and purpose clearly in the first sentence or two of your letter or memo?
- Have you met your reader's purpose?
- Have you avoided jargon and clichés?
- Does your correspondence demonstrate a you-attitude?
- Is your letter or memo clear, concise, complete, and courteous?
- Have you used your chosen format correctly? Checked for correct punctuation and spacing?

LETTER OF INQUIRY

Planning

- In what capacity are you making your request?
- What specifically do you want?
- Why do you need what you are requesting?
- Why are you making your request to this particular source?

Revision

- Does your letter of inquiry identify you, state specifically what you want, establish your need for the information or material you request, tell the recipient why you have chosen him or her as a source, and close graciously?

REPLIES TO LETTERS OF INQUIRY

Planning

- Has the inquirer written to you as a private person or as a member of an organization?
- Can you answer the inquirer's questions point by point?
- Do you have additional information or material that the inquirer would find useful?

Revision

- If you are answering a letter of inquiry, have you answered it as completely as need be? Is there additional information you could refer to or additional material you could send?

LETTERS OF COMPLAINT

Planning

- What specifically is the problem?
- What inconvenience or loss has the problem caused you?
- Do you have all the product information you need?
- What is the adjustment you want?
- How can you motivate the recipient to make a fair adjustment?

Revision

- Do you have a firm but courteous tone in your letter of complaint? Are you specific about the problem and about the adjustment you seek? Have you motivated the recipient to grant your request?

LETTERS OF ADJUSTMENT

Planning

- Is the letter of adjustment good news or bad news?
- If good news, what circumstances caused the problem? Are there special problems that must be handled? What adjustment has been asked for? What adjustment can you offer?
- If bad news, what are the circumstances for refusing the request for adjustment? Is there a substitute that can be offered in place of the requested adjustment?

Revision

- Is your letter of adjustment courteous? Is the you-attitude evident?
- If your letter of adjustment is a bad-news letter, have you explained your refusal in an honest, detailed way? Have you expressed the refusal clearly but courteously? Have you offered a substitute adjustment? Have you opened and closed the letter in a friendly way?

CORRESPONDENCE OF THE JOB HUNT

Planning

- Have you researched the potential employer? Have you analyzed yourself? Do you know your strengths and weaknesses, your skills, and your qualifications for the jobs you seek? Do you have clear career objectives?
- For your letter of application:
  - Do you have the needed names and addresses?
  - What position do you seek?

Revision

- Do your letter of application and resume reflect adequate preparation and self-analysis?
- Are your letter of application and resume completely free of grammatical and spelling errors? Are they well designed and good looking?
- For the letter of application:
  - Have you the right tone, self-confident without arrogance?

Planning

How did you learn of this position?
Why are you qualified for this position?
What interests you about the company?
What can you do for the organization that it needs?
Can you do anything to make an interview more convenient for the employer?
How can the employer reach you?

- For your resume:
  Do you have all the details needed of your past educational and work experiences? Dates, job descriptions, schools, majors, degrees, extracurricular activities, and so forth?
  How will you prepare your resume? Type and photocopy? Commercial printing? Word processing?
  Which resume form will suit your experience and capabilities best? Chronological? Functional? Targeted? Why?
  In a functional resume, which categories would best suit your experience and capabilities?
  Do you know enough about the job sought to use a targeted resume?
  Do you have three references and permission to use their names?
- What follow-up letters do you need? No answer to your application? Follow up an interview? Respond to a job refusal? Accept or refuse a job?

Revision

Does your letter show how you could be valuable to the employer? Will it raise the employer's interest?
Does your letter reflect a sure purpose about the job you seek?
Have you highlighted the courses and work experience that best suit you for the job you seek?
Have you made it clear you are seeking an interview and made it convenient for the employer to arrange one?

- For the resume:
  Have you chosen the resume type that best suits your experiences and qualifications?
  Have you held your resume to one page?
  Have you put your educational and work experience in reverse chronological order?
  Have you given your information in phrases rather than complete sentences?
  Have you used active verbs to describe your experience?
  Does the personal information presented enhance your job potential?
  Do you have permission to use their names from your references?
  If you are using a functional resume, do the categories reflect appropriately your capabilities and experience?
  Has your targeted resume zeroed in on an easily recognizable career objective? Do the listed achievements support the capabilities listed?
- Are the follow-up letters you have written gracious in tone? Have you followed up every interview with a letter? Does that letter invite further communication in some way?
- Does your letter of acceptance of a job show an understanding of the necessary details, such as when you report to work?

Revision

- Does your letter of refusal make clear that you appreciate the offer and thank the employer for time spent with you?

LETTER AND MEMORANDUM REPORTS

Planning

- What is the occasion, subject, purpose, and audience for the report?
- Would lists, tables, and graphs help your discussion?
- What conclusions, recommendations, and decisions do you have to report?
- What action should be the outcome of the report?

Revision

- Is your letter or memorandum report well introduced and summarized? Do your headings reflect the content of the report? Have you used lists, tables, and graphs if they would help? Are your conclusions and recommendations clearly stated? Is any action needed or desired clearly stated?

Exercises

1. Write an unsolicited letter of inquiry to some company asking for sample materials or information. If you really need the information or material, mail the letter, but do not mail it as an exercise.
2. Write a letter to some organization applying for full- or part-time work. Brainstorm and work out in rough form the three kinds of resumes: chronological, functional, and targeted. Choose the one that suits your purposes best, and work it into final form to accompany your letter. It may well be that you are seeking work and can write your letter with a specific organization in mind.
3. Imagine that you are working for a firm that provides a service or manufactures a product you know something about. Someone has written the firm a letter of inquiry asking about the service or product. Your task is to answer the letter.
4. Think about some service or product that has recently caused you dissatisfaction. Find out the appropriate person or organization to write to and write that person or organization a letter of complaint.
5. Swap your letter of complaint written for Exercise 4 for the letter of complaint written by another member of the class. Your assignment is to answer the other class member's letter. You may have to do a little research to get the data you will need for your answer.

6. Write a memo to some college official or to an executive at your place of work. Many of the papers you have written earlier in your writing course are probably suitable for a memorandum format. Or you could choose some procedure, such as college registration, and suggest a new and better procedure. Perhaps your memo could be to an instructor suggesting course changes. A look at Figure 4–7 on page 66 could also suggest a wide range of topics and approaches that you could use in a memo.

CHAPTER 14

Instructions

SITUATIONAL ANALYSIS FOR INSTRUCTIONS

POSSIBLE COMPONENTS OF INSTRUCTIONS

ACCESSIBLE FORMAT

READER CHECKS

Instructing others to follow some procedure is a common task on the job. Sometimes the instructions are given orally. But when the procedure is done by many people or is done repeatedly, written instructions are a better choice. Instructions may be quite simple—as in Figure 14–1—or exceedingly complex—a bookshelf full of manuals. They may be highly technical, dealing with operating machinery or programming computers, for example. Or they may be executive or business oriented, for example, explaining how to complete a form or how to route memorandums through a company. The task is not to be taken lightly. A Shakespearean scholar who had also served in the British Army wrote the following:

> The most effective elementary training [in writing] I ever received was not from masters at school but in composing daily orders and instructions as staff captain in charge of the administration of seventy-two miscellaneous military units. It is far easier to discuss Hamlet's complexes than to write orders which ensure that five working parties from five different units arrive at the right place at the right time equipped with the proper tools for the job. One soon learns that the most seemingly simple statement can bear two meanings and that when instructions are misunderstood the fault usually lies with the original order.[1]

To help you write instructions, we discuss the following in this chapter: situational analysis for instructions, the possible components of instructions, creating an accessible format, and checking with your readers.

Situational Analysis for Instructions

In preparing to write instructions, follow the situational analysis we describe on pages 15–20 in Chapter 2, "Composing." In addition, pay particular attention to the answers to these questions.[2]

What Is the Purpose of My Instructions?

Be quite specific about the purpose of your instructions. Keep your purpose in mind because it will guide you in choosing your content and in arranging and formatting that content. State your purpose in writing, like this:

> To instruct the plant managers, the corporate treasurer, and the plant accountant in the steps they need to follow to establish a petty cash fund.

tionship between cholesterol and atherosclerosis, that is, hardening of the arteries. Understanding the theory helps the readers to understand the guidelines for cholesterol levels set out in the rest of the section and motivates them to follow the guidelines. The entire section is written on a very personal level: What does this theory mean for the reader? Through the use of format, graphics, questions, and simple language, the writers of the brochure make the theory quite accessible for the intended audience.

The section uses some unfamiliar terms, such as *atherosclerosis,* which are defined in the glossary mentioned in the last paragraph of the introduction. However, the authors would have done their readers a kindness by mentioning the glossary again the first time it is needed and giving its page numbers. Remember our advice on page 59 about directing your readers. Locating a glossary for them is a good example of such direction.

As our two excerpts illustrate, many diverse items of information can be placed in a theory or principles section. Remember, however, that the major purpose of the section is to emphasize the principles that underlie the actions later described in the how-to instructions. In this section you're telling your readers *why*. Later you'll tell them *how*. Theory is important, but don't get carried away with it. Experts in a process sometimes develop this section at too great length, burying their readers under information the readers don't need and obscuring more important information that they do need. Make this section, if you have it at all, only as full and as complex as your analysis of purpose and readers demands.

List of Equipment and Materials Needed

In a list of equipment and materials, you tell your readers what they will need to accomplish the process. A simple example would be the list of cooking utensils and ingredients that precedes a recipe. Sometimes in straightforward processes, or with knowledgeable audiences, the list of equipment is not used. Instead, the instructions tell the readers what equipment they need as they need it: "Take a rubber mallet and tap the hubcap to be sure it's secure."

When a list is used, frequently, each item is simply listed by name. Sometimes, however, your audience analysis may indicate that more information is needed. You may want, for instance, to define and describe the tools and equipment needed, as is done in Figure 14–3. If you think your readers are really unfamiliar with the tools or equipment being used, you may even give instruction in its use, as in Figure 14–4. If the equipment cannot be easily obtained, you'll do your readers a service by telling them where they can find the hard-to-get items. As always, your audience analysis determines the amount and kind of information presented.

Basic Tools

You'll need a few basic tools for most home maintenance jobs, and some special tools for special jobs. Some are expensive, and are not needed very often. Is there a place where you can borrow or rent those?

Here are some basic tools and materials you may need for doing simple repairs on the outside of your house.

Nail Set

A *nail set* is a small metal device used to sink the heads of nails slightly below the surface you are driving them into (fig. 1).

Fig. 1

Squares

The *framing square* is a handy measuring tool for lining up materials evenly and making square corners. It is usually metal (fig. 2).

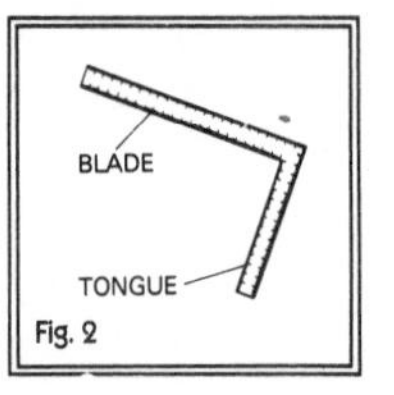

Fig. 2

The *try square* is smaller and is also used for lining up and squaring material. One side is made of wood and is not marked to measure with (fig. 3).

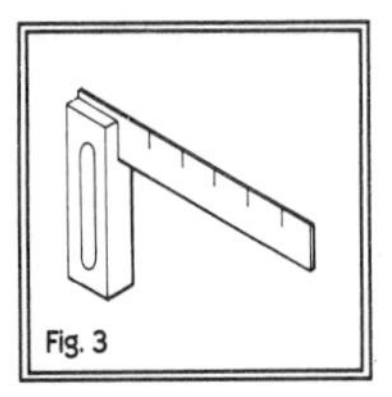
Fig. 3

Miter Box

With a *miter box,* you can saw off a piece of board at an exact angle. It may be of wood, to use with a separate saw (fig. 4). Or it may be steel, with the saw set in the steel box (fig. 5).

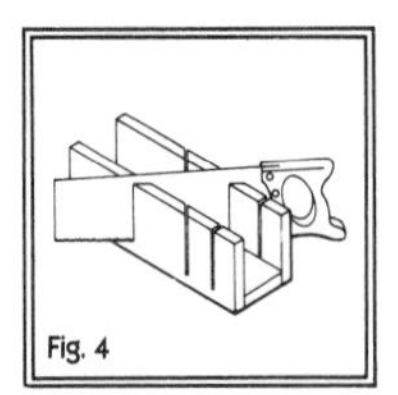
Fig. 4

Masonry Trowels and Jointer

The *trowel* is used to build or repair masonry walls, sidewalks, etc. It has a flat, thin, steel blade set into a handle. The "brick trowel" is the larger and is used for mixing, placing, and spreading mortar. The smaller "pointing trowel" is used to fill holes and repair mortar joints (fig. 6). This process is called "pointing."

The *jointer* is another masonry tool, used to finish joints after the wall is laid (fig. 7). Finish joints are made on the outside of a masonry wall to make it more waterproof and to improve appearance. The "V" and "concave" joints are the most weather tight. A different type of jointer is needed for each type of joint used.

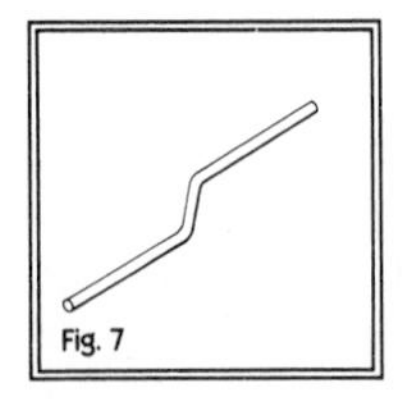
Fig. 7

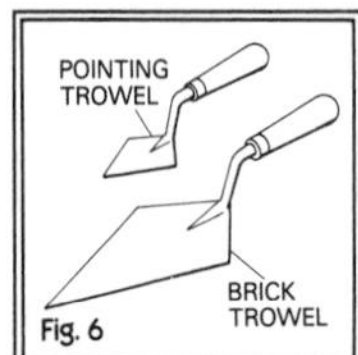

Fig. 6

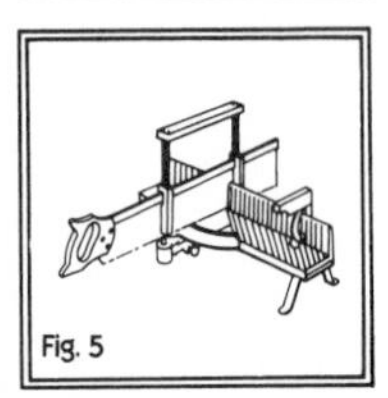
Fig. 5

Figure 14–3 List of Tools
Source: U.S. Department of Agriculture, *Simple Home Repairs: Outside* (Washington, DC: GPO, 1986) 4.

Description of the Mechanism

Instructions devoted to the operation and maintenance of a specific mechanism usually include a section describing the mechanism. Also, when it is central in some process, the mechanism is frequently described. In such sections, follow the principles for technical description given on pages 175–181. Break the mechanism into its component parts, and describe how they function.

To illustrate how this is done, we reproduce several sections from the *Macintosh SE Owner's Guide*. In Figure 14–5, you can see the introduction to the Chapter "Inside the Macintosh SE." Through the use of a list and a graphic, the introduction breaks the mechanism into its component parts and previews the rest of the chapter. The introduction also explains that while the information in the chapter is not needed to operate the Macintosh SE it may be useful in other ways. Writers of computer manuals know that their readers skim and scan their way through the manual. Here you have an invitation to read or not read the chapter, depending on your needs and interest.

In Figure 14–6, from the same chapter, we show you an excerpt that describes the Macintosh mouse, the device that allows a user to move a pointer around the Macintosh screen. The description describes the function of the mouse and explains how it fulfills that function. A graphic accompanies the prose. Mechanism descriptions are generally accompanied by numerous illustrations. Sometimes exploded views of the mechanism are provided, as in Figure 14–7. We hasten to add that in this context "exploded" means the mechanism is drawn in such a way that its component parts are separated and are thus easier to identify. Figure 14–7 makes the concept quite clear.

Warnings

We live in an age of litigation. People who hurt themselves or damage their equipment when following instructions in the use of that equipment frequently sue for damages. If they can prove to a court's satisfaction that they were not sufficiently warned of the dangers involved, they will collect huge sums of money. Because of this, warnings have increasingly become an important part of instructions, even, as Figure 14–8 shows, in the operation of something as simple as an electric can opener.

If they are extensive enough, the warnings may be put into a separate section, as they are in Figure 14–8. But often they are embedded in the how-to instructions. In either case, be sure they are prominently displayed in some manner that makes them obvious to the reader. You may surround them with boxes, print them in type different and larger than the surrounding text, print them in a striking color, or mark them with a symbol of some sort. Frequently, you will use some combination of these devices.

Not only must you make the warnings stand out typographically, you

(Text continued on page 447.)

NAILS, SCREWS, AND BOLTS

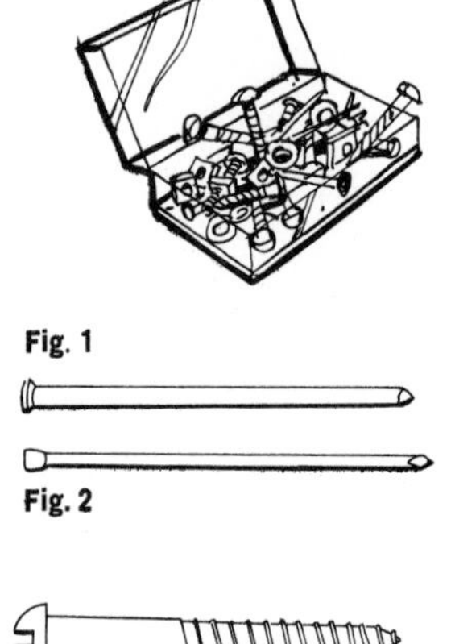

Nails, screws, and bolts each have special uses. Keep them on hand for household repairs.

NAILS come in two shapes.

Box nails have large heads. Use them for rough work when appearance doesn't matter. (Fig 1.)

Finishing nails have only very small heads. You can drive them below the surface with a nail set or another nail, and cover them. Use them where looks is important, as in putting up panelling or building shelves. (Fig. 2)

SCREWS are best where holding strength is important. (Fig 3.) Use them to install towel bars, curtain rods, to repair drawers, or to mount hinges. Where screws work loose, you can refill the holes with matchsticks or wood putty and replace them.

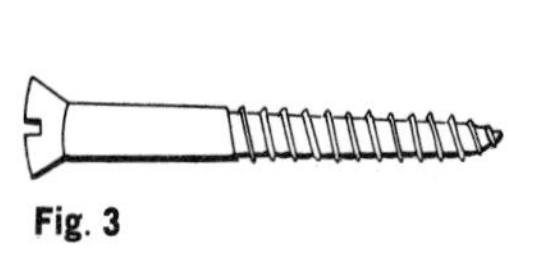

Use **molly screws** or **toggle bolts** on a plastered wall where strength is needed to hold heavy pictures, mirrors, towel bars, etc.

Molly screws have two parts (Fig. 4). To install, first make a small hole in the plaster and drive the casing in even with the wall surface. Tighten screw to spread casing in the back. Remove screw and put it through the item you are hanging, into casing, and tighten.

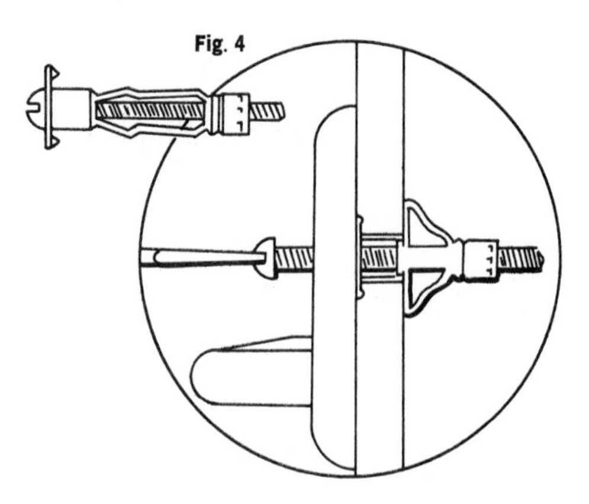

Figure 14–4 Instruction in Equipment Use

Source: U.S. Department of Agriculture, *Simple Home Repairs: Inside* (Washington, DC: GPO, 1989) 22–23.

Toggle bolts. (Fig. 5) Drill a hole in the plaster large enough for the folded toggle to go through. Remove toggle. Put bolt through towel bar or whatever you are hanging. Replace toggle. Push toggle through the wall and tighten with a screwdriver.

Plastic anchor screws (Fig. 6) should be used where you want to attach something to a concrete wall. To install, first make a small hole in the wall and drive casing in even with the wall surface. Put screw through item and into the casing, and tighten.

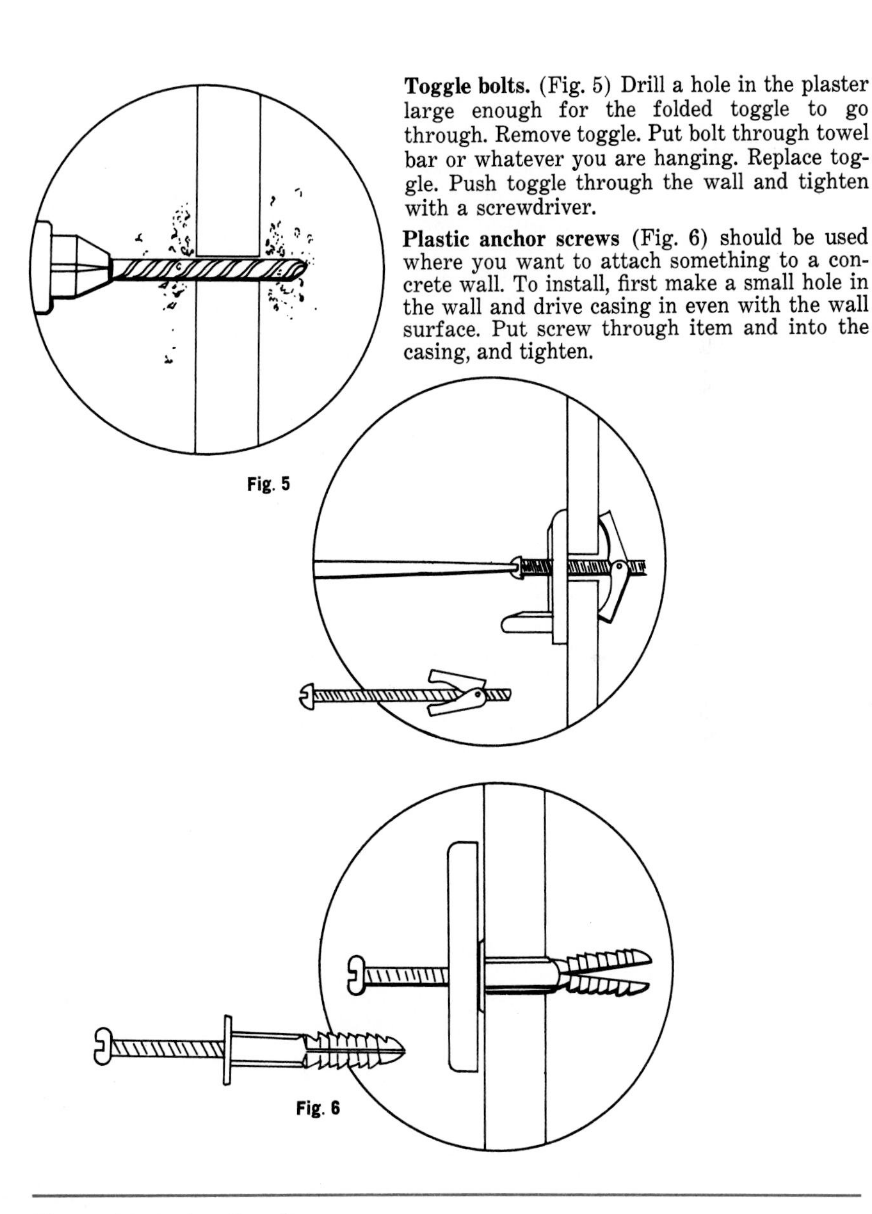

Figure 14–4 *(continued)*

Figure 14–5 Introduction to Mechanism Description
Source: Apple Computer, Inc., *Macintosh SE Owner's Guide* (Cupertino, CA: 1988) 32–34.

Inside the Macintosh SE

If you used the training disk, *Your Apple Tour of the Macintosh SE,* or went through the tutorial in the *Macintosh System Software User's Guide,* you've already learned the basic skills you'll use to work with your Macintosh SE computer. The next two chapters expand on what you've learned.

This chapter briefly introduces you to the more important parts of the Macintosh SE—the basic hardware components of your computer system. In Chapter 3, you'll read about how they work together with software.

You don't need to know the material in this chapter or in Chapter 3 to use your Macintosh SE. But a basic familiarity with how a Macintosh works will help you understand advanced concepts that you may encounter in using system software and applications. It may also come in handy when considering hardware options and peripheral devices for your computer.

Here's a brief summary of what you'll read about in this chapter:

- the microprocessor
- RAM and ROM
- the display
- disk drives and disks
- the mouse and the keyboard

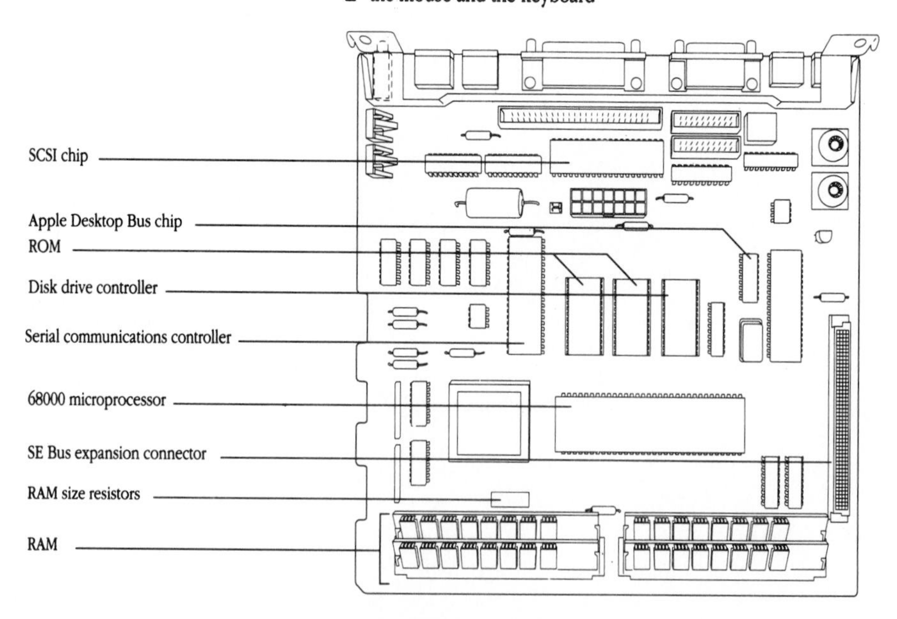

Macintosh SE main circuit board

The mouse

The Macintosh mouse is a hand-operated device that lets you easily control the location of the pointer on your screen and make selections and choices with the mouse button. Coupled with the graphic elements of the Macintosh User Interface—icons, windows, pull-down menus, and so on—the mouse makes ordinary operation of the system almost effortless: you view your work on the screen and interact with it merely by pointing with the mouse and clicking the mouse button.

The Macintosh mouse contains a rubber-coated steel ball that rests on the surface of your working area. When the mouse is rolled over that surface, the ball turns two rotating axles inside the mouse, each connected to an interrupter wheel. These wheels contain precisely spaced slots through which beams of light are aimed at detectors. The wheels track vertical and horizontal motions. As the axles turn the wheels, the light beams shining through their slots are interrupted; the detectors register the changing optical values, and a small integrated circuit inside the mouse interprets them and signals the operating system to move the pointer on your screen accordingly. (There's also a third axle that helps balance the ball and keep it rolling smoothly.)

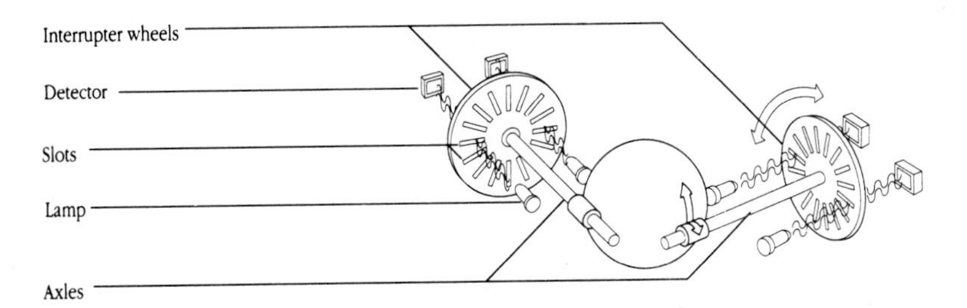

The mouse registers relative movement only; the operating system can tell how far the mouse has moved and in which direction, but not the mouse's absolute location. That's why you can pick up the mouse and move it to another place on your table or desk surface and the pointer will not move. You can adjust the speed with which the pointer on the screen responds to the mouse's movements by using the Control Panel desk accessory in the Apple menu.

Figure 14–6 Description of the Macintosh Mouse
Source: Apple Computer, Inc., *Macintosh SE Owner's Guide* (Cupertino, CA: 1988) 37–38. Copyright © 1988 Apple Computer, Inc. Reprinted by permission.

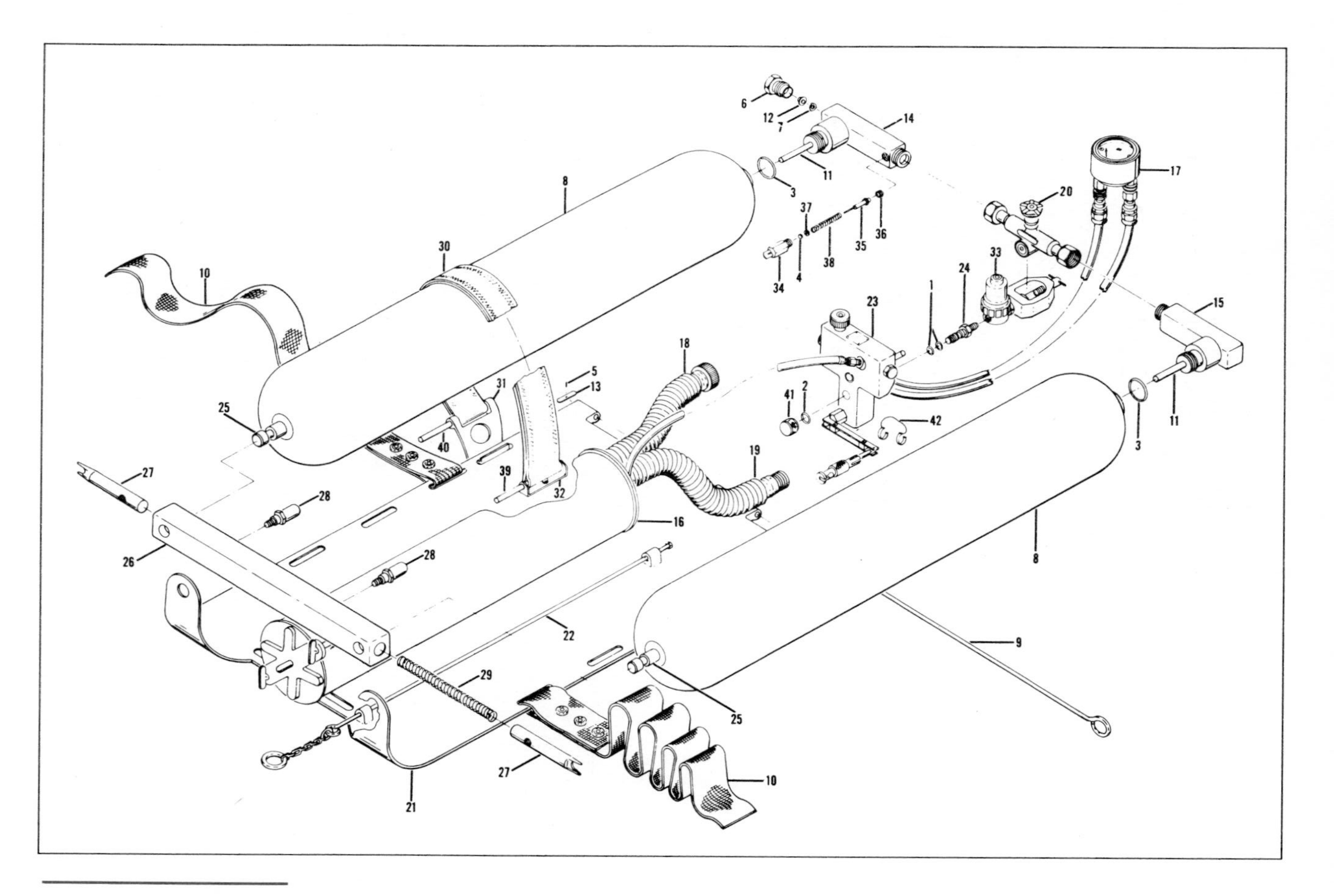

Figure 14–7 Exploded View of Equipment Assembly

IMPORTANT SAFEGUARDS

When using electrical appliances, basic safety precautions should always be followed, including the following:

1. Read all instructions.
2. To protect against risk of electrical shock, do not put power unit in water or other liquid.
3. Close supervision is necessary when any appliance is used by or near children.
4. Unplug from outlet when not in use, before putting on or taking off parts, and before cleaning.
5. Avoid contacting moving parts.
6. Do not operate any appliance with a damaged cord or plug or after the appliance malfunctions, or is dropped or damaged in any manner. Return appliance to the nearest authorized service facility for examination, repair or electrical or mechanical adjustment.
7. The use of attachments not recommended or sold by the appliance manufacturer may cause fire, electric shock or injury.
8. Do not use outdoors.
9. Do not let cord hang over edge of table or counter, or touch hot surfaces.
10. Do not open pressurized (aerosol-type) cans or cans of flammable liquids, such as lighter fluids.

SAVE THESE INSTRUCTIONS

Figure 14–8 Warning Section
Source: Reprinted by permission of Underwriters Laboratories, Northbrook, Illinois.

must use language that is absolutely clear about the hazards involved. Any lack of clarity can result in a preventable accident almost certainly followed by a costly lawsuit against your employer.

Although no terminology is completely agreed upon, three levels of warning have been widely accepted, designated by the words *caution, warning,* and *danger.*[6]

Caution Use the term **caution** to alert the reader that not following the instructions exactly may lead to a wrong or inappropriate result. However, in a caution, no danger to people or equipment is involved. Figure 14–9 shows how a caution might be used to advise a technician to follow the steps of a procedure in proper order.

Check Valve Test

- Place the mouthpiece shut-off valve in the **Diving** position.
- Place the mouthpiece in your mouth, squeeze the inhalation hose closed, and attempt to inhale through the mouthpiece. If it is possible to inhale with the inhalation hose closed off, the check valve is missing or defective.

CAUTION

If the mouthpiece shut-off valve is in the **Open** position, the test will incorrectly indicate a defective or missing check valve.

Figure 14–9
A Caution Message

Warning Use the term ***warning*** to alert the reader to faulty procedures that may damage equipment, as in the warning from a compact disk player manual shown in Figure 14–10. The exclamation point inside the triangle in Figure 14–10 is a commonly accepted symbol to attract the reader's attention and to stress the importance of the message. You will see it used on all three levels of warnings.

Danger Use the term ***danger*** for the highest level of warning: a warning to prevent injury or death. Obviously, you must make such danger messages stand out typographically and write them with utter clarity. Figure 14–11 presents a good example.

How-to Instructions

The actual instructions on how to perform the process obviously lie at the heart of any set of instructions. For sample sets of how-to instructions,

Figure 14–10
A Warning Message

⚠ **WARNING** ⚠

- Do not use force to open or close the disk tray. Force may result in a damaged tray.
- Place nothing but a compact disk in the tray. Inserting objects other than disks in the tray may result in a damaged tray.

⚠ **DANGER** ⚠

- **Use no oil.**
- Oil coming in contact with a high-pressure connection in diving equipment may result in an **explosion.**
- To prevent serious injury or death, **use no oil.**

Figure 14–11
A Danger Message

see Figures 14–12 through 14–15. The same general principles apply to all how-to instructions. What we now tell you about writing them is well illustrated by these samples.

Style When writing how-to instructions, one of your major concerns is to use a clear, understandable style. To achieve this, write your instructions in the active voice and imperative mood. *Turn the mouse upside down and rotate the plastic dial counterclockwise as far as it will go.* The imperative mood is normal and acceptable in instructions. It's clear and precise and will not offend the reader. By using the format shown in Figure 14–12, you can use the imperative mood even when several people with distinct tasks have to carry out the procedure. In the format shown, the headings to the left identify the responsible actor, allowing the imperative mood to be used in the right action column. It's an efficient system. (For more on the active voice and imperative mood, see pages 183–184.)

Notice also that the sets of how-to instructions we've shown you all use a list format. Each numbered step usually contains only one instruction and at the most two or three closely related instructions. The list format keeps each step in a series clear and distinct from every other step. Listing has several other advantages as well:

- It makes it obvious how many steps there are.
- It makes it easy for readers to find their place on the page.
- It allows the reader to use the how-to instructions as a checklist.

Use familiar, direct language and avoid jargon. Tell your readers to *check* things or to *look them over.* Don't tell them to *conduct an investigation.* Tell your readers to *use* a wrench, not to *utilize* one. Fill your instructions with readily recognized verbs like *adjust, attach, bend, cap, center, close, drain, install, lock, replace, spin, turn,* and *wrap.* For more on good style, see Chapter 6, "Achieving a Readable Style."

If your how-to instructions call for calculations, include sample cal-

DELUXE CHECK PRINTERS, INC. **STANDARD OPERATING PROCEDURES**

Procedure C-9
Establishing, Changing, or Eliminating the Petty Cash Fund
Accounts Payable and Purchasing Manual - C

SUMMARY: The petty cash fund is a fixed cash fund reserved for minor expenditures of $50 or less. This procedure explains how to establish, change, or eliminate the petty cash fund.

NOTE: When a petty cash fund is established, the plant manager should assign responsibility to no more than two cash drawer custodians, with one individual having primary responsibility. The accounts payable clerk must not be a custodian of the petty cash fund.

See Procedure C-17 to disburse petty cash. See Procedure C-18 to replenish the petty cash fund. See Appendix J for petty cash fund controls.

RESPONSIBILITY	ACTION
Plant manager	1. Request authorization from corporate treasurer for one of the following: • establish petty cash fund • change amount of existing petty cash fund • eliminate existing petty cash fund
Corporate treasurer	2. Review request and approve or disapprove and notify plant accountant of decision.
Plant accountant	3. Notify plant manager of decision.
	4. If establishing fund or increasing existing fund, have check prepared from Account 1030 (Regular Cash Account) for authorized amount, payable to cash, debiting Account 1010 (Petty Cash Account) on check voucher.
	4a. Place check in check cashing fund box and withdraw authorized amount of cash.
	4b. Place cash in petty cash fund, and notify plant manager that petty cash fund is established or increased.
	-OTHERWISE-
	5. If decreasing or eliminating existing fund, use Daily Report of Cash, form A-30-Q (Exhibit 24), to credit Account 1010.

Rewritten by: Kathy Huebsch

Figure 14–12 Standard Operating Procedures
Source: Reprinted by permission of Deluxe Check Printers, Inc.

culations to clarify them for the reader. See Figure 14–13 for an example of such calculations.

Graphics Be generous with graphics. Word descriptions and graphics often complement each other. The words tell *what* action is to be done. The graphics show *where* it is to be done and often also show the *how*. Our samples demonstrate well the relationship between words and graphics. Note that graphics are often annotated to allow for easy reference to them. As Figures 14–14 and 14–15 demonstrate, graphics can be used to show the worker, or at least the worker's hands, actually performing the job.

Arrangement When writing performance instructions, arrange the process being described into as many major routines and subroutines as needed. For example, a set of instructions for the overhaul and repair of a piece of machinery might be broken down as follows:

- Disassembly of major components
- Disassembly of components
- Cleaning
- Inspection
- Lubrication
- Repair
- Reassembly of components
- Testing of components
- Reassembly of major components

Notice that in this case the steps are in chronological order. Our samples also demonstrate chronological order.

If there are steps that are repeated, it is sometimes a legitimate practice to tell the reader to "Repeat steps 2, 3, and 4." But whether you would do so depends upon your analysis of the reader's situation. Visualize your reader. Maybe he or she will be perched atop a shaky ladder, your instructions in one hand, a tool in the other. Under such circumstances, the reader will not want to be flipping pages around to find the instructions that need to be repeated. You will be wiser and kinder to print once again all the instructions of the sequence. But if the reader will be working in a comfortable place with both feet on the ground, you would probably be safe enough saying, "Repeat steps. . . ."

Such reader and situation analysis can help you make many similar decisions. Suppose, for example, that your readers are not expert technicians, and the process calls for them to use simple test equipment. In such a situation, you should include the instructions for operating the test equipment as part of the routine you are describing. On the other hand, suppose your readers are experienced technicians following your instruc-

Sample Calculations

The first house plan discussed in Chapter 8 (Figure 8.3) is analyzed. Orlando weather data are used. The calculations follow.

Form for Calculating Window Areas in Naturally Ventilated Houses

Project ______________________

Analyst ______________________

1. House conditioned floor area = 1334 (1) ft^2
2. Average ceiling height = 8 (2) ft
3. House volume
   = 1334 (Step 1) x 8 (Step 2) = 10,672 (3) ft^3
4. Design air change rate/hr = (recommended value is 30) 30 (4) ACH
5. Required airflow rate, cfm
   = 10,672 (Step 3) x 30 (Step 4) ÷ 60 = 5336 (5) cfm
6. Design month = may (6)

   (Recommended months = May for Florida and the Gulf Coast
   = June for more northern southeast cities)
7. Nearest city location with weather data = Orlando (7)
8. From weather data in Appendix B, determine windspeed (WS) and wind direction (WD) for design month
   - 8a. WS = 8.8 (8a) mph
   - 8b. WD = SE (8b)
9. From prevailing wind direction and building orientation, determine incidence angle on windward wall. Incidence angle = about 10 (9) degrees
10. From Table A1, determine inlet-to-site 10 meter windspeed ratio = 0.35 (10)

Figure 14–13 Sample Calculations
Source: Subrato Chandra, Philip W. Fairey III, and Michael M. Houston, *Cooling with Ventilation* (Washington, DC: GPO, 1986) 64–65.

11. Determine windspeed correction factors

11a. For house location and ventilation strategy, determine terrain correction factor from Table A2 = 0.67 (11a)

11b. For neighboring buildings, determine neighborhood convection factor from Table A3 = 0.77 (11b)

Assume h = 8 ft, =
g = 24 ft

11c. If sizing windows for the second floor or for house on stilts, use a height multiplication factor of 1.15. Otherwise, use 1.0. Selected value = 1.0 (11c)

12. Determine overall windspeed correction factor

= 0.67 (Step 11a) x 0.77 (Step 11b) x 1.0 (Step 11c) = 0.52 (12)

13. Determine site windspeed in ft/min

= 8.8 (Step 8a) x 0.52 (Step 12) x 88 = 403 (13) ft/min

14. Determine window inlet airspeed

= 403 (Step 13) x 0.35 (Step 10) = 141 (14) ft/min

15. Determine net aperture inlet area

= 5336 (Step 5) ÷ 141 (Step 14) = 37.8 (15) ft^2

16. Determine total inlet + outlet area, insect screened. Assumes fiberglass screening with a porosity of 0.6

= 2 x 37.8 (Step 15) x 1.67 = 126 (16) ft^2

17. Since typical window or door framing is about 20% of the gross area, determine gross total operable area required as

= 1.25 x 126 (Step 16) = 156 (17) ft^2

18. Determine gross operable area as a % of floor area

= 158 (Step 17) ÷ 1334 (Step 1) x 100 = 11.8 (18) %

Note: This gross operable area requirement can be met the by same area of windows if the windows are 100% operable (awning, casement, hopper, etc.). The window area required will be twice this value if single-hung or sliding windows are used which have only 50% of the area as operable.

Figure 14–13 *(continued)*

The mouse

Be careful not to drop the mouse or let it hang from a table by its cable. Use common sense in treating it as carefully as you can.

The surface your mouse moves on should be as smooth, clean, and dust-free as possible. And give the mouse itself an occasional cleaning.

Here's how to clean the mouse:

1. **Turn the mouse upside down and rotate the plastic dial counterclockwise as far as it will go.**
2. **Holding one hand over the ball and dial to catch them, turn the mouse back right side up. The dial and the ball will drop into your hand.**

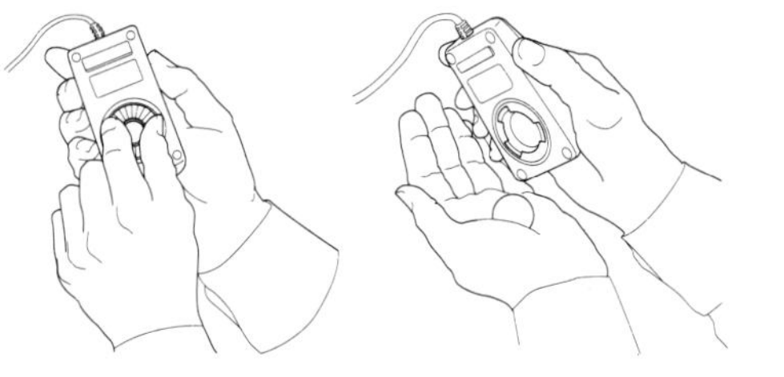

3. **Inside the case are three plastic rollers, similar to those on a tape recorder. Using a cotton swab moistened with alcohol or tape head cleaner, gently wipe off any oil or dust that has collected on the rollers, rotating them to reach all surfaces.**
4. **Wipe the ball with a soft, clean, dry cloth. (Don't use tissue or anything that may leave lint, and don't use a cleaning liquid.)**

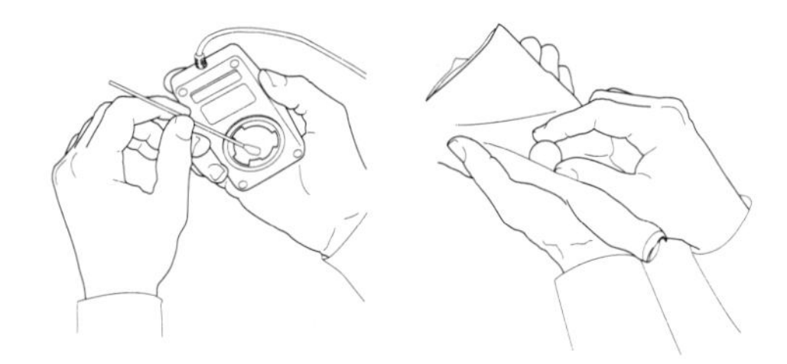

5. **Blow gently into the case to remove any dust that has collected there.**
6. **Put the ball back into its case and, lining up the indicator on the dial with the O on the back of the case, reinsert the dial and turn it clockwise as far as it will go.**

Figure 14–14 How-to Instructions: Mouse
Source: Apple Computer, Inc., *Macintosh SE Owner's Guide* (Cupertino, CA: 1988) 64–65. Copyright © 1988 Apple Computer, Inc. Reprinted by permission.

tions at a comfortable workbench with a well-stocked library of manuals nearby. Then you can assume that they know how to operate any needed test equipment, or you can refer them to another manual that describes how to operate the test equipment.

Only the instructions in Figure 14–15 have a conclusion of sorts, labeled in this case "Your Reward." For the most part, instructions have no conclusions. They simply end with the last instruction. On occasion, particularly when writing for lay people, you might wish to close with a summary of the chief steps of the process or, perhaps, a graceful close (see page 302). However, such endings are not general practice.

Tips and Troubleshooting Procedures

Many sets of instructions contain sections that either give the reader helpful tips on how to do a better job or that provide guidance when trouble occurs.

Tips You may present tips in a separate section, as illustrated in Figure 14–16. Or, just as likely, you may incorporate them into the how-to instructions, as in this excerpt on setting flexible tile. In the excerpt, the last sentence in instructions 1, 2, 3, and 5 gives the reader a tip that should make the task go more easily:

1. Remove loose or damaged tile. A warm iron will help soften the adhesive.
2. Scrape off the old adhesive from the floor or wall. Also from the tile if you're to use it again.
3. Fit tiles carefully. Some tile can be cut with a knife or shears, others with a saw. Tile is less apt to break if it's warm.
4. Spread adhesive on the floor or wall with a paint brush or putty knife.
5. Wait until adhesive begins to set before placing the tile. Press tile on firmly. A rolling pin works well.[7]

Troubleshooting Procedures You may incorporate troubleshooting procedures into your how-to instructions, as in this excerpt:

> Tighten screws in the hinges. If screws are not holding, replace them one at a time with a longer screw. Or insert a matchstick in the hole and put the old screw back.[8]

Perhaps more often, troubleshooting procedures will have a section of their own, as in Figures 14–17 and 14–18. Both figures illustrate a typical format, a three-column chart with headings such as *Problem, Possible Cause,* and *Possible Remedy.* Notice that the chart in Figure 14–17 uses graphics to illustrate the problem, an excellent technique you should use where possible. Notice also that the remedies are given as instructions in the active voice, imperative mood.

The example in Figure 14–18 gives page references when appropriate to guide the reader to additonal information, also an excellent idea.

Glossary

If your audience analysis tell you that your reader will not comprehend all the terminology you plan to use in your instructions, you'll need to provide definitions. If you need only a few definitions, you can define terms as you use them. You can even provide graphic definitions as in the definition of the "Underwriters' knot" in Figure 14–15.

If you must provide many definitions, you'll probably want to provide a glossary as a separate section. Figure 14–19 illustrates a typical layout and style. Arrangement is alphabetical, and the first sentence in each definition is a fragment. The rest of the definition is in complete sentences. See also pages 164–169, where we discuss definitions, and pages 286–287, where we discuss glossaries.

(Text continued on page 460.)

ELECTRIC PLUGS—Repair or Replace

YOUR PROBLEM

- Lamps or appliances do not work right.
- A damaged plug is dangerous.
- It's hard to hire help for small repairs.

WHAT YOU NEED

- New plug—if your old one cannot be used. (Buy one with a UL label)
- Screwdriver
- Knife

HOW-TO

1. Cut the cord off at the damaged part. (Fig 1.)
2. Slip the plug back on the cord. (Fig 2.)
3. Clip and separate the cord. (Fig. 3)
4. Tie Underwriters' knot. (Fig. 4)
5. Remove a half-inch of the insulation from the end of the wires. **Do not cut any of the small wires.** (Fig. 5)

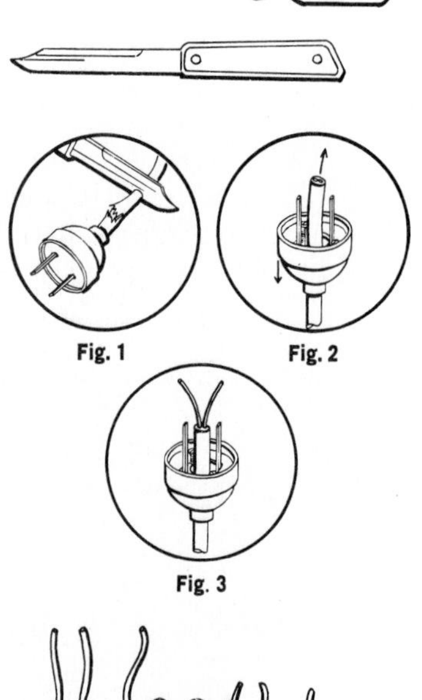

Fig. 4

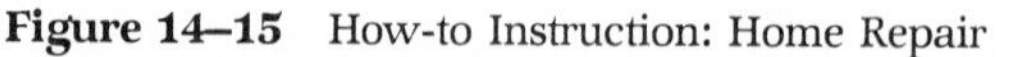

Figure 14–15 How-to Instruction: Home Repair

Source: U.S. Department of Agriculture, *Simple Home Repairs: Inside* (Washington, DC: GPO, 1989) 3–4.

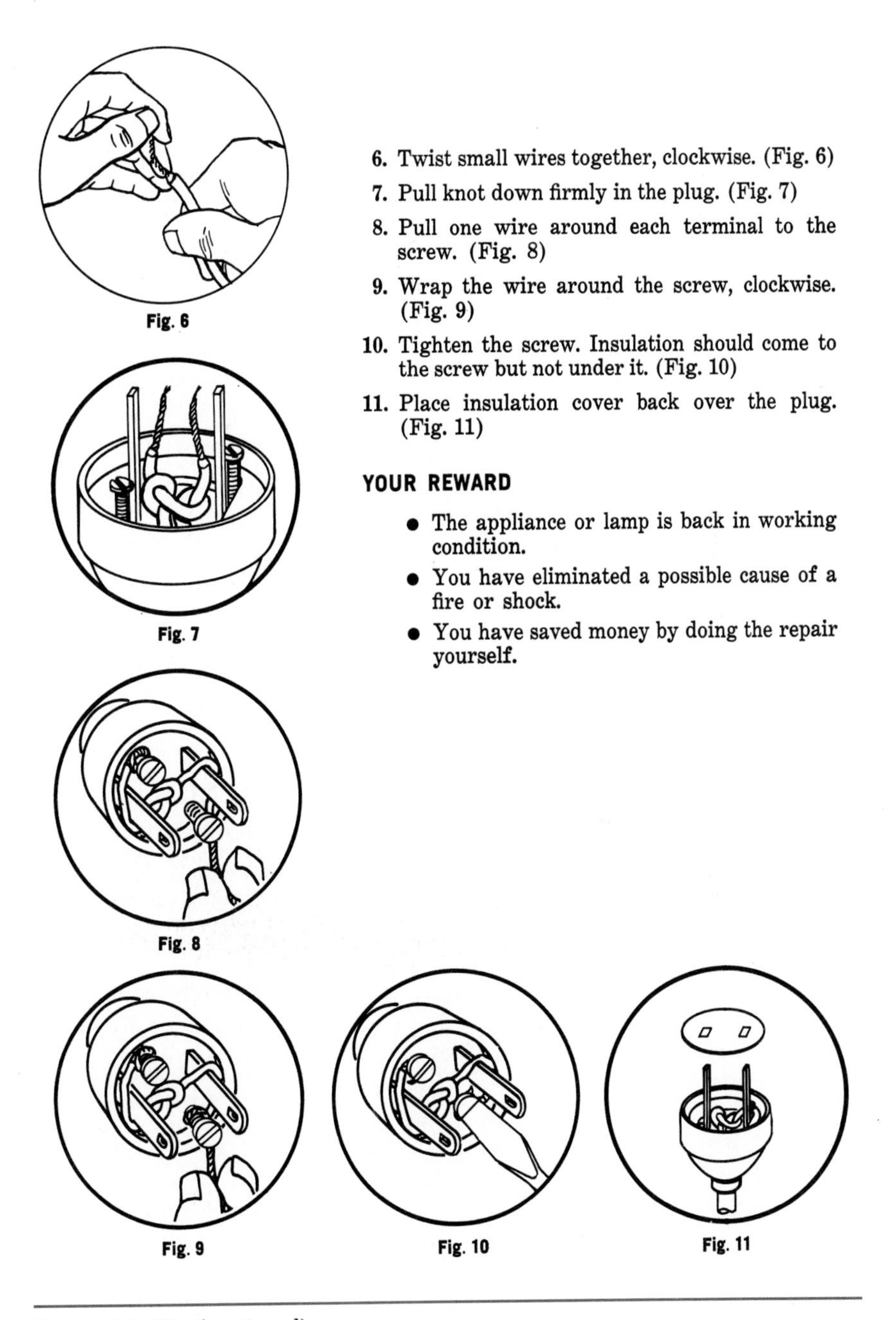

6. Twist small wires together, clockwise. (Fig. 6)
7. Pull knot down firmly in the plug. (Fig. 7)
8. Pull one wire around each terminal to the screw. (Fig. 8)
9. Wrap the wire around the screw, clockwise. (Fig. 9)
10. Tighten the screw. Insulation should come to the screw but not under it. (Fig. 10)
11. Place insulation cover back over the plug. (Fig. 11)

YOUR REWARD

- The appliance or lamp is back in working condition.
- You have eliminated a possible cause of a fire or shock.
- You have saved money by doing the repair yourself.

Figure 14–15 *(continued)*

Figure 14–16 Tips
Source: U.S. Department of Health and Human Services, *Eating to Lower Your Blood Cholesterol* (Washington, DC: GPO, 1989) 23–24.

New Ways To Make Sauces and Soups

Sauces, including gravies and homemade pasta sauces, and many soups often can be prepared with much less fat. Before thickening a sauce or serving soup, let the stock or liquid cool – preferably in the refrigerator. The fat will rise to the top and it can easily be skimmed off. Treat canned broth-type soups the same way.

For sauces that call for sour cream, substitute plain low-fat yogurt. To prevent the yogurt from separating, mix 1 tablespoon of cornstarch with 1 tablespoon of yogurt and mix that into the rest of the yogurt. Stir over medium heat just until the yogurt thickens. Serve immediately. Also, whenever you make creamed soup or white sauces, use skim or 1% milk instead of 2% or whole milk.

New Ways To Use Old Recipes

There are dozens of cookbooks and recipe booklets that will help you with low-fat cooking. But there is no reason to stop using your own favorite cookbook. The following list summarizes many of the tips. Using them, you can change tried and true recipes to low-saturated fat, low-cholesterol recipes. In some cases, especially with baked products, the quality or texture may change. For example, using vegetable oil instead of shortening in cakes that require creaming will affect the result. Use margarine instead; oil is best used only in recipes calling for **melted** butter. Substituting yogurt for sour cream sometimes affects the taste of the product. Experiment! Find the recipes that work best with these substitutions.

Instead of	Use
1 tablespoon butter	1 tablespoon margarine or ¾ tablespoons oil
1 cup shortening	⅔ cup vegetable oil
1 whole egg	2 egg whites
1 cup sour cream	1 cup yogurt (plus 1 tablespoon cornstarch for some recipes)
1 cup whole milk	1 cup skim milk

Low-Fat Cooking Tips

Your kitchen is now stocked with great tasting, low-saturated fat, low-cholesterol foods. But you may still be faced with the temptation to fix your favorite higher fat meats, rich soups, and baked breads and cookies. The suggestions below will help you to reduce the amount of total and saturated fats in these foods.

New Ways To Prepare Meat, Poultry, Fish, and Shellfish

When you prepare meats, poultry, and fish, remove as much saturated fat as possible. Trim the visible fat from meat. Remove the skin and fat from the chicken, turkey, and other poultry. And, if you buy tuna or other fish that is packed in oil, rinse it in a strainer before making tuna salad or a casserole, or buy it packed in water.

Changes in your cooking style can also help you remove fat. Rather than frying meats, poultry, fish, and shellfish, try broiling, roasting, poaching, or baking. Broiling browns meats without adding fat. When you roast, place the meat on a rack so that the fat can drip away.

Finally, if you baste your roast, use fat-free ingredients such as wine, tomato juice, or lemon juice instead of the fatty drippings. If you baste turkeys and chickens with fat use vegetable oil or margarine instead of the traditional butter or lard. Self-basting turkeys can be high in saturated fat–read the label!

TROUBLE-SHOOTING

Most baler operating problems are caused by improper adjustment or delayed service. This chart is designed to help you when a problem develops, by suggesting a probable cause and the recommended solution.

Apply these suggested remedies carefully. Make certain the source of the trouble is not some place other than where the problem exists. A thorough understanding of the baler is a must if operating problems are to be corrected satisfactorily. Refer to the operator's manual for detailed repair procedures.

TROUBLE-SHOOTING CHART

PROBLEM	POSSIBLE CAUSE	POSSIBLE REMEDY
Knotter Difficulties — Twine Baler		
 KNOT IN TWINE OVER BALE	Tucker fingers did not pick up needle twine or move it into tying position properly.	Adjust tucker fingers. Adjust needles or twine disk. Check twine disk and twine-box-tension. Install plungerhead extensions.
	Hay dogs do not hold end of bale.	Free frozen hay dogs. Replace broken hay-dog springs. Reduce feeding rate. Install plungerhead extensions.
 TWINE BROKEN IN KNOT	Extreme tension on twine around billhook during tying cycle causes twine to shear or pull apart.	Loosen twine-disk-holder spring. Smooth off all rough surfaces and edges on billhook.

Figure 14–17 Troubleshooting Chart
Source: Reprinted by permission of Deere and Company.

ELECTRICAL SYSTEM

PROBLEM	POSSIBLE CAUSE	POSSIBLE REMEDY	PAGE REFERENCE
Battery will not charge	Loose or corroded connections.	Clean and tighten connections.	66
	Sulfated or worn-out battery.	Check electrolyte level and specific gravity.	67
	Loose or defective alternator belt.	Adjust belt tension or replace belt.	52
"CHG" indicator glows with engine running	Low engine speed.	Increase speed.	
	Defective battery.	Check electrolyte level and specific gravity.	67
	Defective alternator.	Have your John Deere dealer check alternator.	

Figure 14–18 Troubleshooting Chart
Source: Reprinted by permission of Deere and Company.

Glossary

1. **Atherosclerosis** - A type of "hardening of the arteries" in which cholesterol, fat, and other blood components build up on the inner lining of arteries. As atherosclerosis progresses, the arteries to the heart may narrow so that oxygen-rich blood and nutrients have difficulty reaching the heart.

2. **Carbohydrate** - One of the three nutrients that supply calories (energy) to the body. Carbohydrate provides 4 calories per gram—the same number of calories as pure protein and less than half the calories of fat. Carbohydrate is essential for normal body function. There are two basic kinds of carbohydrate—simple carbohydrate (or sugars) and complex carbohydrate (starches and fiber). In nature, both the simple sugars and the complex starches come packaged in foods like oranges, apples, corn, wheat, and milk. Refined or processed carbohydrates are found in cookies, cakes, and pies.
   - **Complex carbohydrate** - Starch and fiber. Complex carbohydrate comes from plants. When complex carbohydrate is substituted for saturated fat, the saturated fat reduction helps lower blood cholesterol. Foods high in starch include breads, cereals, pasta, rice, dried beans and peas, corn, and lima beans.
   - **Fiber** - A nondigestible type of complex carbohydrate. High-fiber foods are usually low in calories. Foods high in fiber include whole grain breads and cereals, whole fruits, and dried beans. The type of fiber found in foods such as oat and barley bran, some fruits like apples and oranges, and some dried beans may help reduce blood cholesterol.

3. **Cholesterol** - A soft, waxy substance. It is made in sufficient quantity by the body for normal body function, including the manufacture of hormones, bile acid, and vitamin D. It is present in all parts of the body, including the nervous system, muscle, skin, liver, intestines, heart, etc.
   - **Blood cholesterol** - Cholesterol that is manufactured in the liver and absorbed from the food you eat and is carried in the blood for use by all parts of the body. A high level of blood cholesterol leads to atherosclerosis and coronary heart disease.

Figure 14–19 Glossary
Source: U.S. Department of Health and Human Services, *Eating to Lower Your Blood Cholesterol* (Washington, DC: GPO, 1989) 27.

Accessible Format

Your major goal in setting up your format in instructions should be to make the information accessible for your readers. Our Part III, "Document Design in Technical Writing," is especially helpful in this regard.

The theory section shown in Figure 14–2 on pages 437–438 demonstrates excellent accessibility. The type is large and readable, and the format is especially helpful for those readers who may scan the document. The headings standing apart to the left of the print allow the reader to scan quickly, looking for points of interest. Also, headings phrased as questions are more likely to arrest the attention of scanning readers and attract them to read the text. Curiosity is put to work. They may want to know the answers to the questions.

The graphic of the narrowed artery and the table showing cholesterol levels highlight the two key points in the section. The scanning reader who stops only long enough to absorb the information in the two graphics will at least know the principal danger of high cholesterol and what a desirable cholesterol level is.

Look now at Figure 14–20, a government document instructing its readers in how to file a form to establish their relationship with alien relatives who may wish to immigrate to the United States. The document is an example of inaccessible format. Both the headings and the print are small. The page is cluttered and intimidating.

The headings are not worded in a way to lead readers to the information they seek. Terms like "Eligibility," "Documents previously submitted," and "Documents in general" are probably meaningful to the person who wrote them but not to the typical reader of these instructions. The format violates most of the principles discussed in Chapter 10, "Document Design." Furthermore, the style of the instructions violates most of the principles discussed in Chapter 6, "Achieving a Readable Style."

Now look at Figure 14–21, which is the same document after it has been revised and given a new format to make it accessible. Certain things are immediately obvious. The print is bigger and there is more white space. The headings are more meaningful and informative. They are phrased from the reader's point of view. They have become the kinds of questions that someone approaching this process might reasonably ask: "Who can file?" and "For whom can you file?" have replaced "Eligibility." Such new headings lead and inform the readers rather than confuse them. The format and style of the instructions now show a knowledge and application of the principles discussed in Chapters 6 and 10. The result is a readable document.

Finally, when a set of instructions runs more than several pages, you should furnish a table of contents (TOC) to help your readers find their way and also to provide an overview of the instructions. The headings in the TOC should duplicate those in the instructions. Figure 14–22 shows a useful TOC in which the headings are meaningful and informative to the readers. (See also pages 282–283.)

Reader Checks

When you are writing instructions, check frequently with the people who are going to use them. Bring them a sample of your theory section. Discuss it with them. See if they understand it. Does it contain too much theory

(Text continued on page 466.)

☆ U.S. GOVERNMENT PRINTING OFFICE: 1984-437-621

(PLEASE TEAR OFF HERE BEFORE SUBMITTING PETITION)

U.S. Department of Justice
Immigration and Naturalization Service

PETITION TO CLASSIFY STATUS OF ALIEN RELATIVE
FOR ISSUANCE OF IMMIGRANT VISA

READ INSTRUCTIONS CAREFULLY. FEE WILL NOT BE REFUNDED.

Not all of these instructions relate to the type of case which concerns you. Please read carefully those which do relate. Failure to follow instructions may require return of your petition and delay final action.

1. **Eligibility.** A petition may be filed by a citizen or a lawful permanent resident of the United States to classify the status of alien relatives as follows:

   a. By citizen of the United States: Except as noted in paragraph 2, a citizen of the United States may submit a petition on behalf of a spouse or sons and daughters (regardless of age or marital status). A United States citizen at least 21 years of age may submit a petition for a parent, brother, or sister. If the petition is for a son or daughter who is married or at least 21 years of age, or both, or for a brother or sister, do not submit petitions for the beneficiary's spouse or unmarried children under 21 years of age. If the petition is approved, the beneficiary's spouse and unmarried children under 21 years of age, if accompanying or following to join him/her, will automatically be eligible for the same preference status.

   b. By a lawful permanent resident alien: Except as noted in paragraph 2, an alien lawfully admitted to the United States for permanent residence may submit a petition on behalf of a spouse or an unmarried child regardless of age. However, if a lawful permanent resident alien is married to a citizen and wishes to petition for an unmarried child, such alien should consult the nearest office of the Immigration and Naturalization Service for advice as to whether it would be preferable, or *necessary*, for the United States citizen spouse to submit the petition instead. If the petition is for an unmarried son or daugther, do not submit petitions for the beneficiary's unmarried children under 21 years of age. If the petition is approved, the beneficiary's unmarried children under 21 years of age, if accompanying or following to join him/her, will automatically be eligible for the same preference status.

2. **Petitions which cannot be approved.** Approval cannot be given to a petition on behalf of—

   a. A parent, brother, or sister, unless the petitioner is a United States citizen and at least 21 years of age.

   b. An adoptive parent, unless the relationship to the United States citizen petitioner exists by virtue of an adoption which took place while the child was under the age of 16, and the child has thereafter been in the legal custody of, and has resided with, the adopting parent or parents for at least 2 years. While the legal custody must be after the adoption, residence occurring prior to the adoption can satisfy the residence requirement.

   c. A stepparent, unless the marriage creating the status of stepparent occurred before the citizen stepchild reached the age of 18 years.

   d. An adopted child, unless the child was adopted while under the age of 16 and has thereafter been in the legal custody of, and has resided with, the adopting parent or parents for at least 2 years. While the legal custody may be after the adoption, residence occurring prior to the adoption can satisfy the residence requirement.

   e. A stepchild, unless the child was under the age of 18 years at the time the marriage creating the status of stepchild occurred.

   f. A wife or husband by reason of any marriage ceremony where the contracting parties thereto were not physically present in the presence of each other, unless the marriage shall have been consummated.

   g. A grandparent, grandchild, nephew, niece, uncle, aunt, cousin, or in-law.

3. **Supporting documents.** The following documents must be submitted with the petition:

   a. To prove United States citizenship of petitioner (where petition is for relative of a citizen).

   (1) If you are a citizen by reason of birth in the United States, submit your birth certificate. If your birth certificate is unobtainable, see "Secondary Evidence" below for submission of document in place of birth certificate.

   (2) If you were born outside the United States and became a citizen through the naturalization or citizenship of a parent or husband, and have not been issued a certificate of citizenship in your own name, submit evidence of the citizenship and marriage of such parent or husband, as well as termination of any prior marriages. Also, if you claim citizenship through a parent, submit your birth certificate and a separate statement showing the date, port, and means of all your arrivals and departures into and out of the United States. (Do not make or submit a photostat of a certificate of citizenship.)

   (3) If your naturalization occurred within 90 days immediately preceding the filing of this petition, of if it occurred prior to September 27, 1906, the naturalization certificate must accompany the petition. Do not make or submit a photostat of such certificate.

   b. To prove family relationship between petitioner and beneficiary.

   (1) If petition is submitted on behalf of a wife or husband, it must be accompanied by a certificate of marriage to the beneficiary and proof of legal termination of all previous marriages of both wife and husband.

   (2) If a petition is submitted by a mother on behalf of a child (regardless of age), the birth certificate of the child, showing the name of the mother, must accompany the petition. If the petition is submitted by a father or stepparent on behalf of a child (regardless of age), certificate of marriage of the parents, proof of termination of their prior marriages, and birth certificate of the child showing the names of the parents thereon, must accompany the petition.

   (3) If petition is submitted on behalf of a brother or sister, your own birth certificate and the birth certificate of the beneficiary, showing a common mother, must accompany the petition. If the petition is on behalf of a brother or sister having a common father and different mothers, marriage certificate of your parents, and proof of termination of their prior marriages must accompany the petition.

   (4) If petition is submitted on behalf of a mother, your own birth certificate, showing the name of your mother, must accompany the petition. If petition is submitted on behalf of a father or stepparent, your own birth certificate, showing the names of the parents thereon, and marriage certificate of your parents must accompany the petition, as well as proof of termination of prior marriages of your parents.

   (5) If either the petitioner or the beneficiary is a married woman, marriage certifcate(s) must accompa y the petition. However, when the relationship between the petitioner and beneficiary is that of a mother and child (regardless of age), the mother's marriage certificate need not be submitted if the mother's present married name appears on the birth certificate of the child.

   (6) If the petitioner and the beneficiary are related to each other by adoption, a certified copy of the adoption decree must accompany the petition.

   c. Secondary evidence.

   If it is not possible to obtain any one of the required documents or records shown above, the following may be submitted for consideration:

   (1) Baptismal certificate.—A certificate under the seal of the church where the baptism occurred within two months

Form I-130 (Rev. 5-5-83) N

Figure 14–20 Government Instructions before Revision

after birth, showing date and place of the child's birth, date of baptism, and the names of the child's parents.

(2) School record.—A letter from the school authorities having jurisdiction over school attended (preferably the first school), showing the date of admission to the school, child's date of birth or age at that time, place of birth, and the names and places of birth of parents, if shown in the school records.

(3) Census Record.—State or federal census record showing the name(s) and place(s) of birth, and date(s) of birth or age(s) of the person(s) listed.

(4) Affidavits.—Written statements sworn to or affirmed by two persons who were living at the time, and who have personal knowledge, of the event you are trying to prove—for example, the date and place of birth, marriage, or death. The persons making the affidavits may be relatives and need not be citizens of the United States. Each affidavit should contain the following information regarding the person making the affidavit: his/her full name and address; date and place of birth; relationship to you, if any; full information concerning the event; and complete details concerning how he/she acquired knowledge of the event.

d. Documents and secondary evidence unavailable.

If you are unable to submit required evidence of birth, death, marriage, divorce or adoption because the event took place in a foreign country which does not record such events, and secondary evidence is unavailable, attach a statement to this effect, setting forth the date and place of each of your entries into the United States. Also attach any letters, photographs, remittances or similar documents which tend to support the claimed relationship and three passport type photographs of yourself.

e. Documents previously submitted.

If your birth abroad, or the birth abroad of any person through whom citizenship is claimed by you, was registered with an American consul, submit with this petition any registration form that was issued. If any required documents were submitted to and retained by the American consul in connection with such registration, or in connection with the issuance of a United States passport or in any other official matter, and you wish to use such documents in support of this petition instead of submitting duplicate copies, merely list such documents in an attachment to this petition and show the location of the consulate. If you wish to make similar use of required documents contained in any Immigration and Naturalization Service file, list them in an attachment to this petition and identify the file by name and number. Otherwise, the documents required in support of this petition must be submitted.

f. Documents in general.

All supporting documents must be submitted in the original. If you desire to have the original returned to you, and if copies are by law permitted to be made, you may submit photostatic or typewritten copies. Photostatic copies unaccompanied by the original may be accepted if the copy bears a certification by an immigration or consular officer that the copy was compared with the original and found to be identical. Any document in a foreign language must be accompanied by a translation in English. The translator must certify that he is competent to translate and that the translation is accurate. (Do not make a copy of a certificate of naturalization or citizenship.)

4. **Preparation of petition.** A separate petition for each beneficiary must be typewritten or printed legibly with pen and ink.

   (If you need more space to answer fully any questions on this form, use a separate sheet(s), identify each answer with the number of the corresponding question, and date and sign each sheet.) **Be sure this petition and attached Form I-130A are legible.**

5. **Submission of petition.** If you are residing in the United States, send the completed petition to the office of the Immigration and Naturalization Service having jurisdiction over your place of residence. If you are residing outside the United States consult the nearest American consulate as to the consular office or foreign officer of the Service designated to act on your petition. If you are a United States citizen petitioning for an immediate relative classification in behalf of your unmarried child, the petition must be submitted in sufficient time for action to be completed on the petition and for the child to obtain a visa and reach the United States before the date on which he/she will be 21 years of age.

6. **Approval of petition.** Upon approval of a petition filed by a United States citizen for his/her alien spouse, unmarried minor child, or parent, an immigrant visa may be issued to the alien without regard to the annual limitation on immigrant visa issuance. In the cases of all other aliens for whom immigrant visa petitions are approved, an immigrant visa number will be required. Availability of an immigrant visa number depends on the volume of demand by aliens in the same visa classification who have an earlier priority date on the visa waiting list.

7. **Fee.** A fee of thirty-five dollars ($35) must be paid for filing this petition. It cannot be refunded regardless of the action taken on the petition. DO NOT MAIL CASH. ALL FEES MUST BE SUBMITTED IN THE EXACT AMOUNT. Payment by check or money order must be drawn on a bank or other institution located in the United States and be payable in United States currency. If petitioner resides in Guam, check or money order must be payable to the "Treasurer, Guam." If petitioner resides in the Virgin Islands, check or money order must be payable to the "Commissioner of Finance of the Virgin Islands." All other petitioners must make the check or money order payable to the "Immigration and Naturalization Service." When check is drawn on an account of a person other than the petitioner, the name of the petitioner must be entered on the face of the check. If petition is submitted from outside the United States, remittance may be made by bank international money order or foreign draft drawn on a financial institution in the United States and payable to the "Immigration and Naturalization Service" in United States currency. Personal checks are accepted subject to collectibility. An uncollectible check will render the petition and any document issued pursuant thereto invalid. A charge of $5.00 will be imposed if a check in payment of a fee is not honored by the bank on which it is drawn.

8. **Penalties.** Severe penalties are provided by law for knowingly and willfully falsifying or concealing a material fact or using any false document in the submission of this petition.

9. **Authority.** The authority for collecting the information requested on this form is contained in 8 U.S.C. 1154(a). Submission of the information solicited is voluntary. The principal purpose for which the information is solicited is to determine the eligibility of the beneficiary for the benefits sought. The information solicited may also, as a matter of routine use, be disclosed to other federal, state, local, and foreign law enforcement and regulatory agencies, the Department of Defense including any component thereof (if either the beneficiary or petitioner has served, or is serving in the Armed Forces of the United States), the Department of State, Central Intelligence Agency, Interpol, and individuals and organizations, during the course of investigation to elicit further information required by this Service to carry out its functions. Failure to provide any or all of the solicited information may result in the denial of the petition.

Figure 14–20 *(continued)*

U.S. Department of Justice
Immigration and Naturalization Service (INS)

Petition for Alien Relative

Instructions

Read the instructions carefully. If you do not follow the instructions, we may have to return your petition, which may delay final action.

1. Who can file?

A citizen or lawful permanent resident of the United States can file this form to establish the relationship of certain alien relatives who may wish to immigrate to the United States. You must file a separate form for each eligible relative.

2. For whom can you file?

A. If you are a citizen, you may file this form for:

1) your husband, wife, or unmarried child under 21 years old
2) your unmarried child over 21, or married child of any age
3) your brother or sister if you are at least 21 years old
4) your parent if you are at least 21 years old.

B. If you are a lawful permanent resident you may file this form for:

1) your husband or wife
2) your unmarried child

NOTE: If your relative qualifies under instruction A(2) or A(3) above, separate petitions are not required for his or her husband or wife or unmarried children under 21 years old. If your relative qualifies under instruction B(2) above, separate petitions are not required for his or her unmarried children under 21 years old. These persons will be able to apply for the same type of immigrant visa as your relative.

3. For whom can you *not* file?

You cannot file for people in these four categories:

A. An adoptive parent or adopted child, if the adoption took place after the child became 16 years old, or if the child has not been in the legal custody of the parent(s) for at least two years after the date of the adoption, or has not lived with the parent(s) for at least two years, either before or after the adoption.
B. A stepparent or stepchild, if the marriage that created this relationship took place after the child became 18 years old.
C. A husband or wife, if you were not both physically present at the marriage ceremony, and the marriage was not consummated.
D. A grandparent, grandchild, nephew, niece, uncle, aunt, cousin, or in-law.

4. What documents do you need?

You must give INS certain documents with this form to show you are eligible to file. You must also give INS certain documents to prove the family relationship between you and your relative.

A. For each document needed, give INS the original and one copy. However, because it is against the law to copy a Certificate of Naturalization, a Certificate of Citizenship or an Alien Registration Receipt Card (Form I-151 or I-551), give INS the original only. **Originals will be returned to you.**

B. If you do not wish to give INS the original document, you may give INS a copy. The copy must be certified by:

1) an INS or U.S. consular officer, or
2) an attorney admitted to practice law in the United States, or
3) an INS accredited representative

(INS still may require originals).

C. Documents in a foreign language must be accompanied by a complete English translation. The translator must certify that the translation is accurate and that he or she is competent to translate.

5. What documents do you need to show you are a United States citizen?

A. If you were born in the United States, give INS your birth certificate.
B. If you were naturalized, give INS your original Certificate of Naturalization.
C. If you were born outside the United States, and you are a U.S. citizen through your parents, give INS:
   1) your original Certificate of Citizenship, or
   2) your Form FS-240 (Report of Birth Abroad of a United States Citizen).
D. In place of any of the above, you may give INS your valid unexpired U.S. passport that was initially issued for at least 5 years.
E. If you do not have any of the above and were born in the United States, see the instructions under 8, below, *"What if a document is not available?"*

6. What documents do you need to show you are a permanent resident?

You must give INS your alien registration receipt card (Form I-151 or I-551). Do not give INS a photocopy of the card.

7. What documents do you need to prove family relationship?

You have to prove that there is a family relationship between your relative and yourself.

In any case where a marriage certificate is required, if either the husband or wife was married before, you must give INS documents to show that all previous marriages were legally ended. In cases where the names shown on the supporting documents have changed, give INS legal documents to show how the name change occurred (for example, a marriage certificate, adoption decree, court order, etc.).

Find the paragraph in the following list that applies to the relative you are filing for.

If you are filing for your:

A. **husband or wife,** give INS:

1) your marriage certificate
2) a color photo of you and one of your husband or wife, taken within 30 days of the date of this petition.

Form I-130 (08-01-85) N

Figure 14–21 Government Instructions after Revision

These photos must have a white background. They must be glossy, un-retouched, and not mounted. The dimension of the facial image should be about 1 inch from chin to top of hair in 3/4 frontal view, showing the right side of the face with the right ear visible. Using pencil or felt pen, lightly print name (and Alien Registration Number, if known) on the back of each photograph.

3) a completed and signed Form G-325A (Biographic Information) for you and one for your husband or wife. Except for name and signature, you do not have to repeat on the G-325A the information given on your I-130 petition.

B. **child** and you are the **mother,** give the child's birth certificate showing your name and the name of your child.
C. **child** and you are the **father or stepparent,** give the child's birth certificate showing both parents' names and your marriage certificate.
D. **brother or sister,** give your birth certificate and the birth certificate of your brother or sister showing both parents' names. If you do not have the same mother, you must also give the marriage certificates of your father to both mothers.
E. **mother,** give your birth certificate showing your name and the name of your mother.
F. **father,** give your birth certificate showing the names of both parents and your parents' marriage certificate.
G. **stepparent,** give your birth certificate showing the names of both natural parents and the marriage certificate of your parent to your stepparent.
H. **adoptive parent or adopted child,** give a certified copy of the adoption decree and a statement showing the dates and places you have lived together.

8. What if a document is not available?

If the documents needed above are not available, you can give INS the following instead. (INS may require a statement from the appropriate civil authority certifying that the needed document is not available.)

A. Church record: A certificate under the seal of the church where the baptism, dedication, or comparable rite occurred within two months after birth, showing the date and place of child's birth, date of the religious ceremony, and the names of the child's parents.
B. School record: A letter from the authorities of the school attended (preferably the first school), showing the date of admission to the school, child's date and place of birth, and the names and places of birth of parents, if shown in the school records.
C. Census record: State or federal census record showing the name, place of birth, and date of birth or the age of the person listed.
D. Affidavits: Written statements sworn to or affirmed by two persons who were living at the time and who have personal knowledge of the event you are trying to prove; for example, the date and place of birth, marriage, or death. The persons making the affidavits need not be citizens of the United States. Each affidavit should contain the following information regarding the person making the affidavit: his or her full name, address, date and place of birth, and his or her relationship to you, if any; full information concerning the event; and complete details concerning how the person acquired knowledge of the event.

9. How should you prepare this form?

A. Type or print legibly in ink.
B. If you need extra space to complete any item, attach a continuation sheet, indicate the item number, and date and sign each sheet.
C. Answer all questions fully and accurately. If any item does not apply, please write "N/A"

10. Where should you file this form?

A. If you live in the United States, send or take the form to the INS office that has jurisdiction over where you live.
B. If you live outside the United States, contact the nearest American Consulate to find out where to send or take the completed form.

11. What is the fee?

You must pay $35.00 to file this form. **The fee will not be refunded, whether the petition is approved or not.** DO NOT MAIL CASH. All checks or money orders, whether U.S. or foreign, must be payable in U.S. currency at a financial institution in the United States. When a check is drawn on the account of a person other than yourself, write your name on the face of the check. If the check is not honored, INS will charge you $5.00.

Pay by check or money order in the exact amount. Make the check or money order payable to "Immigration and Naturalization Service". However,

A. if you live in Guam: Make the check or money order payable to "Treasurer, Guam", or
B. if you live in the U.S. Virgin Islands: Make the check or money order payable to "Commissioner of Finance of the Virgin Islands".

12. When will a visa become available?

When a petition is approved for the husband, wife, parent, or unmarried minor child of a United States citizen, these relatives do not have to wait for a visa number, as they are not subject to the immigrant visa limit. However, for a child to qualify for this category, all processing must be completed and the child must enter the United States before his or her 21st birthday.

For all other alien relatives there are only a limited number of immigrant visas each year. The visas are given out in the order in which INS receives properly filed petitions. To be considered properly filed, a petition must be completed accurately and signed, the required documents must be attached, and the fee must be paid.

For a monthly update on dates for which immigrant visas are available, you may call (202) 632-2919.

13. What are the penalties for submitting false information?

Title 18, United States Code, Section 1001 states that whoever willfully and knowingly falsifies a material fact, makes a false statement, or makes use of a false document will be fined up to $10,000 or imprisoned up to five years, or both.

14. What is our authority for collecting this information?

We request the information on this form to carry out the immigration laws contained in Title 8, United States Code, Section 1154(a). We need this information to determine whether a person is eligible for immigration benefits. The information you provide may also be disclosed to other federal, state, local, and foreign law enforcement and regulatory agencies during the course of the investigation required by this Service. You do not have to give this information. However, if you refuse to give some or all of it, your petition may be denied.

It is not possible to cover all the conditions for eligibility or to give instructions for every situation. If you have carefully read all the instructions and still have questions, please contact your nearest INS office.

Figure 14–21 *(continued)*

Table of Contents

Figure 14–22 Table of Contents
Source: U.S. Department of Health and Human Services, *Eating to Lower Your Blood Cholesterol* (Washington, DC, 1989) i.

or too little? Submit your how-to instructions to the acid test. Let members of the audience for whom the instructions are intended—but who are not familiar with the process—attempt to perform the process following your instructions. Encourage them to tell you where your instructions are confusing. A procedure called **protocol analysis** can be a help at this point. In this procedure, you ask the person following your instructions to speak into a tape recorder, giving his or her observations about the instructions while attempting to follow them. Here is an excerpt from a set of such observations made by someone trying to use a computer manual and on-line help to aid him in a word processing exercise:

> Somehow I've got the caps locked in here. I can't get to the lower case. OK, I'm struggling with trying to come off those capitals. I'm not having any luck. So, what do I need to do. I could press help. See if that gets me anything. Using the keyboard. I'll try that. 2.0. I can't do that because it's in this mode. I'm getting upper case on the numbers so I can't type in the help numbers. So I'll reset to get rid of that. Big problem. Try reset. Merging text, formatting, setting margins, fixing problems. I can't enter a section number because I can't get this thing off lock. Escape. Nothing helps. Well, I'm having trouble here.[9]

Such information pinpoints troublesome areas in instructions. If you were writing instructions that were to be used by many people, protocol analysis would be a worthwhile investment in time and money. In any case, whether you use protocol analysis or not, if your readers can't follow your instructions, don't blame them. Rather, examine your instructions to see where you have failed. Often, you will find you have left out some vital link in the process or assumed knowledge on the part of your readers that they do not possess.

Planning and Revision Checklist

You will find the planning and revision checklists that follow Chapter 2, "Composing" (pages 33–35 and inside the front cover), and Chapter 4, "Writing for Your Readers" (pages 78–79), valuable in planning and revising any presentation of technical information. The following questions specifically apply to instructions. They summarize the key points in this chapter and provide a checklist for planning and revising.

Planning

- What is the purpose of your instructions?
- What is your reader's point of view?
- How and where will your readers use these instructions?
- What content does your reader really need and want?
- How should you arrange your content? Which of the following components should you include as a separate section? Which should you omit or include in another component, for example, theory in the introduction?
  - Introduction
  - Theory or principles of operation: How much theory do your readers really need or want?
  - List of equipment and materials needed: Are your readers famliar with all the needed equipment and material? Do they need additional information?
  - Description of the mechanism: Does some mechanism play a significant role in these instructions?

Revision

- Have you made the purpose of your paper clear to your readers?
- Can your readers scan your instructions easily, finding what they need?
- Do you have sufficient headings?
  - Do your headings stand out?
  - Are they meaningful to your readers?
  - Would it help to cast some as questions?
- Is all terminology unfamiliar to the reader defined somewhere?
- Is your print size large enough for your readers and their location?
- Is all your content relevant? Is it needed or desired by your readers? Have you made it easy for your readers to scan and to skip parts not relevant to them?
- Have you covered any needed theory adequately?
- Do your readers know what equipment and material they will need?
  - Do they know how to use the equipment needed?
  - If not, have you provided necessary explanations?
- Have you provided any necessary mechanism description?

Planning

Warnings: Are there expected outcomes that will be affected by improper procedure? Are there places in the instructions where improper procedure will cause damage to equipment or injury or death to people?

How-to instructions: Can your instructions be divided into routines and subroutines? What is the proper sequence of events for your how-to instructions?

Tips and troubleshooting procedures: Are there helpful hints you can pass on to the reader? What troubles may come up? How can they be corrected?

Glossary: Do you have to define enough terms to justify a glossary?

- What graphics will help your instructions? Do you have them available or can you produce them?

Revision

- **Are your caution, warning, and danger messages easy to see and clear in their meaning? Are you sure you have alerted your readers to every situation where they might injure themselves or damage their equipment?**
- Have you broken your how-to instructions into as many routines and subroutines as necessary?
- Are your steps in chronological order, with no steps out of sequence?
- Are your how-to instructions written in the active voice, imperative mood?
- Have you used a list format with short entries for each step of the instructions?
- Have you used simple, direct language and avoided jargon?
- If needed, have you provided sample calculations?
- Have you used graphics whenever they would be helpful? Are they sufficiently annotated?
- Have you provided helpful tips that may help your readers to do the task more efficiently?
- Have you anticipated trouble and provided troubleshooting procedures?
- If your troubleshooting is covered in a separate section, is the section laid out in a way to clearly distinguish among problem, cause, and remedy?
- Do you have enough definitions to warrant a glossary?
- Are your instructions long enough to warrant a table of contents?
- Have you checked with your readers?
  - Have you allowed a typical reader to attempt the procedure using your instructions?
  - Have you corrected any difficulties such a check revealed?
- Have you checked thoroughly for all misspellings and mechanical errors?

Exercises

1. Writing instructions offers a wide range of possible papers. Short papers might consist of nothing more than an introduction and a set of how-to instructions. Examples—good and bad—of such short instructions can be found in hobby kits and accompanying such things as toys, tents, and furniture that must be put together. Textbook laboratory procedures frequently are examples of a short set of instructions. Using the Planning and Revision Checklist for this chapter, write a short set of instructions. Here are some suggested ideas:

    Developing film
    Drawing a blood sample
    Applying fertilizer
    Setting a bicycle gear
    Completing a form
    Accomplishing some do-it-yourself task around a house
    Replacing a part in an automobile or some other mechanism
    Cleaning a carpet
    Balancing a checkbook

2. Using the Planning and Revision Checklist for this chapter, write a set of instructions that includes at least six of the eight possible components listed on page 484. The components do not have to be in separate sections, but they must be clearly recognizable for what they are. Here are some suggested topics:

    Testing electronic equipment
    Writing (or following) a computer program
    Setting up an accounting procedure for a small business
    Conducting an agronomy field test
    Checking blood pressure
    Painting an automobile

3. Figure 14–23 is a set of instructions for ridding a house of termites. It is usable in its current form, but it could be greatly improved. In a collaborative group examine and discuss the instructions. Using the Revision Checklist for this chapter, decide on ways to improve them. When the discussion is finished, each member of the group should individually prepare a revision. If you are working with a typewriter, use headings such as those shown on page 255. If you have access to word processing, use some typographical variation to make the instructions more accessible.

Termites/Search and Destroy

Your Problem
- Wood materials are threatened by termites.
- Wood materials have deteriorated.
- The house needs to be checked for the presence of termites.
- Preventive action against termites is required.

What You Need
- Flashlight
- Penknife or icepick
- Pick and shovel
- Chemical solution (consult an exterminator to determine the type of chemicals or treatment to use).

How-To: Checking for Termites
Wood decay and damage by insects are threats to the upkeep of the home.

The insects most destructive to wood in buildings are termites. There are two varieties: The "drywood" termite and the "subterranean" or "ground-nesting termite." Both thrive on wood for food.

"Drywood" termites can live without moisture, so that protection against them is very difficult. However, there are not many "drywood" termites in this country.

"Subterranean" or *"ground-nesting" termites* are a serious problem in the southern States. Subterranean termites live in colonies in the ground and require moisture to survive. The worker termites attack damp wood which is in contact with the ground. They may build earthen tunnels from the ground up to the wood. They will sometimes completely eat away the inside of a piece of wood while leaving the outside surface intact.

1. Check for termites at least twice a year.

2. During the spring and summer (termite mating season), call an exterminator to identify large numbers of flying insects that you cannot identify.

3. Look for earthen tunnels in the following locations:
- Along masonry foundation and basement walls
- Around openings where pipes enter walls
- Along the surface of metal pipes (fig. 1).

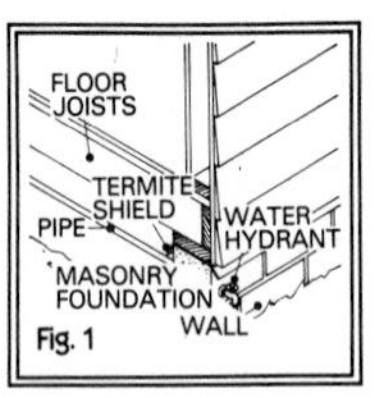

Fig. 1

4. Examine all cracks in slabs, and loose mortar in masonry walls. Check all joints where wood meets with concrete or masonry, at walls, slabs, piers, etc.

5. Inspect all wood and wood structures that are near the ground. Pay special attention to any that touch the house, such as fences, wood trellises, carports, etc. Examine crawl spaces that provide moist conditions.

6. Check windowsills, door thresholds, porches and the underside of stairs. Be on the lookout for peeling and blistering paint.

Figure 14–23 Termite Protection Instructions
Source: U.S. Department of Agriculture, *Simple Home Repairs: Outside* (Washington, DC: GPO, 1986) 38-40.

7. If you suspect that wood has termite damage, probe with a sharp point, such as an icepick or penknife (fig. 2). If the point penetrates the wood to a depth of ½ inch, when you use only hand pressure, it's a good indication of wood damage by termites.

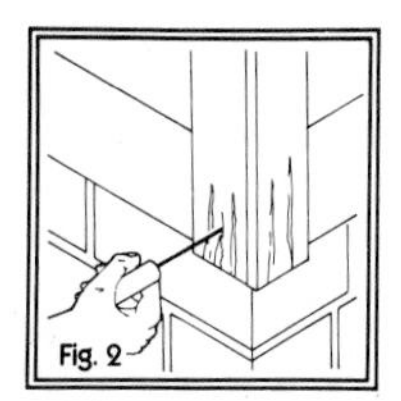
Fig. 2

How-To: Protecting Against Termites

Chemicals needed to control termites are toxic to animals and plant life. There is also danger of contaminating the water supply. The chemicals should be applied with extreme caution and preferably by an experienced person.

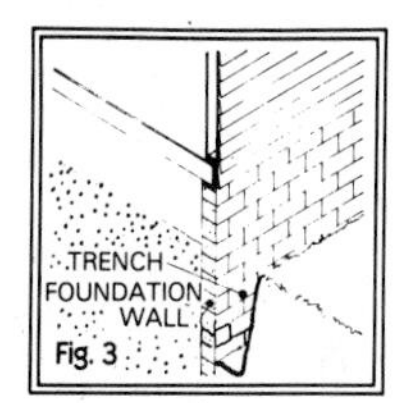

Fig. 3

1. The following procedure should be followed when chemical treatments are necessary for an existing building:

(a) Dig a trench, approximately 1 foot wide and 3 feet deep, adjacent to the foundation wall (fig. 3).

(b) Prepare a solution of the insecticide. (Consult your County Extension Office or local exterminator regarding the recommended type and mixing instructions.)

(c) Pour the insecticide against the exposed wall surface and into the trench as it is backfilled. The solution should be applied to all other locations where wood and masonry meet at a joint. It should also be applied to other areas that have earth floors.

(d) Use extreme caution with these chemicals since they will also be poisonous to humans and pets. If a chemical is used inside the house, the room or space must be well ventilated and vacated for a period of time.

2. All surface water should be directed away from the building, allowing no water to accumulate at the foundations.

3. Cover the earth of unpaved basements with plastic film 4 mil or heavier.

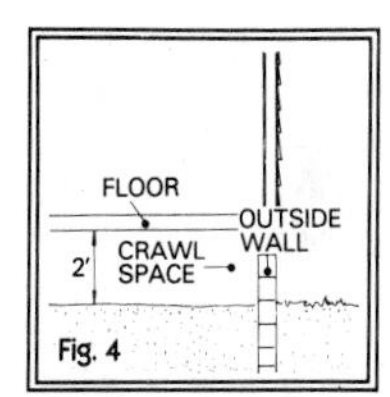

Fig. 4

4. Keep crawl spaces well ventilated. A house of 1,000 square feet should have at least 8 vents, 16 inches x 8 inches, open at all times. Crawl spaces should be at least 2 feet in height (fig. 4). Keep the space clear of wood scraps.

5. Untreated wood should not come closer than 6 inches to the ground.

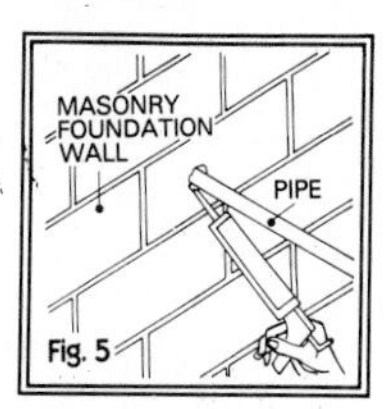

Fig. 5

6. Using caulking compound, seal all openings where pipes pass through foundation walls or other walls of the house (fig. 5). Also, seal any cracks or points of loose mortar in masonry walls.

7. If there is a termite shield around the foundation it should be straightened and turned down (at least 2 inches) at approximately a 45-degree angle (fig. 6).

8. Make sure all scraps of lumber or stumps are removed when a building project is complete.

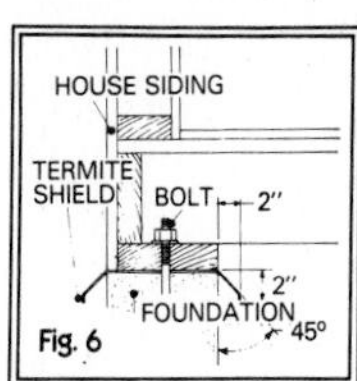

Fig. 6

Your Benefits

- Controlling and preventing wood deterioration in your home.
- Preventing costly repairs later.
- Assuring an attractive appearance of the wood in your home.

Figure 14–23 *(continued)*

CHAPTER 15

Proposals

PROPOSALS DEFINED

BASIC PLAN OF PROPOSALS

PROJECT SUMMARY

PROJECT DESCRIPTION

Introduction
Rationale and Significance
Plan of Work
Facilities and Equipment

PERSONNEL QUALIFICATIONS

BUDGET

APPENDIXES

URGE TO ACTION

PROPOSAL FOR A STUDENT REPORT

In this chapter, we define proposals, give a general plan for a proposal and describe how to prepare each of its parts, and discuss proposals for student reports.

Proposals Defined

All projects have to begin somewhere and with someone. In the professional world, the starting point is often a proposal. In simplest terms, a proposal is an offer to provide a service or a product to someone in exchange for money. For example, Professor X of the university's Sociology Department proposes to the U.S. Department of Health and Human Services (HHS) that she research the characteristics of children in single-parent families. In her proposal she describes her planned research and explains its benefits. She states her qualifications to conduct the research and details the costs of the project. HHS will review Professor X's proposal and, based upon HHS's needs and the credibility of the proposal, will grant or deny the funding requested.

In another example, Company Y announces that it intends to buy 50 new delivery trucks. It describes, that is, specifies, what the trucks must be like in size, carrying capacity, power, and so forth. Truck dealers read the announcement. If they have trucks that meet Company Y's specifications, they prepare a proposal offering the 50 trucks at a certain price. As part of their proposal, they show how their trucks meet or exceed the specifications. Company Y will buy the trucks that best meet their specifications at the best price.

In business, industry, and government, proposals are an everyday necessity. As in the first example, a person, company, or agency that desires additional work may offer to render service to another. Or, as in the second example, the original suggestion may come from the person, company, or agency that would like someone else to do a job that needs doing or to furnish some product it needs. In either case, a proposal is required.

The flowchart in Figure 15–1 depicts the typical routes of a proposal with its inception through award of contract.

Block 1 indicates the "unsolicited situation," in which one offers, unasked, to perform work for another. Block 1a indicates the opposite or "solicited situation," in which someone requests services or products from someone else. Scanning across the flowchart, you see the various stages that ensue until either a work contract is awarded or the proposed project dies out before any productive work is done.

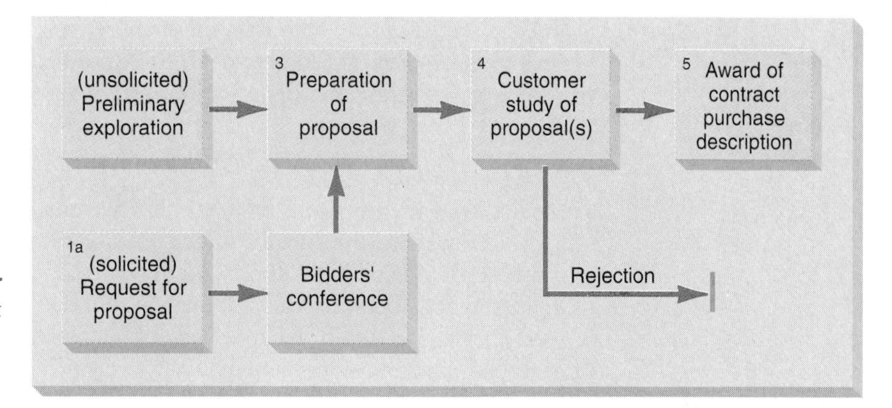

Figure 15–1 Flowchart of Activities from Proposal to Award of Contract

Companies, institutions, and agencies solicit work to be done by outsiders. They may spread the news informally by word of mouth. Officially and formally, they may issue a Request for Proposal (RFP) and mail it to likely prospects. Many U.S. Government RFPs are printed in the *Federal Register,* where they are eagerly scanned by university and private research laboratories. Corporations may place their RFPs in the *Commerce Business Daily*. The RFP may resemble a want ad or it may be a bound document of a hundred pages or more.

Organizations who are interested in the project outlined in the RFP respond with a proposal in the hope they will win out against the competition. The interim is a time of anxious waiting. During times of normal prosperity, the qualified competitor stands, perhaps, one chance in ten of winning out in the competition. Happy is the contractor who is awarded the contract. Happier still is the person who can live with the contract and gain a profit from it.

Refer again to Figure 15–1, specifically to block 1, at which point the course of the **unsolicited proposal** begins.

The term *unsolicited,* as applied to proposals, may be misleading. There is a great deal of exchange of information within the business community. The word gets around that X Company might be interested in having someone else conduct a research project for it. Those who are privileged to hear the word take it back to their own Y Company. Heads get together, inquiries are discreetly made, and a "feeler" letter may go to X Company. As a result, a proposal may be prepared on the basis of slim information. Again, chances are low that this kind of unsolicited proposal will be accepted, but the chance is worth taking if Y Company has personnel to spare.

The customary procedure for unsolicited proposals is as follows:

1. An individual or group desiring to do a task for pay prepares an unsolicited proposal and sends it to an appropriate sponsor—a company, a private foundation, or a government agency.

2. The sponsor, who will pay for the task if approved, considers the merits of the proposal, the qualifications of the people involved, and the benefits that might result if the task has a productive outcome.
3. If the benefits seem worth the money, the proposal is funded.

When and if you reach the stage that an unsolicited proposal seems to be in order, go to the library to dig up all the supporting data you can find. Identify one or more sponsors who might be interested. Consult with knowledgeable persons in your organization. Write a letter to the proposed sponsor to obtain its specifications for proposals. Then settle down for weeks of work to create an attractive and persuasive proposal. The chances may be worth your investment of time and energy.

Because the "unsolicited proposal" brings all proposal functions into play, we limit the rest of our discussion to it. Recognize, however, that where our advice on proposal preparation necessarily has to be somewhat general, the directions in an RFP may be quite specific. Figure 15–2 reproduces the guidelines laid down in an RFP from the National Agricultural Pesticide Impact Assessment Program (NAPIAP). When you have clear guidance, such as that in Figure 15–2, follow it to the letter. For

Figure 15–2 Guidelines for Preparing Proposals for the National Agricultural Pesticide Impact Assessment Program

Guidelines for Preparing NAPIAP Proposals

The proposal should be concise and show a clear relationship to *Selection Criteria for NAPIAP Related Research.* The following outline is recommended as guidance in the preparation of proposals, which should be kept to a maximum of three pages. Send five copies to the Chief, USDA Forest Service, Attention: Director of FPM, P.O. Box 96090, Washington, DC 20090-6090.

1. Title Page
   a. Study title
   b. Principal investigator(s)
   c. Name(s) of performing organization(s) and address(es)
   d. Forest Service contact or liaison and telephone number
   e. Date of submission
   f. Approval signatures of appropriate officials
2. Objectives (include risk and/or benefit data sought)
3. Justification Statement
4. Expected Accomplishments
5. Research Approach (method, statistical design, etc.)
6. Research Timetable
7. Personnel Support (Show all personnel involved in the research.)
8. Cost Estimates (Show estimated annual costs by source of funds. Include indirect costs and contributed funds where appropriate.)

example, in presenting a proposal to the NAPIAP, you would use the headings from their outline exactly as shown. Failure to follow RFP guidelines usually will result in a rejected proposal.

Whether the proposal is solicited or unsolicited, we can safely say that a proposal has one major objective: To get you (or your company) accepted to provide a service or a product. Essentially, a **proposal** is an argument in which every item of content, every point of diction, and every matter of display and format must be designed to accomplish this objective. (Because proposals are arguments, you may wish to review Chapter 9, "Arguing.")

Basic Plan of Proposals

Short proposals of several pages will probably be presented as a letter or memo (see Chapter 13, "Correspondence"). Longer proposals may be presented in a formal report format and be accompanied by a letter of transmittal (see pages 276–277). Remember that a proposal reaches your prospective client's desk as a "package." Once the wrappings are removed, the contents will trigger a reaction. The initial reaction will be visual and tactile. In larger reports, a sturdy binding, an attractive cover, clean printing, and standard format will place the reader in a receptive mood.

With the use of desktop publishing word processing equipment and software (see Chapter 10, "Document Design"), almost anyone can turn out attractive proposals. However, don't allow the possibilities in desktop publishing to tempt you into an overelaborate system requiring a great variety of typefaces. Avoid also conspicuous overdressing, such as handmade paper and deckled edges. But there is a happy medium between too much and too little. Neatness, good layout, commercial-quality printing, and attention to detail may tip the scale of judgment in your favor. Whether large or small, the proposal should be organized for selective reading through the use of headings (see pages 248–257).

In one common format, the parts of a proposal consist of the following:

Project Summary
Project Description
    Introduction
    Rationale and Significance
    Plan of Work
    Facilities and Equipment
Personnel
Budget
Appendixes

This basic plan is illustrated in Figure 15–3, a proposal written by a student in a technical writing class. The plan can accommodate modest

(Text continued on page 483.)

4062 Hoven Street • St. Paul, MN 55108
(612) 624-7750

20 September, 1990

Ms. Nancy Johnson, Chair
Wyoming Lake Citizens Committee
Marinette County Courthouse
Wyoming Lake, WI 54126

Proposal for a Feasibility Study of Rehabilitating Wyoming Lake

Dear Ms. Johnson:

Thank you for approving my submission of this proposal for a feasibility study of rehabilitating Wyoming Lake. My proposal begins with a project summary followed by a project description, a statement of my qualifications, and a budget.

Project Summary

Objectives of proposed work

Plan of work

Relevance to Committee

Early research shows that Wyoming Lake is clogged by over 2-million cubic yards of sediment that effectively destroy the lake's potential as a recreational resource. This letter proposes that the Wyoming Lake Citizens Committee commission a study to ascertain the feasibility of dredging the lake and restoring its recreational use. The study would analyze the sediment deposits, investigate needed dredging equipment and methods, and ascertain feasibility of a dredging project by using the criteria of time, cost, and community benefits and problems. Using the report, the Committee would have a basis upon which to proceed in its decisions concerning possible implementation of a dredging project.

Project Description

Subject and purpose of proposal

In 1988, the Wyoming Lake Citizens Committee concerned by the increasing deterioration of Wyoming Lake, commissioned a study of the sediment deposits in the lake. Since the completion of that study, no further steps have been taken in the lake's rehabilitation. Each year the sediment deposits grow thicker, further impeding the use of the lake and increasing the ultimate cost of their removal. I propose that your committee commission me to carry out a study to evaluate the feasibility of removing the sediment from Wyoming Lake. I will investigate and

Figure 15–3 Student Proposal

Page 2
Ms. Nancy Johnson
20 September 1990

report on the equipment and methods needed for removal of the deposits. As a major outcome of the study, I will present an evaluation of the feasibility of a dredging project based upon the criteria of cost, time-to-completion, and community benefits and problems. If the dredging project proves feasible and is completed, it will restore Wyoming Lake as an important recreational resource to the citizens of the area.

Implications of proposed work

Definition of problem

Rationale and significance. Charles J. Anderson's report, "Sediment Deposits in Wyoming Lake," commissioned by the Wyoming Lake Citizens Committee in 1988, shows the extent of the lake's problems. Mr. Anderson reports that there are over 2-million cubic yards of sediment. As a result, in over 5 miles of shoreline there are few areas that can be used for swimming or wading. Swimming is no longer allowed in the public park at the south end of the lake.

Background of problem

Need for a solution

Description of solution and its benefits

Wyoming Lake is typical of many lakes left in Wisconsin and Minnesota when the great glaciers of the ice age retreated some 11,000 years ago. Like many such lakes it is subject to sedimentation as silt washes into it from the surrounding area. Left alone, in time it will become a bog and ultimately a completely filled-in area. The longer the problem remains, the most costly a solution becomes. At some point, the cost will become prohibitive, and the lake will be lost as a resource for water sports. However, with timely intervention, the sediment can be removed and the lake kept clear. Rehabilitation projects restoring sediment clogged lakes have been carried out successfully by several nearby cities.

Feasibility of solution

The experience of the neighboring cities indicates that a rehabilitation project is technologically feasible. Modern dredging equipment is available that will remove the sediment deposits of the size and type found in Wyoming Lake. More pertinent perhaps are the social and political questions of how long will the project take, how much will it cost, what problems will it create for the citizens of Wyoming Lake, and what benefits will the citizens receive in return for the cost of the project. Should the benefits outweigh the cost and problems created, the project would seem to be feasible. Again the experience of neighboring cities indicates the likely feasibility of the project on these social and political terms.

Figure 15–3
(continued)

Continued

Page 3
Ms. Nancy Johnson
20 September 1990

Scope and task breakdown

Plan of work. The proposed study breaks down into these three tasks:

1. Analyzing the location, depth, and content of the sediment deposits.
2. Investigating the equipment and methods needed for removal of the deposits.
3. Applying the criteria of time, cost, and community benefits and problems.

Methods to be used

The Anderson report provides adequate information as to the location and depth of the sediment deposits. I'll refine his research by taking sediment samples from three locations in the lake. The samples will be sent to the soils laboratory of the University of Wisconsin–Madison for analysis. The results of the analysis will suggest possible uses for the removed sediments.

To ascertain equipment and methods, I'll correspond with dredging companies that work in the Midwest. I'll give them estimates of the amount of sediment to be removed and ask them to comment on the methods, equipment, time-to-completion, and cost. I'll also correspond with the Wisconsin Department of Natural Resources (DNR). They should be able to provide information concerning legal restrictions and procedures involved in such projects. I'll also query the DNR concerning possible state and federal assistance.

Time and work schedule

Research on all three tasks outlined will be carried out more or less simultaneously and should take eight weeks. Allowing an additional two weeks for me to organize and write my findings, I should be able to have a final report within ten weeks after you authorize work to begin.

Products of the project

As products of the project, I will furnish to you two monthly progress reports and a final report. The final report will be a feasibility study that reports and analyzes the data gathered in the study. The report will make a recommendation as to the feasibility of the project. Further, if the project proves feasible, the report will recommend methods and equipment for carrying it out. The report will estimate the cost and time-to-completion of the project. It will detail what are likely to be the major problems to be

Figure 15–3
(continued)

Page 4
Ms. Nancy Johnson
20 September 1990

encountered and possible solutions to them. It will discuss the benefits that may result from the project.

The report will not provide a detailed blueprint on how to proceed with the project, but it will provide a basis for the Wyoming Lake Citizens Committee to make such decisions as are needed.

Facilities and equipment

Facilities and equipment. No expensive facilities or equipment are needed to carry out this research. I will provide a car and a boat as needed at nominal cost. Equipment for sampling the lake's sediment can be borrowed at no cost from the Wyoming Lake Maintenance Department.

Personnel Qualifications

Personnel and qualifications

I am a senior student in agricultural engineering at the University of Minnesota, where I carry a 3.3 grade point average based upon a 4.0 scale. I am familiar through my course work with dredging equipment and methods. In the summer of 1988, I assisted Charles Anderson in his study of the Wyoming Lake sediment deposits. In the summer of 1989, as an intern with the Wisconsin DNR, I assisted in several eutrophication studies of Wisconsin lakes. (Eutrophication is the process by which lakes become more favorable to plant life than to animal life, leading to their becoming clogged with plants.) I am familiar, therefore, with the problems of Wisconsin lakes in general and with Wyoming Lake in particular.

The people listed below will provide references for me if requested:

Professor Arnold Flikke
Department of Agricultural Engineering
University of Minnesota
St. Paul, MN 55108

Dr. Charles Murphy
Department of Natural Resources
Madison, WI 53702

Figure 15–3
(continued)

Continued

Budget

Page 5
Ms. Nancy Johnson
20 September 1990

Budget

In this project budget I indicate estimated expenses for several items. Where estimates are indicated, the cost may be less than estimated and will not be more. Costs are as follows:

Use of my boat and motor	$ 15
Use of my car @ 20¢ a mile (estimated)	$ 80
Cost of soil sample analysis	$ 10
Mailing and telephone expense (estimated)	$ 30
Fee	$450
TOTAL	$585

Payment is due within 30 days of the report's acceptance by the Wyoming Lake Citizens Committee.

Sincerely yours,

Barbara J. Buschatz

Barbara J. Buschatz

Figure 15–3
(continued)

proposals that run in length from a page or two to proposals for millions of dollars that may fill a five-foot shelf. The headings and annotations in Figure 15–3 correspond to the description of a proposal that follows in this chapter. By scanning the headings and annotations, you can find examples of every component we describe. By seeing these components in context, you can better understand how all these parts of a proposal work together to achieve the desired result—acceptance of the proposal.

Students, while probably not ready for the large proposals often found in industry and government, can use proposals in two ways. First, they can follow the general proposal format described in this chapter and by simulating a real proposal situation produce the letter or memo proposals frequently used in industrial and governmental settings. The proposal in Figure 15–3 is such a simulation. That is, the student author, by starting with a real situation and projecting her imagination, did all the things she would have to do in working for hire as a researcher. Despite the fictional trappings, the proposal is realistic and professional.

Second, on a still smaller scale the student can submit a proposal for a report to a teacher. We discuss this smaller proposal at the end of this chapter and illustrate it in Figure 15–4. But now we discuss the parts of a typical proposal.

Project Summary

Remember that proposals are read by busy people. Whether they are scientists or business people, when reading a proposal they are fulfilling an executive, decision-making role: Should the proposal be rejected or accepted? Therefore, treat them like executives and put a summary of the key factors of your proposals up front (see pages 291–292). Include in your summary the objectives of your proposed work. Point out why the work is relevant to the reader or the subject matter area. Clearly and succinctly summarize the plan of work. Because you can't write a good summary until you have completed the rest of the proposal, the summary should be the last thing you write.

Project Description

The reviewer of your proposal will judge your proposal largely by the material in your project description. The project description itself breaks down into four parts: introduction, rationale and significance, plan of work, and facilities and equipment.

MEMORANDUM

Date 27 January 1991

To Professor Victoria Mikelonis
Department of Rhetoric
University of Minnesota

From David M. Zellar DMZ

Subject Proposal for a Feasibility Report

Subject and purpose

Professor Milton Weller, Head of the Department of Entomology, Fisheries and Wildlife, wishes to purchase a portable tape recorder for the Department. The recorder will be used on field trips, primarily to record bird vocalizations. I have volunteered to help the Department select an appropriate recorder.

Professor Weller and I have discussed the criteria to be used in comparing possible recorders. We agreed that for ease of operation the tape recorder should be a cassette recorder and not reel-to-reel. Other criteria to be used are the following:

Scope

Frequency response: Bird vocalizations range from 100 to 18000 Hz. Therefore, the recorder should have a frequency response to match.

Size and weight: Because the recorder is to be used on field trips, it should be small and light. Professor Weller and I agreed that 20 pounds was probably a reasonable limit for maximum weight.

Battery operation: The recorder will be used in places where electricity is not available. Because it will be used on extended field trips of a week or more, even rechargeable batteries would not be practical. Therefore, the recorder must operate on dry cell batteries.

Maintenance requirements: Ideally, the recorder should be available for purchase locally, and the availability of local maintenance is a must.

Cost: Professor Weller, because of budgetary restrictions, has set a maximum limit of $500 for cost.

At least two tape recorders, the Marantz Superscope and the AIWA TPR-945 seem to fit the criteria. Undoubtedly, investigation will turn up others.

Figure 15–4 Proposal for a Review

Page 2
Professor Mikelonis
27 January 1991

Time and task breakdown

Task and Time Breakdown

Professor Weller wishes to buy the recorder as soon as possible. Furthermore, to use the project as a feasibility report in your class, I must complete it in the next six weeks. Therefore, I propose the following task and time breakdown:

1. Visit stores where tape recorders are sold. Interview salespeople, gather literature on available recorders, and examine and record with potential selections (2 weeks).
2. Perform my analysis of available recorders, comparing them by our criteria (1 week).
3. Write and turn in to you a progress report that includes preliminary findings and an organizational plan for the final report, with a copy to Professor Weller (1 week).
4. Write a preliminary draft of the report and discuss it with you (1 week).
5. Write the final copy of the report and submit it to you no later than 10 March 1991.

Resources available

Resources Available

For information about bird vocalizations, I'll use Birds of North America by Robbins, Bruun, Zim, and Singer. Professor Weller will continue to serve as a resource concerning criteria.

There are over one-hundred stereophonic and high fidelity equipment dealers in the Minneapolis–St. Paul area. I have dealt with at least ten of them on various occasions. All recorders of the caliber to be considered in this study have literature available that furnishes accurate information as to their size, weight, and capabilities.

Qualifications

Qualifications

For the last eight years I have owned and operated stereophonic and high fidelity equipment. I'm familiar with equipment terminology and can interpret sales literature with no difficulty.

I am a senior in Wildlife. Having been on two extended field trips—one a week long, the other two weeks—I understand the techniques and problems of capturing bird vocalizations and of carrying and caring for field equipment.

Figure 15–4
(continued)

Introduction

In the introduction you set down everything needed to inform your readers about the objectives of your proposed work. Make the subject and purpose—the *what* and the *why*—of your work completely clear. When proposal readers cannot tell immediately where your work is heading or what it will accomplish, they lose interest rapidly. The introduction should point out, briefly, the *so-what* as well as the *why*. That is, in addition to telling what you hope to accomplish with the work, explain the implications of the work. In so doing, you are telling the readers how the work will be valuable and, therefore, why it is worth paying for. You will cover this same point much more thoroughly in the rationale and significance section.

Rationale and Significance

In your rationale and significance section, you bring your sales ability to play. It's at this section that the readers look to see if they are going to get their money's worth. In general, a proposal presents a plan to solve some problem that the people with the money have. Therefore, consider the following items for inclusion in this section:

- Definition of the problem
- Immediate background of the problem
- Need for solution to the problem
- Description of the solution
- Benefits that will come from the solution
- Feasibility of the solution

Definition of the Problem People and companies often suffer for years from a problem they have never defined and may not suspect they have. Exorbitantly high telephone bills, delays in invoicing and shipping, poor inventory control, poor customer relations, ineffective advertising, high maintenance costs—such "facts of life" may steadily grow worse through inattention and lack of recognition. An outsider, not having been blinded by day-to-day living with the problem, is usually in a far better position to detect its presence and to formulate a description of it.

Depending on the scale of the proposal, you should spend from a paragraph to several pages in defining, locating, and describing the problem you propose to solve. By this means you may "shock" your intended client into sudden and full awareness that a problem really does exist. However, you should guard against overstatement and overdramatization, because these techniques can boomerang.

Immediate Background of the Problem Before we can devise solutions, we must first discover what generated the problem in the first place. For example, trash collection and disposal is a generally recognized problem today. Populous cities along the Atlantic Coast transport their refuse hundreds of miles, where it is dumped in landfills in lightly populated areas. Citizens near these dumping grounds are increasingly raising their voices in protest. How did we get into this sorry plight? Discovering the origin and development of the problem may lead to means for its solution. Have Americans become an increasingly wasteful people? How much of the blame can be placed on marketing practices that emphasize bottles, cans, and cartons? We don't know, but we must find out.

A historical review can be quite enlightening and helpful in proving your grasp of the problem. However, your job is not to write history but to solve an existing problem. Therefore, keep the historical review brief and concentrate on the recent past. Be selective. Make the main points stand out.

Need for Solution to the Problem Why does the problem need to be solved? Why can't things stay as they are? Given the inertia in many organizations, the inclination to let well enough alone, you must be particularly persuasive in answering this question. The answers often relate to money, health, and social responsibility. For example, in the 1920s and 1930s, automobile exhausts were discharging contaminants into our atmosphere just as they are today—and at a higher rate per vehicle mile. But it was not until the 1950s that a few clear-sighted individuals pointed with alarm to the extent of air pollution from this source alone. Respiratory irritation and pulmonary congestion, smog that has killed the frail and aged, soiling of buildings, and grime that settles everywhere cost us millions of dollars every year. The cost in health, comfort, and cleanliness cannot be stated in dollars, yet Congress did not pass a comprehensive clean air act until 1990.

For the most part, people are not really aware of local problems they have grown up with and become accustomed to. In a proposal dealing with such a problem, you may have to shock them into acknowledging that the problem is a crucial one that will adversely affect their health or well-being. Spell out the consequences of ignoring the problem, as the Surgeon General of the United States has done with cigarette smoking.

Description of the Solution Having convinced your readers of the need for the solution, you must now present the solution. Describe it briefly. Perhaps it's a plan to heat public buildings with pelletized garbage. The Surgeon General might call for better education about the effects of smoking. Don't go into great detail at this point; your plan of work that comes later will provide the specifics of the solution proposed.

Benefits That Will Come from the Solution The benefits that will come from the solution are the positive side of the need for the solution. The benefits are the implications, the so-whats of your solution. Don't assume that your readers will see these benefits for themselves. To be persuasive, you have to spell them out.

In this section state as specifically as you can some of the good things that will happen if the solution is adopted. With respect to the trash disposal problem, for example, you might stress that avoiding the long-distance shipping of trash would save millions of dollars a year of tax money. Freight cars and trucks would be released for more profitable work. On the whole, the solution might prove to be income producing rather than tax consuming.

Feasibility of the Solution Will the proposed solution really solve the problem? Is it feasible? Arguments for the feasibility of solutions are to be found in physics and chemistry, engineering, psychology and social science, mathematical models, and analyses. Such arguments are meant to demonstrate that the solution can be made to work; that is, the solution is both practical and appropriate in terms of present knowledge and attainments.

When you argue for the feasibility of your proposed solution, you must consider the two meanings of **feasible.** A feasible action or undertaking is one that

- Can be accomplished in a purely practical sense (that is, not precluded by physical, economic, or similar concerns).
- Is suitable or appropriate (that is, generally conceded to be desirable as well as practical).

Suppose you wish to determine whether it would be feasible to provide all students at a certain university access to microcomputers and word processing programs to be used for writing assignments. Furthermore, these programs would check the students' spelling, punctuation, diction, and sentence style. First, the project would have to be assessed in practical terms. Could the university equip itself with enough microcomputers in view of purchase costs, maintenance costs, student population, room space, need for security, and so on?

Your research might establish that these practical obstacles are insurmountable, in which case there would be no reason to go any further. If the school can neither afford nor install the microcomputers, their desirability becomes irrelevant.

Suppose, however, that the practical problems appear to be manageable. At this point, you need to consider the more subtle issue of desirability. For instance, if the word processing programs were readily available to students, would they come to depend on them so much that they would forget the basic rules of mechanics and style? Would they forget how to spell?

Many alumni and teachers might raise this question to illustrate the seeming "unsuitability" of such programs. That response is a natural one, based on the tradition that tedious mental work is essential to learning anything. But suppose your research indicates that the word processing programs are excellent teaching tools that enhance rather than detract from the students' writing abilities—that the students actually learn the rudiments of spelling, punctuation, diction, and style faster when using word processing than by older methods. Furthermore, word processing could enable students who did have a sound grasp of these fundamentals to progress much more rapidly toward advanced concepts. You might then conclude that word processing is not only practical but appropriate. (For a more detailed treatment of feasibility, see Chapter 17, "Feasibility Reports.")

Plan of Work

After you have discussed the problem and its solution, you have to tell your readers how you are going to carry out the work. In your plan of work, you will give your readers information on the following:

- scope
- methods to be used
- task breakdown
- time and work schedule

You may also include statements concerning

- likelihood of success
- products of the project

Scope The scope statement sets boundaries and states what is to be done within these boundaries. You may state that you will sample public opinion or disregard the question entirely. You may adopt existing techniques and modify them or create entirely new techniques. You may limit study to estimates of technical feasibility, or you may agree to demonstrate its suitability. You may study all possible sites within your state, only one site, or a selected few. In other words, the scope statement establishes the depth, breadth, and means of your approach. Also, for everyone's protection, you may include in it negative observations covering what you will not do.

Methods to Be Used When you propose a solution to a problem, you must demonstrate that you have and understand the methodology required to carry it out. Giving your readers a plan of attack, a method of operation, a systematic analysis of intended procedures will attest to your practical good sense and know-how. Frequently, in research projects you may be proposing methods that have been created for earlier research similar to your own. When such is the case, you may save space and time by referring to such methods and providing references to sources, usually scholarly journals, where complete details can be found. Be sure to refer, however, only to sources that your readers can easily obtain and comprehend.

Task Breakdown Break your plan of work down into specific tasks that you will perform. For example, frequently, beginning tasks may consist of initial exploration: a search of the literature, correspondence and interviews, and so on. In projects that extend over a long period of time, reporting progress will certainly be one of the tasks. (See Chapter 16, "Progress Reports.")

Often, one or several tasks may have to be completed before others can be started. You cannot, for example, assemble equipment before you have collected parts, analyze data before you have acquired them, or report progress before you have made any.

By breaking the project into its component tasks, you assure your readers that the total job will be done in an orderly manner and be completed on schedule.

Time and Work Schedule With few exceptions, work proposals and agreements specify the calendar period within which projects are to be completed. The period may extend for days, months, or years. Further, the schedule may stipulate that portions of the work are to be done in a stated order and are to be completed by a given date. For complicated projects, time and effort must be carefully allocated. It is common practice to prepare a time-based flowchart showing activities and their intended duration and to provide this chart as a graphic. (See pages 341–343.) Even limited projects, such as student research problems, require planning for efficient performance to guarantee that work will be completed by target or deadline dates.

Likelihood of Success At first thought, you may feel that you would be foolhardy to write a proposal concerning any problem that you could not guarantee to solve. But understand that every problem exists simply because it has not been solved. If the solution were self-evident, then the problem would very probably have been solved before your arrival on the scene. Furthermore, some problems—cancer, racial tensions, and drug abuse—must be attacked, and are being attacked, even though the leads are slim and the prospects for early and full solution are dim. The universality and severity of such problems warrant the expenditure of effort even though the solutions may not be found in our lifetime.

Given infinite time, money, and means, we can reasonably hope to solve all problems that now afflict humanity. But when you prepare a proposal, you must set bounds on time, money, and means. To protect yourself and your professional reputation against charges of misrepresentation and nonfulfillment, promise no more than you can reasonably expect to accomplish.

Products of the Project Stating what you are to produce throughout a project and at its termination does commit you to definite performance and productiveness. Sometimes you may find it burdensome and costly to satisfy all points of your agreement. However, if you succeed in dodging this issue and do not specify what you will produce and supply, you may be placing yourself in a still worse situation. Your customer may construe your proposal as promising delivery of a working model or prototype equipment or as promising follow-up maintenance and consulting services free of extra cost. Therefore, a clear and specific statement of products to be delivered, although it commits you, also sets bounds on what you are to deliver. For the moral here, read the guarantee on the next roll of photographic film you buy. You will see that the manufacturer limits its obligation to replacing the original roll of film—and only if the original was defective in manufacture, labeling, or packaging. If you leave the lens cap on your camera, you, not the film manufacturer, will have to stand the loss.

Facilities and Equipment

In the section discussing facilities and equipment, you tell your readers what you will need to carry out your proposal. For the simplest of projects you may need only a pad of writing paper and a ballpoint pen. But many projects, apparently simple on first consideration, may shortly prove to require more than you bargained for. Facilities and equipment needed may include laboratory space, secretarial support, a technical library, use of a car, surveying instruments, testing apparatus, materials for constructing models, photographic equipment—and so on. On a still larger scale, you might need a building, a truck, the use of a light aircraft, and computers.

Before you decide to prepare a proposal, you must realistically determine the facilities and equipment the proposed work will require. Some you may already have. For those items you do not already possess, all or some of the cost of acquiring them must be included in the cost and financing of the project. If you already have the items, some or all of their original cost should be paid off, amortized, during the life of the project proposed.

Personnel Qualifications

In the section on personnel qualifications, you name the people who will carry out the proposal and give their qualifications for the job. A lack of experience in the people offering a proposal makes prospective employers pause, whereas successful accomplishment in the past promises success in the future. Whenever possible, therefore, you should cite earlier successes with similar problems when you prepare the new proposal. The past and present problems need not be identical—but they should overlap and have major points in common. You should include dates, contract numbers, names and addresses, references to published research, and so forth, so that your prospective client can verify your statements.

Many persons in research, industry, and government enjoy national and international reputations for their knowledge and performance. Institutions seek to attract celebrated "names." If you qualify, or if people in your organization qualify, you would do well to include their names and biographical sketches in a proposal. Resumes much like those prepared for the job hunt may be suitable (see pages 404–411). Known names tend to be regarded as a known quantity. However, if you cannot guarantee the availability of such persons for the project, do not use wording that can be construed as a firm promise. Otherwise, your contract may be broken on the grounds of misrepresentation.

To support performance records, you would do well to include the names and addresses of individuals, companies, and agencies that are able and willing to testify on behalf of the people listed. Satisfied customers can write letters that spell the difference between a hit and a near miss with proposals. Be sure, however, that they hold you in favor and will respond promptly and generously to inquiries concerning you. Obtain their permission well in advance of using them as references.

Budget

Practices differ in handling money matters in connection with proposals. Sometimes the cost accounting is included in the proposal proper, usually labeled as "Budget." Sometimes it accompanies the proposal as an addendum or rider. And sometimes it is handled in a separate financial and legal contract. In any case, you must never commit yourself to working for others without a binding agreement covering dollar amounts, hours of labor, fees and profits, timing, and method of payment. Penalty clauses for nonperformance or late performance, limits on fees and profits as a percentage of the whole, and delayed payment practices should be scrutinized closely. If you are not adept in these matters, go to a qualified financial expert or attorney.

If you do set up a budget, classify your costs in some coherent way that clearly shows where the money is going by using headings such as these:

Materials	Test Equipment	Administrative Expense
Labor	Travel	Fee

If your proposal stretches over several years, give a year-by-year breakdown of expenses.

In small proposals, the detailed itemization and cost breakdown of your major entries—such as the kinds and cost of test equipment and travel—will be presented right in the budget, as in Figure 15–3. In larger proposals, only the totals may be presented in the budget, with reference to detailed itemizations in an appendix.

Appendixes

Think carefully before you put anything into an appendix. There are decided advantages and disadvantages. An advantage may be that the ease of reading the proposal is increased if detailed analyses, budget itemizations, personnel resumes, and so forth are put into an appendix. The disadvantage may be that important material placed in an appendix may be overlooked. Also, material indiscriminately thrown into an appendix may add bulk to your proposal but not credibility. It's a judgment call.

In any event, appendixes are frequently used in proposals. Sometimes they include previously prepared exhibit literature that companies and experienced consultants keep on hand. They may include biographical sketches, descriptions of earlier projects, company organization charts, and statements concerning company policy, security clearances held, employment practices, and physical location and setting.

Urge to Action

Every proposal involves sales functions. A good product, of course, sells itself, but it does no harm to advertise its merits also. Therefore, at one or more places in a proposal, particularly at the end of the proposal or in the transmittal letter, you may wish to include several sentences whose purpose is to trigger acceptance of the proposed project. For example, if you have proposed the purchase of microcomputers and word processing software, you may wish to state that adoption of the proposal this spring would make it possible to have the machines in operation by fall.

You may stress the need and the benefits. You may give evidence of your sincere intentions by requesting an interview or by stating your willingness to modify the proposal. Don't overdo this urging, however. Many an overzealous salesperson has killed a sale by talking too long and too loud after the customer was willing to buy.

Proposal for a Student Report

In a proposal for a report, the student addresses not a client but the instructor to propose a report the student intends to write. The student should tailor the proposal for a report to fit the proposed report and the instructor's needs but, in general, should include the following information:

- Subject, purpose, and scope for the proposed report
- Task and time breakdown
- Resources available
- Qualifications for doing the report

Figure 15–4 illustrates a typical proposal for a report.

Planning and Revision Checklist

You will find the planning and revision checklists that follow Chapter 2, "Composing" (pages 33–35 and inside the front cover), and Chapter 4, "Writing for Your Readers" (pages 78–79), valuable in planning and revising any presentation of technical information. The following questions specifically apply to proposals. They summarize the key points in this chapter and provide a checklist for planning and revising.

Planning

- Is the proposal solicited or unsolicited?
  - If solicited, have you read and understood all the requirements set down by the RFP?
  - If unsolicited, have you obtained the sponsor's proposal specifications? Have you read and understood them?
- Who are your readers? Do they have technical competence in the field of the proposal? Is it a mixed audience, some technically educated, some not?

Revision

- Does your proposal have good design and layout? Does its appearance suggest the high quality of the work you propose to do?
- Does the project summary succinctly state the objectives and plan of the proposed work? Does it show how the proposed work is relevant to the readers' interests?
- Does the introduction make the subject and purpose of the work clear? Does it briefly point out the so-whats of the proposed work?
- Have you defined the problem thoroughly?

Planning

- What is the problem the proposed work is designed to remedy?
    - What is the immediate background of the problem?
    - Why does the problem need to be solved?
- What is your proposed solution to the problem?
    - What benefits will come from the solution?
    - Is the solution feasible, that is, both practical and appropriate?
- How will you carry out the work proposed?
    - Scope?
    - Methods to be used?
    - Task breakdown?
    - Time and work schedule?
- Do you want to make statements concerning the likelihood of success or failure and the products of the project?
- What facilities and equipment will you need to carry out the project?
- Who will do the work? What are their qualifications for doing the work? Can you obtain references for past work accomplished?
- How much will the work cost? Consider such things as materials, labor, test equipment, travel, administrative expense, and fee. Who will pay for what?
- Will you need to include an appendix? Consider such things as biographical sketches, descriptions of earlier projects, and employment practices.
- Will the proposal be better presented in a report format or a letter or memo format?
- Do you have a student report to propose? Consider including the following in your proposal:
    - Subject, purpose, and scope of report
    - Task and time breakdown
    - Resources available
    - Your qualifications for doing the report

Revision

- Is your solution well described? Have you made its benefits and feasibility clear?
- Will your readers be able to follow your plan of work easily? Have you protected yourself by making clear what you will do and what you will not do? Have you been careful not to promise more results than you can deliver?
- Have you carefully considered all the facilities and equipment you will need?
- Have you presented the qualifications of project personnel in an attractive but honest way? Have you asked permission from everyone you plan to use as a reference?
- Is your budget realistic? Will it be easy for the readers to follow and understand?
- Do all the items in the appendix lend credibility to the proposal?
- Have you included a few sentences somewhere that urge the readers to take favorable action on the proposal?
- Have you satisfied the needs of your readers? Will they be able to comprehend your proposal? Do they have all the information they need to make a decision?

Exercises

1. Check to see if your school has an office whose task is to help the faculty obtain grants. Such an office will almost certainly have outdated RFPs that it may be willing to let you have. When you have obtained several RFPs, discuss them in class, raising questions such as these about each:
   - What work does the RFP want done or what product furnished? What problem does the RFP present to be solved?
   - Does it specify a length for the proposal?
   - Does the RFP make clear the information proposals must contain?
   - Does it furnish an outline to follow? If so, what does the outline require?
   - Does the RFP require a specific format for the proposal? What is it?
   - Does the RFP make clear the criteria by which submitted proposals will be evaluated and who will do the evaluation?
2. Do any of the RFPs examined in Exercise 1 present a problem for which a class group could collaborate on a proposal? If so, form a group to work on it. Using group discussion techniques as described in Chapter 3, "Writing Collaboratively," examine the problem presented by the RFP. Work together as a group to prepare a proposal that presents a solution to the problem and satisfies the specifications set down in the RFP.
3. The following steps can lead to a proposal prepared either by an individual or a group.
   a. Assume that you have been asked by the Community Planning Commission to identify what you believe to be your town's most urgent problem. Here are some likely possibilities:
      - Traffic and parking congestion
      - Need for new business and industry
      - A deteriorated business district
      - Housing for low-income families
      - Inadequate public transportation
      - Need for a central downtown heating system

   b. List the means you would use to gather information for a proposal to be submitted to the Community Planning Commission. Consider questionnaires, interviews, library research, and the like. Be as specific as possible.

   c. If you were hired to research the problem referred to in a and b, how would you break down the whole project into tasks? How much calendar time do you estimate you would need to complete the investigation? What facilities and equipment would you need? What would be the cost of your project? What section titles do you foresee for the final report?

d. Based on the information you have generated for a, b, and c, simulate a situation in which you can submit a proposal similar to the one in Figure 15–3. Applying your imagination to as realistic a situation as you can find, write a proposal to a client.

4. Write a proposal for a report, as demonstrated in Figure 15–4. Follow your instructor's directions as to what kind of report it should be.

CHAPTER 16

Progress Reports

THE ENDING

Overall Appraisal
Conclusions and Recommendations

ADDITIONAL CONSIDERATIONS

Physical Appearance
Style and Tone
Originality
Accomplishment and Foresight
Exceeding Expectations

Distance, as well as differences in main interests, often separate workers from those who make use of their work, whether that work is research, construction, design, or whatever. A progress report, submitted as a bound report, letter, or memorandum, helps to keep the client in touch with the work being done.

The main and obvious function of any **progress report** is to give the client an accounting of the work that has been done. It explains how you have spent your hours and the client's money and what you have accomplished as a result of the investment. Though this purpose is dominant, we must not lose sight of four other purposes that are discharged by a progress report:

- It enables the client to check on progress, direction of development, emphasis of the investigation, and general conduct of the work. Thus, the client can alter the course of the work before too much time and money have been invested.
- It enables workers to estimate work done and work remaining with respect to the total time and effort available.
- It compels workers to evaluate their work and focus their attention.
- If the work includes a final report, as a research project or a feasibility study would, it provides a sample report that helps both the client and the workers to decide upon the tone, content, and plan of the final report.

Several different arrangements are used to report progress. A popular plan for a year-long project is shown in Figure 16–1. A progress letter is sent at the close of work each month, except at the end of a month closing the first, second, and third quarters of the year. At the end of these first three quarters, a formal bound report of progress is sent to the customer. This quarterly report recapitulates the main contents of the preceding monthly letters and adds the work done during the month since the last letter.

Toward the end of a project (about eleven months into a year's project), when affairs often have reached a critical stage, a preliminary copy of the proposed final report may be sent either in addition to or instead of the eighth monthly letter. Usually the prose has yet to be edited and polished; some or all of the illustrations may be roughly sketched or omitted entirely. The submission of this report enables the client to react and criticize and thereby get a final report more to his or her liking (see Figure 16–1).

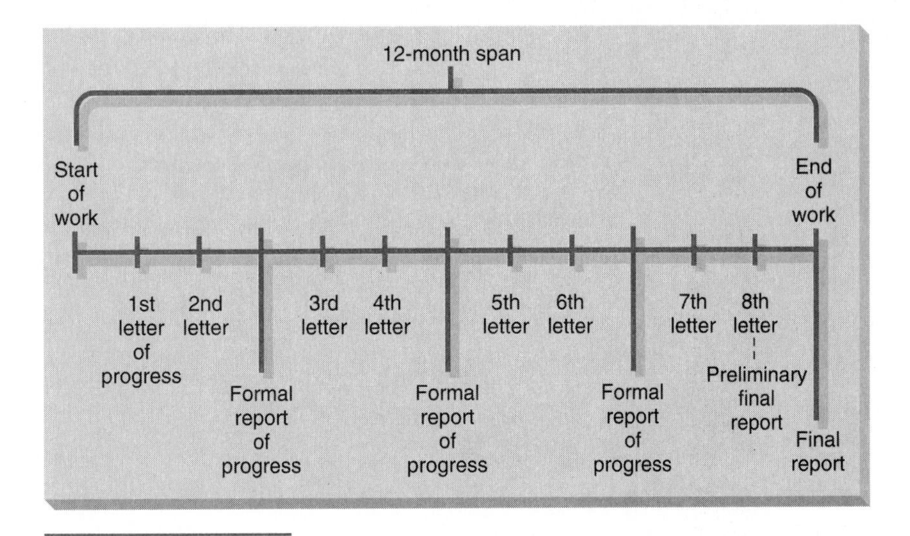

Figure 16–1 Two-Level Method of Reporting Progress

Progress reports, like most pieces of a writing, have a beginning, a middle, and an end. The following plan for a progress report shows what these three parts may contain:

Beginning

- Introduction
- Project description

Middle

- Summary of the work done in the preceding period(s) included in all project reports after the first
- Work done in period just closing
- Work planned for next work period
- Work planned for periods thereafter

End

- Overall appraisal of work to date
- Conclusions and recommendations concerning work

Recognize that the plan shown here is for a full-scale report. More modest progress reports may collapse some of the parts together. For example, the project description may be part of the introduction or, if there are no changes in it, omitted altogether. "Work planned for next period" and "Work planned for periods thereafter" may be lumped together under a heading like "Work remaining." Conclusions and recommendations may be integrated into the overall appraisal. But, no matter how, all these elements should be presented.

In Figure 16–2, we show you a sample progress report. It is linked both to the sample proposal on pages 478–482 and the excerpts from a student feasibility report used as examples in Chapter 17, "Feasibility Reports." The sample progress report is a modest one that collapses some of the suggested parts of a full-scale progress report together. However, it contains all the essential elements of a progress report that we describe in this chapter. Looking through it now will help you to understand the discussion that follows.

The Beginning

In the beginning of a progress report, introduce the report and describe the project.

Introduction

Progress report **introductions** are, in general, typical four-part introductions that make clear the subject, purpose, scope, and plan of the report (see pages 292–297). They should clearly relate the report to the work being done. Also, because progress reports are executive reports, give the readers some idea of your overall progress in the introduction.

Project Description

The **project description** (also known by other names, such as *work statement* and *contractual requirements*) spells out what the workers are required to do and produce. It makes clear the purpose and scope of the project. If changes are made in the contractual agreement, the project description should reflect these changes.

The Middle

The middle section of every progress report must bring together two elements: time and the tasks accomplished or to be accomplished during that time. This fact suggests that the middle portion of the progress report can be organized around either time or tasks.

(Text continued on page 506.)

4062 Hoven Street • St. Paul, MN 55108
(612) 624-7750

November 15, 1990

Ms. Nancy Johnson, Chair
Wyoming Lake Citizens Committee
Marinette County Courthouse
Wyoming Lake, WI 54126

First Progress Report on Feasibility Study of Rehabilitating Wyoming Lake

Dear Ms. Johnson:

On 15 October, after your acceptance of my proposal to study the feasibility of rehabilitating Lake Wyoming by removing its sediment deposits, I started work on all phases of the study. As agreed upon, I am submitting my monthly progress report.

Subject, purpose, scope, and plan of development

This report details the work completed and the work remaining on the three tasks of the study, as described next under Project Description, and concludes with an overall appraisal of the progress made.

Good news for client

Work on the study has progressed well. I also have the good news that with prompt action we may be eligible for major assistance in meeting the cost of the rehabilitation project. See the report for further details on this.

Project Description

Purpose of project

Purpose. The purpose of the Lake Wyoming Study is to evaluate the feasibility of removing sediment from Wyoming Lake to rehabilitate the lake and restore it as a recreational resource.

Scope and task breakdown

Scope. The study breaks down into three major tasks:

1. Analyzing the location, depth, and content of the sediment deposits.
2. Investigating the equipment and methods needed for removal of the deposits.
3. Applying the criteria of time, cost, and community benefits and problems.

Figure 16–2 Sample Progress Report

Continued

Page 2
Ms. Nancy Johnson
November 15, 1990

Combination plan

Location, Depth, and Content of the Deposits

Work completed. Adequate data on location and depth of the sediment deposits already exist in a report commissioned by the Wyoming Lake Citizens Committee in 1988. This report—Charles J. Anderson's "Sediment Deposits in Wyoming Lake"—provides detailed maps to the location of the sediment deposits and reports the results of soundings to determine deposit depths. Mr. Anderson's work indicates that the problem is a major one. The deposits cover over 1.2 million square yards of lake bottom. To rehabilitate the lake may require the removal of up to 2 million cubic yards of sediment.

Work remaining. In order to determine the content of the sediment deposits I will in the next week take sediment samples from three locations in the lake. These samples will be sent to the soils laboratory at the University of Wisconsin–Madison for analysis. The potential use of the removed deposits will depend to some extent on the results of this analysis.

Equipment and Methods

Work completed. I have written to the Madison Dredging Company, Madison, Wisconsin; The Capital Dredging Company, St. Paul, Minnesota; and Mississippi Dredges, Inc., St. Louis, Missouri. I have given these companies estimates of the volume of sediment to be removed and asked them to comment on the method, equipment, time to completion, and approximate cost of the project.

Details of good news

Telephone conversations with James H. Olson, Wisconsin Department of Natural Resources (DNR), have provided useful information concerning the legal restrictions and procedures involved in the project. Mr. Olson will send me a bulletin that should contain complete information regarding these matters. He also said that if we act promptly the project might qualify as a pilot project and be eligible for state and federal assistance for up to half its cost.

Notice of possible change of direction

Work remaining. The possibility exists that the City of Wyoming Lake might be better off buying the necessary dredging equipment and contracting the job on its own, rather than having a dredging firm do the job. I intend to explore this possibility with several firms that sell equipment.

Figure 16–2
(continued)

Page 3
Ms. Nancy Johnson
November 15, 1990

Time, Cost, and Community Benefits and Problems

Work completed. Some progress has been made in this area. Mr. Olson of the DNR has provided a list of city managers in the state whose cities have completed rehabilitation projects similar to ours. The replies from the dredging company correspondence should be helpful concerning time and costs. Also, Mr. Olson told me that public communication and proper coordination with the DNR often pose problems. He suggested that early appointment of a project coordinator would help with both these problems.

Work remaining. This week I'll begin correspondence with the city managers whose names were given to me by Mr. Olson. I'll ask them questions concerning time, cost, problems, and benefits. Using the replies received from them, I'll design questionnaires to be administered to a random selection of city residents to ascertain what they see as benefits, problems, and problem solutions.

Overall appraisal

Overall Appraisal

Good progress has been made. Data are available on location and depth of sediment deposits. Work has begun to determine the content of the deposits. Expected letters from major dredging contractors and dredge manufacturers should provide information on equipment, method, time, and costs. Conversations with the DNR indicate that legal restrictions and procedures do not present insurmountable problems. Planned correspondence with city managers should provide a basis for judging benefits and problems.

Exceeding expectations

I will call you next week concerning the possibility that prompt action on our part may make the city eligible for state and federal assistance for up to 50% of the cost of the project. With such a possibility, we may want to move our timetable forward.

Work has progressed well enough that I would consider an earlier submission of my final report than originally planned.

Sincerely,

Barbara J. Buschatz

Barbara J. Buschatz

Figure 16–2
(continued)

Time Plan

If the **time plan** is used for the overall plan of a progress report, the reader understands what time period is being discussed at any point in the report. Let us look at four possible headings for proof of this point:

- Work Previously Done
- Work Done in the Period Just Closing
- Work Scheduled for the Next Period
- Work Scheduled for Periods Thereafter

This often-used plan has a dynamic character. It gives the recital a cumulative effect. And it offers few problems in obtaining coherent flow from section to section.

We have commented that the time plan tends to be dynamic—that is, it gives the impression that something is being accomplished. Having said this, we should also remark that the time plan may lead to windy generalization and too much emphasis on procedure. But, with all its possible faults, it is widely used and generally has proven effective.

Task Plan

The project description of the progress report often gives the task breakdown of the entire project. These tasks may be performed at different times, by separate individuals, working at different locations. If this, or something close to it, is the working arrangement, then it may be convenient to organize the progress report on the **task plan.** It is a task to prepare, administer, and tabulate a questionnaire. It is a task to install air conditioners or design a new wing for a hospital. It is a task to conduct library or empirical research. Once you have clearly defined and described your tasks, it is a simple matter to arrange the body of your report around those tasks, as is done in the sample progress report in Figure 16–2.

- Location, Depth, and Content of the Deposits
- Equipment and Methods
- Time, Cost, and Community Benefits and Problems

The task plan is realistic and objective. The customer and project supervisor can readily and firmly estimate the amount of work done and remaining on each task. But, by the same token, it throws the glaring spotlight of unfavorable publicity on tasks that are not going well. Also, if the tasks are to be done in sequence, with little or no overlapping, then a given progress report may have solid prose to devote to only one of the tasks, with the other tasks being essentially blank for the time being.

Therefore, the task plan seems most appropriate when several or all tasks are being performed concurrently.

Combination Plan

Clearly, you can use time and task plans in combination, the exact arrangement depending upon where you wish to put the emphasis. You could, for instance, choose between these arrangements:

I. Work Done in Preceding Periods
  A. Location, Depth, and Content of the Deposits
  B. Equipment and Methods
  C. Time, Cost, and Community Benefits and Problems
II. Work Done in Period Just Closing
  A. Location, Depth, and Content of the Deposits
    [etc.]

* * *

I. Location, Depth, and Content of Deposits
  A. Work Done in Preceding Periods
  B. Work Done in Period Just Closing
  C. Work Scheduled for Next Period
  D. Work Scheduled for Periods Thereafter
II. Equipment and Methods
  A. Work Done in Preceding Periods
    [etc.]

The sample progress report in Figure 16–2 is based upon the second of these two combination plans. We do not, however, offer any one plan or combination of plans as being generally superior to others. We suggest that you collect the information you wish to include in the progress report and then select the plan that will best hold and present the information. And you should always be prepared to try a different plan if the first choice proves unwieldy.

The Ending

End your progress report by appraising the work to date and offering any conclusions and recommendations you may have.

Overall Appraisal

In the ending of the progress report, you draw things together for your readers. You summarize and review your progress. You provide answers to the kinds of questions executives are likely to ask. Are you on time or ahead of time? Are you running into unexpected problems? How are you solving them? Is the scope of the work changing? If so, how and why? Is a consultation needed between you and the client? Are costs running higher than expected? How much higher and why? Would some new approach or procedure be more efficient and less costly? Is something unexpected and significant showing up in the research? Does it throw past findings into doubt? Are materials called for in construction specifications no longer available? What can be substituted for them?

Conclusions and Recommendations

The overall appraisal is essentially an executive summary (see pages 291–292) and is, therefore, likely to contain your conclusions and recommendations. However, if you wish for some reason to emphasize your conclusions and recommendations, you may wish to give them a section of their own.

Remember what you know about executive reading habits from Chapter 4, "Writing for Your Readers." Because executives read selectively, they may skip the middle of the report and come directly to the ending. Therefore, this section of the progress report should be able to stand on its own. Also, you may wish to consider moving the overall appraisal and conclusions and recommendations nearer the front of the report, perhaps placing them right after the introduction. In modest letter and memorandum reports, you might consider making them part of the introduction.

Additional Considerations

The total impression a progress report conveys to the reader is, of course, important. But this total impression is the integrated product of more particular things, such as the following.

Physical Appearance

To put the reader–client in a receptive frame of mind, a progress report must be physically attractive. By this we do not mean expensive and glamorous, but rather neat, appropriate, and well-designed. Reports that exceed letter or memorandum length should have a protective cover both front and back. The title page should be tasteful and uncluttered. The print (or typing) should be clean and legible. However, the cost of all the progress letters and reports arriving from a project should not exceed a very small percentage (perhaps 5%) of the total funding of the project.

After all, we are paid to make progress and not to linger lovingly over the reporting of progress.

Style and Tone

Progress reports are a project's emissaries. If these emissaries seem tired, confused, and unhappy, what then is the customer to think of the workers "back home"? Progress reports, therefore, should read with vigor, firmness, and authority—one might risk optimism. Yet their forcefulness must lie in more than artful writing. Generalizations must be bolstered by recitations of detailed factual accomplishment. Snags, problems, and delays should be honestly discussed, but the accent should be on positive accomplishments. A neat balance between these two aspects of a project will prevent progress reports from reading like either a trail of disaster or an outpouring of giddy optimism. Excess in either direction will always have its day of reckoning.

Originality

The first progress report often leaves much to be desired. For one thing, the project just recently got under way, and whatever progress has been made cannot yet be crystallized. For another thing, the first progress report, lacking precedent, sometimes seems tentative and experimental. The next two or three progress reports usually represent a substantial improvement over the first, for they have accomplishments to report and they profit from earlier experience. However, after the third or fourth progress reports of a series, the reports tend to hit a plateau or go downhill. A feeling of ennui and repetition may be disturbing to authors. This slacking off must be prevented at any cost. Bringing in new blood to the writing staff may help. A staff review of the project may help. But a clear recognition of the need to maintain reading interest, verve, and originality is a necessity. Keep out of the rut. Do not simply warm over last month's progress report like the proverbial Sunday roast. Take a fresh look at the whole problem of reporting progress.

Accomplishment and Foresight

Your client will find it heartening to learn all you have accomplished on his or her behalf. Past performance is probably the most reassuring promise of future performance. Yet investigators should not seem to be moving blindly into the future work periods, like an automobile driver about to run off the margin of his only road map. Therefore, progress reports, while stressing what has been done, should give adequate attention to plans for the future. For the next work period, the plans should be firm and detailed; for work periods thereafter, the plans may understandably be less specific. Showing your plans reveals you to be a professional and also makes it possible for your client to suggest modifications. Getting your client into the act usually works to everyone's benefit.

Exceeding Expectations

A progress report should give your customer a pleasant but mild surprise. Notification that a new task has been started a few days before its scheduled beginning, three or four graphic aids, a technical appendix, and some noticeable improvements in format are useful ways of cheering your customer without undue labor or expense on your part. On the other hand, avoid sudden and excessive novelty, such as using multiple overlays and color printing that have no precedent in previous progress reports of the project. Again, do not increase the length of the previous progress reports by more than, say, 20%, for your customer may suspect that you are trying to "cover up" and divert attention. In a research project, do not include a set of firm conclusions or ultimate recommendations in a progress report, for you will thereby steal the fire from your final report. At the worst, your client may collapse the project before its normal completion date.

Planning and Revision Checklist

You will find the planning and revision checklists that follow Chapter 2, "Composing" (pages 33–35 and inside the front cover), and Chapter 4, "Writing for Your Readers" (pages 78–79), valuable in planning and revising any presentation of technical information. The following questions specifically apply to progress reports. They summarize the key points in this chapter and provide a checklist for planning and revising.

Planning

- Do you have a clear description of your project available, perhaps in your proposal?
- Do you have the project tasks clearly defined? Do all the tasks run in sequence or do some run concurrently? In general, are the tasks going well or badly?

Revision

- Does your report present an attractive appearance?
- Does the plan you have chosen show off your progress to its best advantage?
- Is your tone authoritative with an accent on the positive?

Planning

- Which would show off your work to its best advantage, a time plan, a task plan, or some combination of the two?
- What items need to be highlighted in your summary and appraisal?
  - Are there problems to be discussed?
  - Can you suggest solutions for the problems?
  - Is your work ahead or behind schedule?
  - Are costs running as expected?
  - Do you have some unexpected good news you can report?

Revision

- Have you supported your generalizations with fact?
- Does your approach seem fresh or tired?
- Do you have a good balance between work accomplished and work to be done?
- Can your summary and appraisal stand alone? Would they satisfy an executive reader?

Exercises

1. The following progress report is poorly organized and formatted. Reorganize it and rewrite it. Use a letter format, but furnish a subject line and the headings readers need to find their way in the report.

Forestry Research Associates
222 University Avenue
Madison, Wisconsin 53707

June 30, 1991

Mr. Lawrence Campbell, Director
Council for Peatlands Development
420 Duluth Street
Grand Forks, ND 58201

Dear Mr. Campbell:

Well, we have our Peatland Water-Table Depth Research Project underway. This is our first progress report. As you know, by ditching peatlands, foresters can control water-table depths for optimum growth of trees on those peatlands. Foresters, however, don't have good data on which water-table depths will encourage optimum tree growth. This study is an attempt to find out what those depths might be. We have to do several things to obtain the needed information. First, we have to measure tree growth on plots at varying distances from existing ditches on peatland. Then, we have to establish what the average water-table depth is on the plots during the growing season of June, July, August, and September. To get meaningful growth and water-table depth figures, we have to gather these data for three growing seasons. Finally, we have to correlate average water-table depth with tree growth. The correlation should establish the water-table depth that promotes optimum tree growth. Knowing that, foresters can recommend appropriate average water-table depths.

When the snow and ice went out in May, we were able to establish 14 tree plots on Northern Minnesota peatlands. Each plot is one-fortieth of a hectare. The distances of the plots from a drainage ditch vary from 0 to 100 meters. The plots have mixed stands of black spruce and tamarack. During June, we measured height and diameter at breast height (DBH) of a random selection of trees on each plot. We marked the measured trees so that we can return to them for future measurements. We will measure them again in September of this year and in June and September of the next two years.

While we were measuring the trees, we began placing two wells on each plot. The wells consist of perforated, plastic pipes driven eight feet into the mineral soil that underlies the plot. We should have all the wells in by the end of next week. We'll measure water-table depths once a month in July, August and September. In the next two years, we'll measure water-table depths in June, July, August, and September. We will also measure water-table depths any time there is a rainfall of one inch or more on the plots.

We have our research well underway, and we're right on schedule. We have made all our initial tree measurements and will soon obtain our first water-table depth readings.

By the way, the entire test area seems to be composed of raw peat to a depth of about twenty cm with a layer of well-decomposed peat about a meter thick beneath that. However, to be sure there are no soil differences that would introduce an unaccounted for variable into our calculations, we'll do a soil analysis on each of the 14 plots next summer. We will do this additional work at no added cost to you. During the cold-weather months, between growing seasons, we'll prepare water-table profiles that will cover each plot for each month of measurement. At the completion of our measurements in the third year, we'll correlate these profiles with the growth measurements on the plots. This correlation should enable us to recommend a water-table depth for optimum growth of black spruce and tamarack. We have promised two progress reports per growing season and one each December. Therefore, we will submit our next report on September 30.

Sincerely,

Robert Weaver
Principal Investigator

2. Write a progress report addressed to a client for some project you are currently working on, perhaps the project you developed for either Exercise 2 or 3 in Chapter 15. Use a memorandum or letter format.

CHAPTER 17

Feasibility Reports

PREPARATION OF THE FEASIBILITY REPORT

Introduction
Discussion
The Summation

A feasibility report documents the results of a feasibility study. It defines the study in terms of its objectives and the criteria that determine whether something is feasible or not. Feasibility is determined by the answers to questions concerning technological possibility, economic practicality, social desirability, ecological soundness, and so forth. The **feasibility report** presents, interprets, and summarizes the data relevant to feasibility. It presents the conclusions of the study and recommends actions to be taken. Before we get into the preparation of a feasibility report, let's examine first the feasibility study of which the report is the end product. Although exactly how you would conduct a feasibility study would depend on your discipline and your level of expert knowledge, we can, in general terms, describe the conduct of a study for you. (See also pages 488–489.)

Logic of the Feasibility Study

A feasibility study always involves a choice among options. The options may involve doing something or not doing it. For example, should the Department of Transportation require active restraint devices in all automobiles? Or given the decision that something should be done, the choice may lie among the options available to do it: Should the Department of Transportation require air bags activated by impact or seat-belt restraints activated when a car door is shut?

At all levels of human activity, from the individual engrossed in personal and domestic problems to the highest level of policy making in government, we live in a society where such decision making goes on:

- Homeowners may discuss whether to replace the wornout furnace with a new conventional furnace or to switch to a heat pump.
- The town council may debate whether to install a downtown heating plant and sell heat to local businesses.
- A company may study the feasibility of manufacturing a new product.

- The state legislature may argue about where to locate a hazardous waste dump.
- The governments of the United States, Canada, and Mexico may study the feasibility of a 300-billion dollar water system to bring water from Alaska and the Yukon to the rest of Canada and the United States and to Mexico.

Such questions—whether domestic and local or high level and global—are not easily answered. In an attempt to get the best possible answers, we study the problem, defining it and considering our options. We gather information that we hope will help us reach a wise decision. The homeowners with the wornout furnace may go to the library to read about heat pumps. They may call a friend who has one. They may write away for government documents and talk to salespeople.

Gradually, the homeowners form criteria to judge the option of the conventional furnace against that of the heat pump. The criteria may involve cost, both initial and operating, longevity of the appliance, safety, comfort, and so forth. When the criteria are formed and enough information is at hand, the homeowners make an analysis. They run comparisons applying the criteria to the data. The heat pump costs more initially, but it costs less to operate. Will the lower operating cost offset the higher initial cost? The heat pump can also serve as an air conditioner. How much is that worth? And so forth. After a time, they reach some conclusions; that is, they form opinions based upon their evidence and their reasoning that will guide them to their decision. Although it's unlikely that the homeowners will write a feasibility report, they are engaging in what we can recognize as a feasibility study. At any level, a feasibility study involves the same steps the homeowners have followed:

- Setting the purpose and scope of the study
- Gathering and checking information
- Analyzing data
- Reaching conclusions
- Arriving at a decision or recommendation

As in all creative mental exercises, there will be a good deal of back and forth movement in following these steps, but all the steps should be taken.

We discuss how to gather, check, and analyze your information in Chapter 5, "Gathering and Checking Information." We discuss conclusions and recommendations in Chapter 11, "Design Elements of Reports," on pages 299–302 and later in this chapter on pages 532–534. Because formulating purpose and scope is so critical to the success of a feasibility study, we discuss that at some length here.

Purpose

Before you do any research, define the precise purpose of all the work you will do. Usually, a single sentence is ideal for the purpose statement:

> The purpose of this investigation is to determine whether X Company of Old Town should establish a branch plant in New Town.

An announcement such as this may seem easy and self-evident, perhaps superfluous. But many investigators have floundered around and eventually bogged down simply because they did not clearly and consciously formulate the objective toward which they were striving.

If you do not know what you are trying to accomplish with a feasibility study, you have no sensor to tell you when you are on or off the right track. Get on the right track at the start, and keep on it by constantly reminding yourself of the purpose.

Here are some additional examples of purpose statements:

- The purpose of this study is to determine the feasibility of using particle board to sheathe the interior of house boats.
- The purpose of this investigation is to select the best methods for instructing elementary schoolchildren about the effects of toxic pollution on water quality.
- Our primary objective is to decide which of several microcomputers would be the best choice for the Department of Mechanical Engineering to purchase.

Inexperienced investigators are sometimes inclined to resent and reject the notion that their work should have a stateable and stated purpose. They are satisfied, rather, "to learn all they can about such-and-such a subject." This kind of open-ended purpose is not suitable for a feasibility study. Here an extended illustration will help to clarify our point.

One student investigator proposed to her instructor in technical writing that she be permitted to investigate the fossil spore content of a cranberry bog. "What bog do you have in mind?" her instructor asked. The student looked pained but under pressure disclosed that she had no particular bog in mind. Her instructor immediately sent her to the Department of Geology to learn the name and location of a bona fide cranberry bog in the vicinity of the campus. When she returned with an answer, both student and instructor were pleased, for visible headway had been made.

"Now," the instructor asked, "why do you want to investigate the fossil spore population of this bog?" The student looked more pained than before. After some moments of cogitation, she volunteered this information: "Well, I want to find out what kinds of fossil spores are in this bog and how many there are."

"But what are you going to determine?" her instructor pressed.

"Why, just what I said," the student replied, "what kinds are there and how many."

"I'm afraid that won't do," said her instructor. "The effort would be interesting and informative, I'm sure, but you are really describing an activity rather than stating the purpose of that activity. The real question hinges on what you will do, or recommend that someone else do, as a result of what you learn."

It was a sultry August afternoon. Minds were working none too well, and tempers were brittle. But by 5:15 the student was at last equipped with an acceptable purpose, as follows:

> The purpose of my project is to decide whether Mrs. Rose's cranberry bog contains a fossil spore population large enough to justify the acquisition of exploration rights over the bog by the Department of Geology for field research by its graduate students.

Equipped with this hard purpose, the student would not waste her time beating around cranberry bushes and plowing knee-deep in watery bogs. Rather, she was committed to making a decision, settling a question. In other words, she was now furnished with an outcome to be accomplished instead of a nebulous description of a messy activity.

Scope

Once the purpose of a feasibility study has been clearly and exactly decided, a fairly tight logical process has been set in motion. Certain other decisions follow with near inevitability. Ordinarily, the next major decision hangs on this question: Given my stated purpose, what must I do to accomplish it? The term *scope* is often applied to this concept. Whatever it is called, the decision consists of determining the actions to be taken, the range of data to be delved into, the bounds to be set to the problem, the criteria to be used, the emphasis to be made. To illustrate, let us use a previous example of a purpose statement to find what it implies in terms of scope:

Purpose

- Should X Company of Old Town establish a branch plant in New Town?

Scope

- Does X Company now have, or can it develop, enough business in New Town to justify a branch there?
- Does New Town offer adequate physical facilities, utilities, and other services for plant operation there—office space, transportation, communications, and so forth?
- Can the required staff be obtained, whether by local hiring or moving personnel into the area, or both?
- Are local business practices and codes, tax structure, and so forth favorable for conducting business there?
- What effect, for better and worse, would opening a branch plant in New Town have upon overall company organization, operations, policy, financial condition?

Necessary as it is to generate and compile the chief items of the scope statement early in an investigation, we would be dishonest to pretend either that the process is easy and forthright or that first efforts will produce a scope statement that will hold unchanged throughout the investigation. Devising a scope statement requires insight, foresight, and hindsight—in other words, speculative thinking. But the ability to create serviceable scope statements improves with practice and with experience gained in the subject matter.

In any case, as an investigator you should not remain blindly committed to your initial statement but should reexamine it from time to time in the light of the information you gather. Look for holes, overlaps, superfluous items, and the like. Frequently, a person unacquainted with the study is in a far better position than you to spot shortcomings and illogicalities in the statement. Therefore, someone outside the study should be asked to review and react to the list of scope items you compile.

For the most part, feasibility studies are conducted by experts in the field of the study or, in many cases, teams of experts from several fields. For example, environmental impact studies are specialized forms of feasibility studies that decide whether some new project, perhaps a new highway, is environmentally sound or not. The study may bring together civil engineers, wildlife biologists, soil scientists, archaeologists, and so forth, all working in a collaborative effort.

Preparation of the Feasibility Report

When the study is complete, the results, conclusions, and recommendations have to be reported to the ultimate users of the study. Generally speaking, the users are not experts in the field of the study. The users of an environmental impact study may be citizen groups and state legislators. In industry, the users of feasibility studies will be the executives charged with the decision-making responsibility. All of these diverse audiences are, in general, acting as executives, and reports for them should be written in the manner we describe as suitable for executives (see pages 64–69). As the writer of a feasibility report, you should write in plain language, avoiding technical jargon when possible. Give necessary definitions and background information. Use suitable graphics. Emphasize consequences and function over methodology and theory. Interpret your data and state clearly the conclusions and recommendations that your best professional judgment leads you to.

Feasibility reports do not have definite formats. They may include all or some of the following elements:

- Letter of transmittal or preface
- Title page
- Table of contents
- List of illustrations
- Glossary of terms
- Executive summary
- Introduction
- Discussion
- Factual summary
- Conclusions
- Recommendations
- Appendixes
- References

How many of these elements you include will depend upon audience factors and the length and complexity of the report. For example, a long report aimed at a narrow audience of several people should have a letter of transmittal. A long report for a more general audience would have a preface instead. A short feasibility report of only several pages may be cast as a memorandum or letter and essentially consist of only an executive summary, discussion, conclusions, and recommendations.

We discuss all the elements you may need for a feasibility report in Chapter 11, "Design Elements of Reports," and in Chapter 13, "Correspondence." Here we single out the key elements—introduction, discussion, factual summary, conclusions, and recommendations—for brief discussion and exemplification. In our discussion of these key elements, we'll draw upon a student report entitled "The Feasibility of Removing Sediment Deposits from Wyoming Lake."[1] We reproduce the table of contents from that report in Figure 17–1. A glance at it now will help to orient you to the parts as we discuss them.

Figure 17–1 Table of Contents for Sample Report

TABLE OF CONTENTS

ANNEXES

Introduction

Depending upon the length and complexity of the report, the introduction may range from a single paragraph to several pages. However, the introduction should be limited to essential preliminaries such as the following:

- Subject, purpose, and scope of the feasibility study
- Reasons for conducting the study
- Identification and characteristics of the person or company performing the study (if not given earlier in a preface or letter of transmittal)
- Definition and historical background of the problem studied (If this can be given briefly. If not, consider doing this in the discussion.)
- Any limitations imposed upon the study
- Procedures and methods employed in the study (briefly!)
- Acknowledgment of indebtedness to others (if not given in an earlier preface or letter of transmittal)
- Preview of the report that follows

If all of these topics are separately treated, the introduction might run to considerable length. Some of these topics can be combined and handled very briefly. Others can be omitted altogether if they are not pertinent to the report that follows. Certainly, for a brief report of some ten pages or so, the introduction should seldom exceed one page.

Read over the following introduction to our student feasibility report:

Problem definition

Occasion for report

Purpose of study

Scope of study

Preview of report

> In recent years sediment deposits have built up along most of the shoreline area of Wyoming Lake, making swimming and other recreational activities undesirable and even hazardous. Under contract to the Wyoming Lake Citizens Committee, I have studied the problem and evaluated the feasibility of removing these deposits in order to rehabilitate the lake and restore it as a recreational resource.
>
> The study (1) analyzed the location, depth and content of the deposits; (2) investigated the equipment and methods needed for the removal of the deposits; and (3) using the criteria of time, cost, and benefits and problems to the community, evaluated the feasibility of the project.
>
> This report summarizes the results of the study, draws conclusions from the results, and recommends actions to be taken by the Citizens Committee. A detailed discussion of the results is presented in the annexes to this report.

The prose in the introduction is businesslike and yet reads easily. In the first paragraph the author economically sets forth the rationale of her study. The second paragraph does double duty in that it depicts the author's logical procedure and also previews the organization used in the discussion. The last paragraph makes the transition from the actual study that was conducted to the presentation of the results of that study—that is, to the report the reader has at hand. The author has made a good beginning.

Discussion

The discussion of a feasibility report presents and discusses the results of the study—the information upon which the investigator bases the conclusions and recommendations. In logical progression, the discussion, therefore, precedes the summary, conclusions, and recommendations of the report. In traditional report formats, the logical order is followed. Also, in brief reports presented as letters or memorandums, the logical order is frequently used on the grounds that the reader does not have to wait very long for the conclusions and recommendations. However, in long reports, the discussion is sometimes removed from this central position in the report. Feasibility reports are executive reports, and executives, for the most part, like the key data, conclusions, and recommendations early. Following this executive preference, some feasibility report writers present their reports in this order:

- Introduction
- Factual summary
- Conclusions
- Recommendations
- Discussion (Annexes)

When the discussion is removed from the center of the report and placed near the end, it is sometimes labeled "Annexes." The table of contents to our sample report shown in Figure 17–1 shows the discussion in this annex position, which is the format our student actually chose. Figure 17–2 shows a format with the discussion in a central position. When you use this format, it's wise to place an executive summary near the front of the report to satisfy the executive's need for key information early in the report. (For an example of a feasibility report with the discussion in the central position, see Appendix A, "A Student Report," pages 625–646.) In either central or annex position, the discussion must present enough evidence to justify the conclusions and recommendations reached.

There is no rigid organizational plan to be followed for the discussion itself, but three plans seem to be used more than others:

- Problem–solution–evaluation (see pages 486–489)
- Argument (see Chapter 9, "Arguing")
- Comparison (see pages 204–206)

TABLE OF CONTENTS

Figure 17–2 Plan with Discussion in Central Position

No doubt, the heavy use of these three plans results from the nature of feasibility reports. In them you are either presenting and evaluating the solution to a problem, arguing for your conclusions and recommendations, or comparing two or more alternatives. Whichever plan you choose, remember that you must present a balanced discussion pointing out the risks in your recommendations as well as the benefits.

In our sample report, the author chose a problem–solution–evaluation approach:

A. **LOCATION AND DEPTH OF SEDIMENT DEPOSITS**
The definition of the problem, quite literally, its extent and depth.

B. **LAKE REHABILITATION**
The discussion of the solution to the problem in terms of equipment needed, method, legal restrictions and procedures, and the disposition of the removed deposits.

C. **EVALUATION OF PROJECT**
The application of the criteria of time, cost, and benefits and problems to evaluate the solution. As part of the evaluation, two alternative ways of carrying out the solution are compared.

We reproduce parts of the discussion from our sample feasibility report here for your examination. The first segment is from Annex A, in which the writer defines the problem. In brackets we refer to our reproductions of her figures and tables.

A. Location and Depth of Sediment Deposits

The locations of the sediment deposits in Wyoming Lake are indicated by the shaded areas in Figure 1 [Figure 17–3], designated A, B, C, etc. The accompanying table indicates the square yardage in each area. The mapped areas indicate that a considerable portion of the lake has sediment deposits. In over five miles of shoreline there are few areas that can be used for swimming or wading. The only area that does not contain sediment deposits is near the shoreline along the park at the south end. However, swimming has been discontinued in this area also.

Depths of a selected number of deposits in Wyoming Lake are indicated in Figure 2 [Figure 17–4]. The readings were taken at selected locations around the lake to give an overall picture of the depth of the deposits beneath the water. The depth readings show that deposits are generally 6 to 11 feet deep below 3 to 5 feet of water. The deposits extend out over a large area, in some cases 300 to 400 yards from the shoreline. Removal of 6 feet of these deposits would provide water depth of 9 to 12 feet and eliminate most deposits down to the solid base.

Table 1 [Figure 17–5] provides estimates of the volume of sediment removed and the volume of water obtained by removing 3 or 6 feet of the deposits. The elimination of the large quantity of sediment, coupled with the addition of 260 to 520 million gallons of water, would substantially improve the condition of Wyoming Lake.

In the discussion, enough detail is presented and analyzed so that the readers can comprehend the true extent of the problem. The analysis uses

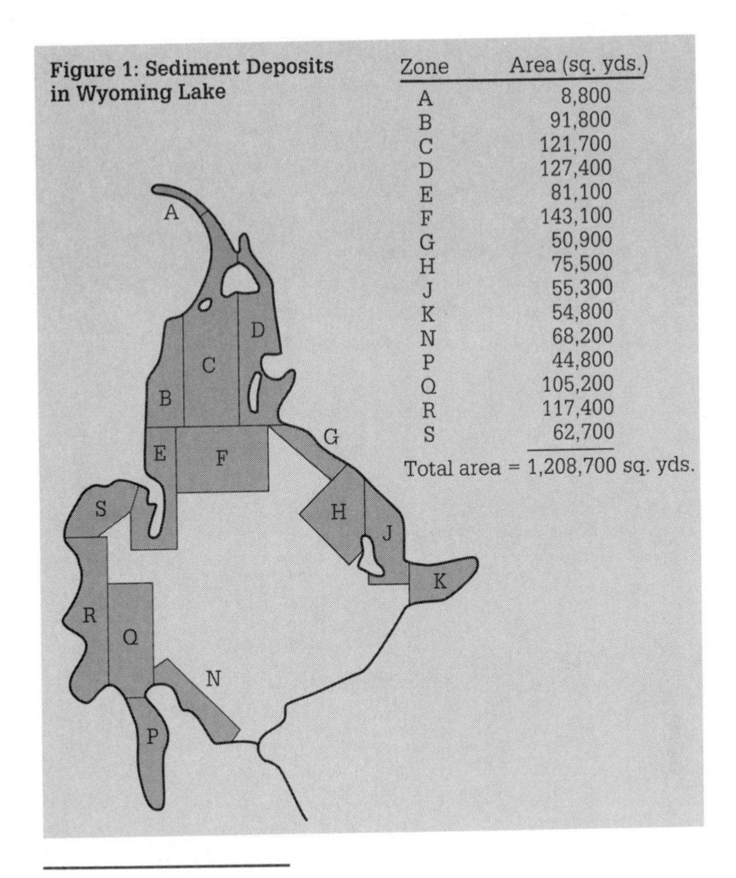

Figure 1: Sediment Deposits in Wyoming Lake
Zone Area (sq. yds.)
A 8,800
B 91,800
C 121,700
D 127,400
E 81,100
F 143,100
G 50,900
H 75,500
J 55,300
K 54,800
N 68,200
P 44,800
Q 105,200
R 117,400
S 62,700
Total area = 1,208,700 sq. yds.
A
B
C
D
E
F
G
H
J
K
N
P
Q
R
S

Figure 17–3 Sample Report, Figure 1.
Source: Figure based on data from the following unpublished report: Charles J. Anderson, *Sediment Deposits in Wyoming Lake* (A report for the Wyoming Lake Citizens Committee, Wyoming Lake, Wisconsin, 1988), 6–8.

the logical techniques of generalizing from particulars and particularizing from generalizations—that is, induction and deduction (see pages 199–204). Recognize that every reader may not want this much detail. That is why our writer has chosen to put these details in the subordinate position of an annex. But for those readers who want or need the details, she has presented them with care. Her use of figures and tables is particularly skillful.

In comparing two alternatives for dredging the lake, the writer presented a good many calculations showing how she arrived at her cost figures. We present one such section for you here. Notice the writer's use of lists and informal tables to make her data easily accessible to the reader.

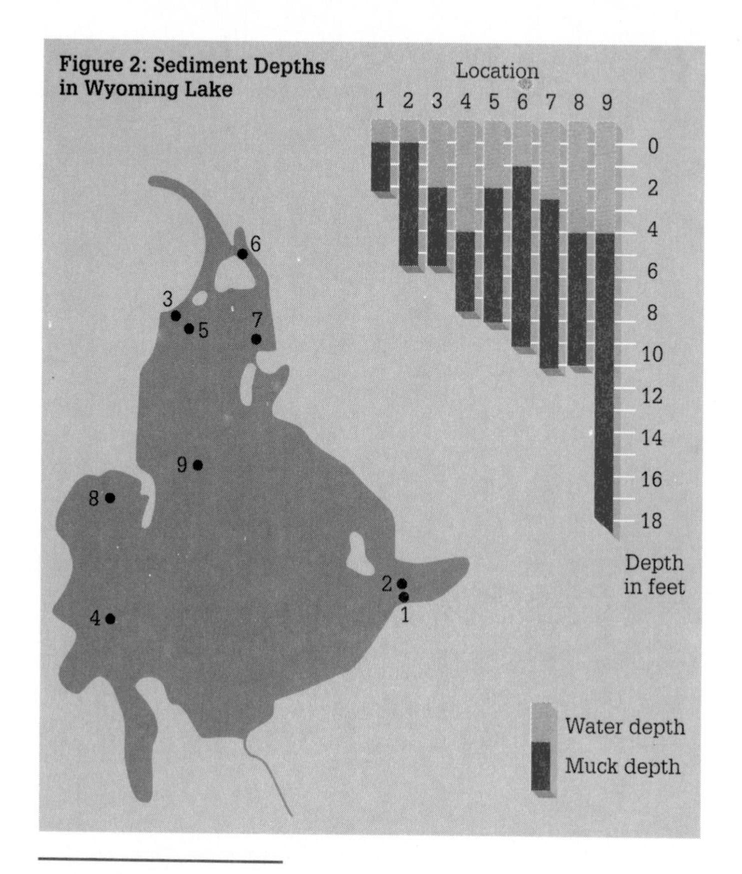

Figure 17–4 Sample Report, Figure 2.
Source: Figure based on data from the following unpublished report: Charles J. Anderson, *Sediment Deposits in Wyoming Lake* (A report for the Wyoming Lake Citizens Committee, Wyoming Lake, Wisconsin, 1988), 8.

TABLE 1. Sedimentation Volumes in Wyoming Lake

Depth of Sediment Removed	Volume of Sediment Removed	New Water Depth	Water Added to Lake Body
3 feet	1.2 million cubic yards	6 to 8 feet	262 million gallons
6 feet	2.4 million cubic yards	9 to 11 feet	524 million gallons

Figure 17–5 Sample Report, Table 1

Purchase of dredging equipment. Estimates of the cost of dredging equipment suitable for Wyoming Lake are based on the following operational requirements:

Maximum distance pumped	6,000 feet
Maximum discharge elevation	82 feet
Total material removed	2,400,000 cubic yards
Maximum digging depth	12 feet

Based upon the above requirements, the proposed equipment consists of the following:

1	14-inch Hydraulic dredge	$210,000
1	14-inch Booster pump	$ 80,000
6,000 feet	16-inch Discharge pipe	$ 50,000
	Pontoons, connectors, and joints	$ 50,000
	Total Equipment Cost	$390,000

The booster pump is required because of the long distance to the farthest deposit areas and the necessity of pumping to an elevation of 82 feet.

The 16-inch discharge pipes are specified to reduce the frictional loss in the pipe, so that long distances and high elevations can be accommodated.

The Summation

As a lawyer does near the end of a trial, the feasibility report author has to sum up the evidence, draw conclusions, and make recommendations. In the discussion all the pertinent and concrete evidence has been presented. But what does it all add up to? The informed and dedicated readers might be able to draw their own conclusions, but the author occupies a position of special advantage: complete familiarity with the material and its ramifications. The author must draw on that special position to report the outcome of the study clearly and succinctly.

In its full logical form, the outcome of a feasibility study is reported in three stages:

Factual Summary	Conclusions	Recommendations
Two or more facts given in meaningful context.	Interpretive generalizations, implications.	Actions urged as a consequence of conclusions.

In short feasibility reports, these three parts may be used in various combinations. For example, all three might be combined in what is essentially an executive summary. (See pages 291–292.) In another arrangement, the factual summary may be combined with the conclusions while the recommendations stand alone. No matter how you mix and match these three parts, in summing up your findings, be sure to present all three elements: facts, conclusions, and recommendations. In the next three sections, we discuss how to present these elements.

Factual Summary The term **factual summary** distinguishes this type of summary from those that contain conclusions and recommendations as well as data. Creating a good factual summary is one of the hardest challenges you face as a writer. In meeting that challenge, keep these principles in mind:

- You can summarize only that information appearing elsewhere in the report. A factual summary must not introduce new or additional information.
- Make every statement in a factual summary a genuine assertion of fact. Keep opinions out of factual summaries. Unless this rule is observed, you build an unsure foundation for the conclusions that follow.

No: Every American home should have a telephone. Who says so? This statement sets up an arbitrary standard of the author's own making. In a factual summary, avoid preferential and opinionative wording such as *ought, should, better, best.*

Yes: In 1988, 93% of U.S. households had telephone service (*Statistical Abstract of the United States, 1989,* 544.)

- Integrate your facts for the reader. Gather facts together from various places in the report and put them together in a meaningful context. In your discussion you may in three separate statments offer these facts:

  Model A retails for $425.
  Model B retails for $385.
  Model C retails for $465.

  In your factual summary bring these three separate statements together in a statement like this one:

  Of the three models tested, model B at $385 costs the least.

You may have to reword, rearrange, and otherwise process your data, but don't change their meaning.

- Make every statement pave the way for one or more conclusions to follow. The only valid reason for summarizing facts is to use them as a foundation for conclusions.

We reproduce for you here the factual summary from our student report. To demonstrate the ratio of summary data to discussion data, we have put in boldface type the two summary statements that relate to the discussion sections we have already shown you on pages 526–529.

Throughout her factual summary, the writer has integrated her data and boiled down her data and calculations to those absolutely essential to support the conclusions that follow.

Factual Summary

Wyoming Lake was once a widely used recreational area. However, over the past decade the lake has stagnated, clogged by sediment, making the lake unfit for many recreational uses.

There are presently 2,400,000 cubic yards of sediment deposits averaging 6 to 11 feet in depth along most of the shoreline of the lake. In order to remove these deposits and rehabilitate the lake, certain steps have to be taken. First, the city of Wyoming Lake has to apply for a dredging permit from the state Department of Natural Resources, Division of Resource Development, and from the U.S. Army Corps of Engineers. After permission is granted, the lake can be dredged. With prompt application on the part of the city, the dredging project could be designated a pilot project by the Department of Natural Resources and receive a 50% subsidy.

Deposit areas where the removed sediment can be used as fill are available within pumping distance from the lake. These areas can actually benefit from this fill because the lake bottom sediment tests out to be about the same as most agricultural soils. Thus, the deposits would improve subsoils and scalped areas and would be suitable for filling and reclaiming nearby abandoned gravel pits.

A hydraulic dredge is used for dredging operations. Two alternatives for accomplishing the dredging were considered: the city's contracting with a dredging firm and the city's buying a dredge and doing the job itself.

Informal bids for the project received from dredging firms fell into a range from $720,000 to $1,200,000.

If the city bought a dredge and did the job itself, **the cost of the dredge and associated equipment would be $390,000.** The operating cost for the two years estimated to be necessary for completing the project would be approximately $304,000. The total of purchase price and operating costs is $694,000, a cost lower than the lowest bid. Also, the dredge would be available for future projects or resale at approximately 60% of its $210,000 purchase price.

Other cities in the state that have dredged their lakes have found that property values and tax revenues have increased when their lakes once more provided a healthy recreational environment in which to swim, fish, and water ski.

Problems that other cities have encountered include paying for the project, poor communication with the public during the project, and gaining approval of the dredging work by the Department of Natural Resources as the work proceeds. Bond sales, local assessment, and state and federal assistance are all available as means of paying for the project. Appointment of an experienced project coordinator is a way of solving the remaining problems.

Conclusions **Conclusions** act as the intermediate step or bridge between the facts in the factual summary and the recommendations. They are the inferences and implications you draw from your data through the use of induction and deduction. They are the answers you should be able to give if someone looks at your data and asks, not unkindly, "So what?"

Conclusions can be presented in normal essay style or in a series of short, separate statements. Perhaps for reasons of succinctness and to make the logic of the writer's reasoning more evident, the presentation of short statements is an often-used style. It is the style used by our student writer. However you present your conclusions, you should arrange them in a clear and premeditated order that reveals your thinking process to the reader. Do not insist that the order of events in the factual summary be exactly paralleled in the conclusions. Do not insist that each statement in the factual summary be represented by one and only one conclusion. The truth is that a single conclusion often embraces two or more factual summary statements—and sometimes not consecutive ones at that.

We can give you no foolproof guidelines to the number of conclusions you should have. Too few and your logic will not be adequately revealed. Too many and you run the risk of confusing your readers. However, we can point out that a prime way to make an error in writing conclusions is to assume that something that is obvious to you will be obvious to your readers. Remember that you have been living with your material for a long time and know it well. Not knowing the territory as well as you do, your readers may miss implications that you should point out. Do your readers a favor and wrap things up neatly for them.

If your readers can read your introduction, factual summary, conclusions, and recommendations and feel that you have justified your recommendations, you have probably presented your material well. The readers should not be left wondering why you chose alternative A over alternative B. Ideally, your last conclusion should serve as a stepping-off platform for the recommendations that follow.

We present the conclusions drawn by our student writer. In numbers in parentheses following each conclusion, the author refers the readers to pages in the annexes where complete support for the conclusions can be found—an excellent practice. As we frequently emphasize, help your readers to read your report selectively.

Conclusions

1. Wyoming Lake will become completely useless as a recreational resource if its sediment deposits are not removed (pp. 1–5).
2. Suitable dredging technology for removing the sediment within two years and environmentally sound areas for depositing the removed sediment are both available (pp. 5, 9–11).
3. The total cost of the city purchasing a dredge and the associated equipment and running the dredging project itself is lower than the lowest bid for contracting the project with a dredging company (pp. 11–14). Therefore, the city serving as its own contractor seems the sounder choice of the two.
4. The appointment of an experienced project coordinator should forestall potential problems of communication with the public and poor coordination with the Department of Natural Resources (p. 17).
5. Standard techniques for funding the project such as bond sales and tax assessments are available. State and federal assistance is also available (p. 17).
6. With prompt action, state and federal assistance for up to 50% of the cost may be available (p. 11).
7. The city should benefit from a healthy recreational environment, and property values should increase, resulting in increased tax revenues (p. 15).
8. The project, given the right decisions and prompt action, is feasible. The first step is application for approval of the project by the State Department of Natural Resources and the Army Corps of Engineers.

Recommendations In contrast to conclusions, a **recommendation** is an action statement. That is, it recommends that the report users take some proposed action or refrain from taking it. *Always be sure that the first (or only) recommendation discharges the purpose set forth early in the report.*

If the purpose is to determine whether X Company should establish a branch in Memphis, Tennessee, and the author has determined the plan is feasible, the recommendation (first or only) should read something like the following:

> X Company should establish a branch plant in Memphis, Tennessee.

Any further recommendations usually implement the first. If the first recommendation favors establishment of a branch plant, succeeding recommendations may detail the necessary steps to carry out the plan. If the first recommendation opposes the establishment of a branch, succeeding recommendations may propose alternative plans or urge that the whole question be reinvestigated several years hence.

Recommendations are opinions, professional judgments. No logic, no array of evidence—nothing under the sun—can make them anything else. Let them therefore be sound and well considered.

The recommendations offered by the author may or may not be accepted by the customer or report user. However, for the feasibility study report, the author is obliged to arrive at one or more recommendations. If the research has been properly conducted, then these recommendations should carry great weight.

Our student-writer's recommendation section consists of one primary recommendation followed by five implementing recommendations.

Recommendations

Primary recommendation

City government and the Wyoming Lake Citizens Committee should take the necessary steps to rehabilitate Wyoming Lake as follows:

Implementing recommendations

- Apply promptly to the Department of Natural Resources and the Army Corps of Engineers for permission to begin dredging.
- Take necessary steps to qualify the rehabilitation project as a pilot project eligible for state and federal assistance at the 50% level.
- Appoint a project coordinator.
- Appoint a committee to investigate and report on the best method of funding the project.
- When funding is assured and all permissions secured, purchase a dredge and the associated equipment and begin operations under the leadership of the project coordinator.

When as the writer you have stated your recommendations, you have completed your objectives as an investigator. Your task is over. For an example of a complete feasibility report, see Appendix A.

Planning and Revision Checklist

You will find the planning and revision checklists that follow Chapter 2, "Composing" (pages 33–35 and inside the front cover), and Chapter 4, "Writing for Your Readers" (pages 78–79), valuable in planning and revising any presentation of technical information. The following questions specifically apply to feasibility reports. They summarize the key points in this chapter and provide a checklist for planning and revising.

Planning

- What is the purpose of your feasibility study? State it in one sentence.
- What is the scope of your study?
  - What are your options?
  - What information do you need?
  - What are your criteria?
- Who is your client? What is your client's technical level?
- What will be the approximate length of your feasibility report?
- What format do you plan for your report? What report elements will it require?
- Which of these items will you need in your introduction?
  - Purpose and scope of study?
  - Reasons for study?
  - Information about investigator?
  - Definition and background of problem?
  - Study limitations?
  - Study methodology?
  - Acknowledgments?
  - Report preview?
- Will you organize your report with factual summary, conclusions, and recommendations first or last? If last, do you need an executive summary at the beginning?
- Will you use one of these organizational plans?
  - Problem–solution–evaluation?
  - Argument?
  - Comparison?

Revision

- Is your report suitable for an executive audience? Have you done the following?
  - Organized and formatted the report for selective reading?
  - Used plain language?
  - Provided needed definitions and background?
  - Used suitable graphics?
  - Emphasized consequences and function over methodology and theory?
  - Clearly stated conclusions and recommendations near the front of the report, either in separate sections or an executive summary?
- Does your introduction clearly state the purpose and scope of the study and preview the report?
- Does your discussion support the conclusions and recommendations reached?
- Does your summation sum up the evidence, draw conclusions, and make recommendations?
- Do the data and reasoning in your summation make the recommendations obvious, even inevitable?
- Does your first recommendation discharge the purpose of the study?

Planning

- Are there tables and graphs that would help your reader?
- What are the key facts that need to go in your factual summary? What facts can be brought together to lead to meaningful conclusions?
- What are the key conclusions you can draw from your data?
- Which of your options do your conclusions support?
- What recommendations do your conclusions lead to?

Exercises

1. The feasibility study is a necessary prelude to any important action. In your local newspaper identify any news items that call for feasibility studies before action. Is your community or school considering any actions that should be preceded by feasibility studies? Are there topics in your discipline that would lend themselves to feasibility studies?
2. Using the material you gathered for Exercise 1, prepare a list like the one shown here, but substitute your own subject matter following the four colons:

    General field: Meteorology
    Specific topic: Short-term weather forecasting
    Purpose: To determine the feasibility of making 24-hour forecasts for specific areas to expedite road crews for highway storm treatment
    Client: Richard Ferguson, Road Maintenance Engineer at state Department of Transportation

3. Using the list you prepared for Exercise 2, determine the scope of your treatment (see pages 519–520). Rough out the areas of information you will need and the methods you will use to gather the information (see Chapter 5, "Gathering and Checking Information").
4. Using the proposal for a report shown in Figure 15–4 on pages 484–485 as a model, submit to your teacher a proposal for the feasibility study and report described in Exercises 2 and 3 above.

5. Using the planning checklists following Chapter 2 and this chapter, prepare an organizational plan for your report. Begin with an updated purpose, scope, and audience statement. Decide on an appropriate format for the report. Will it be a memorandum, letter, or a full report? If a full report, which elements of formal reports will you include? How many and what kinds of graphics do you anticipate using in your report? Be prepared to justify your choices in a discussion with your teacher.
6. Write, revise, and edit the report planned in Exercise 5.

CHAPTER 18

Empirical Research Reports

RESULTS

DISCUSSION

A FINAL WORD

Can American caterpillars be crossbred with oriental silkworms? Is "baby talk" good for children's language development? Can wine be made from milk? What is a feasible way to store heat gathered by solar collectors? Can optical fibers be used as acoustic sensors?

To get answers to questions of this kind, you can do two things:

- You can find, usually through library research, the answers that previous researchers have obtained.
- You can obtain first-hand answers for yourself, by the direct empirical methods of experimentation and observation.

Of course, you can—and probably should—use these two methods in combination. However, in studies and reports of the kind treated in this chapter, the emphasis falls upon empirical methods.

An illustration may help to clarify our point. Suppose that we have a chunk of glass, crude and irregular, dumped out of the ladle and unmolded. We desire to find the impact strength of the chunk of glass, that is, how many pounds of force will be required to shatter it. We may approach our solution in two ways:

- We may read up on the chemical makeup of the chunk of glass. We may measure its geometric properties. We may pass white light through it to obtain a reading of its internal structure. By turning to suitable handbooks we may then estimate the minimum impact force required to shatter the chunk.
- We can whack the chunk with a hammer, hitting harder and harder until it shatters. A pressure gauge or similar accessory will tell us how hard we had to hit to get the result we wanted. This is the pragmatic test, pure and simple.

The **empirical research** study places the emphasis on the second approach.

Empirical research is very common in our daily lives. Suppose a homeowner has had a new picture window installed, 5 feet high and 12 feet wide. Since buying a home six years ago, he has become a faithful reader of the "Household Hints" column in his morning newspaper. He tries to recall what the contributors to the column have had to say about cleaning large windows. What did his parents do? He recalls a mishmash of advice: Start at the bottom. Start at the top. Use a circular cleaning motion. Use broad horizontal strokes. Never clean with the sun on the glass. Never clean the inside when it's dark outside; you'll be sorry in the morning. Squeaky glass is clean glass. Scouring powders will dull the surface. For guaranteed results, call your Handi-Dandi Window Cleaning Service. Helpful or not, this body of "theory" runs through his head as the time comes to clean the new picture window.

What cleaning agents should he use: A commercial window spray? Detergent and water? A powder such as Belle de Jour? Ammonia and vinegar diluted with water? And what should he use to wipe it? His cabinet under the sink holds an extensive array of likely tools: natural sponges, synthetic sponges, rolls of cotton gauze, paper towels, old Turkish towels, torn bed sheets, and chamois skins.

Though he may never state his objectives in so many words, he senses them keenly. He wants to keep the cleaning costs down. He wants a window that is clear and sparkling clean. Washing windows is not his favorite pastime, so he wants the glass to stay clean as long as possible. And he wants to get the job done quickly and efficiently, avoiding spillage, fuss, and excitation of any allergies he may have.

Being a conscientious homeowner, he now tries out the various cleaning agents, wiping materials, and application techniques. It's a long haul. Every cleaning session is followed by hard stares at the window and by stocktaking and debate. He concludes that every method is less than perfect but that some are better than others.

During coffee breaks at work, he may report to his colleagues on his efforts. He may explain the problem he faced, summarize what "Household Hints" said about window cleaning, describe his cleaning apparatus and methods, detail which methods left streaks and which didn't, and, finally, tell how his experiments helped him decide which method was best. In his informal conversation, he has followed, quite naturally, what has come to be the standard format for reports of empirical research:

- Introduction and Literature Review
- Materials and Methods
- Results
- Discussion

When put into final form, the empirical research report may also contain the usual elements found in reports, such as tables of content, abstracts, and references. Where the research is reported—usually either in a journal article or a student thesis—will determine the exact format. Because we cover the additional parts in Chapter 11, "Design Elements of Reports," we deal here only with the sections in the preceding list. Before we do, however, we have a few words about audience adaptation in empirical research reports.

Audience Adaptation

To illustrate the characteristics of empirical research reports, we draw on several well-written research reports. We have been careful to select passages that you should be able to read regardless of your specialization. On several occasions, however, we do define terms that you, as a nonspecialist, might not know. We place these definitions in brackets to distinguish them from the authors' work. Our need to define terms demonstrates a major point about audience adaptation in empirical research reports. Experts write these reports for their fellow experts. In these reports, you are free to use your professional vocabulary. You may use the standard vocabulary and standard knowledge of your field and expect your audience to understand you. In fact, your audience would be annoyed if you took time to define terms or explain concepts known to them.

However, there are times, even in reports written for your fellow experts, when you may be moving to the fringe of what is standard knowledge. If you use a new term or a highly specialized one, you will have to define it, using the definition techniques we describe on pages 164–169. Nor is there any reason in writing for experts to set aside the concepts of good style discussed in Chapter 6, "Achieving a Readable Style." A heavy, pretentious style, full of long convoluted sentences is a bad style, no matter the audience.

For further help in adapting for an expert audience, you may wish to review pages 69–72 in Chapter 4, "Writing for Your Readers."

Introduction and Literature Review

When research reports are presented in journal articles, most often the introduction and literature review are integrated. The major function of this integrated section is to describe the subject, scope, significance, and objectives of the research.

The literature review, as the name implies, reviews the scientific literature pertinent to the research being reported. In it, the investigator defines the problem being investigated as a way of leading to a statement of objectives. The investigator may also use the literature review to explain a choice of materials or methodology, or show the rationale for the investigation.

Because space in journals is expensive, the integrated introduction and literature review is held to information absolutely necessary to the investigation. However, when a research report is presented as a student thesis rather than an article, the literature review is often quite extensive and presented separately from the introduction. When such is the case, its purpose is not only to introduce the research but to demonstrate the investigator's mastery of a certain subject matter. In writing a thesis, always check with your advisor to determine the type of literature review required and the subject matter coverage desired.

Sample Introduction and Literature Review

Read over now the introduction and literature review of the article, "Sex Determining Temperatures in Turtles: A Geographic Comparison."

Subject: temperature as a sex determinant in turtles

The sex of many turtles is determined by the incubation temperature of the egg (Bull, 1980). Laboratory studies of six genera of turtles, for example, show that a developmental temperature of 25 C produces all males, 31 C or higher produces all females, and survival is sufficiently high in some studies to conclude that this is not due to differential death of the sexes (Pieau, 1975; Yntema, 1976; Bull and Vogt, 1979; Yntema and Mrosovsky, 1980). The laboratory studies have been corroborated by field studies, suggesting the nest temperature is the sex determining agent in these species (Bull and Vogt, 1979, and unpubl.).

Significance of subject

Temperature-dependent sex determination is unique because a variety of factors interact to determine the sex ratio: (1) maternal behavior in choosing a nest site, (2) the zygote's [the fertilized egg] response to temperature in becoming male or female, and (3) environmental effects on the temperature of the nesting area. As a result of the environmental component to sex determination, temporal and spatial variation of the environment may cause the sex ratio to vary. On the one hand, this sex ratio variation may jeopardize the evolutionary stability of temperature-dependent sex determination (Bull, 1980, 1981; Bulmer and Bull, 1982; see also Charnov and Bull, 1977). Here, however, we consider how the sex ratio evolves in response to the environment, rather than evolution of the sex determining mechanism per se. Long term environmental changes may lead to temporary sex ratio biases, but from Fisher (1930), selection acting on maternal behavior and/or zygotic sex determination would be expected to restore the equilibrium sex ratio.

Scope of research

Objective stated as hypothesis

A straightforward prediction of this hypothesis is that, in the warmer parts of a species range, (i) mothers should choose relatively cooler nests, and/or (ii) embryos should develop as male/female at higher temperatures than in the cooler localities. We have investigated the latter possibility by comparing sex determining temperatures among turtles from different geographic areas.[1]

As is typical of a scientific research article, this article is documented by parenthetical references (see pages 313–321). Notice how the authors use the literature cited to define the nature and scope of the problem. They state the precise objective of the research clearly in the form of a hypothesis: ". . . in the warmer parts of a species' range . . . embryos should develop as male/female at higher temperatures than in the cooler localities." We know now what question the researchers have asked of nature. We will expect before the article is finished to have an answer to the question: To know whether the hypothesis has been proved or disproved.

Statement of Objectives

The objectives need not always be stated as a hypothesis. Frequently, they are expressed as questions, as in this passage from another report:

Objectives as questions

> We then seek to answer the following questions raised by the model: 1) Is the spatial arrangement of nectar rewards within the inflorescence [pattern of flowers on a stalk or in a cluster] exploited by foraging bees in the manner predicted by our model? 2) Does the plant's pattern of nectar rewards elicit bee behavior which promotes pollen transfer?[2]

Sometimes, the writer simply states what was studied and the purpose for the study as in this statement from "Sexual Selection in a Brentid Weevil":

Objective as simple statement

> I report here the results of a study of the influence of body size on mate preference and on success in intrasexual competition in a natural aggregation of the neotropical brentid weevil, *Brentus anchorago* L. (Coleoptera, Curculionoidea, Brentidae).[3]

No matter how you present your objectives, be sure to present them with absolute clarity. No doubt should exist in the reader's mind concerning what you were up to in your research.

Choice of Materials or Methodology

The weevil article demonstrates how the literature review is used to explain the choice of materials used in the investigation:

Brentids make good candidates for studies of sexual selection and individual variation because most species of the family exhibit pronounced sexual dimorphism [differences between the sexes in characteristics such as size and color] (Muizon, 1960; Haedo Rossi, 1961; Damoiseau, 1967, 1971). The males generally possess greater body length, a stouter rostrum [beak or snout], and more powerful mandibles [jaws], one of which may be grossly enlarged (Darwin, 1871). Within each sex there is impressive phenotypic variation in body size, especially in males, which fight one another with snout and mandible for access to females (Wallace, 1869; Meads, 1976). The most size-variable brentid may be *B. anchorago:* after examining a large series of this species, Sharp (1895) commented that "the variation in length is enormous, and perhaps not equalled in the case of any other species of Coleoptera, small males being only 10–11 mm long, while large examples of the same size attain 52 mm." Such variation in size is common within a single aggregation, and is important in male mating success, in female choice, and in patterns of mating in the aggregation as a whole.[4]

Similarly, the investigator could use the literature review to explain or justify a choice of methodology.

Rationale for the Investigation

Often, the rationale for the investigation lies in past research. That is, past research may not have solved a problem adequately, or perhaps it was faulty in some way. There may be many reasons, and the investigator can use the literature review as a medium to express the reason or reasons for the research that he or she has conducted. The literature review of an article concerning Israeli agricultural extension publications gives a lapse of time in research as the rationale for the investigation being reported:

Molho found that in the Kibbutzim, 40 percent of the farmers read all, and the other 60 percent read part of the extension publications in the branch in which they specialized (27). Among Moshav farmers, 27 percent did not read at all extension publications, and the rest read part of them. Since this research was conducted twenty years ago, it became high time to re-evaluate Moshav farmers' reading habits, and to check if the initial difficulties in reading still existed.[5]

Verb Tense in Literature Reviews

Choosing proper verb tense is frequently a problem in writing a literature review. It will help to keep these principles in mind. When referring to the actual work that researchers have done in the past, use the past tense. When referring to the knowledge their research produced, if the knowledge is still considered to be true, use the present tense. Thus, the literature review usually mixes together past and present tense, as shown in this passage in which we have italicized the verb forms:

Past tense for actual work done

Present tense for current knowledge

> The best evidence for this theory *was provided* by Paulson (1974). He *pinned* live females of a number of species of coenagrionid damselflies to sedge stems and *presented* them to males of the various species. He then *scored* whether the male *approached* and *attempted* to clasp the female, and whether he *was* successful in clasping her. . . . Thus, for most of the species, reproductive isolation *is* normally *maintained* by the inability of males to clasp non-conspecific females with their anal appendages, presumably because their differently shaped and sized appendages *are* not compatible with the thoracic structure of the non-conspecific females.[6]

Materials and Methods

The major criterion by which you can measure the success of a materials and methods (M & M) section is simply stated: Using this section, an experienced researcher in the discipline should be able to duplicate the research. For a second criterion, an experienced researcher should be able to evaluate the research by using this section. If these criteria are not met, the M & M section fails.

M & M sections follow a fairly definite pattern. They may contain the following parts:

- Design of the investigation
- Materials
- Procedures
- Methods for observation and interpretation

Not every M & M section will contain all these parts, but every one should contain all of the parts needed to meet the criteria of duplication and evaluation.

Design of the Investigation

Give your readers an overview of your investigative design before you plunge them into the details. The overview need not, usually should not, be elaborate. Our turtle researchers give it in one sentence and, indeed, include in the same sentence information about the materials used:

The effect of temperature on sex determination was studied in turtles of the subfamily Emydinae, genera *Graptemys* (map turtles), *Pseudemys* (sliders), and *Chrysemys* (painted turtles), from populations in the northern U.S. (Wisconsin) and southern U.S. (Alabama, Mississippi, and Tennessee) (Table 1).[7]

TABLE 1. Sex ratio as a function of incubation temperature in turtles. (Data presented as percentage male, with the sample size in parentheses.)

		Temperature (degrees C)						
Species	**Loc**[1]	**28.0**	**28.3**	**29.0**	**29.3**	**29.5**	**30.0**	**30.6**
Chrysemys picta	TN	2 (41)	9 (11)	0 (12)	—	0 (5)	0 (16)	0 (14)
Chrysemys picta	WI	98 (94)	—	63 (38)	—	0 (7)	0 (56)	0 (22)
Graptemys geographica	WI	100 (26)	—	33 (6)	—	—	0 (28)	—
G. ouachitensis	WI	100 (93)	—	83 (64)	—	—	1 (89)	—
G. pseudogeographica	TN	100 (7)	96 (25)	0 (5)	28 (47)	16 (25)	0 (5)	0 (22)
G. pseudogeographica	WI	100 (70)	100 (14)	92 (24)	58 (57)	33 (15)	11 (82)	0 (17)
G. pulchra	MS	100 (17)	—	0 (4)	—	—	0 (14)	—
Pseudemys scripta	AL	100 (21)	—	37 (16)	—	—	0 (17)	—
Pseudemys scripta	TN	—	92 (36)	—	—	30 (40)	—	5 (42)

[1] Locality—AL: Alabama, Perry Co., Lakeland Farms; MS: Mississippi, Greene Co., Chickasawhay River; TN: Tennessee, Obion Co., Reelfoot Lake; WI: Wisconsin, Vernon Co., Mississippi River.

Materials

Materials can be human, animal, vegetable, or mineral. They are whatever you used by way of subjects, material, or equipment to do your research. In a report for the social sciences, instruments such as questionnaires would be described in this section. Remember that your descriptions of your materials have to be accurate enough that your readers could obtain similar materials. In the case of animals and plants, this usually means using their scientific as well as their common names. If you have a good deal of necessary information about your subjects or materials, use a table to display some of it. All these attributes are exhibited in the preceding passage quoted from the turtle research.

Sometimes the descriptions of materials can be precise, indeed, as in this passage:

> The organism used in this study was the wild type Chicago strain of the confused flour beetle, *Tribolium confusum* Duval, maintained by Prof. Thomas Park at the University of Chicago for over four decades on a single-sifted flour medium.[8]

Equipment used throughout the experiment can be described at this point, as in this passage:

> A bank of five parallel G.E. G8T5 germicidal lamps were used to generate ultraviolet light predominantly at 254 nm. The uncovered dishes to be irradiated were placed on a rotating platform 82 cm from the light source. A 10 cm diameter aperture midway between the light source and the rotating platform was used to collimate the incident light and to reduce shielding by the sides of the culture dishes.[9]

Equipment used only at specific times in the experiment will more likely be described at the appropriate place in the procedure, as in this sentence: "Thirty young adults drawn from a common stock were placed in a rearing jar."[10]

Procedures

In the procedures part, you describe for your readers step by step how you did your investigation. The description should be as complete as necessary, but remember you are writing for an expert audience. When you are working with a procedure or equipment common in the discipline, you do not need to describe it in detail. For instance, in the sentence just quoted the writers do not define or describe a "rearing jar." However, if you anticipate that your readers might have some question about why you conducted some step as you did, take time to explain.

You can save a great many words by referring to procedures described elsewhere rather than repeating the information found in the original source. This is an excellent practice so long as you don't refer your readers to journals unobtainable either by reason of their geographic location or their obscurity. All of these techniques are evident in these excerpts from our turtle report:

Succinct description of methodology

Eggs were removed from gravid [full of ripe eggs] females or from fresh nests and placed under controlled incubation temperatures within two weeks of removal, usually within one week. . . .

In nearly all cases eggs from each clutch were divided between three different incubation temperatures in the range 28–31 C and incubated at constant temperature throughout development; preliminary work on *Graptemys* had indicated that this range included male determining as well as female determining temperatures. The temperatures 28.3, 29.5, and 30.6 C were studied at Carnegie; 28.0, 29.0, 29.3, and 30.0 C were studied at Wisconsin, with separate clutches. Since incubators were precise only to within ±0.2 C degrees of the intended temperature, there was little to be gained from further subdividing this temperature range into narrower intervals.

Reference to procedure described elsewhere

Hatchling sex was diagnosed by inspection of gonads under a dissecting microscope (Bull and Vogt, 1979); specimens have been deposited in the Carnegie museum. All species studied here are known to develop as male at 25 C and female at 30–31 C with sufficiently high survival that the possibility of differential mortality of the sexes can be ruled out as the cause of the sex ratio biases (Bull and Vogt, 1979, and unpubl.).[11]

Methods for Observation and Interpretation

Finally, tell your readers how you observed your materials during the investigation and how you interpreted your results. Because these methods of observation and interpretation are often quite standardized, this part can frequently be quite short, as is this passage from our weevil report:

Reference to procedure described elsewhere

The behavioral sampling techniques I used are described in Altmann (1974). Sequences of male and female behavior that occurred around drilling females were dictated into a tape recorder, using Focal-Animal sampling. Focal animals were selected sequentially along the log to reduce repeat samplings to a minimum. Periodically I scanned the entire log and recorded the lengths of all weevils involved in copulations, fights, and guardings by an All Occurrence of Some Behaviors sampling technique. All major types of behavior were videotaped, and the verbal descriptions of motor patterns were prepared from these tapes.[12]

Use of standard sampling technique

Voice in Materials and Methods Sections

Notice that in the last passage quoted the author uses the first-person *I*. However, also notice that in most of the passages we have quoted from M & M sections the authors have seldom used the first person and have used far more passive voice than active voice sentences. In most cases, it is either obvious that the researchers performed the steps described, or it is unimportant who performed them. Under such circumstances, passive voice is as good a choice or perhaps even a better choice than active voice. But don't fear using active voice and first person when they seem appropriate to you. Most modern stylebooks encourage such practices, and an occasional *I* or *we* reminds your readers that real people are at work.

Also, remember that passive voice used carelessly creates a great many dangling modifiers: "After drawing the blood, the calf was returned to the pen." Here the case of the blood-drawing calf can be cleared up with a judicious use of active voice: "After drawing the blood, I returned the calf to the pen." (See pages 125–126 and 183–184.)

Results

Because your results section answers the questions you have asked, it is the most important section of your report. Nonetheless, it is often the shortest section of an empirical report. It often takes a great deal of work to gain only a few bits of knowledge.

Begin your results section with an overview of what you have learned. The first sentence or two in the results section should be like a lead in a newspaper story, where the main points of the story are quickly given, to be followed in later paragraphs by the details. In a results section, the following details, if at all possible, are presented in tables and graphs. You do not then need to restate such details. But you may want to refer to key data, both to emphasize their significance and to help your readers comprehend your tables and graphs. The following excerpts from our turtle report demonstrates these attributes:

Reference to table
An important result stated and explained

The hatchling sex ratios obtained are shown in Table 1 [see our page 547]. A striking result in all populations is that there is an abrupt change from virtually all males to all females over a narrow range of temperatures. In every sample except Tennessee *C. picta*, all or nearly all turtles are male at 28 C, female at 30 C and above. (Tennessee *C. picta* are nearly all females at 28 C and above, all males at 25 C, so there is an abrupt change in this population as well.) This abrupt change in sex ratio is referred to as a threshold (Bull, 1980), and the temperature producing a sex ratio (proportion male) of 1/2 is defined here as the *threshold temperature*. . . .

A result key to disproving hypothesis

The samples at 29 C offer the most extensive set of interspecific comparisons for a temperature within the range of intermediate sex ratios. Among these there is no indication that threshold temperatures are higher in southern species, and some comparisons even suggest that threshold temperatures can at most be only marginally higher in the southern species. . . .

Another result disproving hypothesis

Even if threshold temperatures are higher in most of these southern populations (having escaped detection because of the small samples or possible experimental errors), in several cases the maximum difference is of the order of 0.4 C degrees.[13]

At this point, probably even a nonbiologist could tell whether the turtle researchers' hypothesis is proved or disproved. If not, the discussion should tell us.

Discussion

The discussion interprets the results. It answers questions such as these:

- Do the results really answer the questions raised?
- Are there any doubts about the results? Why? Was the methodology flawed? How could it be improved?
- Were the research objectives met?
- Was the hypothesis proved or disproved?
- How do the results compare with results from previous research? Are there areas of disagreement? Can disagreements be explained?
- What are the implications for future work?

Though the discussion section may cover a lot of ground, keep it tightly organized around the answers to the questions that need to be asked. Our turtle researchers did exactly that as these excerpts from their discussion show:

Restatement of hypothesis

According to our hypothesis, the southern populations should have higher threshold temperatures, perhaps by as much as 2 C, to compensate for the presumed warmer nests.

Implication of negative data

The data overwhelmingly indicate that (1) threshold temperatures do not increase to nearly the same magnitude as climatic temperatures between the Wisconsin and southern populations. In many cases threshold temperatures are at most only slightly higher (0.3–0.4 C) in the southern population, if indeed higher. We therefore must reject the hypothesis that differences in threshold temperature between populations are of the same magnitude and direction as differences in climatic temperature. Furthermore, (2) in the two northern intraspecific comparisons, the threshold temperatures differ in the opposite direction as climatic temperatures, and not even one North–South interspecific comparison provides significant evidence of a higher threshold in the southern population. We therefore reject even the less stringent form of our hypothesis—that, regardless of the magnitude of difference, climatic and threshold temperatures merely covary in the same direction. Of course data on additional populations may show these results to be atypical of turtles, but even so, these results require explanations in themselves.

Rejection of hypothesis

Further evidence against hypothesis

If threshold temperatures do not increase with ambient temperature, there might be greater female-based sex ratios in the South than in the North. Preliminary evidence is to the contrary, however. In a study of three southern U.S. turtle communities, McCoy and Vogt (unpubl.) observed an adult sex ratio of nearly 1/2 in *Pseudemys* and female-based sex ratios in several species of *Graptemys*. However, Vogt (1980) observed even more heavily female-based sex ratios in two species of *Graptemys* at the Wisconsin site, and the sex ratio biases were documented using a variety of collecting techniques (Vogt, 1980a). Thus although southern localities are warmer then northern ones and threshold temperatures do not compensate, sex ratios are not obviously more female-based in the South.

Three explanations for negative results, all with implications for future research

We consider three possible explanations for a lack of a positive association between environmental temperature and threshold temperature in these turtles. (1) The characteristics we diagnose as male and female in hatchlings may be irrelevant to sex in adults, if the sex phenotype changes later in life. Although numerous observations support a hypothesis of constant sex phenotype (Yntema, 1976, 1981; Bull and Vogt, 1979; Vogt, 1980), unequivocal data are lacking, and this should be entertained as a possibility. (2) The threshold temperatures in this study may not accurately reflect sex determination in nests. Nest temperatures fluctuate on a daily basis, and the fluctuations possibly affect sex determination in a way not evident from threshold temperatures. . . .

Perhaps the most likely explanation for the lack of correspondence between climatic temperature and threshold temperature is that (3) nest temperatures do not correspond to climatic temperatures, either because differences in nesting behavior compensate (overcompensate) for the climatic differences or because climatic data are misleading when extrapolated to subsurface soil temperatures. If maternal behavior compensates, this means that southern turtles choose nest sites which are relatively cool, whereas northern turtles choose nest sites which are relatively warm in their local environment. This has not been studied directly. However, southern turtles begin nesting a few weeks earlier than northern ones (Vogt, unpubl.), and this will partly compensate for the climatic differences, provided the southern turtles also cease nesting earlier or at the same time as the northern ones. Comparison of June temperatures for the southern localities with July temperatures for the northern ones removes only about half the difference in climatic temperatures, and this is likely an overestimate of the magnitude of compensation from early laying. Other means of maternal behavior compensating for climatic differences are possible as well, such as changes in nest position relative to shading vegetation. And it is further possible that the female has only a limited range of nest habitats available, so that nest temperature depends more upon the habitat than on ambient temperature. The resolution of why these threshold temperatures do not (strongly) accord with climatic temperatures clearly requires supplementation with data of nest temperatures in the different populations.[14]

A Final Word

In this chapter we have given you general advice about reporting empirical research. If you are to become a professional in any field that requires such reporting, doing it well will be of vital importance to you. Therefore, we strongly urge you to examine representative journals and student theses in your discipline. Observe closely their format and style. Most journals have a section labeled something like "Information for Contributors." This section will inform you about manuscript preparation and style. Often, it will refer you to the style manual, such as the *Council of Biology Editors Style Manual,* that governs the journal. Likely, your library will have a copy of the manual you need. Check it out and read it carefully. It will supplement what you have learned here.

Planning and Revision Checklist

You will find the planning and revision checklists that follow Chapter 2, "Composing" (pages 33–35 and inside the front cover), and Chapter 4, "Writing for Your Readers" (pages 78–79), valuable in planning and revising any presentation of technical information. The following questions specifically apply to empirical research reports. They summarize the key points in this chapter and provide a checklist for planning and revising.

Planning

- What is the subject, scope, and significance of your research?
- What were your objectives? How can you best state your objectives? As hypotheses? As questions? As a statement of purpose?
- What do want to accomplish with your literature review? Definition of the research problem? Explanation of choice of materials and methods? Rationale for investigation?
- Is your report going to be a journal article or a thesis? If a thesis, have you consulted with your thesis advisor about it?
- Do you have the following well in mind for your materials and methods section?
  - Design of the investigation?
  - Materials?
  - Procedures?
  - Methods for observation and interpretation?
- Are all your results in? Can some of them be tabulated or displayed in charts or graphics?
- Which of these questions need to be answered in your discussion section?
  - Do the results really answer the questions raised?
  - Are there any doubts about the results? Why? Was the methodology flawed? How could it be improved?
  - Were the research objectives met?
  - Was the hypothesis proved or disproved?
  - How do the results compare with results from previous research? Are there areas of disagreement? Can disagreements be explained?
  - What are the implications for future work?

Revision

- Will your reader know the subject, scope, significance, and objectives of your investigation? Are your objectives stated absolutely clearly?
- Have you used past and present tense appropriately in your literature review?
- Would an experienced researcher in your field be able to use your materials and methods section either to duplicate your investigation or to evaluate it?
- Have you used active voice and passive voice appropriately in your materials and methods section? If you have used passive voice, have you avoided dangling modifiers?
- Do the first few sentences of your results section present an overview of the results? Have you used tables and graphs when appropriate?
- Have you kept your discussion tightly organized around the questions that needed answering?

Exercises

1. Empirical research is primarily concerned with fact finding and interpretation. Do you see any similarity between empirical research and feasibility studies (Chapter 17)? In what major respects are they different?

2. Referring to Chapter 5 of this book, "Gathering and Checking Information," determine how the methods used to gather information are affected by the nature and purpose of the investigation. How do the techniques discussed in Chapter 5 relate to empirical research?
3. Choose a research problem in your discipline, perhaps with the help of an instructor in that field. Research the literature in the problem sufficiently so that you can formulate an empirical study to deal with some aspect of the problem. Then write an introduction and literature review and a materials and methods section for the study. Submit your work to both your writing teacher and the teacher in the discipline.
4. Choose an empirical research report published in a journal in your discipline. With the help of this chapter and the style manual that governs the journal, if one is available to you, analyze the report's format, style, organization, and content. How closely does it follow the principles of this chapter and the style manual? Present your analysis as a written report for your writing instructor.

PART V
Oral Reports

Oral Reports

CHAPTER 19

Oral Reports

PRESENTATION

Physical Aspects of Speaking
Audience Interaction

VISUAL AIDS

Purpose
Visual Aids Criteria
Visual Content
Visual Presentation Tools

Oral reports are a major application of reporting technical information. You will have to report committee work, laboratory experiments, and research projects. You will give reports at business or scholarly meetings. You will instruct, if not in a teacher–student relationship, perhaps in a supervisor–subordinate relationship. You may have to persuade a group that a new process your section has devised is better than the present process. You may have to brief your boss about what your department does to justify its existence. In this chapter we discuss preparing and presenting your oral report, with a heavy emphasis on the ways in which you can provide visual support.

Preparation

In many ways, preparing an oral report is much like composing a written report. The situational analysis is virtually identical to the situational analysis described in Chapter 2, "Composing." You have to consider your purpose and audience, discover your material, and arrange your material. You will not in most cases write out your material beyond an outline stage. However, you should rehearse its delivery several times, which is akin to writing and revising. In addition to these tasks, you will have others peculiar to the speech situation.

Find out as much as you can about the conditions under which you will speak. Inquire about the size of the room you will speak in, the time allotted for the speech, and the size of the audience. If you have to speak in a large area to a large group, will a public address system be available to you? Find out if you will have a lectern for your notes. If you plan to use visual aids, inquire about the equipment. Does the sponsoring group have projectors to show 35-mm slides or transparencies? Many speakers have arrived at a hall and found all of their vital visual aids worthless because projection equipment was not available. Find out if there will be someone to introduce you. If not, you may have to work your credentials as a speaker into your talk. Consider the time of day and day of the week. An audience listening to you at 3:30 on Friday afternoon will not be nearly as attentive as an audience earlier in the day or earlier in the week. Feel free to ask the sponsoring group any of these questions. The more you know beforehand, the better prepared and therefore the more comfortable you will be.

Delivery Techniques

There are four basic **delivery techniques,** but you really need to think about only two of them. The four are (1) impromptu, (2) speaking from memory, (3) extemporaneous, and (4) reading from a manuscript. Impromptu speaking involves speaking "off the cuff." Such a method is too risky for a technical report, where accuracy is so vital.

In speaking from memory, you write out a speech, commit it to memory, and then deliver it. This gives you a carefully planned speech, but we cannot recommend it as a good technique. The drawbacks are (1) your plan becomes inflexible; (2) you may have a memory lapse in one place that will unsettle you for the whole speech; (3) you think of words rather than thoughts, which makes you more artificial and less vital; and (4) your voice and body actions become stylized and lack the vital spark of spontaneity.

We consider the best methods to be extemporaneous speech and the speech read from a manuscript, and we will discuss these in more detail.

The Extemporaneous Speech

Unlike the impromptu speech, with which it is sometimes confused, the extemporaneous speech is carefully planned and practiced. In preparing for an extemporaneous speech, you go through the planning and arranging steps described in Chapter 2. But you stop when you complete the outline stage. You do not write out the speech. Therefore, you do not commit yourself to any definite phraseology. In your outline, however, include any vital facts and figures that you must present accurately. You will want no lapses of memory to make you inaccurate in presenting a technical report.

Before you give the speech, practice it, working from your outline. Give it several times, before a live audience preferably, perhaps a roommate or a friend. As you practice, fit words to your outlined thoughts. Make no attempt to memorize the words you choose at any practice session, but keep practicing until your delivery is smooth. When you can go through the speech without faltering, you are ready to present it. When you practice a speech, pay particular attention to timing. Depending upon your style and the occasion, plan on a delivery rate of 120–180 words per minute. Nothing, *but nothing,* will annoy program planners or an audience more than to have a speaker scheduled for 30 minutes go for 40 minutes or an hour. The long-winded speaker probably cheats some other speaker out of his or her allotted time. Speakers who go beyond their scheduled time can depend upon not being invited back.

We recommend that you type your outline. Use capitals, spacing, and underlining generously to break out the important divisions. Use boldface type if you have word processing capability. But don't do the entire outline in capitals. That would make it hard to read. As a final refinement, place your outline in a looseleaf ring binder. By so doing you can be sure that it will not become scattered or disorganized.

There are several real advantages of the extemporaneous speech over the speech read from manuscript. With the extemporaneous speech you will find it easier to maintain eye contact with your audience. You need only glance occasionally at your outline to keep yourself on course. For the rest of the time you can concentrate on looking at your audience.

You have greater flexibility with an extemporaneous speech. You are committed to blocks of thought but not words. If by looking at your audience you see that they have not understood some portion of your talk, you are free to rephrase the thought in a new way for better understanding. If you are really well prepared in your subject, you can bring in further examples to clarify your point. Also, if you see you are running overtime, you can condense a block by leaving out some of your less vital examples or facts.

Finally, because you are not committed to words, you retain conversational spontaneity. You are not faltering or groping for words, but neither are you running by your audience like a well-oiled machine.

The Manuscript Speech

Most speech experts recommend the extemporaneous speech above reading from a manuscript. We agree in general. However, speaking in a technical situation often requires the manuscript speech. Papers delivered to scientific societies are frequently written and then read to the group. Often, the society will later publish your paper. Often, technical reports contain complex technical information or extensive statistical material. Such reports do not conform well to the extemporaneous speech form, and you should plan to read them from a manuscript.

Planning and writing a speech is little different from writing a paper. However, in writing your speech try to achieve a conversational tone. Certainly, in speaking you will want to use the first person and active voice. Remember that speaking is more personal than writing. Include phrases like "it seems to me," "I'm reminded of," "Just the other evening, I," and so forth. Such phrases are common in conversation and give your talk extemporaneous overtones. Certainly, prefer short sentences to long ones.

Type the final draft of your speech. Just as you did for the extemporaneous speech outline, be generous with capitals, spacing, and underlining. Plan on about three typed pages per five minutes of speech. Put your pages in order and place them in a looseleaf binder.

When you carry your written speech to the lectern with you, you are in no danger of forgetting anything. Nevertheless, you must practice it, again preferably aloud to a live audience. As you practice, remember that because you are tied to the lectern, your movements are restricted. You will need to depend even more than usual on facial expression, gestures, and voice variation to maintain audience interest. Do not let yourself fall into a sing-song monotone as you read the set phrases of your written speech.

Practice until you know your speech well enough to look up from it for long periods of good eye contact. Plan an occasional departure from your manuscript to speak extemporaneously. This will aid you to regain the direct contact with the audience that you so often lose while reading.

Arranging Content

For the most part you will arrange your speech as you do your written work. However, the speech situation does call for some differences in arrangement and even content, and we will concentrate on these differences. We will discuss the arrangement in terms of introduction, body, and conclusion.

Introduction

A speech introduction should accomplish three tasks: (1) create a friendly atmosphere for you to speak in, (2) interest the audience in your subject, and (3) announce the subject, purpose, scope, and plan of development of your talk.

Be alert before you speak. If you can, mingle and talk with members of the group to whom you are going to speak. Listen politely to their conversation. You may pick up some tidbit that will help you to a favorable start. Look for bits of local color or another means to establish a common ground between you and the audience. When you begin to speak, mention some member of the audience or perhaps a previous speaker. If you can do it sincerely, compliment the audience. If you have been introduced, remember to acknowledge and thank the speaker. Unless it is a very formal occasion, begin rather informally. If there is a chairperson and a somewhat formal atmosphere, we recommend no heavier beginning than, "Mr. Chairman (or Madame Chairwoman), ladies and gentlemen."

Gain attention for your subject by mentioning some particularly interesting fact or bit of illustrative material. Anecdotes are good if they truly tie in with the subject. But take care with humor. Avoid jokes that really don't tie in with the subject or the occasion. Forget about risqué stories.

Some speakers try to startle their audience, but there are pitfalls to this method. Shortly after World War II, one of the authors of this text began a speech by pretending to be Adolf Hitler for a few minutes. (The purpose of the speech was to expose the wickedness of the big lie in propaganda.) He raged and ranted and denounced Wall Street, Franklin Roosevelt, and the Jews. The audience froze into resentment, and he never got them back. When he stopped his play acting and explained why he had imitated Hitler, most of the audience never even heard him—they were already too angry.

Be careful also about what you draw attention to. Do not draw attention to shortcomings in yourself, your speech, or the physical surroundings. Do not begin speeches with apologies.

Announce your subject, purpose, scope, and plan of development in a speech just as you do in writing. (See pages 292–297.) If anything, giving your plan of development is more important in a speech than in an essay. Listeners cannot go back in a speech to check on your arrangement the way that a reader can in an essay. So, the more guideposts you give an audience, the better. No one has ever disputed this old truism: (1) Tell the audience what you are going to tell them. (2) Tell them. (3) Tell them what you just told them. In instructional situations, some speakers provide their audiences with a printed outline of their talk.

Body

When you arrange the body of a speech, you must remember one thing: A listener's attention span is very limited. Analyze honestly your own attention span—be aware of your own tendency to let your mind wander. You listen to the speaker for a moment, and then perhaps you think of lunch, of some problem, or an approaching appointment. Then you return to the speaker. When you become the speaker, remember that people do not hang on your every word.

What can you do about the problem of the listener's limited attention span? In part, you solve it by your delivery techniques. We will discuss these in the next section of this chapter. It also helps to plan your speech around intelligent and interesting repetition.

Begin by cutting the ground you intend to cover in your speech to the minimum. Build a five-minute speech around one point, a 15-minute speech around two. Even an hour-long talk probably should not cover more than three or four points.

Beginning speakers are always dubious about this advice. They think, "I've got to be up there for 15 minutes. How can I keep talking if I only have two points to cover? I'll never make it." Because of this fear they load their speeches with five or six major points. As a result, they lull their audience into a state of somnolence with a string of generalizations.

In speaking, even more than in writing, **your main content should be masses of concrete information—examples, illustrations, little**

narratives, analogies, and so forth—supporting just a few generalizations. As you give your supporting information, repeat your generalization from time to time. Vary its statement, but cover the same ground. The listener who was out to lunch the first time you said it may hear it the second time or the third. You use much the same technique in writing, but you intensify it even more in speaking.

We have been using the same technique here in this chapter. We began this section on the speech body by warning you that a listener's attention span is short. We reminded you that your listening span is short; same topic but a new variation. We asked you what you can do about a listener's limited span; same topic with only a slight shift. In the next paragraph we told you not to make more than two points in a 15-minute speech. We nailed this point down in the next paragraph by having a dubious speaker say, "I've got to be up there for 15 minutes. How can I keep talking if I have only two points to cover?" In the paragraph just preceding this one we told you to repeat intelligently so that the reader "who was out to lunch the first time you said it may hear it the second time or the third." Here we were slightly changing an earlier statement that "You listen to the speaker for a moment, and then perhaps you think of lunch. . . ." In other words, we are aware that the reader's attention sometimes wanders. When you are paying attention we want to catch you. Try the same technique in speaking, because the listener's attention span is even more limited than the reader's.

Creating suspense as you talk is another way to generate interest in your audience. Try organizing a speech around the inductive method. That is, give your facts first and gradually build up to the generalization that they support. If you do this skillfully, using good material, your audience hangs on, wondering what your point will be. If you do not do it skillfully or use dull material, your audience will tune you out and tune into their private worlds.

Another interest-getting technique is to **relate the subject matter to some vital interest of the audience.** If you are talking about water pollution, for example, remind the audience that the dirtier their rivers get, the more tax dollars it will eventually take to clean them up.

Visual aids often increase audience interest. Remember to keep your graphics big and simple. No one is going to see typewritten captions from more than three or four feet away. Stick to big pie and bar graphs. If you have tables, print them in letters from two to three inches high. If you are speaking to a large group, put your graphic materials on transparencies and project them on a screen. Prepare your transparencies with care. Don't just photocopy typed or printed pages or graphics from books. No one behind the first row will see them. To work, letters and numbers on transparencies should be at least twice normal size. Large-type typewriters are available. Word processing makes large type fonts available.

If you need an assistant to help you project visual aids, get one or bring one with you.

Do not display a visual aid until you want the audience to see it. While the aid is up, call your listeners' attention to everything you want them to see. Take the aid away as soon as you are through with it. If using a projector, turn it off whenever it is not in use. Be sure to key every visual aid into your speaker's script. Otherwise, you may slide right by it. (See also the section, "Visual Aids," pages 572–582.)

Conclusion

In ending your speech, as in your written reports, you have your choice of several closes. You can close with a summary, or a list of recommendations including a call for some sort of action, or what amounts to "Good-bye, it's been good to talk to you." As in the introduction, you can use an anecdote in closing to reinforce a major point. In speaking, never suggest that you are drawing to a close unless you really mean it. When you suggest that you are closing, your listeners perk up and perhaps give a happy sigh. If you then proceed to drag on, they will hate you.

Second, remember that audience interest is usually highest at the beginning and close of a speech. Therefore, you will be wise to provide a summary of your key points at the end of any speech. Give your listeners something to carry home with them.

Presentation

After you have prepared your speech you must present it. For many people giving a speech is a pretty terrifying business. Before speaking they grow tense, have hot flashes and cold chills, and experience the familiar butterflies in the stomach. Some people tremble before and even during a speech. Try to remember that these are normal reactions, for both beginning and experienced speakers. Most people can overcome them, however, and it is even possible to turn this nervous energy to your advantage.

If your stage fright is extreme, or if you are the one person in a hundred who stutters, or if you have some other speech impediment, seek clinical help. The ability to communicate ideas through speech is one of humanity's greatest gifts. Do not let yourself be cheated. Winston Churchill had a speech impediment as a child. Some of the finest speakers we have ever had in class were stutterers who admitted their problem and worked at it with professional guidance. Remember, whether your problems are large or small, the audience is on your side. They want you to succeed.

Physical Aspects of Speaking

What are the physical characteristics of good speakers? They stand firmly but comfortably. They move and gesture naturally and emphatically but avoid fidgety, jerky movements and foot shuffling. They look directly into the eyes of people in the audience, not merely in their general direction. They project enthusiasm into their voices. They do not mumble or speak flatly. We will examine these characteristics in detail—first movement and then voice.

Movement A century ago a speaker's movements were far more florid and exaggerated than they are today. Today we prefer a more natural mode of speaking, closer to conversation than oratory. To some extent, electronic devices such as amplifying systems, radio, and television have brought about this change. However, you do not want to appear like a stick of wood. Even when speaking to a small group or on television (or, oddly enough, on the radio) you will want to move and gesture. If you are speaking in a large auditorium, you will want to broaden your movements and gestures. From the back row of a 2,500-seat auditorium, you look about three inches tall.

Movement during a speech is important for several reasons. First, it puts that nervous energy we spoke of to work. The inhibited speaker stands rigid and trembles. The relaxed speaker takes that same energy and puts it into purposeful movement.

Second, movement attracts attention. It is a good idea to emphasize an idea with a pointing finger or a clenched fist; and a speaker who comes out from behind the lectern occasionally and walks across the stage or toward the audience awakens audience interest. The speaker who passively utters ideas deadens the audience.

Third, movement makes you feel more forceful and confident. It keeps you, as well as your audience, awake. This is why good speakers while speaking over the radio will gesture just as emphatically as though the audience could see them.

What sorts of movements are appropriate? To begin with, **movement should closely relate to your content.** Jerky or shuffling motions that occur haphazardly distract an audience. But a pointing finger combined with an emphatic statement reinforces a point for an audience. A sideward step at a moment of transition draws attention to the shift in thought. Take a step backward and you indicate a conclusion. Step forward and you indicate the beginning of a new point. Use also the normal descriptive gestures that all of us use in conversation: gestures to indicate length, height, speed, roundness, and so forth.

For most people, **gesturing is fairly normal.** They make appropriate movements without too much thought. Some beginning speakers, however, are body inhibited. If you are in this category, you may have to cultivate movement. In your practice sessions and in your classroom speeches, risk artificiality by making gestures that seem too broad to you. Oddly enough, often at the very point where your gestures seem artificial and forced to you, they will seem the most natural to your audience.

Allow natural gestures to replace nervous mannerisms. Some speakers develop startling mannerisms and remain completely oblivious of them until some brave but kind soul points them out. Some that we have observed include putting eyeglasses off and on; knocking a heavy ring over and over on the lectern; fiddling with a pen, pointer, chalk, cigar, microphone cord, ear, mustache, nose, you name it; shifting from foot to foot in time to some strange inner rhythm; and pointing with the elbows while the hands remain in the pants pockets. Mannerisms may also be vocal. Such things as little coughs or repeating comments such as "OK" or "You know" to indicate transitions may become mannerisms.

Listeners are distracted by such habits. Often they will concentrate on the mannerisms to the exclusion of everything else. They may know that a speaker put her eyeglasses on and off 22 times but not have the faintest notion of what she said. If someone points out such mannerisms in your speaking habits, do not feel hurt. Instead, work to remove the mannerisms.

Movement includes facial movement. Do not be a deadpan. Your basic expression should be a relaxed, friendly look. But do not hesitate to smile, laugh, frown, or scowl when such expressions are called for. A scowl at a moment of disapproval makes the disapproval that much more emphatic. Whatever you do, do not freeze into one expression, whether it be the stern look of the man of iron or the vapid smile of a smoker in a magazine ad.

Voice **Your voice should sound relaxed, free of tension and fear.** In a man, people consider a deep voice to be a sign of strength and authority. Most people prefer a woman's voice to be low rather than shrill. If you do not have these attributes, you can develop them to some extent. Here we must refer you to some of the good speech books in our bibliography (Appendix C), where you will find various speech exercises described. If, despite hard work, your voice remains unsatisfactory in comparison with the conventional stereotypes, do not despair. Many successful speakers have had somewhat unpleasant voices and through force of character or intellect directed their audiences to their ideas and not their voices.

Many beginning speakers speak too fast, probably because they are anxious to be done and sit down. A normal rate of speech falls between 120 and 180 words a minute. This is actually fairly slow. Generally, you will want a fairly slow delivery rate. **When you are speaking slowly, your voice will be deeper and more impressive.** Also, listeners have trouble following complex ideas delivered at breakneck speed. Slow up and give your audience time to absorb your ideas.

Of course, **you should not speak at a constant rate, slow or fast. Vary your rate.** If you normally speak somewhat rapidly, slowing up will emphasize ideas. If you are speaking slowly, suddenly speeding up will suggest excitement and enthusiasm. As you speak, change the volume and pitch of your voice. Any **change in volume,** whether from low to loud or the reverse, will draw your listener's attention and thus emphasize a point. The same is true of a **change in pitch.** If your voice remains a flat monotone and your words come at a constant rate, you deprive yourself of a major tool of emphasis.

Many people worry about their accent. Normally, our advice is *don't.* If you speak the dialect of the educated people of the region where you were raised, you have little to worry about. Some New Englanders, for example, put *r*'s where they are not found in other regional dialects and omit them where they are commonly found. Part of America's richness lies in its diversity. Accents vary in most countries from one region to another, but certainly not enough to hinder communication.

If, however, your accent is slovenly—"Ya wanna cuppa coffee?"—or uneducated, do something about it. Work with your teacher or seek other professional help. Listen to educated speakers and imitate them. **Whatever your accent, there is no excuse for mispronouncing words.** Before you speak, look up any words you know you must use and about which you are uncertain of the pronunciation. Speakers on technical subjects have this problem perhaps more than other speakers. Many technical terms are jawbreakers. Find their correct pronunciation and practice them until you can say them easily.

Audience Interaction

One thing speakers must learn early in their careers is that they cannot count on the audience's hanging on every word. Some years ago an intelligent, educated audience was asked to record its introspections while listening to a speaker. The speaker was an excellent one. Despite his excellence and the high level of the audience, the introspections revealed that the audience was paying something less than full attention. Here are some of the recorded introspections:

> *God, I'd hate to be speaking to this group. . . . I like Ben—he has the courage to pick up after the comments. . . . Did the experiment backfire a bit? Ben seems unsettled by the introspective report. . . . I see Ben as one of us because he is under the same judgment. . . . He folds his hands as if he was about to pray. . . . What's he got in his pocket he keeps wriggling around. . . . I get the feeling Ben is playing a role. . . . It is interesting to hear the words that are*

emphasized. . . . This is a hard spot for a speaker. He really must believe in this research. . . .

Ben used the word "para-social." I don't know what that means. Maybe I should have copied the diagram on the board. . . . Do not get the points clearly . . . cannot interrupt . . . feel mad . . . More words. . . . I'm sick of pedagogical and sociological terms. . . . Slightly annoyed by pipe smoke in my face. . . . An umbrella dropped. . . . I hear a funny rumbling noise. . . . I wish I had a drink. . . . Wish I could quit yawning. . . . Don't know whether I can put up with these hard seats for another week and a half or not. . . . My head itches. . . . My feet are cold. I could put on my shoes, but they are so heavy. . . . My feet itch. . . . I have a piece of coconut in my teeth. . . . My eyes are tired. If I close them, the speaker will think I'm asleep. . . . I feel no urge to smoke. I must be interested. . . .

Backside hurts . . . I'm lost because I'm introspecting. . . . The conflict between introspection and listening is killing me. Wish I didn't take a set so easily. . . . If he really wants me to introspect, he must realize himself he is wasting his time lecturing. . . . This is better than the two hour wrestling match this afternoon. . . . This is the worst planned, worst directed, worst informed meeting I have ever attended. . . . I feel confirmation, so far, in my feelings that lectures are only 5% or less effective. . . . I hadn't thought much about coming to this meeting, but now that I am here it is going to be O.K. . . . Don't know why I am here. . . . I wish I had gone to the circus. . . . Wish I could have this time for work I should be doing. . . . Why doesn't he shut up and let us react. . . . The end of the speech. Now he is making sense. . . . It's more than 30 seconds now. He should stop. Wish he'd stop. Way over time. Shut Up. . . . He's over. What will happen now? . . .[1]

As some of the comments reveal, perhaps being asked to record vagrant thoughts as they appeared made some members of the audience less attentive than they normally would have been. But most of us know that we have very similar thoughts and lapses of attention while we attend classes and speeches.

Reasons for audience inattention are many. Some are under the speaker's control; some are not. The speaker cannot do much about such physical problems as hard seats, crowded conditions, bad air, and physical inactivity. The speaker can do something about psychological problems such as the listeners' passivity and their sense of anonymity, their feeling of not participating in the speech.

Audience Analysis Even before they begin to speak, good speakers have taken audience problems into account. They have analyzed the audience's education and experience level. They have planned to keep their points few and to repeat major points through carefully planned variations. They plan interesting examples. While speaking they attempt to interest the audience through movement and by varying the speech rate, pitch, and volume.

But good speakers go beyond these steps and analyze their audience and its reactions as they go along. In an extemporaneous speech and even to some extent in a written speech, you can make adjustments based on this audience analysis.

To analyze your audience, you must have good eye contact. You must be looking at Ben, Bob, and Irma. You must not merely be looking in the general direction of the massed audience. Look for such things as smiles, scowls, fidgets, puzzled looks, bored expressions, interested expressions, sleepy eyes, heads nodded in agreement, heads nodded in sleep, heads shaken in disagreement. You will not be 100% correct in interpreting these signs. Many students have learned to smile and nod in all the proper places without ever hearing the instructor. But, generally, such physical actions are excellent clues as to how well you are getting through to your audience.

Reacting to Audience If your audience seems happy and interested, you can proceed with your speech as prepared. If, however, you see signs of boredom, discontent, or a lack of understanding, you must make some adjustments. Exactly what you do depends to some extent on whether you are in a formal or informal speaking situation. We will look at the formal situation first.

In the *formal* situation you are somewhat limited. If your audience seems bored, you can quickly change your manner of speaking. Any change will, at least momentarily, attract attention. You can move or gesture more. With the audience's attention gained, you can supply some interesting anecdotes or other illustrative material to support better your abstractions and generalizations. If your audience seems puzzled, you know you must supply further definitions and explanations and probably more concrete examples. If your audience seems hostile, you must find some way to soften your argument while at the same time preserving its integrity. Perhaps you can find some mutual ground upon which you and the audience can agree and move on from there.

Obviously, such flexibility during the speech requires some experience. Also it requires that the speaker have a full knowledge of the subject. If every bit of material the speaker knows about the subject is in the speech already, the speaker has little flexibility. But do not be afraid to adjust a speech in midstream. Even the inexperienced speaker can do it to some extent.

Many of the speaker's problems are caused by the speech situation's being a one-way street. The listeners sit passively. Their normal desires to react, to talk back to the speaker, are frustrated. The problem suggests the solution, particularly when you are in a more informal speech situation, such as a classroom or a small meeting.

In the more *informal* situation, you can stop when a listener seems puzzled. Politely ask him where you have confused him and attempt to clarify the situation. If a listener seems uninterested, give him an opportunity to react. Perhaps you can treat him as a puzzled listener. Or, you can ask him what you can do to interest him more. Do not be unpleasant.

Put the blame for the lack of interest on yourself, even if you feel it does not belong there. Sometimes you may be displeased or shocked at the immediate feedback you receive, but do not avoid it on these grounds. And do not react unpleasantly to it. You will move more slowly when you make speaking a two-way street, but the final result will probably be better. Immediate feedback reveals areas of misunderstanding or even mistrust of what is being said.

In large meetings where such informality is difficult, you can build in some audience reaction through the use of informal subgroups. Before you talk, divide your audience into small subgroups, commonly called **buzz groups.** Use seating proximity as the basis for your division if you have no better one. Explain that after your talk the groups will have a period of time in which to discuss your speech. They will be expected to come up with questions or comments. People do not like to seem unprepared, even in informal groups. As a result, they will be more likely to pay attention to your speech in order to participate well in their buzz groups.

Whether you have buzz groups or not, often you will be expected to handle questions following a speech. If you have a chairperson, he or she will field the questions and repeat them, and then you will answer them. If you have no chairperson, you will perform this chore for yourself. Be sure everyone understands the question. Be sure you understand the question. If you do not, ask the questioner to repeat it and perhaps to rephrase it.

Keep your answers brief, but answer the questions fully and honestly. When you do not have the answer, say so. Do not be afraid of conflict with the audience. But keep it on an objective basis; talk about the conflict situation, not personalities. If someone reveals through his question that he is becoming personally hostile, handle him courteously. Answer his question as quickly and objectively as you can and move on to another questioner. Sometimes the bulk of your audience will grow restless while a few questioners hang on. When this occurs release your audience and, if you have time, invite the questioners up to the platform to continue the discussion. Above all, during a question period be courteous. Resist any temptation to have fun at a questioner's expense.

Visual Aids

Today, good speakers increasingly use visual aids during their talks.

Purpose

You will use a visual aid (1) *to support* and *expand* the content of your message and (2) *to focus the audience's attention* on a critical aspect of your presentation.[2]

Support The first purpose of any visual material is *to support your message*—to enlarge on the main ideas and give substance and credibility to what you are saying. Obviously, the material must be relevant to the idea being supported. Too often a speaker gives in to the urge to show a visually attractive or technically interesting piece of information that has little or no bearing on the subject.

Suppose, for the purposes of our analysis, that you were asked to meet with government people to present a case for your company's participation in a major federal contract. Your visual support would probably include information about the company's past performances with projects similar to the one being considered. You would show charts reflecting the ingenious methods used by the company's development people to keep costs down; performance statistics to indicate your high-quality standards; and your best conception-to-production times to show the audience how adept you are at meeting target dates.

In such a presentation, before an audience of tough-minded officials, you wouldn't want to spend much of your time showing them aerial views of the company's modern facilities or photographs of smiling employees, antiseptic production lines, and the company's expensive air fleet. Such material would hardly support and expand your arguments that the company is used to working and producing on a Spartan budget.

Focus Your second reason for using visual aids is *focus of attention.* A good visual can arrest the wandering thoughts of your audience and bring their attention right down to a specific detail of the message. It forces their mental participation in the subject.

When you are dealing with very complex material, as you often will be, you can use a simple illustration to show your audience a single, critical concept within your subject.

Visual Aids Criteria

What about the visual aids themselves? What makes one better than another for a specific kind of presentation? Before we consider individual visual aids, let's look at the qualities that make a visual aid effective for the technical speaker.

Visibility First, a visual aid must be *visible.* If that seems so obvious that it hardly need be mentioned at all, it may be because you haven't experienced the frustration of being shown something the speaker feels is important—and not being able to read it, or even make out detail. To be effective, your visual support material should be clearly visible from the most distant seat in the house. If you have any doubts, sit in that seat and look. Remember this when designing visual material: **Anything worth showing the audience is worth making large enough for the audience to see.**

Clarity The second criterion for a good visual is *clarity*. The audience decides this. If they're able to determine immediately what they are seeing, the visual is clear enough. Otherwise, it probably calls for further simplification and condensation. Such obvious mistakes as pictorial material out of focus, or close-ups of a complex device that will confuse the audience, are easy to understand. But what about the chart that shows a relationship between two factors on x- and y-axes when the axes are not clearly designated or when pertinent information is unclear or missing? **Visual material should be immediately clear to the audience, understandable at a glance, without specific help from the speaker.**

Simplicity The third criterion for good visual support is *simplicity*. No matter how complex the subject, the visual itself should include no more information than absolutely necessary to support the speaker's message. If it is not carrying the burden of the message, it need not carry every detail. Limit yourself to *one* idea per visual—mixing ideas will totally confuse an audience, causing them to turn you off midsentence.

When using words and phrases on a visual, limit the material to key words that act as visual cues for you and the audience. If a visual communications expert wished to present the criteria for a good visual he might *think* something like this:

A good visual must manifest visibility.
A good visual must manifest clarity.
A good visual must manifest simplicity.
A good visual must be easy for the speaker to control.

What would he show the audience? If he knows his field as well as he should, he'll offer the visual shown in Figure 19–1.

Figure 19–1 Criteria for Visual Aids

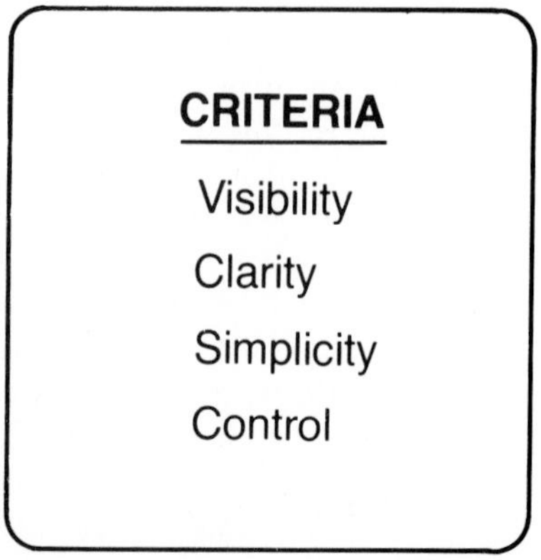

CRITERIA

Visibility
Clarity
Simplicity
Control

The same information is there. The visual is being used appropriately to provide emphasis while the speaker supplies the ideas and the extra words. The very simplicity of the visual has impact and is likely to be remembered by the audience.

Control The fourth quality a visual aid needs is *control,* speaker control. You should be able to add information or delete it, to move forward or backward to review, and, finally, to *take it away* from the audience to bring their attention back to you.

Some very good visual aids can meet the other criteria and prove almost worthless to a speaker because they cannot be easily controlled. The speaker, who must maintain a flow of information and some kind of rapport with his audience while he is doing it, can't afford to let his visual material interfere with his task. Remember, visual material is meant to *support* you as a speaker, not to replace you.

Visual Content

So far we have discussed the *why* of visual support material. The remaining two questions of concern to you are: What do I use? How do I use it? Let's consider them in that order, applying the criteria already established as we go. Types of visual material can be roughly classified in six categories: (1) graphs, (2) tables, (3) representational art, (4) words and phrases, (5) cartoons, and (6) hardware. Graphs, tables, and representational art are discussed in detail in Chapter 12, "Graphical Elements of Reports." You'll want to apply the suggestions made there to the visual materials you use when speaking. In the next few pages, we'll discuss using words and phrases, cartoons, and hardware.

Words and Phrases There will always be circumstances in which you will want to emphasize key words or phrases by visual support, as in Figure 19–1. This type of visual can be effectively used in making the audience aware of major divisions or subdivisions of a topic, for instance.

There is danger, however, in the overuse of words—too many with too much detail. Some speakers tend to use visuals as a "shared" set of notes for their presentation, a self-limiting practice. Audiences who are involved in reading a long, detailed piece of information won't recall what the speaker is saying.

With technical presentations, there is still another problem with the use of words. Too often, because they may be parts of a specialized vocabulary, they do more to confuse the audience than increase their understanding. Such terms should be reserved for audiences whose technical comprehension is equal to the task of translating them into meaningful thoughts.

Cartoons Cartooning is no more than illustrating people, processes, and concepts with exaggerated, imaginative figures—showing them in whatever roles are necessary to your purpose. (See Figure 19–2.) Not only does it heighten audience interest, but cartooning can be as specific as you want it to be in terms of action or position.

Some situations in which you might choose to use cartooned visual material are these:

- When dealing with subjects that are sensitive for the audience.
- When showing people-oriented action in a stationary medium—any visual aid outside the realm of motion pictures or video.

The resourceful speaker will use cartoons to help give additional meaning to other forms of visual support. The use of cartoons as elements in a block diagram tends to increase viewer interest.

Cartooning, like any other technique, can be overdone. There are circumstances in which the gravity of the situation would suggest that you consider only the most formal kinds of visual support material. On other occasions, cartooning may distract the audience or call too much attention to itself. Your purpose is not to entertain but to communicate.

Hardware After all this analysis of visual support material, you may wonder if it wouldn't be somewhat easier to show the real thing instead.

Certainly, there will be times when the best visual support you can have is the actual object. Notably, the introduction of a new piece of equipment will be more effective if it is physically present to give the audience an idea of its size and bulk. If it is capable of some unique and important function, it should definitely be seen by the audience. (The greatest difficulty with the use of actual hardware is control. The device that is small enough for you to carry conveniently may be too small to be seen from the audience.)

Even when the physical presence of a piece of equipment is possible, it is important to back it with supplementary visual materials. Chances are the audience will not be able to determine what is happening inside

Figure 19–2 Cartoon Used to Illustrate a Computer Process

the machine, even if they understand explicitly the principle involved. With this in mind, you will want to add information with appropriate diagrams, graphs, and scale drawings.

In this discussion of visual support, the points of visibility, clarity, simplicity, and control have been stressed over and over. The reason for this is their importance to the selection and use of visual support by the technical communicator. In the end, it is you who can best decide which visual support form is required by your message and your audience.

You are also faced with the choice of visual tools for presenting your visual material. The next section will deal with popular visual tools, their advantages, disadvantages, and adaptability to the materials we've already discussed.

Visual Presentation Tools

The major visual tools are these:

- Chalkboards
- Charts
- Slides
- Movies/videos
- Overhead projection
- Computer presentations

In the next few pages we'll discuss them individually, with an eye on the advantages each one offers the technical speaker.

The Chalkboard Anyone who has attended school in the past half century is familiar with the chalkboard. It has been a standard source of visual support for much longer than that and often the only means of presenting visual information available to the classroom teacher.

As a visual aid, it leaves something to be desired. In the first place, preparing information on a chalkboard, especially technical information in which every sliding scrawl can have significance, takes time. And after the material is in place, it cannot be removed and replaced quickly.

Second, the task of writing on a surface that faces the audience requires that you turn your back toward them while you write. And people don't respond well to backs. They want you to face them while you're talking to them.

Add to these problems the difficulties of moving a heavy, semiperma-nent chalkboard around, and you begin to wonder why anyone bothers.

Low cost and simplicity are the reasons. The initial cost of a chalkboard is higher than you may think; but the cost of erasers and chalk is minimal. In spite of its drawbacks, a chalkboard is also easy to use. It may take time, but there's nothing very complicated about writing a piece of infor-mation on a chalkboard. This simplicity, of course, gives it a certain flexibility, makes it essentially a spontaneous visual aid on which speakers can create their visual material as they go.

There are specific techniques for using a chalkboard that make it a more effective visual tool and help overcome its disadvantages. Let's consider them one at a time.

- Plan ahead. Unless there is a clear reason for creating the material as you go, prepare your visual material before the presentation. Then cover it. Later, you can expose the information for the audience at the appropriate point in your speech.
- Be neat and keep the information simple and to the point. If your material is complex, find another way of presenting it.
- Prime the audience. Before showing your information, tell them what they are going to see and why they are going to see it.

This last point is especially important when you are creating your visual support as you go. Priming your listeners will allow you to maintain the flow of information and, at the same time, prepare them mentally to understand and accept your information.

Charts Charts take a couple of forms. The first is the individual **hardboard chart,** rigid enough to stand by itself and large enough to be seen by the audience from wherever they might be seated in the room. It is always prepared before the presentation, sometimes at considerable cost.

The second chart form is the **flip chart,** a giant-sized note pad that may be prepared before or during the presentation. When you have completed your discussion of one visual, you simply flip the sheet containing it over the top of the pad as you would the pages of a tablet.

The two types of charts have a common advantage. Unlike the chalkboard, they allow for reshowing a piece of information when necessary—an important aid to speaker control.

The following techniques will help you use charts more effectively during your presentation. They're really rules of usage, to be followed each time you choose this visual form for support.

- Keep it simple. Avoid complex, detailed illustrations on charts. A three-by-five-foot chart is seldom large enough for detailed visibility.
- Ask for help. Whenever possible, have an assistant on one side of your charts to remove each one in its turn. This avoids creating a break in your rapport with the audience while you wrestle with a large cardboard chart or a flimsy flip sheet.
- Predraw your visuals with a very light-colored crayon or chalk. During the presentation, you can simply draw over the original lines in darker crayon or ink. This allows you to create an accurate illustration a step at a time for clarity.

- Prime the audience. Tell them what they are going to see and why before you show each visual, for the same reason you would do it with the chalkboard.

Slides The 35-mm slide, with its realistic color and photographic accuracy, has always been a popular visual tool for certain types of technical presentations. Where true reproduction is essential, no better tool is available.

Modern projectors have two notable advantages over their predecessors. First, the introduction of slide magazines has made it possible for you to organize your presentation and keep it intact. Second, remote controls allow you to operate the projector—even to reverse the order of your material—from the front of the room.

To use slide projectors effectively, however, you must turn off the lights in the room. Any time you keep your audience in the dark, you risk damaging the direct speaker–listener relationship on which communication hinges. In a sense, it takes the control of the presentation out of your hands. Long sequences of slides tend to develop a will and a pace of their own. They tire an audience and invite mental absenteeism.

There are ways to handle the built-in problems of a slide presentation, simple techniques that can greatly increase audience attention and the effectiveness of your presentation.

- When using slides in a darkened room, light yourself. A disembodied voice in the dark is little better than a tape recorder; it destroys rapport and allows the audience to exit into their own thoughts. To minimize this effect, arrange your equipment so you may stay in the front of the room and use a lectern light or some other soft, nonglaring light to make yourself *visible* to the audience.
- Break the presentation into short segments of no more than five or six slides.
- Always tell the audience what they're going to see, and what they should look for.

Everything considered, slides are an effective means of presenting visual material. But like any visual tool, they require control and preparation on your part. The important thing to remember is that they are there only to support your message—not to replace you.

Movies and Videos Whenever motion and sound are important to the presentation, movies and videos are the visual forms available to the speaker that can accomplish the effect. Like slides, they also provide an exactness of detail and color that can be critical to certain subjects. There is really no other way an engineer could illustrate the tremendous impact aircraft tires receive during landings, for instance. The audience would understand the subject only if they were able to view, through the eye of the camera, the distortion of the rubber when the plane touches down.

But movies and videos *are* the presentation. They cannot be considered visual support material in the sense of the term developed in this book. They simply replace the speaker as the source of information, at least for their duration. If they become the major part of the presentation, the speaker is reduced to an announcer with little more to do than introduce and summarize their content.

This makes movies and videos the most difficult visual forms to control. Yet they can be controlled and, if they are to perform the support functions we've outlined, they must be. Some effective techniques are given here.

- Prepare the audience. Explain the significance of what you are about to show them.
- If a film or video is to be used, it should make up only a small part of the total presentation.
- Whenever possible, break the film or video into short three- or four-minute segments. Between segments you can reestablish rapport with the audience by summarizing what they have seen and refocusing attention on the important points in the next segment.

Overhead Projection Throughout this discussion of visual support, we've intentionally stressed the importance of maintaining a good speaker–audience relationship. It's an essential in the communication process. And it's fragile. Any time you turn your back to the audience, or darken the room, or halt the flow of ideas for some other reason, this relationship is damaged.

The overhead projector effectively eliminates all of these rapport-dissolving problems. The image it projects is bright enough and clear enough to be used in a normally lighted room without noticeable loss of visibility. And just as important from your point of view, it allows you to remain in the front of the room *facing* your audience throughout your presentation. The projector itself is a simple tool, and like all simple tools it may be used without calling attention to itself.

Visual material for the overhead projector is prepared on transparent sheets the size of typing paper. The methods for preparing these transparencies have become so simplified and made so inexpensive that the overhead has become a universally accepted visual tool in both the classroom and industry.

Perhaps the most important advantage of the overhead projector is the total speaker control it affords. With it, you may add information or delete it in a variety of ways, move forward or backward to review at will, and *turn it off* without altering the communicative situation in any way. It is the last capability that makes the overhead projector unique among visual tools. By flipping a switch, you can literally "remove" the visual material from the audience's consideration, bringing their attention back to you and what you are saying. Because the projector is used in a lighted room, this on-and-off process seldom distracts the audience or has any effect on the speaker–audience relationship.

There are three ways to add information to a visual while the audience looks on—an important consideration when you want your listeners to receive information in an orderly fashion. In order of their discussion, they are (1) overlays, (2) revelation, and (3) writing on the visual itself.

The **overlay technique** (Figure 19–3) combines the best features of preparing your visuals in advance and creating them at the moment of their need. It is the simple process of beginning with a single positive transparency and adding information with additional transparencies by "overlaying" them, placing each one over the first so they may be viewed by the audience as a single, composite illustration. Ordinarily, no more than two additional transparencies should be used this way, but it's possible to include as many as four or five.

The technical person, who must usually present more complex concepts a step at a time to ensure communication, can immediately see the applications of such a technique.

The technique of **revelation** is simpler. (See Figure 19–4). It is the process of masking off the parts of the visual you don't want the audience to see. A plain sheet of paper will work. By laying it over the information you want to conceal for the moment, you can block out selected pieces of the visual. Then when you're ready to discuss this hidden information,

Figure 19–3 Using Overlays with an Overhead Projector

Figure 19–4 The Technique for Revelation

you simply remove the paper. The advantage is clear enough. If you don't want the audience to read the bottom line on the page while you're discussing the top line, this is the way to control their attention.

Writing information on an overhead-projection transparency (Figure 19–5) is nearly as easy as writing on a sheet of paper at your desk. Several felt-tipped pens available for this purpose may be used to create visual material in front of the audience. Often you can achieve your

Figure 19–5 Writing on Transparencies

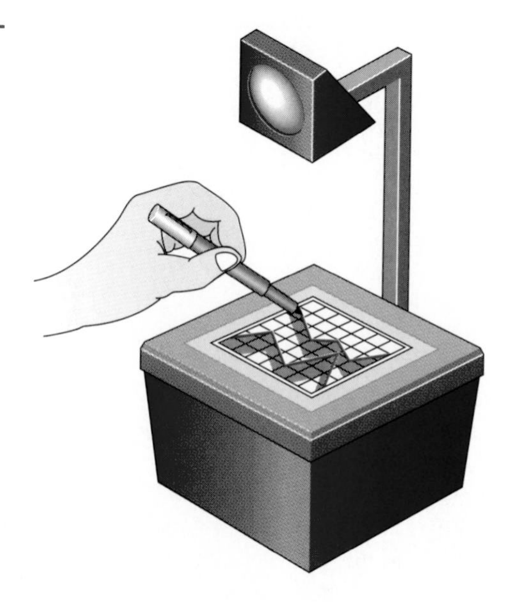

purposes by simply underlining or circling important parts of your visual—a means of focusing audience attention on the important aspects of your message.

A final way of directing audience attention with overhead-projection transparencies is simply to **use your pencil as a pointer.** The "profile" shadow of the pencil will be seen on the screen, directing the audience's attention to the proper place.

Computer Presentations As is to be expected, microcomputers play an increasingly important role in preparing and presenting graphics during talks. To begin with, computer equipment and programs are available to create transparencies and 35-mm slides of any graphic you can produce on your computer screen. You, therefore, can create for your talks computer graphics of the kind we discuss in Chapter 12, "Graphical Elements of Reports."

However, you can obtain even more versatility in your speech graphics by using computer projectors that display whatever is on your computer screen on a projection screen. The computer projector allows you to use your computer during your talk to create and display graphics and data as you need them. In essence, the **computer projector** is an overhead projector linked with a computer's capability for creating graphics.

Computer presentations allow you great control of your visual material. However, don't be misled by the great capability of the computer into creating graphics that are too complex to be readily readable. As with all graphics, keep your computer-generated graphics visible, clear, and simple. Use them not to impress your listeners but to focus their attention and to support and clarify your message.

Planning and Revision Checklist

You will find the planning and revision checklists that follow Chapter 2, "Composing" (pages 33–35 and inside the front cover), and Chapter 4, "Writing for Your Readers" (pages 78–79), valuable in planning and revising any presentation of technical information. The following questions specifically apply to oral reports. They summarize the key points in this chapter and provide a checklist for planning and revising.

Planning

- What are the conditions under which you will speak?
- What equipment is available to you?
- Which delivery technique will be more appropriate? Extemporaneous? Manuscript?

Revision

In one sense, you can't revise a talk you have already given, unless, of course, you will have an opportunity to repeat it somewhere. But you can use revision techniques in your practice sessions. Most of the questions asked under

Planning

- If you are speaking extemporaneously, have you prepared a speech outline to guide you?
- If reading from a manuscript, have you introduced a conversational tone into your talk? Is your typed manuscript easy to read from?
- Do you have a good opening that will interest your audience and create a friendly atmosphere?
- Have you limited your major points to fit within your allotted time?
- Does your talk contain sufficient examples, analogies, narratives, and data to support your generalizations? Have you repeated key points?
- Can you relate your subject matter to some vital interest of your audience?
- Which visual aids do you plan to use?
  - Graphs?
  - Tables
  - Representational art?
  - Words and phrases?
  - Cartoons?
  - Hardware?
- Which tools of presentation will you use?
  - Chalkboard?
  - Charts?
  - Slides?
  - Movies?
  - Overhead projection?
  - Computer presentations?
- Have you prepared your graphics? Do they successfully focus the listener's attention and augment and clarify your message? Do they meet the four criteria that govern good graphics?
  - Visibility
  - Clarity
  - Simplicity
  - Control
- Do you have a good ending ready, perhaps a summary of key points or an anecdote that supports your purpose?
- Have you rehearsed your talk several times?

Revision

"Planning" lend themselves to that. Also, you can critique your speeches, looking for ways to improve your delivery techniques in future speeches. These next questions lend themselves to that. It helps if you have someone in the audience to give you friendly but honest answers to all the questions asked under planning and revision.

- Did your gestures support your speech? Did they seem normal and relaxed? Did you avoid nervous mannerisms?
- Was your speech rate appropriate? Did you vary rate, pitch, and volume occasionally? Could everyone hear you?
- Did you pronounce all your words correctly?
- Did you have good interaction with your audience? Were they attentive or fidgety?
- Did your talk fit comfortably into the time allotted for it?
- Did the questions that followed your talk indicate a good understanding of it? Did the questions indicate friendliness or hostility to your key points?
- Were you sufficiently informed to answer the questions raised?
- Were there any indications that members of your audience could not see or readily comprehend any of your graphics?

Exercises

1. Deliver a speech in one of the following situations:
   - **a.** You are an instructor at your college. Prepare a short extemporaneous lecture on a technical subject. Your audience is a class of about 20 upperclassmen.
   - **b.** You are the head of a team of engineers and technicians that has developed a new product or process. Your job is to persuade a group of senior managers from your own firm to accept the process or product for company use. Assume these managers to have a lay person's knowledge about your subject. Speak extemporaneously.
   - **c.** You are a known expert on your subject. You have been invited to speak about your subject at the annual meeting of a well-known scientific association. You are expected to write out and read your speech. You are to inform the audience, which is made up of knowledgeable research scientists and college professors from a diversity of scientific disciplines, about your subject or to persuade them to accept a decision you have reached.
2. Change one of your written reports into an oral report. Deliver the report extemporaneously. Prepare several visual aids to support major points.

PART VI
Handbook

Any living language is a growing, flexible instrument with rules that are constantly changing by virtue of the way it is used by its live, independent speakers and writers. Only the rules of a dead language are unalterably fixed.

Nevertheless, at any point in a language's development, certain conventions of usage are in force. Certain constructions are considered errors that mark the person who uses them as uneducated. It is with these conventions and errors that this handbook primarily deals. We also include a section on outlining. To make the handbook easy to use as a reference, we have arranged it in alphabetical order. Each convention and error dealt with has an abbreviated reference tag. The tags are reproduced on the back endpapers along with some of the more important proofreading symbols. If you are in a college writing course, your instructor may use some combination of these tags and symbols to indicate revisions needed in your reports.

Abbreviations *(Ab)*

Although most people are familiar with the kinds of abbreviations we encounter in everyday conversation and written material, from "mph" to "Mon." to "Dr.," technical abbreviations are something else again. Each scientific and professional field generates hundreds of specialized terms, and many of these terms are often abbreviated for the sake of conciseness and simplicity.

Thus, before deciding to use technical abbreviations in an article or report, you must first consider your audience—lay people or scientists, executives or engineers? Only readers with a technical background will be able to interpret the specialized shorthand for the field in question. When in doubt, then, avoid all but the most common abbreviations. If you must use a technical term, spell it out in full the first time it appears and include both the abbreviation and a definition in parentheses after it. You can then safely use the abbreviation if the term crops up again in your report.

Standard technical and scientific abbreviations include the following:

absolute	abs
acre or acres	(spell out)

alternating current (as adjective)	a-c
atomic weight	at. wt
barometer	bar.
Brinell hardness number	Bhn
British thermal units	Btu or B
meter	m
square meter	m^2
microwatt or microwatts	mu w or μW
miles per hour	mph
National Electric Code	NEC
per	(spell out)
revolutions per minute	rpm
rod	(spell out)
ton	(spell out)

The system implied by these illustrative abbreviations can be described by a brief set of rules.

(1) Use the same (singular) form of abbreviation for both singular and plural terms:

cu ft	either cubic foot or cubic feet
cm	either centimeter or centimeters

But there are some common exceptions:

no.	number
nos.	numbers
p.	page
pp.	pages
ms.	manuscript
mss.	manuscripts

(2) Use lowercase letters except for letters standing for proper nouns or proper adjectives:

ab	*but*	Btu or B
mph	*but*	Bhn

(3) For technical terms, use periods only after abbreviations that spell complete words. For example, *in* is a word, and the abbreviation for inches could be confused with it. Therefore, use a period:

ft	*but*	in.

abs *but* bar.
cu ft *but* at. wt

(4) Remember the hyphen in the abbreviations a-c and d-c when you use them as adjectives:

This a-c motor can be converted to 28 volts dc.

(5) Spell out many short and common words:

acre rod per ton

(6) In compound abbreviations, use internal spacing only if the first word is represented by more than its first letter:

rpm *but* cu ft
mph *but* mu w

(7) With few exceptions, form the abbreviations of organization names without periods or spacing:

NEC ASA

(8) Abbreviate terms of measurement only if they are preceded by an arabic expression of exact quantity:

55 mph *and* 20-lb anchor

But:
We will need an engine of greater horsepower.

Acronyms *(Acro)*

Acronyms are formed in two ways. In one way, the initial letters of each word in some phrase are combined to form a word. An example would be WYSIWYG, an acronym for the computer phrase "What you see is what you get." In a second way, some combination of initial letters or several letters of the words in the phrase are combined. An example would be *radar* for *ra*dio *d*etection *a*nd *r*anging.

Technical writing uses acronyms freely, as in this example from a description of a computer program that performs statistical analysis:

> It has good procedural capabilities, including some time-series-related plots and ARIMA forecasting, but it doesn't have depth in any one area. Although it has commands to create EDA displays, these graphics are static and are printed with characters rather than with lines.[1]

Use acronyms without explanation only when you are absolutely sure your readers know them. If you have any doubts at all, at least provide the words from which the acronym stems. If in doubt about that being enough, provide a definition of the complete phrase. In the case of the paragraph just quoted, the computer magazine in which it was printed provided a glossary giving both the complete phrases and definitions:

ARIMA (auto-regressive integrated moving-average): a model that characterizes changes in one variable over time. It is used in time-series analysis.

EDA (exploratory data analysis): The use of graphically based tools, particularly in initial states of data analysis, to inspect data properties and to discover relationships among variables.[2]

Acronyms can be daunting to those unfamiliar with them. Even when you have an audience that knows their meaning, a too-heavy use of acronyms can make your writing seem lumpish and uninviting.

Apostrophe *(Apos)*

The apostrophe has three chief uses: (1) to form the possessive, (2) to stand for missing letters or numbers, and (3) to form the plural of certain expressions.

Possessives

Add an apostrophe and an *s* to form the possessive of most singular nouns, including proper nouns, even when they already end in an *s* or another sibilant such as *x:*

man's
spectator's
jazz's
Marx's
Charles's

Exceptions to this rule occur when adding an apostrophe plus an *s* would result in an *s* or *z* sound that is difficult to pronounce. In such cases, usually just the apostrophe is added:

Xerxes'
Moses'
conscience'
appearance'

To understand this exception, pronounce *Marx*'s and then a word like *Moses*'s or *conscience*'s. (Note, also, that though a proper noun such as *Marx* may be italicized its possessive *s* remains in roman type.)

To form plurals into the possessive case, add an apostrophe plus *s* to words that do not end in an *s* or other sibilant and an apostrophe only to those that do:

men's
data's
spectators'
agents'
witnesses'

To show joint possession, add the apostrophe and *s* to the last member of a compound or group; to show separate possession, add an apostrophe and *s* to each member:

Gregg and Klymer's experiments astounded the class.
Gregg's and Klymer's experiments were very similar.

Of the several classes of pronouns, only the indefinite pronouns use an apostrophe to form the possessive.

Possessive of Indefinite Pronouns	Possessive of Other Pronouns
anyone's	my (mine)
everyone's	your (yours)
everybody's	his, her (hers), its
nobody's	our (ours)
no one's	their (theirs)
other's	whose
neither's	

Missing Letters or Numbers

Use an apostrophe to stand for the missing letters in contractions and to stand for the missing letter or number in any word or set of numbers where for one reason or another a letter or number is omitted:

can't, don't, o'clock, it's (it is), etc.
We were movin' downriver, listenin' to the birds singin'.
The class of '49 was Colgate's best class in years.

Plural Forms

An apostrophe is sometimes used to form the plural of letters and numbers, but this style is gradually dying, particularly with numbers.

6s and 7s (but also 6's and 7's)
A's and *b*'s

Brackets *(Brackets)*

Brackets are chiefly used when a clarifying word or comment is inserted into a quotation:

> "The result of this [disregard by the propulsion engineer] has been the neglect of the theoretical and mathematical mastery of the engine inlet problem."
> "An ideal outlet require [sic] a frictionless flow."
> "Last year [1990] saw a partial solution to the problem."

Sic, by the way, is Latin for *thus.* Inserted in a quotation, it means that the mistake found there is the original writer's, not yours. Use it with discretion.

Capitalization *(Cap)*

We provide the more important rules of capitalization. For a complete rundown see your college dictionary.

Proper Nouns

Capitalize all proper nouns and their derivatives:

Places

America American Americanize Americanism

Days of the Week and Months

Monday Tuesday January February

But not the seasons:

winter spring summer fall

Organizations and Their Abbreviations

American Kennel Club (AKC)
United States Air Force (USAF)

Capitalize *geographic areas* when you refer to them as areas:

> The Andersons toured the Southwest.

But do not capitalize words that merely indicate direction:

> We flew west over the Pacific.

Capitalize the names of *studies* in a curriculum only if the names are already proper nouns or derivatives of proper nouns or if they are part of the official title of a department or course:

Department of Geology
English Literature 25
the study of literature
the study of English literature

Note: Many nouns (and their derivatives) that were originally proper have been so broadened in application and have become so familiar that they are no longer capitalized: *boycott, macadam, spoonerism, italicize, platonic, chinaware, quixotic.*

Literary Titles

Capitalize the first word, the last word, and every important word in literary titles:

But What's a Dictionary For
The Meaning of Ethics
How to Write and Be Read

Rank, Position, Family Relationships

Capitalize the titles of rank, position, and family relationship unless they are preceded by *my, his, their,* or similar possessive pronouns:

Professor J. E. Higgins
I visited Uncle Timothy.
I visited my uncle Timothy.
Dr. Milton Weller, Head, Department of Entomology

Colon *(Colon)*

The colon is chiefly used to introduce quotations, lists, or supporting statements. It is also used between clauses when the second clause is an example or amplification of the first and in certain conventional ways with numbers, correspondence, and bibliographical entries.

Introduction

Place a colon before a quotation, a list, or supporting statements and examples that are formally introduced:

Mr. Smith says the following of wave generation:

> The wind waves that are generated in the ocean and which later become swells as they pass out of the generating area are products of storms. The low pressure regions that occur during the polar winters of the Arctic and Antarctic produce many of these wave-generating storms.

The various forms of engine that might be used would operate within the following ranges of Mach number:

M-0 to M-1.5	Turbojet with or without precooling
M-1.5 to M-7	Reheated turbojet, possibly with precooling
M-7 to M-10 +	Ramjet with supersonic combustion

Engineers are developing three new engines: turbojet, reheated turbojets, and ramjets.

Do not place a colon between a verb and its objects or a linking verb and the predicate nouns.

Objects

The engineers designed turbojets, reheated turbojets, and ramjets.

Predicate nouns

The three engines the engineers are developing are turbojets, reheated turbojets, and ramjets.

Do not place a colon between a preposition and its objects:

The plane landed at Detroit, Chicago, and Rochester.

Between Clauses

If the second clause consists of an example or an amplification of the first clause, then the colon may replace the comma, semicolon, or period:

The docking phase involves the actual "soft" contact: the securing of lines, latches, and air locks.

The difference between these two guidance systems is illustrated in Figure 2: The paths of the two vehicles are shown to the left and the motion of the ferry as viewed from the target station is shown to the right.

You may follow a colon with a capital or a small letter. Generally, a complete sentence beginning after a colon is given a capital.

Styling Conventions

Place a colon after a formal salutation in a letter, between hour and minute figures, between the elements of a double title, and between chapter and verse of the Bible:

> Dear Ms. Jones:
>
> 3:45
>
> at 7:15 p.m.
>
> *Mary Russell Mitford: Her LIfe and Writing*
>
> I Samuel 7:14–18

Comma *(C)*

The most used—and misused—mark of punctuation is the comma. Writers use commas to separate words, phrases, and clauses. Generally, commas correspond to the pauses we use in our speech to separate ideas and to avoid ambiguity. You will use the comma often: About two out of every three marks of punctuation you use will be commas. Sometimes your use of the comma will be essential for clarity; at other times you will be honoring grammatical conventions. (See also run-on sentences.)

Main Clauses

Place a comma before a coordinating conjunction *(and, but, or, nor, for, yet)* that joins two main (independent) clauses:

> During the first few weeks we felt a great deal of confusion, but as time passed we gradually fell into a routine.
>
> We could not be sure that the plumbing would escape frost damage, nor were we at all confident that the house could withstand the winds of almost hurricane force.

If the clauses are short, have little or no internal punctuation, or are closely related in meaning, then you may omit the comma before the coordinating conjunction:

> The wave becomes steeper but it does not tumble yet.

In much published writing there is a growing tendency to place two very short and closely related independent clauses (called contact clauses) side by side with only a comma between:

> The wind starts to blow, the waves begin to develop.

Sentences consisting of *three* or more equal main clauses should be punctuated uniformly:

> We explained how urgent the problem was, we outlined preliminary plans, and we arranged a time for discussion.

In general, identical marks are used to separate equal main clauses. If the equal clauses are short and uncomplicated, commas usually suffice. If the equal clauses are long, internally punctuated, or if their separateness is to be emphasized, semicolons are either preferable or necessary.

Clarification

Place a comma after an introductory word, phrase, or clause that might be over-read or that abnormally delays the main clause:

> As soon as you have finished polishing, the car should be moved into the garage. *(Comma to prevent over-reading)*
>
> Soon after, the winds began to moderate somewhat, and we were permitted to return to our rooms. *(First comma to prevent over-reading)*
>
> If the Polar ice caps should someday mount in thickness and weight to the point that their combined weight exceeded the Equatorial bulge, the earth might suddenly flop ninety degrees. *(Introductory clause abnormally long)*

After a short introductory element (word, phrase, or clause) where there is no possibility for ambiguity, the use of the comma is optional. Generally, let the emphasis you desire guide you. A short introductory element set off by a comma will be more emphatic than one that is not.

Nonrestrictive Modifiers

Enclose or set off from the rest of the sentence every nonrestrictive modifier, whether a word, a phrase, or a clause. How can you tell a nonrestrictive modifier from a restrictive one? Look at these two examples:

> *Restrictive*
>
> A runway *that is not oriented with the prevailing wind* endangers the aircraft using it.
>
> *Nonrestrictive*
>
> The safety of any aircraft, *whether heavy or light,* is put in jeopardy when it is forced to take off or land in a crosswind.

The restrictive modifier is necessary to the meaning of the sentence. Not just any runway but "a runway that is not oriented with the prevailing wind" endangers aircraft. The writer has *restricted* the many kinds of runways he could talk about to one particular kind. In the nonrestrictive example, the modifier merely adds descriptive details. The writer does not restrict *aircraft* with the modifier but rather simply makes the meaning a little clearer.

Restrictive modifiers cannot be left out of the sentence if it is to have the meaning the writer intends; nonrestrictive modifiers can be left out.

Nonrestrictive Appositives

Set off or enclose every nonrestrictive appositive. As used here the term *appositive* means any element (word, phrase, or clause) that parallels and repeats the thought of a preceding element. According to this view, a verb may be coupled appositively with another verb, an adjective with another adjective, and so on. An appositive is usually more specific or more vivid than the element that it is an appositive to; an appositive makes explicit and precise something that has not been clearly enough implied.

Some appositives are restrictive and, therefore, are not set off or enclosed.

> *Nonrestrictive*
>
> A crosswind, *a wind perpendicular to the runway,* causes the pilot to make potentially dangerous corrections just before landing.
>
> *Restrictive*
>
> In some ways, Mr. Bush *the President* had to behave differently from Mr. Bush *the Vice President.*

In the nonrestrictive example, the appositive merely adds a clarifying definition. The sentence makes sense without it. In the restrictive example, the appositives are essential to the meaning. Without them we would have, "In some ways, Mr. Bush had to behave differently from Mr. Bush."

Series

Use commas to separate members of a coordinate series of words, phrases, or clauses if *all* the elements are not joined by coordinating conjunctions:

> Instructions on the label state clearly how surfaces should be prepared, how the contents are to be applied, and how to clean and polish the mended article.
>
> To mold these lead figures you will need a hot flame, a two-pound block of lead, the molds themselves, a file or a rasp, and an awl.

Under the microscope the sensitive, filigree-like mold appeared luminous and transparent and faintly green.

Other Conventional Uses

Date

On November 26, 1950, all the forces of nature seemed to combine to wreak havoc upon the Middle Atlantic States.

Note: When you write the month and the year without the day, it is common practice to omit the comma between them—as in June 1990.

Geographical Expression

During World War II Middletown, Pennsylvania, was the site of a huge military airport and supply depot.

Title after Proper Name

A card in yesterday's mail informed us that Carleton Williams, M.D., Pediatrician, would soon open new offices on Beaver Street, State College.

Noun of Direct Address

Clifton, do you suppose that we can find our way back to the cabin before nightfall?

Informal Salutation

Dear Jane,

Dangling Modifier *(DM)*

Many curious sentences result from the failure to provide the modifier something to modify:

Having finished the job, the tarpaulins were removed.

In this example it seems as though the tarpaulins have finished the job. As is so often the case, a passive voice construction has caused the problem (see pages 125–126). If we recast the sentence in active voice, we remove the problem:

Having finished the job, the workers removed the tarpaulins.

Dash *(Dash)*

In technical writing, you will use the dash almost exclusively to set off parenthetical statements. You may, of course, use commas or parentheses for the same function, but the dash is the most emphatic separator of the three. You may also use the dash to indicate a sharp transition. On a typewriter you make the dash with two hyphens. You do not space between the words and the hyphens or between the hyphens themselves.

Typewriter

The first phase in rendezvous--sighting and recognizing the target--is so vital that we will treat it at some length.

Typeset

The target must emit or reflect light the pilot can see—but how bright must this light be?

Diction *(D)*

For good diction, choose words that are accurate, effective, and appropriate to the situation. Many different kinds of linguistic sins can cause faulty diction. Poor diction can involve a choice of words that are too heavy or pretentious: *utilize* for *use, finalize* for *finish, at this point in time* for *now,* and so forth. Tired old clichés are poor diction: *with respect to, with your permission, with reference to,* and many others. We talk about such language in Chapter 6, particularly in the section on pomposity (pages 129–132).

Sometimes the words chosen are simply too vague to be accurate: *inclement weather* for *rain, too hot* for *600° C.* See the section "Specific Words" (pages 127–128) in Chapter 6 for more on this subject.

Poor diction can mean an overly casual use of language when some degree of formality is expected. One of the many synonyms for *intoxicated,* such as *bombed, stoned,* or *smashed,* might be appropriate in private conversation but totally wrong in a police or laboratory report.

Poor diction can reflect a lack of sensitivity to language—to the way one group of words relates to another group. Someone who writes that "The airlines are beginning a crash program to solve their financial difficulties" is not paying attention to relationships. The person who writes that the "Steelworkers' Union representatives are getting down to brass tacks in the strike negotiations" has a tin ear, to say the least.

Make your language work for you, and make it appropriate to the situation.

Ellipsis *(Ell)*

Use three spaced periods to indicate words omitted within a quoted sentence, four spaced periods if the omission occurs at the end of the sentence:

> "As depth decreases, the circular orbits become elliptical and the orbital velocity . . . increases as the wave height increases."
>
> "As the ground swells move across the ocean, they are subject to headwinds or crosswinds. . . ."

You need not show an ellipsis if the context of the quotation makes it clear that it is not complete:

> Wright said the accident had to be considered a "freak of nature."

Exclamation Point *(Exc)*

Place an exclamation point at the end of a startling or exclamatory sentence.

> On 9 January 1952, the *S.S. Pennsylvania* sank with no survivors in the North Pacific in a storm that produced 50-knot winds for 33 hours and waves up to 48 feet high!

With the emphasis in technical writing on objectivity, you will seldom use the exclamation point.

Fragmentary Sentence *(Frag)*

Most fragmentary sentences are either verbal phrases or subordinate clauses that the writer mistakes for a complete sentence.

A verbal phrase has in the predicate position a participle, gerund, or infinitive, none of which functions as a complete verb:

> Norton, *depicting* the electromagnetic heart. (participle)
>
> The *timing* of this announcement about Triptycene. (gerund)
>
> Braun, in order *to understand* tumor cell growth. (infinitive)

When your fragment is a verbal phrase, either change the verb to a complete verb or repunctuate the sentence so that the phrase is joined to the complete sentence of which it is actually a part.

Fragment

Norton, depicting the electromagnetic heart. She made a mockup of it.

Rewritten

Norton depicted the electromagnetic heart. She made a mockup of it.

Norton, depicting the electromagnetic heart, made a mockup of it.

Subordinate clauses are distinguishable from phrases in that they have complete subjects and complete verbs (rather than verbals) and are introduced by relative pronouns (who, which, that) or by subordinating conjunctions (because, although, since, after, while).

The presence of the relative pronoun or the subordinating conjunction is a signal that the clause is not independent but is part of a more complex sentence unit. Any independent clause can become a subordinate clause with the addition of a relative pronoun or subordinating conjunction.

Independent Clause

Early transistors were only the size of a pea.

Subordinate Clause

Although early transistors were only the size of a pea.

Repunctuate subordinate clauses so that they are joined to the complex sentence of which they are a part.

Fragment

Although early transistors were only the size of a pea. They were enormous compared to modern transistors.

Rewritten

Although early transistors were only the size of a pea, they were enormous compared to modern transistors.

Various kinds of elliptical sentences minus a subject and/or a verb do exist in English, for example, "No!" "Oh?" "Good shot." "Ouch!" "Well, now." These constructions may occasionally be used for stylistic reasons, particularly, the representation of conversation, but they are seldom needed in technical writing. If you do use such constructions, use them sparingly. Remember that major deviations from normal sentence patterns will probably jar your readers and break their concentration upon your thought, the last thing that any writer wants.

Hyphen *(Hyphen)*

Hyphens are used to form various compound words and in breaking up a word that must be carried over to the next line.

Compound Numbers

See "Numbers."

Common Compound Words

Observe dictionary usage in using or omitting the hyphen in compound words.

governor-elect	acid resistant
ex-treasurer	Croesus-like
Russo-Japanese	drill-like
pro-American	self-interest

But:

glasslike	wrist watch
neophyte	sweet corn
newspaper	weather map
newsstand	sun lamp
housewife	prize fight

Compound Words as Modifiers

Use the hyphen between words joined together to modify other words:

a half-spent bullet
an eight-cylinder engine
their too-little-and-too-late methods

Be particularly careful to hyphenate when omitting the hyphen may cause ambiguity:

two-hundred-gallon drums
two hundred-gallon drums
a pink-skinned Buddha

Sometimes you have to carry a modifier over to a later word, creating what is called a *suspended hyphen:*

> GM cars come with a choice of four-, six-, or eight-cylinder engines.

Word Division

Use a hyphen to break a word that must be carried over to the next line. Words compounded of two roots or a root and an affix are divided at the point of union:

self-/important	wind/jammer
cross-/pollination	desir/able
bladder/wort	anti/dote
summer/time	manage/able

Note: The first two words in this list are always spelled with a hyphen; the remaining words use the hyphen only when the word is divided at the end of a line.

In general, noncompound words of more than one syllable are divided between any two syllables but only between syllables:

soph/o/more	con/clu/sion
sat/is/fac/tion	sym/pa/thy

A syllable of one letter is never set down alone; a syllable of only two letters is seldom allowed to stand alone unless it is a prefix or a suffix, and then only if it is pronounced as spelled:

hello/	elec/tro/type
method/	de/mand
pilot/	ac/cept
many/	walked/
saga/	start/ed

If a consonant is doubled because a suffix is added, include the second consonant with the suffix:

spin/ner	slip/ping
stir/ring	slot/ted

But:

stopped/	pass/ing
lapped/	stall/ing

Because *-ped* is not pronounced as a syllable, it should not be carried over. In words like *passing* and *stalling,* both consonants belong to the root, and, therefore, only the suffix *-ing* is carried over.

Italicization *(Ital)*

Italic print is a distinctive typeface, like this sample: *Scientific American.* When you type or write, you represent italics by underlining, like this:

Scientific American

Foreign Words

Italicize foreign words that have not yet become a part of the English language:

We suspected him always of holding some *arrière pensée.*

Karl's everlasting *Weltschmerz* makes him a depressing companion.

Also italicize Latin scientific terms.

cichorium endivia (endive)

Percopsis omiscomaycus (trout-perch)

But do not italicize Latin abbreviations or foreign words that have become a part of the language:

etc. bourgeois
vs. status quo

Your collegiate dictionary will normally indicate which foreign words are still italicized and which are not.

Words, Letters, and Numbers Used as Such

The words *entrance* and *admission* are not perfectly interchangeable

Don't forget the *k* in *picnicking.*

His *9*s and *7*s descended below the line of writing.

Titles

Italicize the titles of books, plays, magazines, newspapers, ships, and artistic works:

Webster's New World Dictionary *The Free Press*
Othello *S.S. Pennsylvania*
Scientific American *Mona Lisa*

Misplaced Modifier *(MM)*

As in the case of dangling modifiers, curious sentences result from the modifier's not being placed next to the element modified:

> An engine may crack when cold water is poured in unless it is running.

Probably, with a little effort, no one will misread this example; but, undeniably, it says that the engine will crack unless the water is running. Move the modifier and the sentence is clear:

> Unless it is running, an engine may crack when cold water is poured in.

It should be apparent from the preceding examples that a modifier may be in the wrong position to convey one meaning but in the perfect position to convey a different meaning. In the next example, the placement of *for three years* is either right or wrong. It is in the right position to modify *to work* but in the wrong position to modify *have been trying*.

> I have been trying to place him under contract to work here for three years. (three-year contract)

As the examples should suggest, correct placement of modifiers sometimes amounts to more than mere nicety of expression. It can mean the difference between stating falsehood and truth, between saying what you mean and saying something else.

Numbers *(Num)*

There is a good deal of inconsistency in the rules for handling numbers. It is often a question as to whether you should write the number as a word or as a figure. We will give you the general rules. Your instructor or your organization may give you others. As in all matters of format, you must satisfy whomever you are working for at the moment. Do, however, be internally consistent within your reports. Do not handle numbers differently from page to page of a report.

Numbers as Words

Generally, you write all numbers nine and under, and rounded-off large numbers, as words:

> six generators
> about a million dollars

However, when you are writing a series of numbers, do not mix up figures and words. Let the larger numbers determine the form used:

five boys and six girls

But:

It took us 6 months and 25 days to complete the experiment.

Numbers as Sentence Openers

Do not begin sentences with a figure. If you can, write the number as a word. If this would be cumbersome, write the sentence so as to get the figure out of the beginning position:

Fifteen months ago, we saw the new wheat for the first time.
We found 350 deficient steering systems.

Compound Number Adjectives

When you write two numbers together in a compound number adjective, spell out the first one or the shorter one to avoid confusing the reader:

Twenty 10-inch trout
100 twelve-volt batteries

Hyphens

Two-word numbers are hyphenated on the rare occasions when they are written out:

Eighty-five boxes

or:

Eighty-five should be enough.

Numbers as Figures

The general rule here is to write all exact numbers over nine as a figure. This rule probably holds more true in technical writing, with its heavy reliance on numbers, than it does in general writing. However, as we noted, rounded-off numbers are commonly written as words. The precise figure could give the reader an impression of exactness that might not be called for.

Certain conventional uses call for figures at all times.

Dates, exact sums of money, time, address

1 January 1992 or January 1, 1992

$3,422.67 *but* about three thousand dollars
1:57 p.m. *but* two o'clock
660 Fuller Road

Technical units of measurement

6 cu ft
4,000 rpm

Cross-references

See page 22.
Refer to Figure 2.

Fractions

When a fraction stands alone, write it as an unhyphenated compound:

two thirds
fifteen thousandths

When a fraction is used as an adjective, you may write it as a hyphenated compound. But if either the numerator or the denominator is hyphenated, do not hyphenate the compound. More commonly, fractions used as adjectives are written as figures.

two-thirds engine speed
twenty-five thousandths
3/4 rpm

Outlining

As illustrated in the accompanying sample outline, an outline has a title, purpose statement, audience statement, and body. We have annotated the sample outline to point out major outlining conventions. Following the sample outline, we provide other major outlining principles.

DESALINATION METHODS FOR AIR FORCE USE

Purpose: To choose a desalination method for Air Force bases located near large bodies of salt water.

Audience: Senior officers

First level, use capital roman numerals

Second level, use capital letters

Third level, use arabic numerals

Fourth level, use lowercase letters

Capitalize only first letter of entry and proper nouns

Use no punctuation after entries

I. Statement of the problem
    A. Need for a choice
    B. Choices available
        1. Electrodialysis
        2. Reverse osmosis
    C. Sources of data
        1. Air force manuals
        2. Expert opinion
            a. Journals
            b. Interviews
II. Explanation of criteria
    A. Cost
    B. Purity
    C. Quantity
III. Electrodialysis
    A. Theory of method
    B. Judgment of method
        1. Cost
        2. Purity
        3. Quantity
IV. Reverse osmosis
    A. Theory of method
    B. Judgment of method
        1. Cost
        2. Purity
        3. Quality
V. Choice of method

- Make all entries grammatically parallel. (See the entry for "Parallelism.") Do not mix noun phrases with verb phrases, and so forth. A formal outline with a hodgepodge of different grammatical forms will seem to lack—and, in fact, may lack—logic and consistency.

Incorrect	*Correct*
I. The overall view	I. The overall view
II. To understand the terminal phase	II. The terminal phase
III. About the constant-bearing concept	III. The constant-bearing concept

- Never have a single division. Things divide into two or more; so, obviously, if you have only one division, you have done no dividing. If you have a "I" you must have a "II." If you have an "A" you must have a "B," and so forth.

Incorrect	*Correct*
I. Visual capabilities	I. Visual capabilities
A. Acquisition	A. Acquisition
II. Interception and closure rate	B. Interception and closure rate
III. Braking	II. Braking

- Do not have entries for your report's introduction or conclusion. Outline only the body of the report. Of course, the information in your purpose statement belongs in your introduction, and perhaps the information about audience belongs there as well.
- Use substantive statements in your outline entries. That is, use entries such as "Reverse osmosis" or "Judgment of method" that suggest the true substance of your information. Do not use cryptic expressions such as "Example 1" or "Minor premise."

Many word processing programs have an outlining feature. This feature has two advantages. First, you can choose the outlining scheme you want, for example, I, A, B, 1, 2. The program automatically writes the appropriate numbers or letters for you—and changes them when you change the outline. Second, you can write your text into the outline.

Parallelism *(Paral)*

When you link elements in a series, they must all be in the same grammatical form. Link an adjective with an adjective, a noun with a noun, a clause with a clause, and so forth. Look at the italicized portion of the sentence below:

> A good test would *use small amounts of plant material, require little time, simple to run, and accurate.*

The series begins with the verbs *use* and *require* and then abruptly switches to the adjectives *simple* and *accurate.* All four elements must be based on the same part of speech. In this case, it's simple to change the last two elements:

> A good test would use small amounts of plant material, require little time, *be simple to run,* and *be accurate.*

Always be careful when you are listing to keep all the elements of the list parallel. In the following example, the third item in the list is not parallel to the first two:

> The process has three stages: (1) the specimen is dried, (2) all potential pollutants are removed, and (3) atomization.

The error is easily corrected:

> The process has three stages: (1) the specimen is dried, (2) all potential pollutants are removed, and (3) the specimen is atomized.

When you start a series, keep track of what you are doing, and finish the series the same way you started it. Nonparallel sentences are at best awkward and off-key. At worst, they can lead to serious misunderstandings.

Parentheses *(Paren)*

Parentheses are used to enclose supplementary details inserted into a sentence. Commas and dashes may also be used in this role, but with some restrictions. You may enclose a complete sentence or several complete sentences within parentheses. But such enclosure would confuse the reader if only commas or dashes were used for the enclosure:

> The violence of these storms can scarcely be exaggerated. (Typhoons and hurricanes generate winds over 75 miles an hour and waves 50 feet high.) The study. . . .

Lists

Parentheses are also used to enclose numbers or letters used in listing:

> This general analysis consists of sections on (1) wave generation, (2) wave propagation, (3) wave action near a shoreline, and (4) wave energy.

Punctuation of Parentheses in Sentences

Within a sentence place no mark of punctuation before the opening parenthesis. Place any marks needed in the sentence after the closing parenthesis:

> If a runway is regularly exposed to crosswinds of over 10 knots (11.6 mph), then the runway is considered unsafe.

Do not use any punctuation around parentheses when they come between sentences. Give the statement *inside* the parentheses any punctuation it needs.

Period *(Per)*

Periods have several conventional uses.

End Stop

Place a period at the end of any sentence that is not a question or an exclamation:

Find maximum average daily temperature and maximum pressure altitude.

Abbreviations

Place a period after abbreviations:

M.D.	etc.
Ph.D.	Jr.

However, some style guides now call for no periods in academic abbreviations such as BA, MD, and PhD.

Decimal Point

Use the period with decimal fractions and as a decimal point between dollars and cents:

.4	$5.60
.05%	$450.23

Pronoun–Antecedent Agreement *(P/ag)*

Pronoun–antecedent agreement is closely related to verb–subject agreement. For example, the problem area concerning the use of collective nouns explained in verb–subject agreement is closely related to the proper use of pronouns. When a collective noun is considered singular, it takes a singular pronoun as well as a singular verb. Also, such antecedents as *each, everyone, either, neither, anybody, somebody, everybody,* and *no one* take singular pronouns as well as singular verbs:

> *Each* of the students had *his* assignment ready.

However, our sensitivity about using male pronouns exclusively when the reference may be to both men and women makes the choice of a suitable pronoun in this construction difficult. Many people might object to the use of *his* as the pronoun in the preceding example. Do not choose to solve the problem by introducing a grammatical error, as in this example of incorrect usage:

> *Each* of the students had *their* assignment ready.

The use of male and female pronouns together is grammatically correct, if a bit awkward at times:

> *Each* of the students had *his or her* assignment ready.

Perhaps the best solution, one that is often applicable, is to use a plural antecedent that allows the use of a neutral plural pronoun, as in this example:

All the students had *their* assignments ready.

The same problem presents itself when we use such nouns as *student* or *human being* in their generic sense, that is, when we use them to stand for all students or all human beings. If used in the singular, such nouns must be followed by singular pronouns:

The *student* seeking a loan must have *his or her* application in by 3 September.

Again, the best solution is to use a plural antecedent:

Students seeking loans must have *their* applications in by 3 September.

See also the entry for sexist usage.

Pronoun Form *(Pron)*

Almost every adult can remember being constantly corrected by parents and elementary school teachers in regard to pronoun form. The common sequence is for the child to say, "Me and Johnny are going swimming," and for the teacher or parent to say patiently, "No, dear, 'Johnny and I are going swimming.' " As a result of this conditioning, all objective forms are automatically under suspicion in many adult minds, and the most common pronoun error is for the speaker or writer to use a subjective case pronoun such as *I, he,* or *she* when an objective case pronoun such as *me, him,* or *her* is called for.

Whenever a pronoun is the object of a verb or the object of a preposition, it must be in the objective case:

It occurred to my colleagues and *me* to check the velocity data on the earthquake waves.

Just between *you* and *me,* the news shook Mary and *him.*

However, use a subjective case pronoun in the predicate nominative position. This rule slightly complicates the use of pronouns after the verb. Normally, the pronoun position after the verb is thought of as objective pronoun territory, but when the verb is a linking verb (chiefly the verb "to be") the pronoun is called a "predicate noun" rather than an object and is in the subjective case.

It is *she.*

It was *he* who discovered the mutated fruit fly.

Question Mark *(Ques)*

Place a question mark at the end of every sentence that asks a direct question:

> What is the purpose of this report?

A request that you politely phrase as a question may be followed by either a period or a question mark:

> Will you be sure to return the experimental results as soon as possible.

> Will you be sure to return the experimental results as soon as possible?

When you have a question mark with quotation marks, you need no other mark of punctuation:

> "Where am I?" he asked.

Quotation Marks *(Quot)*

Use quotation marks to set off short quotations and certain titles.

Short Quotations

Use quotation marks to enclose quotations that are short enough to work into your own text (normally, less than three lines):

> According to Dr. Stockdale, "Ants, wonderful as they are, have many enemies."

When quotations are longer than three lines, set them off by single spacing and indenting them. See the entry for colon for an example of this style. Do not use quotation marks when quotations are set off and indented.

Titles

Place quotation marks around titles of articles from journals and periodicals:

> Nihei's article "The Color of the Future" appeared in *PC World.*

Single Quotes

When you must use quotation marks within other quotation marks, use single marks (the apostrophe on your keyboard):

"Do you have the same trouble with the distinction between 'venal' and 'venial' that I do?" asked the copy editor.

Punctuation Conventions

The following are the American conventions for using punctuation with quotation marks:

Commas and Periods Always place commas and periods inside the quotation marks. There are no exceptions to this rule:

> G. D. Brewer wrote "Manned Hypersonic Vehicles."

Semicolons and Colons Always place semicolons and colons outside the quotation marks. There are no exceptions to this rule:

> As Dr. Damron points out, "New technology has made photographs easy to fake"; therefore, they are no longer reliable as courtroom evidence.

Question Marks, Exclamation Points, and Dashes Place question marks, exclamation points, and dashes inside the quotation marks when they apply *to the quote only or to the quote and the entire sentence at the same time.* Place them outside the quotation marks when they apply to the entire sentence only.

> *Inside*
>
> When are we going to find the answer to the question, "What causes clear air turbulence?"
>
> *Outside*
>
> Did you read Minna Levine's "Business Statistics"?

Run-on Sentence *(Run-on)*

A run-on sentence is two independent clauses (that is, two complete sentences) put together with only a comma or no punctuation at all between them. Punctuate two independent clauses placed together with a period, semicolon, or a comma and a coordinating conjunction (*and, but, for, nor,* or *yet*). Infrequently, the colon or dash is used also. (There are some exceptions to these rules. See the entry for comma.) The following three examples are punctuated correctly, the first with a period, the second with a semicolon, the third with a comma and a coordinating conjunction:

Check the hydraulic pressure. If it reads below normal, do not turn on the aileron boost.

We will describe the new technology in greater detail; however, first we will say a few words about the principal devices found in electronic circuits.

Ground contact with wood is particularly likely to cause decay, but wood buried far below the ground line will not decay because of a lack of sufficient oxygen.

If the example sentences had only commas or no punctuation at all between the independent clauses, they would be run-on sentences.

Writers most frequently write run-on sentences when they mistake conjunctive adverbs for coordinating conjunctions. The most common conjunctive adverbs are *also, anyhow, besides, consequently, furthermore, hence, however, moreover, nevertheless, therefore,* and *too.*

When a conjunctive adverb is used to join two independent clauses, the mark of punctuation most often used is a semicolon (a period is used infrequently), as in this correctly punctuated sentence:

Ice fish are nearly invisible; however, they do have a few dark spots on their bodies.

Often, the sentence will be more effective if it is rewritten completely, making one of the independent clauses a subordinate clause or a phrase.

Run-on sentence

The students at the university are mostly young Californians, most of them are between the ages of 18 and 24.

Rewritten

The students at the university are mostly young Californians between the ages of 18 and 24.

Semicolon *(Semi)*

The semicolon lies between the comma and the period in force. Its use is quite restricted. (See also the entry for run-on sentences.)

Independent Clauses

Place a semicolon between two closely connected independent clauses that are not joined by a coordinating conjunction (*and, but, or, nor, for,* or *yet*):

> The expanding gases formed during burning drive the turbine; the gases are then exhausted through the nozzle.

When independent clauses joined by a coordinating conjunction have internal punctuation, then the comma before the coordinating conjunction may be increased to a semicolon:

> The front lawn has been planted with a Chinese Beauty Tree, a Bechtel Flowering Crab, a Mountain Ash, and assorted small shrubbery, including barberry and cameo roses; but so far nothing has been done to the rear beyond clearing and rough grading.

Series

When a series contains commas as internal punctuation within the parts, use semicolons between the parts:

> Included in the experiment were Peter Moody, a freshman; Jesse Gatlin, a sophomore; Burrel Gambel, a junior; and Ralph Leone, a senior.

Sexist Usage *(Sexist)*

Conventional usages often discriminate against both men and women, but particularly against women. For example, a problem often arises when someone is talking about some group in general but refers to members of the group in the singular, as in the following passage:

> The modern secretary has to be a master of electronic equipment. She has to be able to run a microcomputer and fix a fax machine. On the other hand, her boss still doodles letters on yellow pads. He has yet to come to grips with all the electronic gadgetry in today's office.

This paragraph makes two groundless assumptions: That all secretaries are female and all executives are male. Neither assumption, of course, is valid.

Similarly, in the past, letters began with "Dear Sir" or "Gentlemen." People who delivered mail were "mailmen" and those who protected our streets were "policemen." History books discussed "man's progress" and described how "man had conquered space."

However, of late, we have recognized the unfairness of such discriminatory usages. Most organizations now make a real effort to avoid sexist usages in their documents. How can you avoid such usages once you appreciate and understand the problem?

Titles of various kinds are fairly easy to deal with. *Mailmen* have become *mail carriers; policemen, police officers; chairmen, chairpersons* or simply *chairs;* and so forth. We no longer speak of "man's progress" but of "human progress."

The selection of pronouns when dealing with groups in general sometimes presents more of a problem. One way to deal with it is to move from the singular to the plural. You can speak of *secretaries/they* and *bosses/they,* avoiding the choice of either a male or female pronoun.

You can also write around the problem. You can convert a sentence like the following one from a sexist statement to a nonsexist one by replacing the *he* clause with a verbal phrase such as an infinitive or a participle:

> The diver must close the mouthpiece shut-off valve before he runs the test.
>
> The diver must close the mouthpiece shut-off valve before running the test.

If you write instructions in a combination of the second person (you) and the imperative mood, you avoid the problem altogether:

> You must close the mouthpiece shut-off valve before you run the test.

At times, using plural forms or second person or writing around the problem simply won't work. In an insurance contract, for example, you might have to refer to the policyholder. It would be unclear to use a plural form because that might indicate two policyholders when only one is intended. When such is the case, writers have little recourse except to use such phrases as *he or she* or *he/she.* Both are a bit awkward, but they have the advantage of being both precise and nonsexist.

You can use the search program in your word processing program to find sexist language in your own work. Search for male and female pronouns and *man* and *men.* When you find them, check to see if you have used them in a sexist way or nonsexist way. If you have used them in a sexist way, correct the problem, but be sure not to introduce inaccuracy or imprecision in doing so.

See also the entry for pronoun–antecedent agreement and page 385 in Chapter 13, "Correspondence."

Spelling Error *(Sp)*

The condition of English spelling is chaotic and likely to remain so. George Bernard Shaw once illustrated this chaos by spelling fish, as *ghoti.* To do so he took the *gh* from *rough,* the *o* from *women,* and the *ti* from *condition.* If you have a spelling checker in your word processing program, it will help you avoid many spelling errors and typographical errors. Do remember, though, that a spelling checker will not catch the wrong word correctly spelled. That is, it won't warn you when you used *to* for *too.* You may obtain help from the spelling section in a collegiate dictionary where the common rules of spelling are explained. You can also buy, rather inexpensively, books that explain the various spelling rules and provide exercises to fix the rules in your mind.

Verb Form *(Vb)*

Improper verb form includes a wide variety of linguistic errors ranging from such nonstandard usages as "He seen the show" for "He saw the show" to such esoteric errors as "He was hung by the neck until dead" for "He was hanged by the neck until dead." Normally, a few minutes spent with any collegiate dictionary will show you the correct verb form. College-level dictionaries list the principal parts of the verb after the verb entry.

Verb–Subject Agreement *(V/ag)*

Most of the time verb–subject agreement presents no difficulty to the writer. For example, in the sentence, "He speaks for us all," only a child or a foreigner learning English might say, "He speak for us all." However, various constructions exist in English that do present agreement problems even for the adult, educated, native speaker of English. These troublesome constructions are examined in the following sections.

Words That Take Singular Verbs

The following words take singular verbs: *each, everyone, either, neither, anybody, somebody.*

Writers rarely have trouble with a sentence such as "No one is going to the game." Problems arise when, as is often the case, a prepositional phrase with a plural object is interposed between the simple subject and the verb, as in this sentence: "Each of the freshmen is going to the game." In this sentence the temptation is to let the object of the preposition, *freshmen,* govern the verb and write "Each of the freshmen are going to the game."

Compound Subject Joined by Or *or* Nor

When a compound subject is joined by *or* or *nor,* the verb agrees with the closer noun or pronoun:

> Either the designer or the *builders are* in error.
>
> Either the builders or the *designer is* in error.

In informal and general usage, one might commonly hear, or see, the second sentence as "Either the builders or the designer are in error." In writing you should hold to the more formal usage of the example.

Parenthetical Expressions

Parenthetical expressions introduced by such words as *accompanied by, with, together with,* and *as well as* do not govern the verb:

> Mr. Roberts, *as well as* his two assistants, *is* working on the experiment.

Two or More Subjects Joined by And

Two or more subjects joined by *and* take a plural verb. Inverted word order does not affect this rule:

> Close to the Academy are Cathedral Rock and the Rampart Range.

Collective Nouns

Collective nouns such as *team, group, class, committee,* and many others take either plural or singular verbs, depending upon the meaning of the sentence. The writer must be sure that any subsequent pronouns agree with the subject and verb:

> The *team is* going to receive *its* championship trophy tonight.
>
> The *team are* going to receive *their* football letters tonight.

Note well: When the team was considered singular in the first example the subsequent pronoun was *its.* In the second example the pronoun was *their.*

Appendixes

A. A Student Report

B. Technical Reference Books and Guides

C. A Selected Bibliography

APPENDIX A

A Student Report

Medical Consulting
635 Fisher Street
St. Paul, MN 57838
(612) 334-5929

26 November 1986

Curt Fenwood, President
Fenwood Company
1455 Iowa Avenue
Minneapolis, MN 55407

Dear Mr. Fenwood:

Statement of transmittal

Occasion for report

Enclosed is Medical Consulting's feasibility report *Accommodating Computerized Chemotherapy Pump Implant Patients at North Star Health Center*, completing the study you requested from us in September.

Statement of subject, purpose, and scope

Recommendation

This report should allow North Star Health Center to make an informed decision regarding CCPI patient-care policies. The number of patients involved, the medical risks for those patients, and the cost and availability of necessary equipment, personnel, and facilities are examined. At this time, providing limited care for CCPI patients seems to be the best option for the Center.

Acknowledgments

I am indebted to Dr. William J. Hrushesky and his staff at University Hospital, who have provided me with extensive information about their pump patient program.

I hope this report will guide North Star Health Center in developing a workable CCPI patient-care policy. If you have any questions, please write or call me at (612) 334-5929.

Sincerely yours,

Leah Isaacson

Leah Isaacson
Consultant

Enclosure

Discreet use of typographical variation

ACCOMMODATING COMPUTERIZED CHEMOTHERAPY PUMP IMPLANT PATIENTS AT NORTH STAR HEALTH CENTER

Prepared for

Curt Fenwood

President

Fenwood Company

by

Leah Isaacson

Medical Consulting

Descriptive abstract

Abstract

This report discusses the feasibility of accommodating computerized chemotherapy pump implant patients at North Star Health Center and, if such care is feasible, to what degree North Star Health Center should be involved in pump maintenance. The report examines the Medtronic 8601H pump, the number of patients involved, and the medical risks for those patients. Four CCPI patient-care options are evaluated on the basis of the personnel, equipment, and facilities necessary to treat those patients. Conclusions and recommendations are reached.

26 November 1986

CONTENTS

LIST OF TABLES

Definitions of key terms in report; all items in parallel grammatical form

GLOSSARY

biorhythmic	having to do with the variations in body function
chemotherapy	putting drugs into the bloodstream to kill cancer cells
external administration (of chemotherapy)	infusing chemotherapy into the bloodstream through an intravenous line inserted in the wrist
Huber needle	a needle similar to a hypodermic needle except that the tip is bent at a 90° angle, used to inject drugs into a chemotherapy pump
metastasis	transmission of cancer cells from the original site to other sites in the body
oncology	the scientific study of tumors
protocol	the accepted procedure for medically treating an illness
pump maintenance	refilling and reprogramming the Medtronic pump
traditonal treatment (of cancer)	treatment administered externally either through an intravenous line or an external pump

-vi-

Statement of problem

Solutions considered

Conclusions

Recommendation

ACCOMMODATING COMPUTERIZED CHEMOTHERAPY PUMP IMPLANT PATIENTS AT NORTH STAR HEALTH CENTER

EXECUTIVE SUMMARY

North Star Health Center (NSHC) must decide whether to accept computerized chemotherapy pump implant (CCPI) patients and, if it does accept them, what level of pump maintenance to offer them. Medical Consulting considered four options:

Option 1, no admittance for CCPI patients, would save any initial investment but would exclude 4,333 potential NSHC clients. If CCPI patient care is later desired, costly overstaffing may result.

Option 2, CCPI patients admitted but no pump maintenance, would include CCPI patients at no additional cost but subject these patients to twice-monthly trips for pump maintenance and programming. Including more CCPI patients at a later date would again result in costly overstaffing.

Option 3, CCPI patients admitted with limited pump maintenance, would require a nurse trained in pump techniques. Because such a nurse would also provide general nursing services and be paid the same salary as a general clinic nurse, no significant cost is involved.

Option 4, CCPI patients admitted with complete pump maintenance, would require the purchase of a Medtronic programmer for $4,000. However, a less expensive model may be available at a later time.

Pump complications are possible but are rare. Most could be handled by NSHC staff. Patients with serious emergency problems could be evacuated by helicopter to St. Luke's Medical Center in Duluth. Option 3 would allow NSHC to service CCPI patients without assuming great risk. If the number of CCPI patients warrants it, NSHC can make the transition to Option 4 easily. Therefore, Medical Consulting recommends Option 3, limited pump maintenance.

Subject and purpose of report

Space-and-a-half spacing used, a readable compromise between single-space and double-space

Background information

Statement of scope and plan of development of report and of criteria

INTRODUCTION

This report discusses the feasibility of accommodating computerized chemotherapy pump implant patients at NSHC and, if such care is feasible, to what extent the Center should be involved in pump maintenance.

When NSHC, located in rural Alice Lake, Minnesota, opens as an oncology treatment center in the summer of 1987, it must be prepared to meet the needs of its clientele with the proper staff, equipment, and facilities. Failure to provide adequate care would expose these patients to undue risk. Because the needs of CCPI patients are very different from the needs of the conventionally treated cancer patient, NSHC needs to develop guidelines concerning CCPI patient care now.

The information presented in this report should allow NSHC to develop a sound policy on CCPI patient care. The Medtronic 8610H pump, the number of future pump patients, and the medical risks for those patients are examined. Then, using the cost and availability of necessary personnel, equipment, and facilities as criteria, the following options for CCPI patient care are evaluated:

1. NSHC will not admit CCPI patients.
2. NSHC will admit CCPI patients but will not do any type of pump maintenance.
3. NSHC will admit CCPI patients and include pump refilling in their care but will not reprogram pumps.
4. NSHC will admit CCPI patients and include all aspects of pump maintenance in their care.

Because computerized chemotherapy pump implants are currently in a period of transition from experimental use to widespread use (the pumps received FDA approval on 5 November 1986), the information presented is subject to change.

However, every effort has been made to ensure that the conclusions and recommendations reached in this study are as accurate as possible.

DISCUSSION

A. Pump Background

Description of the Medtronic pump, the mechanism that is the basis for the study

The Medtronic 8610H pump is a programmable drug administration device (Medtronic n.d., 1). Its ability to dispense drugs over an extended period of time makes it useful in the chronic administration of a variety of drugs. The pump is surgically implanted in the abdominal cavity and is completely internal. Patients can receive drug infusion unhindered by an intravenous (IV) bag or an external pump (Roemeling 1986).

Author–date documentation

Because the pump can be programmed to release drugs in different time-modified patterns, it is especially beneficial in chemotherapy treatment. Since 1983 the pump has been clinically tested on cancer patients at the following institutions: University of Minnesota, University of Chicago, University of Texas, Rush-Presbyterian/St. Luke's Medical Center, and Newark Beth Israel Medical Center.

The toxic effects associated with many chemotherapy drugs can be greatly reduced if the drug is administered in a biorhythmic pattern (Roemeling and Hrushesky 1986, 3). For example, if the body has the greatest tolerance for drug X at 6:00 p.m. and is most sensitive to drug X at 8:00 a.m., the pump can be programmed to release drugs slowly during the morning and release drugs quickly in the evening. Thus, the body can get the majority of the drug when it has the highest tolerance for it.

About the size of a hockey puck, the pump consists of nine components (Medtronic n.d., 3–5):

Components listed for easy access

- **antenna.** Receives programming signals (radio frequency) from the Physician's Programmer and transmits them to the electronic module.

• **electronic module.** Interprets signals from antenna as a pattern of drug infusion, stores this pattern, and makes the pump release drugs according to this pattern.

• **peristaltic pump.** Pumps the chemotherapy drugs into the bloodstream via a catheter.

• **collapsible reservoir.** Stores chemotherapy drugs until they are released into the system.

• **battery.** Powers the pump.

• **titanic connector.** Connects a catheter to the pump; all drugs flow through this point on their way to the bloodstream.

• **self-sealing septum.** Allows access to the pump reservoir; a Huber needle inserted here can withdraw fluid to empty the pump or inject fluid to fill the pump.

• **suture pad.** Anchors pump to abdominal wall; stitches connect suture pad and body wall.

• **catheter.** Attaches to pump at connector and ends in hepatic artery; drugs flow through here to enter the bloodstream.

During chemotherapy, drugs flow from the reservoir to the peristaltic pump. The pump pushes the drugs out through the connector into the catheter. The catheter carries the drugs into the hepatic artery, which delivers the drugs to the liver.

Estimates of future patient population

B. Future Patient Populations

The projected number of pump patients in future years is a very important factor in deciding CCPI patient policy. Too few CCPI patients would make the investment in personnel and equipment to treat them financially infeasible. Conversely, a substantial CCPI patient population could easily offset even a large initial personnel and equipment investment.

Because the pumps will have widespread availability for the first time in 1987, future pump patient populations cannot be calculated simply from last year's pump population plus a population increase. Whereas in the last two years University

First-person point of view used when appropriate

Explanation of data in tables

Calculations presented in tables for easy access

Hospital was one of five pump implant centers nationwide and was limited by the FDA to 150 implants, in future years University Hospital, and *all* area medical centers, will be able to implant all pump-eligible patients.

To reflect this change, I have calculated the number of patients eligible for pump implant in 1987. I have determined pump-eligible populations for the state of Minnesota, the five-state area of Minnesota, Iowa, Wisconsin, North Dakota, and South Dakota, and the nation. In Tables I, II, and III, these figures have been classified by cancer type, using the four primary diagnoses that accounted for 88% of University Hospital Pump Patients plus a miscellaneous category (Rabatin 1986).

The vast majority of pump-eligible patients will be newly diagnosed in 1987 because most patients with earlier diagnoses

TABLE I Minnesota Projected Pump-Eligible Population

TYPE OF CANCER	EST. NEW CASES IN 1987	% ELIGIBLE	PUMP-ELIGIBLE PATIENTS
Colon and rectal	2,400	22%	528
Liver and biliary	238	100%	238
Kidney	350	42%	147
Pancreas	450	30%	135
Miscellaneous			140
Total			1,188

TABLE II Five-State Area Projected Pump-Eligible Population

TYPE OF CANCER	EST. NEW CASES IN 1987	% ELIGIBLE	PUMP-ELIGIBLE PATIENTS
Colon and rectal	8,800	22%	1,936
Liver and biliary	816	100%	816
Kidney	1,500	42%	630
Pancreas	1,470	30%	441
Miscellaneous			510
Total			4,333

TABLE III National Projected Pump-Eligible Population

TYPE OF CANCER	EST. NEW CASES IN 1987	% ELIGIBLE	PUMP-ELIGIBLE PATIENTS
Colon and rectal	140,000	22%	30,800
Liver and biliary	13,600	100%	13,600
Kidney	20,000	42%	8,400
Pancreas	25,000	30%	7,500
Miscellaneous			8,320
Total			68,620

will already be receiving some other type of treatment (Hrushesky 1986). Therefore, I have based my figures on the American Cancer Society (1986) and The National Cancer Institute's estimates of new cancer cases in 1987 (Page and Alsire 1985).

It is important to make the distinction between *new* cancer cases and *total* cancer cases when figuring population increases for successive years. If I had based these calculations on the total incidence of cancer, the 1988 population would equal approximately the 1987 pump-eligible population plus a 10%–20% increase. Because these calculations are based on the number of new cancer cases in 1987, the pump-eligible patient population in 1986 is effectively zero, and the pump-eligible population in 1988 will be over twice that of 1987 (the new 1988 cases plus the old 1987 cases). Furthermore, the 1989 pump-eligible population will be approximately triple the 1987 population (the new 1989 cases plus the 1988 and 1987 cases). Thus, the number of patients eligible to receive pump implants will increase rapidly.

In Tables I, II, and III, the total number of new cancer cases is then multiplied by the percentage of patients who would be eligible for pump implant. The percentages are supplied by Dr. Reinhard Roemeling of University Hospital (1986). According to current protocol, a patient must have cancer originating in the liver or with liver metastasis to be eligible for pump implant.

Dr. Roemeling's percentages are the percentage of patients with each type of cancer meeting these criteria.

The calculations in Tables I, II, and III yield the following 1987 pump-eligible patient populations:

Informal table with end results of calculations

Minnesota	1,888 patients
Five-state area	4,333 patients
United States	68,620 patients

Moreover, the number of pump-eligible patients will increase dramatically if initial findings prove valid and the pump is implemented in treatment of other cancers.

Examination of potential problems and their solutions

C. Patient Risks: Severity and Emergency Care

Operating a health-care facility in a rural setting like Alice Lake has certain drawbacks. NSHC cannot possibly have equipment and staff of its own to handle every possible emergency; yet the nearest complete medical center is over 90 miles away in Duluth, Minnesota. Patients at NSHC might run certain risks because of this distance. Pump-implanted patients may be at greater risk than traditionally treated patients because they require such complex equipment and specially trained personnel.

CCPI patients may experience a variety of complications (Medtronic n.d., 9–10). Severe infection may occur around the pump site, or a pump malfunction may allow chemotherapy drugs to enter the bloodstream unregulated. If the pump leaks, chemotherapy drugs could pool in the surrounding tissues, causing extensive tissue damage. Should the pump shut down or the catheter become clogged, the drugs would no longer reach the bloodstream. I contacted University Hospital on the assumption that NSHC would encounter problems similar to those at University Hospital.

Physicians at University Hospital have experienced only six complications out of 150 total pump patients, as shown in Table IV (Rabatin 1986).

TABLE IV Patient Complications: Treatment and Time to Response

NO. OF PATIENTS	PROBLEM	TREATMENT	TIME TO RESPONSE
4	Blocked catheter	External administration of chemotherapy;	3 days
		nonemergency surgery to open catheter passage	1 month
1	Severe infection of pump site	Surgical removal of pump	3 days
1	Drug leakage into tissues	Empty pump;	4 hours
		drain area of drug[a]	4 hours

[a] Drainage is not always necessary; depending on the amount of drugs that has pooled in the tissue.

Four of the six complications involved blocked catheters, a nonemergency situation. As long as chemotherapy is continued via external administration, the blocked catheter presents no threat to the patient. Of course, the catheter should be cleared or replaced as soon as feasible so that the pump can again function normally. If a patient's catheter became blocked, NSHC could administer external chemotherapy, just as it would to traditionally treated patients, until the patient made arrangements to see his or her own doctor.

Severe infection of the pump site is a much more serious problem. Although these infections usually occur within two weeks of implantation, they can develop at any time (Roemeling 1986). When infections become so severe that they do not respond to antibiotics, the pump must be surgically removed. Although NSHC is not likely to encounter this problem--all clientele will be more than two weeks past implantation--it is a possibility. If this occurs, the patient must be closely supervised until the pump can be removed. The pump should be removed at a medical center familiar with pump procedures within three days. Because ground transportation time from Alice Lake to Duluth is two hours and from Alice Lake to Minneapolis is four hours, air transportation would not be necessary. Emergency care vehicles may not be required if the patient is stable enough to be driven by a friend or family member.

When a pump leaks into the surrounding tissues, the pump must be emptied within one hour. If NSHC is staffed with a nurse trained to fill the pump, this could be done on-site. If not, the patient would need emergency air transportation to St. Luke's Medical Center in Duluth. If the amount of drug in the tissue is sufficient to require draining, there is no on-site treatment. The patient must be transported to St. Luke's by air immediately.

St. Luke's Medical Center has a Life-One helicopter designed to handle these kinds of emergencies and any others that may arise. Because Life-One access to NSHC would reduce the risks of *all* patients, by providing quick, emergency access to a complete medical center, NSHC has already planned for a helicopter landing pad.

Risks, advantages, and implications of each option

D. Evaluation of Possible Options

This section evaluates four possible options for CCPI patient care.

1. No Admittance for CCPI Patients

If NSHC does not admit CCPI patients, no additional medical staff, equipment, or facilities are needed. If NSHC does not care for CCPI patients, it need make no additional investments in medical staff, equipment, or facilities. However, this option will exclude 4,333 CCPI patients from the five-state area as possible clientele. Monthly treatment costs are less for CCPI patients than for traditionally treated patients: $375 for CCPI versus $450 for traditional (Bonacci 1985). Because NSHC will charge each client a flat rate to provide all care--common practice for such centers--it will incur higher costs by treating only traditionally treated patients. Should the number of CCPI patients increase, NSHC would need to provide CCPI patient care at a later date. Hiring additional medical personnel trained in pump techniques at this later time would probably result in overstaffing.

2. No Pump Maintenance

In Option 2, NSHC would admit CCPI patients but would not be involved in any type of pump maintenance. Under this option, at least one member of NSHC's medical staff would have to be qualified to recognize pump complications. Pump infections are not difficult to spot, but tissue toxicity caused by leaking pumps is. Extensive first-hand experience with pump patients is not necessary, but familiarity with the appearance of tissue toxicity is. Because this person would be part of NSHC's regular staff, there would be no additional cost. Many experienced oncologists could fulfill this need.

Caring for CCPI patients without doing pump maintenance presents a different set of advantages and disadvantages. The CCPI patients would be admitted, but they would have to leave NSHC every two weeks to have their pumps filled and reprogrammed elsewhere. NSHC would need to include at least one nurse or physician on its staff who could recognize pump complications. This person wouldn't require a higher salary, but finding him or her might require some additional interviewing. Some costs would be cut by the lower monthly treatment costs of CCPI patients, although this could be offset by the shortened stays of CCPI patients, who would have to seek complete care elsewhere. As in option 1, converting to a full-care CCPI facility at a later time could result in overstaffing.

3. Limited Pump Maintenance

Option 3 calls for NSHC to include emptying and refilling patient pumps in its CCPI patient-care regimen. The pumps require filling at two-week intervals, which can be done by a clinic nurse trained in pump technique. The procedure, similar to flushing a Hickman catheter, does not require extensive training; however, it does require hands-on experience. Persons capable of filling the pumps would also be familiar with pump complications. A nurse with this training would command a salary of $2,500 per month. However, the same nurse would perform clinic functions besides filling pumps. Because the salary range of a pump-trained nurse is comparable to that of a general clinic nurse, the additional personnel cost is minimal.

Under this option the CCPI patients involved could now have their pumps filled on-site. Although the pumps are designed to be reprogrammed every two weeks, patients often have them reprogrammed only monthly without risk (Hrushesky 1986). Thus, many CCPI patients could stay at NSHC for one month without interruption. NSHC would probably save significantly with the lower treatment cost of CCPI patients. With the addition of pump-trained personnel at NSHC, some CCPI patient emergencies could be handled at Alice Lake.

4. Complete Pump Maintenance

Becoming involved in all aspects of pump maintenance calls for a staff with more specialized training. Someone must reprogram the pumps every two weeks using the Physician's 8800M Programmer. This person could be the same nurse who refills the pumps or a separate technician at a salary of $1,500 per month. It would be most cost-effective to train the same nurse who refills the pumps to program them, rather than hiring an additional person. Depending on the number of CCPI patients, this nurse could also provide conventional nursing services. Medtronic provides the specialized training needed when the Physician's Programmer is purchased. The Physician's Programmer presently costs about $4,000, but a smaller (and presumably less expensive) portable model is under development (Pierce 1986). A standard three-prong plug 220-volt power outlet for the programmer must be available in the clinic area. In addition, all the requirements of Option 3 must be met.

Under this option, CCPI patients' needs would be met

completely by NSHC. Their stay at Alice Lake would be uninterrupted. Money would be saved on monthly treatment costs by treating CCPI patients, and the pump-trained personnel at Alice Lake could handle many emergencies themselves, using the air transport system only for critical cases. In addition, if, as suggested by Roemeling and Hrushesky (1986), the pump in the future is used in treatment of other cancers, NSHC would be prepared to meet the needs of an expanded CCPI patient population.

Key facts put into meaningful context

FACTUAL SUMMARY

Implanted computerized chemotherapy pumps have been tested on cancer patients since 1983 and have shown very positive results in reducing toxic effects. The Medtronic 8610H pump will have widespread availability for the first time in 1987, the same year NSHC will open for business. Deciding what policy NSHC will adopt toward CCPI patients depends on the number of patients involved, the medical risks those patients would be exposed to, and the cost and availability of the personnel, equipment, and facilities required for CCPI patient care.

Approximately 4,333 patients will be eligible for pump implants in the five-state area of Minnesota, Wisconsin, Iowa, North Dakota, and South Dakota in 1987. If things go as expected, most of those patients will receive pump implants.

Many pump complications are possible, but few have actually been experienced. Most of the complications University Hospital encountered were nonemergencies that NSHC could handle even with a nonspecialized staff. Complications requiring immediate emergency treatment are rare, but NSHC patients with such complications could be evacuated by helicopter to a complete medical center. However, some member of NSHC must be familiar with pump complications if CCPI patients are present. Personnel trained in pump techniques would enable NSHC to handle some emergencies at Alice Lake.

Results of evaluating the four options are summarized as follows:

Options listed for easy comparison

Option 1. No admittance for CCPI patients would save any initial investment but would exclude 4,333 potential NSHC clients. If CCPI patient care is later desired, costly overstaffing may result.

Option 2. No pump maintenance would include CCPI patients at no additional cost but would subject all CCPI clients to twice-monthly trips to their own physicians to have their pumps refilled and reprogrammed. Including more CCPI patient services at a later date would again result in costly overstaffing.

Option 3. Limited pu· p maintenance would include CCPI patients but would require a nurse trained in pump techniques. Because this nurse could provide general nursing services as well and would be paid no more than a general clinic nurse, no significant cost is involved. Patients would need their pumps reprogrammed elsewhere monthly but could spend that month at NSHC uninterrupted.

Option 4. Complete pump maintenance would provide complete pump maintenance for CCPI patients. Medtronic at no cost would train the nurse who fills the pumps to program them, but the Physician's 8800M Programmer would have to be purchased for approximately $4,000. However, a less expensive model is being developed.

CONCLUSIONS

Conclusions presented as separate statements

1. Increasing numbers of cancer patients will be CCPI patients.
2. CCPI patients at NSHC will not be at great risk.
3. A later decision to care for CCPI patients might mean hiring additional pump-trained staff who would handle only CCPI patients, resulting in overstaffing.
4. Pump-trained personnel could be hired now in place of general

Culminating conclusion leading to recommendation

Recommendation fulfilling purpose of study

clinic personnel to perform both pump duties and general duties; thus there would be no additional personnel costs.

5. The $4,000 Physician's 8800M Programmer is a significant investment, but a less expensive Physician's Programmer may be available soon.

6. Option 3 would allow NSHC to be involved in CCPI patient care without assuming a great risk. If the number of CCPI patients warrants it, NSHC can make the transition to Option 4 easily. The Physician's Programmer can be purchased, possibly for less, and Medtronic could train the pump nurse to program the pumps.

RECOMMENDATION

NSHC should implement Option 3 by including, as part of the general medical staff, a nurse with experience refilling computerized chemotherapy pumps.

WORKS CITED

American Cancer Society. 1986. *1986 Cancer Facts and Figures.* New York: American Cancer Society.

Bonacci, Michelle. Memo from Creighton Cancer Center, Houston, TX, to Medtronic, Inc., Minneapolis, MN, 6 Oct. 1985.

Hrushesky, William J. Personal interview. 4 Nov. 1986.

Medtronic. N.d. *Medtronic Model 8800M Physician's Programmer Manual.* Draft ed. Minneapolis: Medtronic, Inc.

Page, Harriet S., and Ardyce J. Alsire. 1985. *Cancer Rates and Risks.* 3rd ed. NIH Publication no. 85–691. Washington, DC: National Cancer Institute.

Pierce, Joanna. Personal interview. 28 Oct. 1986.

Rabatin, Jeffery. Personal interview. 13 Nov. 1986.

Roemeling, Reinhard R. Personal interview. 5 Nov. 1986.

Roemeling, Reinhard R., and William J. Hrushesky. 1986. "Principles of Chemotherapy and Immunotherapy for Head and Neck Cancer." Unpublished study.

APPENDIX B

Technical Reference Books and Guides

Prepared by Donald J. Barrett
Chief Reference Librarian
United States Air Force Academy

Even an average library makes a staggering amount of technical information available to its users. You can gain ready access to that information only through reference books and reference guides. A **reference book,** such as a dictionary, encyclopedia, or atlas, consolidates a good deal of technical information in one location. With a **reference guide** you can find the reference books, periodicals, and reports published in any specific field from agriculture to zoology. This appendix is a guide to both field-specific reference books and field-specific reference guides. The preceding list of subjects covered in this appendix shows the extent of what is available to you and can help you locate the particular works you may need.

Book Guides

You can find information on books in science and technology in standard guides to book publication and general reviewing tools. A few tools devoted specifically to technical publications do exist, but most cover only a small portion of each year's book production, so, in this case, you must use the more general tools. Several of the major guides are listed here and will be found in most libraries.

Book Review Digest. 1905–.

Index to selected book reviews, mostly from general periodicals. Gives publication data, brief descriptive notes, exact citations to reviews in about 80 periodicals. Subject and title indexes.

Books in Print and ***Subject Guide to Books in Print.*** Annual.

Guide to book availability (in print) from 3,600 American publishers.

Books in Print: Author and title lists—give author, title, usually date of publication, edition, price, and publisher.

Subject Guide to Books in Print: Books entered under Library of Congress subject heading, then by author, with title and other bibliographic data.

Cumulative Book Index. 1928–.

World list of all books published in the English language. Author, title, and subject listings arranged in a dictionary sequence. Main entry (fullest information) under author. Gives author, title, edition, series, pagination, price, publisher, date, Library of Congress card number.

Library of Congress Catalogs: Subject Catalog. 1950–.

Cumulative list by subject of works represented by Library of Congress printed cards for publications printed 1945 or later. Since 1983, issued in microfiche only.

New Technical Books. 1915–.

Selectively annotated titles by New York Public Library staff. Classed subject arrangement, includes table of contents, annotation for each book.

Proceedings in Print. 1964–.

Announcement journal for availability of proceedings of conferences. Full citation, price, subject and agency indexes.

Technical Book Review Index. 1935–88.

Very good guide to reviews appearing in scientific, technical, and trade journals, with bibliographic data and exact references to sources of reviews.

United Nations Documents Index. 1950–.

Contains checklists of documents and publications issued by various UN agencies, with subject and author indexes cumulated annually.

Reference Books

Each subject or academic field frequently has its own literature, often ranging from encyclopedias to indexes, dictionaries, biographical tools, and so on. The amount of literature specific to one subject may vary greatly. When you first approach a field, one principal guide, Sheehy's, is available to assist you in becoming familiar with its literature. The current edition of this guide should be found in every major collection:

Guide to Reference Books. Eugene P. Sheehy, ed. 10th ed. Chicago: American Library Association, 1986. Biennial suppls.

Encyclopedias

The encyclopedia, although considered too general by some specialists, is often extremely useful to the researcher in getting under way and learning a field. Several of the encyclopedias for specific subjects are in fact quite detailed and scholarly. The editors and contributors to a well-written special encyclopedia will often be experts in their fields. Of course, the latest developments in a subject would be available only in periodical and report literature. Still, the general works are of great value toward understanding a subject, and they often include bibliographic citations to aid in further research.

Dictionary of Organic Compounds: The Constitution and Physical, Chemical, and Other Properties of the Principal Carbon Compounds and Their Derivatives, Together with Relevant Literature References. Ian M. Heilbron, ed. 5th rev. ed. 7 vols. New York: Chapman and Hall, 1982. Plus annual suppls. 1985–.

Alphabetical list of compounds, large number of cross-references.

Encyclopaedic Dictionary of Physics. 9 vols., 5 suppls. New York: Pergamon, 1961–.

Scholarly work, alphabetically arranged, articles generally under 3,000 words, most with bibliographies. Includes articles on general, nuclear, solid state, molecular, chemical, metal, and vacuum physics. Index, plus a multilingual glossary in six languages.

Encyclopedia of the Biological Sciences. Peter Gray, ed. 2nd ed. New York: Van Nostrand Reinhold, 1970.

Contains 800 articles covering the broad field of the biological sciences as viewed by experts in their developmental, ecological, functional, genetic, structural, and taxonomic aspects. Bibliographies, biographical articles, illustrations, and diagrams are helpful features.

Encylopedia of Chemical Technology. Raymond Kirk and Donald Othmer, eds. 24 vols., index, and suppl. 3rd ed. New York: Wiley-Interscience, 1978–84.

Main subject is chemical technology; about half the articles deal with chemical substances. There are also articles on industrial processes. A bibliography is included for each product, as well as information on properties, sources, manufacture, and uses.

Encyclopedia of Materials Science and Engineering. Michael Bever, ed. 8 vols. Cambridge, MA: MIT P, 1986.

Over 1,550 articles to assist in understanding design and development of new processes. Alphabetical topical arrangement. Index volume includes a systematic outline, author citation index, subject index, and materials information sources.

Encyclopedia of Physical Science and Technology. 15 vols. San Diego: Academic Press, 1987. Supplemental yearbooks.

Covers all aspects of the physical sciences, including electronics, lasers, and optical technology. Over 500 articles, many illustrations, tables, and bibliographic citations.

Encylopedia of Polymer Science and Engineering. 2nd ed. 17 vols. New York: Wiley, 1985–88.

Articles designed to present a balanced account of all aspects of polymer science and technology, with bibliographies included.

McGraw-Hill Encyclopedia of Science and Technology. 6th ed. 20 vols., annual suppls. New York: McGraw, 1987.

Main set includes 7,600 articles, kept current by annual supplements. Covers the basic subject matter of all the sciences and their major applications in engineering, agriculture, and other technologies. Separate index volume. Has many diagrams and charts, and complicated subjects are treated in clear and readable language. Contributors identified in index volume.

Van Nostrand's Scientific Encyclopedia. 7th ed. 2 vols. New York: Van Nostrand Reinhold, 1989.

Includes articles on both basic and applied sciences. Defines and explains over 17,000 terms, arranged alphabetically with extensive cross-references.

Subject Guides

Bibliographers and librarians have gathered research suggestions and bibliographies for many fields into published guides. It should be remembered that the rapidly changing literature in many subjects partially outdates any guide. Therefore, Sheehy's *Guide to Reference Books* and other tools listing current books and indexes should be consulted to supplement these guides.

Chemical Publications, Their Nature and Use. Melvin G. Mellon. 5th ed. New York: McGraw, 1982.

Describes publications by nature and sources. Identifies primary, secondary, and tertiary sources and evaluates their use by subject. Chapters on manual searching techniques and computer searching of data bases.

Geologic Reference Sources: A Subject and Regional Bibliography to Publications and Maps in the Geological Sciences. Dederick Ward and Marjorie Wheeler. 2nd ed. Metuchen, NJ: Scarecrow, 1981.

Subject bibliography, most items not annotated. Subject and geographic indexes. Section on geologic maps.

Guide to Basic Information Sources in Engineering. Ellis Mount. New York: Wiley, 1976.

Annotated list by type material and subject. Author and title index.

Guide to U.S. Government Scientific and Technical Resources. Rao Aluri. Littleton, CO: Libraries Unlimited, 1983.

Covers access to federally sponsored research, reports, patents, translations, and data bases.

How to Find Chemical Information: A Guide for Practicing Chemists, Educators, and Students. Robert E. Maizell. 2nd ed. New York: Wiley, 1986.

Extensive coverage on use of *Chemical Abstracts*, on-line data bases, patents, and standard literature.

Information Sources in Physics. Dennis F. Shaw, ed. 2nd ed. London: Butterworths, 1985.

Subject chapters, bibliographic essays, detailed examination of the literature.

Information Sources in Science and Engineering. C. D. Hurt. Littleton, CO: Libraries Unlimited, 1988.

Chapters on the literature of individual disciplines, annotated citations to over 2,000 titles.

Information Sources in the Life Sciences. H. V. Wyatt. 3rd ed. Boston: Butterworths, 1987.

General essays on searching, plus eight chapters on specific subjects. Worldwide literature coverage.

Science and Engineering Literature: A Guide to Reference Sources. Harold R. Malinowsky. 3rd ed. Littleton, CO: Libraries Unlimited, 1980.

Selected evaluative list of basic reference sources, arranged by major subjects such as physics, chemistry, astronomy, and the like.

Science and Technology: An Introduction to the Literature. Denis Grogan. 4th ed. London: Binbley, 1982.

Student guide to structure of the literature of science and technology.

Science Information Sources: A Universal and International Guide. Herman de Jaeger. 2nd ed. Nijmegen, Holland: Association Scientifique et Technique, 1974.

Selective list by subject of scientific and technical information sources.

Scientific and Technical Information Sources. Ching-Chih Chen. 2nd ed. Cambridge, MA: MIT P, 1986.

Arranged by type of publication. Lists materials for each subject field. Annotated evaluative entries.

Technical Information Sources: A Guide to Patent Specifications, Standards, and Technical Reports Literature. Bernard Houghton. 2nd ed. Hamden, CT: Linnet, 1972.

British emphasis; covers use of patents as source of technical information, use of specifications and reports.

Use of Chemical Literature. R T. Bottle. 3rd ed. London: Butterworths, 1979.
Subject arrangement, guides to resources and services evaluated.

Use of Engineering Literature. K. W. Mildren, ed. London: Butterworths, 1976.
Discusses forms of literature and services with 22 chapters of various engineering fields.

Use of Mathematical Literature. A. R. Dorling, ed. London: Butterworths, 1977.
Describes general literature and use of particular tools, then provides critical accounts by subject experts in their fields. Author and subject indexes.

Using the Biological Literature: A Practical Guide. Elisabeth B. Davis. New York: Marcel Dekker, 1981.
Broad subject chapters; bibliographic lists with brief subject content notes.

Using the Chemical Literature: A Practical Guide. Henry Woodburn. New York: Marcel Dekker, 1974.
Practical guide to use of chemical literature; not a comprehensive bibliography of sources.

Bibliographies

The literature of a field may often be compiled into bibliographies in connection with other publications, and in some cases as an indication of the work of an agency or company. A guide to bibliographies and examples of other types of compilations are given here.

Bibliographic Index: A Cumulative Bibliography of Bibliographies. 1937–.
Alphabetical subject list of separately published bibliographies and bibliographies appearing in books, pamphlets, and periodicals.

Bibliography of Agriculture. 1942–.
Classified bibliography of current literature received in the National Agricultural Library, with cumulative annual subject and author indexes.

Bibliography and Index of Geology. 1969–.
Index produced by Geological Society of America. Arranged in broad subject categories, with author and subject indexes. Formerly, *Bibliography and Index of Geology Exclusive of North America* and *Bibliography of North American Geology.*

Chemical Titles. 1960–.
Author and keyword indexes to titles from 700 journals in pure and applied chemistry. A computer-produced bibliography.

Dissertation Abstracts International: Abstracts of Dissertations Available on Microfilm or as Xerographic Reproductions. 1952–.
Compilation of abstracts of doctoral dissertations from most American universities. Since 1966, Part B has been devoted to the sciences and engineering.

Index of Selected Publications of the RAND Corporation. 1946–.
Coverage includes unclassified publications of the corporation. Abstracts, listed by subject and author, describe content and indexes.

Science Citation Index. 1961–.
Computer-produced index that provides access to related articles by indicating

sources in which a known article by an author has been cited. Not ideally suited to subject searching.

Translations Register-Index. 1967–77.

Lists new translations received in the Special Libraries Association translations center, plus those available from government and commercial sources.

Vertical File Index. 1932–.

Subject and title index to selected pamphlet material in all subjects of interest to the general library.

Biographies

Identification of authors and significant figures in the scientific and technical areas is frequently a problem. Some of the most notable biographical sources are commented on here. A check of Sheehy's *Guide to Reference Books* will reveal many more directories in almost every major subject field.

Biography Index. 1946–.

Index to biographical material in books and magazines. Alphabetical, with index by profession and occupation.

American Men and Women of Science: Physical and Biological Sciences. 17th ed. 8 vols. New York: Bowker, 1989–90.

Standard biographical set for 130,000 personages in the sciences.

Dictionary of Scientific Biography. 14 vols., suppl., and index. New York: Scribner's, 1970–80.

Comprehensive: covers historical and current persons in science field.

Who's Who in America: A Biographical Dictionary of Notable Living Men and Women. Chicago: Marquis, 1899–. Biennial.

The standard dictionary of contemporary bibliographical data. Regional volumes cover persons not of national prominence.

Who's Who in Frontiers of Science and Technology, 1985. 2nd ed. Chicago: Marquis Who's Who, 1985.

Covers scientists who have distinguished themselves by research in fields representing either new directions in traditional fields or research areas using advanced technologies.

Who's Who in Science in Europe: A New Reference Guide to West European Scientists. 5th ed. 3 vols. New York: Gale Research Service, 1987.

More than 40,000 entries including natural and physical sciences.

Dictionaries

Definition of terms for the student and scholar is a problem in the sciences, as in any field. A few general guides are available in addition to glossaries for a single field.

Chambers Science and Technology Dictionary. Peter M. B. Walker, ed. 4th ed. New York: Cambridge UP, 1988.

Successor to *Chambers Technical Dictionary*, completely revised. Dictionary of concise definitions.

Concise Chemical and Technical Dictionary. Harry Bennett, ed. 4th ed. New York: Chemical Publishing, 1986.

Contains 50,000 brief definitions, including chemical formulas.

A Dictionary of Physical Sciences. John Daintith, ed. New York: Rowman, 1983.

Includes some diagrams and cross references.

McGraw-Hill Dictionary of the Life Sciences. Daniel N. Lapedes, ed. New York: McGraw, 1976.

Provides vocabulary of the biological sciences and related disciplines. Over 20,000 terms, useful appendixes.

McGraw-Hill Dictionary of Physics and Mathematics. Daniel N. Lapedes, ed. New York: McGraw, 1978.

More than 20,000 terms, containing both basic vocabulary and current specialized terminology. Illustrated.

McGraw-Hill Dictionary of Scientific and Technical Terms. Sybil P. Parker, ed. 3rd ed. New York: McGraw, 1984.

Gives almost 100,000 definitions, amplified by 2,800 illustrations. Each definition identified with the field of science in which it is primarily used.

Modern Science Dictionary. A Hechtlinger. 2nd ed. Palisade, NJ: Franklin, 1975.

Briefly defined terms for beginning science students.

Commercial Guides

Access to materials from companies working on a specific product can be facilitated by the use of product association and company address information. The standards are an example of tools that many industries must use to satisfy a contractor's requirements.

Annual Book of ASTM Standards, with Related Material. Philadelphia: American Society for Testing and Materials. Annual.

Approximately 50 parts including index. Contains 4,900 ASTM Standards and Tentatives in effect at the time of publication. An example of an essential reference book in industrial technology.

MacRae's Blue Book. 5 vols. Chicago: MacRae's Blue Book. Annual.

Buying directory for engineering products from over 50,000 companies.

Thomas Register of American Manufacturers and Thomas Register Catalog File. 23 vols. New York: Thomas. Annual.

Products lists, alphabetical listing of manufacturers, trade names, catalogs.

Periodicals

In almost any current research, the latest developments in a field will be published in the current periodicals and professional journals. Your first task as a researcher may be to determine what periodicals are published in a given field. Next, you may want to determine what indexing or abstracting services give access to a specific journal. The guides are the most significant in assisting you to locate this type of information.

Ulrich's International Periodicals Directory: Now Including Irregular Serials and Annuals. 28th ed. 3 vols. New York: Bowker, 1989. Biennial, with suppl.
Alphabetical subject list of over 111,000 serials, published in all languages. Alphabetical title index. Indicates coverage of titles in periodical indexes and abstracting services.

World List of Scientific Periodicals Published in the Years 1900–1960. 4th ed. 3 vols. London: Butterworths, 1963–65. Suppl., 1960–68, 1970.
Lists more than 60,000 titles of periodicals concerned with the natural sciences and technology.

Periodical Indexes

A periodical index provides ready access to articles appearing in professional journals and general periodicals. Each index should be examined for inclusion principles, entry format, and any peculiarities unique to that index. The principal indexes for your consideration cover both general and specific fields, and the primary newspaper index to the *New York Times* is also listed.

Agricultural Index, Subject Index to a Selected List of Agricultural Periodicals and Bulletins. 1916–64.
Detailed alphabetical subject index. Continued as *Biological and Agricultural Index.*

Air University Library Index to Military Periodicals. 1949–.
Subject index to significant articles in approximately 70 military and aeronautical periodicals not covered in readily available commercial indexes.

Applied Science and Technology Index. 1958–.
Subject index to periodicals in aeronautics, automation, chemistry, construction, electricity and electrical communication, engineering, geology and metallurgy, industrial and mechanical arts, machinery, physics, transportation, and related subjects. Formerly part of the *Industrial Arts Index.*

Art Index. 1929–.
Author and subject index to fine arts periodicals. Also includes coverage of architecture, graphic arts, industrial design, planning, and landscape design.

Biological and Agricultural Index. 1964–.
Detailed subject index to approximately 190 English language periodicals. Reports, bulletins, and other agricultural agency publications formerly covered in the *Agricultural Index* are no longer covered.

Business Periodicals Index. 1958–.
Subject index to business, financial, and management periodicals, and specific industry and trade journals. Formerly part of the *Industrial Arts Index.*

Current Technology Index. 1981–.
Subject guide to articles in British technical journals, with author index. Formerly *British Technology Index*, 1962–80.

General Science Index. 1978–.

Subject index to 110 periodicals in fields including astronomy, atmospheric sciences, biological sciences, earth sciences, environment and conservation, genetics, oceanography, physics, physiology, and zoology.

Index to U.S. Government Periodicals. 1970–.

Computer-generated guide to 170 selected titles by author and subject.

Industrial Arts Index. 1913–57.

Subject index, split into *Applied Science and Technology Index* and the *Business Periodicals Index.*

New York Times Index. 1851–.

Subject index with precise reference to date, page, and column for each article. Well cross-referenced, brief synopses of articles. Can serve as a guide to locating articles in other unindexed papers.

Public Affairs Information Service Bulletin (PAIS). 1915–.

Very useful index to government, economics, sociology, and so on, covering books, periodicals, documents, and reports. Includes selective indexing of over 1,000 periodicals.

Readers' Guide to Periodical Literature. 1900–.

Best known periodical index. Covers U.S. periodicals of a broad, general nature in all subjects and scientific fields.

Social Sciences Index. 1974–.

Author and subject index to periodicals in fields including economics, environmental science, psychology, planning, and public administration. Covers 300 titles on a more scholarly level than the *Reader's Guide.* Formerly part of the *Social Sciences and Humanities Index* and the *International Index to Periodicals.*

Abstract Services

The abstracting journals assist access to periodical, book, and report literature as does an index. The significant difference is the abstract itself, which frequently gives a better indication of article content and hence makes the abstracting journal significantly more useful than the periodical index.

Abstracts and Indexes in Science and Technology: A Descriptive Guide. 2nd ed. Dolores B. Owen, Metuchen, NJ: Scarecrow, 1985.

Describes approximately 220 abstract and indexing services by subject coverage, arrangement, indexes, abstracts, and other features. Includes information on data bases.

Abstracts of North American Geology. 1966–71.

Abstracts of books, technical papers, maps on the geology of North America. Complements the *Bibliography of North American Geology.*

Air Pollution Abstracts. 1970–76.

Produced by the Air Pollution Technical Information Center of the Environmental Protection Agency. Broad subject arrangement with specific author–subject indexes. Covers periodicals, books, proceedings, legislation, and standards.

ASM Review of Metal Literature. 1944–67.

American Society for Metals abstracting journal of the world's literature concerned with the production, properties, fabrication, and application of metals, their alloys and compounds. Combined to form part of *Metal Abstracts.*

Biological Abstracts. 1926–.

Broad subject coverage of periodicals, books, and papers in all biological fields. Author, keyword, and systematic indexes.

Ceramic Abstracts. 1922–.

Published in 17 subject sections with book, author, and subject indexes.

Chemical Abstracts. 1908–.

Covers chemical periodicals in all languages. Arranged in 80 subject sections; entry includes title, author, publication date, and abstract. Index sections by chemical substance, formula, numbered patent, patent concordance, author, and keyword.

Energy Research Abstracts. 1976–.

Covers all scientific and technical reports, journal articles, conference papers and proceedings, books, patents, theses originated by the Department of Energy, its laboratories and contractors. Classed subject arrangement, with author, title, subject, report, and contract indexes. Succeeds *Nuclear Science Abstracts.*

Engineering Index. 1906–.

Abstracting journal includes coverage of serial publications, papers of conferences and symposia, separate and nonserial publications, some books. Excludes patents. Entries arranged by subject. Separate author index.

Geophysical Abstracts. 1929–71.

Abstracts of current literature pertaining to the physics of the solid earth and to geophysical exploration. Annual author and subject indexes.

A Guide to the World's Abstracting and Indexing Services in Science and Technology. Washington, DC: National Federation of Science Abstracting and Indexing Services, 1963.

List of 1,855 titles originating in 40 countries. Covers the pure and applied sciences including medicine and agriculture. Includes country and subject indexes.

Information Science Abstracts. 1966–.

Classified subject arrangement listing books and periodicals, international in scope. Formerly *Documentation Abstracts.*

International Aerospace Abstracts. 1961–.

Covers published literature in aeronautics and space science and technology. Companion publication to *Scientific and Technical Aerospace Reports.* Arranged in 75 subject categories, each entry gives accession number, title, author, source (book, conference, periodical, etc.), date, pagination, and abstract. Indexed by specific subject, personal author, contract number, meeting paper, and accession number.

Mathematical Reviews. 1940–.

Subject index to mathematical periodicals and books. Arranged by broad subject, with abstracts for most entries. Separate author index, subject classification.

Metallurgical Abstracts. 1934–67.

British publication, combined to form part of *Metals Abstracts.*

Metals Abstracts. 1968–.

Covers all aspects of the science and practice of metallurgy and related fields.

Classed subject arrangement with author index. Publication is a merger of *Metallurgical Abstracts* and the *ASM Review of Metal Literature.*

Meteorological and Geoastrophysical Abstracts. 1950–.

Includes foreign publications, arranged by Universal Decimal Classification subjects. Has separate author, subject, and geographical location indexes.

Mineralogical Abstracts. 1920–.

Classified list of abstracts covering current international literature, including books, periodicals, pamphlets, reports.

Nuclear Science Abstracts. 1947–76.

Covers reports of the U.S. Energy Research and Development Administration (formerly the Atomic Energy Commission), government agencies, universities, industrial and independent research organizations, and worldwide book, journal, and patent literature dealing with nuclear science and technology. Arranged by subject field. Indexes by corporate author, personal author, subject, and report number. Continued by *Energy Research Abstracts.*

Oceanic Abstracts. 1964–.

Covers the worldwide book, periodical, and report literature on the oceans, including pollution, engineering, geology, and oceanography.

Pollution Abstracts. 1970–.

Classed subject listing of books, periodicals, reports, and documents.

Psychological Abstracts. 1927–.

Covers books, periodicals, and reports, arranged by subject with full author and subject indexes, cumulated indexes for 1927–80.

Science Abstracts. Section A—Physics Abstracts. 1898–. ***Section B—Electrical and Electronic Abstracts.*** 1898–. ***Section C—Computer and Control Abstracts.*** 1966–.

Covers books, periodicals, and papers in all languages; sections do not overlap. All sections arranged by subjects. Separate author and conference indexes.

Selected Water Resources Abstracts. 1968–.

Covers water in respect to quality, resources, engineering, and related aspects from books, journals, and reports. Subject and author indexes.

Report Literature

The publication of reports by academic, industrial, and government agencies has been a major development since World War II. Many contracts funded by government agencies have required publication of such reports. The report is now frequently the most recent information on a subject. The bibliographical guides to this type of literature are only recently being developed.

Report Series Codes Dictionary. Eleanor J. Aronson, ed. 3rd ed. Detroit: Gale Research, 1986.

Identifies and explains over 20,000 letter and number codes used in issuing technical reports. Reference notes explain some of the systems used by various agencies. Alphabetical list of designations related to issuing agency and alphabetical agency list to related series codes.

Government Reports Announcements and Index. 1964–.
Formerly titled *Government Reports Announcements* (1971–75), *U.S. Government Research and Development Reports* (1965–71), and *U.S. Government Research Reports* (1954–64). Covers new reports of U.S. government-sponsored research and development released by the Department of Defense and other federal agencies. Arranged by broad subject areas, each entry gives complete bibliographical citation, descriptors, and availability data, and usually an abstract.

Government Reports Index. 1965–75.
Continued as part of *Government Reports Announcements and Index.* Previously issued under other titles. Indexes reports by subject, author, corporate source, and report number.

Scientific and Technical Aerospace Reports. 1963–.
Comprehensive abstracting journal covering worldwide report literature on the science and technology of space and aeronautics. Companion publication to *International Aerospace Abstracts.* Arranged in 74 subject categories. Indexed by subject, corporate source, individual author, contract number, report number, and accession number.

Technical Reports Awareness Circular. 1987–89.
Guide to report literature acquired by the Defense Technical Information center. Successor to the *Technical Abstract Bulletin* published from 1953–86. Some recent issues of the former title are security classified. Well indexed and good availability statements.

Use of Reports Literature. Charles P. Auger, ed. Hamden, CT: Archon, 1975.
Discusses report literature nature, control, and value. Evaluates sources.

Atlases and Statistical Guides

Basic data of interest to the technical researcher on many subjects are found in reliable and frequently updated standard guides. The quality atlas generally contains much more than maps of geographical locations. Statistical guides are of great reference importance to original research in economic, industrial, and social questions.

Commercial Atlas and Marketing Guide. Chicago: Rand McNally. Annual.
In addition to maps, contains much statistical data on trade, manufacturing, business, population, and transportation.

The National Atlas of the United States of America. Washington, DC: U.S. Department of the Interior Geological Survey, 1970.
Outstanding collection of 765 maps, many in color. Covers general reference and special subjects including landforms, geophysical forces, geology, marine features, soils, climate, water, history, and economic sociocultural, and administrative data. Maps, data tables, and diagrams.

Statistical Abstract of the United States. 1878–. Washington, DC: Bureau of the Census. Annual.
Official government standard summary of statistics on the social, political, and economic organization of the United States. Excellent source citations, index.

The World Almanac and Book of Facts. 1868–. New York: Newspaper Enterprise Association. Annual.

The most comprehensive and generally useful of the almanacs of miscellaneous information. Excellent statistical and news summary coverage.

Computerized Information Retrieval

Since the 1970s, bibliographic files have been made available for on-line interactive searching and information retrieval. Over 25 million citations are now available for searching. Normal access is from a local computer terminal to a firm or agency offering access to a data base or a system of data bases. Charges are calculated on the number of minutes a file is in use and the number of citations received. Citations are usually printed off-line and delivered by mail.

A number of individual data base services are now accessible in libraries in compact disk, read-only memory (CD-ROM) versions. With CD-ROM, there is no charge for line access or printing costs.

Three commercial services offer access to a wide spectrum of data bases:

- ORBIT Search Service, Maxwell Online Inc., McLean, VA
- DIALOG, Dialog Information Services, Palo Alto, CA
- BRS, BRS Information Technologies, Latham, NY

From most academic and research libraries, subject searches of over 17 million book titles can also be done via the EPIC service on the OCLC (Online Computer Library Center) System in Columbus, Ohio.

An average search can be expected to cost a minimum of $10–$50, with a key to the cost being good preplanning of search terms and search strategy by an experienced operator. Many major colleges, universities, and information centers offer these services through service bureaus or libraries, with some libraries paying part or all of the costs. Individual access directly through originating firms is also possible. Custom searches of over 1.5 million reports from federal agencies and federally sponsored research are available from National Technical Information Service, U.S. Department of Commerce, Springfield, VA 22161. The cost is about $100 for up to 100 abstracts. Some of the current subjects available in the scientific and technical areas are the following:

Agriculture: AGRICOLA, developed by the National Library of Agriculture

Biological sciences: BIOSIS Previews, prepared by Biosciences Information Service

Business: ABI/INFORM, produced by Data Courier

Chemistry: CA Search, produced by Chemical Abstracts Service of the American Chemical Society
Education: ERIC (Educational Resources Information Center), developed by the National Institute of Education
Engineering: COMPENDEX, produced by Engineering Information
Environmental studies: ENVIROLINE, prepared by Environment Information Center
Geosciences: GEO-REF, produced by the American Geological Institute
Government research and development reports: NTIS, produced by the National Technical Information Service of the Department of Commerce
Mathematics: MATHFILE, produced by the American Mathematical Society
Mechanical engineering: ISMEC, prepared by Data Courier
Metallurgy: METADEX, prepared by the American Society of Metals
Physics: SPIN, Searchable Physics Information Notices, by American Institute of Physics
Pollution and environment: *Pollution Abstracts*, produced by Cambridge Scientific Abstracts
Psychology: PSYCINFO, produced by the American Psychological Association
Science abstracts: INSPEC, produced by the Institution of Electrical Engineers
Science and technology: SCISEARCH, produced by the Institute for Scientific Information

Many other data bases are available for searching, but some have special use restrictions, such as being limited to specific industrial group members.

U.S. Government Publications

Access to government publications, releases, and directives is possible only from a series of federally produced and commercially supplemented catalogs and indexes. The nature of the subject indexing is generally less specific than with nongovernmental periodical indexes and abstracting services. Therefore, more ingenuity on the part of the researcher is generally needed to find pertinent resources.

Monthly Catalog of United States Government Publications. 1895–.

List by agency of publications, printed and processed, issued each month. Subject index only until 1973, then added author and title indexes cumulated annually. Entry gives title, author, publication data, price or availability indication. Superintendent of Documents classification number.

Cumulative Subject Index to the Monthly Catalog of United States Government Publications, 1900–1971. 15 vols. Washington, DC: Carrollton Press, 1973–75.

Covers about 800 thousand publications. Be sure to read the introduction for exclusions and entry policies.

CIS/Index to Congressional Publications. Washington, DC: Congressional Information Service, 1970–.

Part one: Abstracts of congressional publications; Part two: Index of congressional publications and public laws; Part three: Legislative histories. Commercial index giving significant insight into contents of congressional publications, with a detailed subject index.

United States Government Manual. 1935–. Annual. Washington, DC: GPO.

The official handbook of the federal government; describes the purposes and programs of most official and quasi-official agencies, with addresses and lists of current officials.

Federal Register. 1936–.

Daily publication of executive orders, presidential proclamations, and announcements of important rules and regulations of the federal government. Indexed by agency and significant subjects.

Weekly Compilation of Presidential Documents. 1965–.

Makes available transcripts of the president's news conferences, messages to Congress, public speeches and statements and other presidential materials. Indexed.

Code of Federal Regulations. 1949–.

Contains codifications of general and permanent administrative rules and regulations of general applicability and future effect.

Monthly Checklist of State Publications. 1910–.

Records those documents and publications issued by the various states and received in the Library of Congress.

Subject Bibliographies. 1975–.

Lists of publications available in specified subject areas from the Government Printing Office.

American Statistics Index and Abstracts, Annual and Retrospective Edition. A Comprehensive Guide to the Statistical Publications of the U.S. Government. Washington, DC: Congressional Information Service.

Aims to be a master guide and index to all federally produced statistical data. Does not contain the data but describes data and identifies sources.

Bureau of the Census Catalog of Publications, 1790–1972.

An example of a departmental catalog, kept up-to-date by frequent supplements. Many agencies issue such retrospective catalogs.

APPENDIX C

A Selected Bibliography

Technical Writing

Anderson, Paul V. 1987. *Technical Writing: A Reader Centered Approach.* San Diego: Harcourt, Brace Jovanovich.

Blicq, Ron S. 1986. *Technically Write!* 3rd ed. Englewood Cliffs, NJ: Prentice-Hall.

Brusaw, Charles, Gerald J. Alred, and Walter E. Oliu. 1987. *Handbook of Technical Writing.* 2nd ed. New York: St. Martin's.

Burnett, Rebecca E. 1990. *Technical Communications.* 2nd ed. Belmont, CA: Wadsworth.

Cohen, Gerald, and Donald H. Cunningham. 1984. *Creating Technical Manuals.* New York: McGraw.

Day, Robert A. 1988. *How to Write and Publish a Scientific Paper.* 3rd ed. Phoenix: Oryx.

Felker, Daniel B., et al. 1981. *Guidelines for Document Designers.* Washington, DC: American Institutes for Research.

Horton, William K. 1990. *Designing and Writing Online Documentation.* New York: Wiley.

Jordan, Stello, Joseph M. Kleinman, and H. Lee Shimberg, eds. 1971. *Handbook of Technical Writing Practices.* 2 vols. New York: Wiley.

Markel, Michael H. 1988. *Technical Writing: Situations and Strategies.* 2nd ed. New York: St. Martin's.

Michaelson, Herbert B. 1986. *How to Write and Publish Engineering Papers and Reports.* 2nd ed. Philadelphia: ISI.

Mills, Gordon H., and John A. Walter. 1986. *Technical Writing.* 5th ed. New York: Holt.

Pearsall, Thomas E., and Donald H. Cunningham. 1990. *How to Write for the World of Work.* 4th ed. New York: Holt.

Pickett, Nell A., and Ann A. Laster. 1988. *Technical English: Writing, Reading, and Speaking.* 4th ed. New York: Harper.

Souther, James W., and Myron L. White. 1977. *Technical Report Writing.* 2nd ed. New York: Wiley.

Warren, Thomas L. 1985. *Technical Writing.* Belmont, CA: Wadsworth.

Weisman, Herman M. 1985. *Basic Technical Writing.* 5th ed. Columbus, OH: Merrill.

Business Communication

Bowman, Joel P., and Bernadine P. Branchaw. 1988. *Business Report Writing.* 2nd ed. Chicago: Dryden.

Brusaw, Charles T. 1987. *Business Writer's Handbook.* 3rd ed. New York: St. Martin's.

Sigband, Norman B., and Arthur H. Bell. 1989. *Communication for Management and Business.* 5th ed. Glenview, IL: Scott.
Treece, Malra. 1989. *Successful Business Communication.* 4th ed. Boston: Allyn.

Writing in General

Flower, Linda. 1989. *Problem Solving Strategies for Writing.* 3rd ed. San Diego: Harcourt.
Fulwiler, Toby. 1988. *College Writing.* Glenview, IL: Scott.
Kinneavy, James L. 1980. *A Theory of Discourse.* New York: Norton.
Lambuth, David, et al. 1976. *The Golden Book on Writing.* New York: Penguin.
Quiller-Couch, Sir Arthur. 1916. *On the Art of Writing.* New York: Putnam's.
Pearsall, Thomas E., and Donald H. Cunningham. 1988. *The Fundamentals of Good Writing.* New York: Macmillan.
Van Buren, Robert, and Mary Fran Buehler. 1980. *The Levels of Edit.* 2nd ed. Pasadena, CA: JPL, California Institute of Technology.
Williams, Joseph M. 1989. *Style: Ten Lessons in Clarity and Grace.* 3rd ed. Glenview, IL: Scott.
Zinsser, William. 1990. *On Writing Well.* 4th ed. New York: Harper.

Usage

Corbett, Edward P. 1987. *The Little English Handbook.* 5th ed. Glenview, IL: Scott.
Ebbitt, Wilma R., and David R. Ebbitt. 1990. *Index to English.* 8th ed. New York: Oxford.
Follett, Wilson. 1966. *Modern American Usage.* Edited and completed by Jacques Barzun. New York: Hill and Wang.
Fowler, Henry W. 1987. *A Dictionary of Modern English Usage.* Ed. Ernest Gowers. 2nd rev. ed. New York: Oxford UP.
O'Hare, Frank. 1989. *The Modern Writer's Handbook.* 2nd ed. New York: Macmillan.

Style Manuals

Achtert, Walter S., and Joseph Gibaldi. 1985. *The MLA Style Manual.* New York: MLA.
CBE Style Manual Committee. 1983. *Council of Biology Editors Style Manual.* 5th ed. Arlington, VA: Council of Biology Editors.
The Chicago Manual of Style. 1982. 13th ed. Chicago: U of Chicago P.
Publications Manual of the American Psychological Association. 1983. 3rd ed. Washington, DC: American Psychological Association.
U.S. Government Printing Office Style Manual. 1984. Washington, DC: GPO.

Speech

Andrews, James R. 1987. *Public Speaking.* New York: Macmillan.
Ehninger, Douglas et al. 1986. *Principles and Types of Speech Communication.* 10th ed. Glenview, IL: Scott.
Hunt, Gary T. 1987. *Public Speaking.* 2nd ed. Englewood Cliffs, NJ: Prentice-Hall.
Lucas, Stephen E., 1989. *The Art of Public Speaking.* 3rd ed. New York: Random.

Semantics

Condon, John C., Jr. 1985. *Semantics and Communication.* 3rd ed. New York: Macmillan.
Hayakawa, S. I., and Alan R. Hayakawa. 1989. *Language in Thought and Action.* 5th ed. New York: Harcourt.

Logic

Copi, Irving M. 1985. *Informal Logic.* New York: Macmillan.
———. 1986. *Introduction to Logic.* 7th ed. New York: Macmillan.
Toulmin, Steven, et al. 1984. *An Introduction to Reasoning.* New York: Macmillan.

Graphics

Blanchard, Russel W. 1984. *Graphic Design.* Englewood Cliffs, NJ: Prentice-Hall.
Conover, Theodore E. 1990. *Graphic Communication Today.* 2nd ed. St. Paul, MN: West.
Hill, Francis. 1990. *Computer Graphics.* New York: Macmillan.
Hoffman, E. Kenneth. 1990. *Computer Graphics Applications.* Belmont, CA: Wadsworth.
Lefferts, Robert. 1982. *How to Prepare Charts and Graphs for Effective Reports.* New York: Barnes & Noble.
Rehe, Rolf F. 1979. *Typography: How to Make It Legible.* 3rd ed. Carmel, IN: Design Research International.
White, Jan V. 1988. *Graphic Design for the Electronic Age.* New York: Watson-Guptill.

Library Research

Barzun, Jacques, and Henry F. Graff. 1985. *The Modern Researcher.* 4th ed. New York: Harcourt.
Hurt, C. D. 1988. *Information Sources in Science and Technology.* Englewood, CO: Libraries Unlimited.
Mann, Thomas. 1990. *A Guide to Library Research Methods.* New York: Oxford UP.
McCormick, Mona. 1986. *The New York Times Guide to Reference Materials.* New York: NAL.

Chapter Notes

Preface

[1] See Ann Hill Duin, "How People Read: Implications for Writers," and Wayne Slater, "Current Theory and Research on What Constitutes Readable Expository Text," *The Technical Writing Teacher* 15 (1988): 185–93, 195–206.

[2] Walter S. Achtert and Joseph Gibaldi, *The MLA Style Manual* (New York: MLA, 1985).

[3] *Publication Manual of the American Psychological Association*, 3rd ed. (Washington, DC: APA, 1983).

Chapter 1: An Overview of Technical Writing

[1] Paul V. Anderson, "What Survey Research Tells Us about Writing at Work," *Writing in Nonacademic Settings*, eds. Lee Odell and Dixie Goswami (New York: Guilford, 1985) 30.

[2] Anderson 40.

[3] Anderson 54.

[4] Philip W. Swain, "Giving Power to Words," *American Journal of Physics* 13 (1945): 320.

Chapter 2: Composing

[1] Fred L. Luconi, "Artificial Intelligence," *Vital Speeches of the Day* 52 (1986): 605.

[2] Stephen S. Hall, "Aplysia and Hermissenda," *Science 85* May 1985: 33.

[3] Lester Faigley and Thomas P. Miller, "What We Learn from Writing on the Job," *College English* 44 (1982): 562–63.

[4] Lee Odell, Dixie Goswami, Anne Herrington, and Doris Quick, "Studying Writing in Non-Academic Settings," *New Essays in Technical and Scientific Communication: Research, Theory, Practice*, eds. Paul V. Anderson, R. John Brockmann, and Carolyn R. Miller (Farmingdale, NY: Baywood, 1983) 27–28.

[5] For this concept we are indebted to Victoria M. Winkler, "The Role of Models in

Technical and Scientific Writing," *New Essays in Technical and Scientific Communication: Research, Theory, Practice*, eds. Paul V. Anderson, R. John Brockmann, and Carolyn R. Miller (Farmingdale, NY: Baywood, 1983) 111–22.

[6] If you would like to know more about this subject, we suggest you read Walter James Miller, "What Can the Technical Writer of the Past Teach the Technical Writer of Today?" *IRE Transactions on Engineering and Speech* 4 (1961): 69–76.

[7] Blaine McKee, "Do Professional Writers Use an Outline When They Write?" *Technical Communication* 1st Quarter 1972: 10–13.

[8] Lillian Bridwell and Ann Duin, "Looking In-Depth at Writers: Computers as Writing Medium and Research Tool," *Writing On-Line: Using Computers in the Teaching of Writing*, eds. J. L. Collins and E. A. Sommers (Montclair, NJ: Boyton/Cook, 1985) 119.

[9] Eric Brown, "Word Processing and the Three Bears," *PC World* December 1985: 197.

Chapter 3: Writing Collaboratively

[1] For example, see Margaret B. Fleming, "Getting Out of the Writing Vacuum," *Focus on Collaborative Learning*, ed. Jeff Golub (Urbana, IL: NCTE, 1988) 77–84.

[2] For more on this, see Edgar R. Thompson, "Ensuring the Success of Peer Revision Groups," *Focus on Collaborative Learning*, ed. Jeff Golub (Urbana, IL: NCTE, 1988) 109–16.

Chapter 4: Writing for Your Readers

[1] Walter James Miller, "What Can the Technical Writer of the Past Teach the Technical Writer of Today?" *IRE Transactions on Engineering Writing and Speech* 4 (1961): 69–76.

[2] U.S. Bureau of the Census, *Statistical Abstract of the United States, 1989*, 109th ed. (Washington, DC: GPO, 1989) 131.

[3] Alarming Words," *New York Times* 9 Oct. 1986, 12.

[4] This passage and the two following are from Neal E. Carter, "The Political Side of Science," *Vital Speeches of the Day* 52 (1986): 559–60.

[5] Richard Conniff, "Eye on the Storm," *Raytheon Magazine* Fall 1982: 21.

[6] Allen Hammond, "Limits of the Medium," *SIPIscope* 11.2 (1983): 6–7.

[7] See Ann Hill Duin, "How People Read: Implications for Writers," and Wayne Slater, "Current Theory and Research on What Constitutes Readable Expository Text," *The Technical Writing Teacher* 15 (1988): 185–93, 195–206.

[8] Michael Ryan and James Tankard, Jr., "Problem Areas in Science News Writing," *Journal of Technical Writing and Communication* 4 (1974): 233.

[9] *Dietary Salt* (Chicago: Institute of Food Technologists, 1980) 85.

[10] Joseph Callanan and Cameron Foote, "The Light Fantastic," *Raytheon Magazine* Fall 1981: 16.

[11] *The Harvard Medical School Health Letter* July 1978: 1. Copyright © 1978 President and Fellows of Harvard College.

[12] *The Harvard Medical School Health Letter* 1–2.

[13] *The Harvard Medical School Health Letter* 1.

[14] Ryan and Tankard 230.

[15] Daniel B. Felker et al., *Guidelines for Document Designers* (Washington, DC: American Institute for Research, 1981) 41–48.

[16] U.S. Department of Health and Human Services, *Noninsulin-Dependent Diabetes* (Washington, DC: GPO, 1987).

[17] Umberto F. Gianola and Richard R. Shively, "Signal Processor Sorts Sounds from the Sea," *Bell Laboratories Record* May 1980: 167–68.

[18] Mary B. Coney, "The Use of the Reader in Technical Writing," *Journal of Technical Writing and Communication* 8 (1978): 104.

[19] Coney 104.

[20] James W. Souther, "Identifying the Informational Needs of Readers: A Management Responsibility," *IEEE Transactions on Professional Communication* PC-28 (1985): 10.

[21] Souther 10.

[22] Souther 10.

[23] Souther 11.

[24] Thomas N. Huckin, "A Cognitive Approach to Readability," *New Essays in Technical and Scientific Communication,* eds. Paul V. Anderson, R. John Brockmann, and Carolyn R. Miller (Farmingdale, NY: Baywood, 1983) 99.

[25] Lester Faigley, "Nonacademic Writing: The Social Perspective," *Writing in Nonacademic Settings,* eds. Lee Odell and Dixie Goswami (New York: Guilford, 1985) 238.

[26] Thomas E. Pinelli, Virginia M. Cordle, and Raymond F. Vondran, "The Function of Report Components in the Screening and Reading of Technical Reports," *Journal of Technical Writing and Communication* 14 (1984): 89.

[27] David F. Cope, "Nuclear Power: A Basic Briefing," *Mechanical Engineering* 89 (1967): 50.

[28] M. Daily et al., *Application of Multispectral Radar and LANDSAT Imagery to Geologic Mapping in Death Valley* (Pasadena, CA: JPL, California Institute of Technology, 1978) 44.

[29] J. J. Degan, "Microwave Resonance Isolators," *Bell Laboratories Record* Apr. 1966: 123.

[30] J. J. Degan 125.

Chapter 5: Gathering and Checking Information

[1] William H. Whyte, "Small Space Is Beautiful: Design as if People Mattered," *Technology Review* 85 (1982): 40.

[2] Whyte 39.

[3] For a detailed treatment we suggest Earl Babbie, *The Practice of Social Research,* 4th ed. (Belmont, CA: Wadsworth, 1986.)

Chapter 6: Achieving a Readable Style

[1] Raymond K. Neff, "Computing in the University: The Implications of New Technologies," *Perspectives in Computing* Fall 1987: 15.

[2] *Diet and Hyperactivity: Any Connection?* (Chicago: Institute of Food Technologists, 1976) 1–2.

[3] Janice C. Redish and Jack Selzer, "The Place of Readability Formulas in Technical Communication," *Technical Communication* Fourth Quarter 1985: 49.

[4] Francis Christensen, "Notes Toward a New Rhetoric," *College English* Oct. 1963: 7–18.

[5] Daniel B. Felker et al., *Guidelines for Document Designers* (Washington, DC: American Institutes for Research, 1981) 47–48.

[6] As quoted in Felker et al. 64.

[7] As revised in Felker et al. 65.

[8] As quoted in Janice C. Redish, *The Language of Bureaucracy* (Washington, DC: American Institutes for Research, 1981): 1.

[9] CBE Style Manual Committee, *Council of Biology Editors Style Manual: A Guide for Authors, Editors, and Publishers in the Biological Sciences*, 4th ed. (Council of Biology Editors, 1978) 21.

[10] The excerpts from the St. Paul Fire and Marine Insurance Company's old and new Personal Liability Catastrophe Policy are reprinted with the permission of the St. Paul Companies, St. Paul, Minn. 55102.

[11] CBE Style Manual Committee, *Style Manual,* pp. 19–23. Reproduced with permission from *Council of Biology Editors Style Manual,* 4th edition. CBE Style Manual Committee. Council of Biology Editors, 1978.

[12] "Planners Outlaw Jargon," *Plain English* Apr. 1981: 1.

Chapter 7: Informing

[1] Okitsuga Furuya, "Nuclear Power Plant Safety and the Two-Phase Flow Pump," *Scientific Honeyweller* Mar. 1984: 38–39.

[2] U.S. National Aeronautics and Space Administration, *Viking: The Exploration of Mars* (Washington, DC: GPO, 1984) 2–3.

[3] U.S. National Aeronautics and Space Administration, *Voyager at Neptune: 1989* (Washington, DC: GPO, 1989) 16.

[4] Lewis Thomas, "A New Agenda for Science," *SIPIscope* 12.2 (1984): 8.

[5] Thomas 8.

[6] Stephan Wilkinson, "Tiny Keys to Our Electronic Future," *Raytheon Magazine* Winter 1985: 4.

[7] U.S. National Aeronautics and Space Administration, *Exploring the Universe with the Hubble Space Telescope* (Washington, DC: GPO, n.d.) 11.

[8] "If All Our Reporters Were Laid End to End," *The Wall Street Journal* 4 April 1983: 10.

[9] Sir James Jeans, *Stars in Their Courses* (Cambridge: Cambridge UP, 1931) 23–24. Copyright 1931. Reprinted by permission of the publisher.

[10] John A. Lofgren, *Controlling Insect Pests of Shade and Ornamental Trees* (St. Paul, MN: Agricultural Extension Service of the University of Minnesota, 1973) n. pag.

[11] Stephan Wilkinson, "Earth Shakers," *Raytheon Magazine* Winter 1987: 12–15.

Chapter 8: Defining and Describing

[1] Centers for Disease Control, *The Public Health Consequences of Disasters* (Washington, DC: U.S. Department of Health and Human Services, 1989) 33.

[2] U.S. Congress, Office of Technology Assessment, *Biological Effects of Power Frequency Electric and Magnetic Fields* (Washington, DC: GPO, 1989) 19.

[3] F. Richard Stephenson and David H. Clark, "Historical Supernovas," *Scientific American* June 1976: 100.

[4] John S. McNown, "Canals in America," *Scientific American* July 1976: 117.

[5] U.S. National Aeronautics and Space Administration, *Viking: The Exploration of Mars* (Washington, DC: GPO, 1984) 14.

[6] Robin Birley, "A Frontier Post in Roman Britain," *Scientific American* Feb. 1977: 39.

[7] U.S. National Aeronautics and Space Administration, *The Planet Venus* (Washington, DC: GPO, 1978) 4.

[8] U.S. National Aeronautics and Space Administration, *The Voyager Flights to Jupiter and Saturn* (Washington, DC: GPO, 1982) 11.

[9] *The Plow and the Hearth* Fall/Winter Catalog 1986: 27.

[10] U.S. Department of Agriculture, *Wood Handbook* (Washington, DC: GPO, 1974) 17.

[11] *Wood Handbook* 20–24.

[12] Geoffrey A. Cross, "A Bakhtinian Exploration of Factors Affecting the Collaborative Writing of an Executive Letter of an Annual Report," *Research in the Teaching of English* 24 (1990): 176–77.

[13] Thomas M. McCann, "Student Argumentative Writing Knowledge and Ability at Three Grade Levels," *Research in the Teaching of English* 23 (1989): 67.

[14] Stephan Wilkinson, "Tiny Keys to Our Electronic Future," *Raytheon Magazine* Winter 1985: 4.

Chapter 9: Arguing

[1] U.S. National Aeronautics and Space Administration, *Viking: The Exploration of Mars* (Washington, DC: GPO, 1984) 6.

[2] Joel Gurin, "In the Beginning," *Science 80* July/Aug. 1980: 50.

[3] For a more detailed explanation of Toulmin logic, see Steven Toulmin, Richard Rieke, and Allan Janik, *An Introduction to Reasoning,* 2nd ed. (New York: Macmillan, 1984).

[4] The material in this example is adapted from Sharon Begley, Mary Hager, and Larry Wilson, "Is It All Just Hot Air?" *Newsweek* 20 Nov. 1989: 64–66; and Warren T. Brookes, "The Global Warming Panic," *Forbes* 25 Dec. 1989: 96–100.

[5] As repinted in *New York Times* 10 June 1986: 20.

Chapter 10: Document Design

[1] Elizabeth Tebeaux, "The High-Tech Workplace: Implications for Technical Communication Instruction," *Technical Writing: Theory and Practice,* eds. Bertie E. Fearing and W. Keats Sparrow (New York: MLA, 1989) 136.

[2] Robert G. Fuller, "LVs, CDs, and New Possibilities," *Computers in Physics Instruction,* eds. Edward F. Redish and John S. Risley (Redwood City, CA: Addison-Wesley, 1990) 523.

[3] Daniel B. Felker et al., *Guidelines for Document Designers* (Washington, DC: American Institutes for Research, 1981) 79–80.

[4] Jan V. White, *Graphic Design for the Electronic Age* (New York: Watson-Guptill, 1988) 20; Felker et al. 85–86.

[5] Rolf F. Rehe, *Typography: How to Make It Most Legible,* 3rd ed. (Carmel, IN: Design Research International) 34.

[6] M. Gregory and E. C. Poulton, "Even Versus Uneven Right-Margins and the Rate of Comprehension in Reading," *Ergonomics* 13 (1970): 427–34; Rehe.

[7] M. A. Tinker, *Legibility of Print* (Ames: Iowa State UP, 1963); M. A. Tinker, *Bases for Effective Reading* (Minneapolis: U of Minnesota P, 1965); Felker et al. 77–78. Felker et al. recommend 8- to 10-point type. While that guideline works well for many fonts in typeset documents where the resolution from the printer is excellent, 8-point type from most word processing software and desktop printers is not large enough to be read easily.

[8] Philippa J. Benson, "Writing Visually: Design Considerations in Technical Publications," *Technical Communication* Fourth Quarter 1985: 35–39.

[9] Felker et al. 73–76.

[10] Felker et al. 87–88; Rehe 35–36.

[11] J. Foster and P. Coles, "An Experimental Study of Typographical Cuing in Printed Text," *Ergonomics* 20 (1977): 57–66.

[12] Janice C. Redish, Robin M. Battison, and Edward S. Gold, "Making Information Accessible to Readers," *Writing in Nonacademic Settings,* eds. Lee Odell and Dixie Goswami (New York: Guilford, 1985) 129–53.

[13] Linda Flower, John R. Hayes, and Heidi Swarts, "Revising Functional Documents: The Scenario Principle," *New Essays in Technical and Scientific Communication: Research, Theory, Practice,* eds. Paul V. Anderson, R. John Brockmann, and Carolyn R. Miller (Farmingdale, NY: Baywood, 1983) 41–58.

[14] S. M. Glynn and F. J. DiVesta, "Control of Prose Processing Via Instructional and Typographic Cues," *Journal of Applied Psychology* 70 (1978): 595–603; Felker et al. 76.

[15] Jeanne W. Halpern, "An Electronic Odyssey," *Writing in Nonacademic Settings,* eds. Lee Odell and Dixie Goswami (New York: Guilford, 1985) 157.

[16] Halpern 157–201.

[17] Two good books in this area are William K. Horton, *Designing and Writing Online Documentation* (New York: Wiley, 1990); and Joseph S. Dumas, *Designing User Interfaces for Software* (Englewood Cliffs, NJ: Prentice-Hall, 1988).

Chapter 11: Design Elements of Reports

[1] M. Jimmie Killingsworth and Betsy G. Jones, "Division of Labor or Integrated Teams: A Crux in the Management of Technical Communication?" *Technical Communication* Third Quarter 1989: 210.

[2] William E. Splinter, "Center–Pivot Irrigation," *Scientific American* June 1976: 90.

[3] Carolyn J. Mullins, "Once Is Not Enough: WP Files Can Do Extra Duty," *Technical Communication* First Quarter 1982: 20.

[4] Max Weber, "Human Engineering and Its Application to Technical Writing, *Technical Communication* Third Quarter 1972: 2.

[5] *Phthalates in Food* (Chicago: Institute of Food Technologists, 1974) 1.

[6] Carolyn Krause, "Carbon Dioxide and Climate," *Oak Ridge National Laboratory Review* Fall 1977: 40–41.

[7] *Phthalates in Food* 2.

[8] Krause 47.

[9] J. J. Bull, R. C. Vogt, and C. J. McCoy, "Sex Determining Temperatures in Turtles: A Geographic Comparison," *Evolution* 36 (1982): 331.

[10] Walter S. Achtert and Joseph Gibaldi, *The MLA Style Manual* (New York: MLA, 1985).

[11] *Publication Manual of the American Psychological Association,* 3rd ed. (Washington, DC: APA, 1983).

[12] Institute of Food Technologists' Expert Panel on Food Safety and Nutrition and the Committee on Public Information, *Mercury in Food* (Chicago: Institute of Food Technologists, 1973) 1–2.

Chapter 12: Graphical Elements of Reports

[1] The order of our material in this chapter is based upon the order suggested in Dennis E. Minor, "Getting a Handle on Graphics," *Teaching Technical Writing: Graphics: Association of Teachers of Technical Writing Anthology No. 5,* ed. Dixie Lee Hickman, (Assoc. of Teachers of Technical Writing, 1985) 10–16.

Chapter 13: Correspondence

[1] Avery Comarow, "Tracking the Elusive Job," *The Graduate* (Knoxville: Approach 13–30 Corporation, 1977) 42.

[2] Jane L. Anton, Michael L. Russell, and the Research Committee of the Western College Placement Association, *Employer Attitudes and Opinions Regarding Potential College Graduate Employees* (Hayward, CA: Western College Placement Association, 1974) 10.

[3] For example, see Rosemary Ullrich, "The Power of a Positive Job Search," *Business World Women* Fall 1977: 11; Tom Jackson, "10 Musts for a Powerful Resume," *CPC Annual,* 30th ed. (Bethlehem, PA: College Placement Council, 1986) 3: 19–28; and Sandra Grundfest, "A Cover Letter and Resume Guide," *Business Week's Guide to Careers* (New York: McGraw, 1986) 8–13.

Chapter 14: Instructions

[1] G. B. Harrison, *Profession of English* (New York: Harcourt, 1962) 149.

[2] The questions posed here are based on questions presented by Janice C. Redish, Robbin M. Battison, and Edward S. Gold, "Making Information Accessible to Readers," *Writing in Nonacademic Settings,* eds. Lee Odell and Dixie Goswami (New York: Guilford, 1985) 139–43.

[3] Redish, Battison, and Gold 134.

[4] U.S. Department of Health and Human Services, *Eating to Lower Your High Blood Cholesterol* (Washington, DC: GPO, 1989) 1.

[5] U.S. General Services Administration, *Paint and Painting* (Washington, DC: GPO, 1977) 15.

[6] Charles H. Sides, *How to Write Papers and Reports About Computer Technology* (Philadelphia: ISI, 1984) 70.

[7] U.S. Department of Agriculture, *Simple Home Repairs: Inside* (Washington, DC: GPO, 1986) 7–8.

[8] *Simple Home Repairs* 16.

[9] Protocol analysis material furnished by Professor Victoria Mikelonis, U of Minnesota, St. Paul, MN.

Chapter 17: Feasibility Reports

[1] Excerpts from "The Feasibility of Removing Sediment Deposits from Wyoming Lake" reprinted with the permission of Barbara J. Buschatz.

Chapter 18: Empirical Research Reports

[1] J. J. Bull, R. C. Vogt, and C. J. McCoy, "Sex Determining Temperatures in Turtles: A Geographic Comparison," *Evolution* 36 (1982): 326.

[2] Lynn S. Best and Paulette Bierzychudek, "Pollinator Foraging on Foxglove (Digitalis purpurea): A Test of a New Model," *Evolution* 36 (1982): 70.

[3] Leslie K. Johnson, "Sexual Selection in a Brentid Weevil," *Evolution* 36 (1982): 251.

[4] Johnson.

[5] Abraham Blum and Moshe Azencot, "Small Farmers' Habits of Reading Agricultural Extension Publications: The Case of Moshav Farmers in Israel," *Journal of Technical Writing and Communication* 19 (1989): 383.

[6] Hugh M. Robertson and Hugh E. H. Paterson, "Mate Recognition and Mechanical Isolation in Enallagma Damselflies (Odonata: Coenagrionidae)," *Evolution* 36, (1982): 243.

[7] Bull, Vogt, and McCoy 326.

[8] David M. Craig, "Group Selection Versus Individual Selection: An Experimental Analysis," *Evolution* 36 (1982): 272.

[9] Gerald L. Chan and John B. Little, "Further Studies on the Survival of Non-Proliferating Human Diploid Fibroblasts Irradiated with Ultraviolet Light," *International Journal of Radiation Biology* 41 (1982): 360.

[10] Craig 272.

[11] Bull, Vogt, and McCoy 326–27.

[12] Johnson 253.

[13] Bull, Vogt, and McCoy 327–29.

[14] Bull, Vogt, and McCoy 329–30.

Chapter 19: Oral Reports

[1] These introspections were compiled at a session of the National Training Laboratory in Group Development that one of the authors attended in Bethel, Maine.

[2] The material on visual aids has been especially prepared for this chapter by Professor James Connolly of the University of Minnesota.

Part VI: Handbook

[1] Minna Levine, "Business Statistics," *MacUser* April 1990: 128.

[2] Levine 120.

INDEX

* **Note:** Numbers in *italics* refer to graphical representations.

MARKING SYMBOLS

This list of marking symbols refers you to Part VI, the "Handbook," where you can find discussions of style and usage. The list furnishes you with a heading and a page reference.